Bioactive Compounds in Foods and Their By-Products

Bioactive Compounds in Foods and Their By-Products

Guest Editor

Irene Dini

MDPI

Basel • Beijing • Wuhan • Barcelona • Belgrade • Novi Sad • Cluj • Manchester

Guest Editor
Irene Dini
Department of Pharmacy
University of Naples
Federico II
Napoli
Italy

Editorial Office
MDPI AG
Grosspeteranlage 5
4052 Basel, Switzerland

This is a reprint of the Special Issue, published open access by the journal *Molecules* (ISSN 1420-3049), freely accessible at: https://www.mdpi.com/journal/molecules/special_issues/7ZT91Q4334.

For citation purposes, cite each article independently as indicated on the article page online and as indicated below:

Lastname, A.A.; Lastname, B.B. Article Title. *Journal Name* **Year**, *Volume Number*, Page Range.

ISBN 978-3-7258-8335-6 (Hbk)
ISBN 978-3-7258-8336-3 (PDF)
https://doi.org/10.3390/books978-3-7258-8336-3

Contents

About the Editor

Irene Dini

Irene Dini is an Associate Professor of Food Chemistry at the Department of Pharmacy at the University of Naples Federico II, where she studies bioactive compounds in foods, their metabolic functions, and their applications in human nutrition. Her research focuses on the valorization of plant matrices and agro food by products, with particular attention to the identification, characterization, and biological evaluation of phytochemicals with antioxidant, anti inflammatory, and metabolic regulatory properties. Through an interdisciplinary approach that integrates analytical chemistry, food science, and nutritional biochemistry, she promotes innovative and sustainable strategies aimed at improving human well being.

Her scientific activity has received international recognition, and she has been consistently included since 2021 in the global ranking of the top two percent of scientists according to the Stanford classification, reflecting the impact, continuity, and relevance of her contributions. She collaborates with national and international research groups and serves as Guest Editor for scientific initiatives dedicated to bioactive compounds, fermentation, and sustainable food systems. Alongside her research, she is deeply committed to teaching and scientific dissemination, fostering a broader understanding of the relationship between diet, health, and environmental responsibility.

Preface

This Reprint brings together recent scientific advances on bioactive compounds derived from foods and their by products, offering a comprehensive view of how these molecules contribute to human health and sustainable resource management. The volume spans a wide scientific scope, from the characterization and quantification of phenolics, flavonoids, glucosinolates, carotenoids, saponins, and dietary fibers to the study of their biological activities and technological applications. A central motivation for assembling this work is the growing need to valorize agro food residues, transforming them into functional ingredients capable of supporting nutrition, wellness, and environmental responsibility.

The contributions included here illustrate how innovative processing, fermentation, enzymatic treatments, and green extraction methods can enhance the availability, stability, and efficacy of bioactive compounds. They also highlight the importance of plant biodiversity and the potential of underutilized matrices as sources of high value functional molecules.

This Reprint is intended for researchers, food technologists, nutrition scientists, and professionals working in the fields of functional foods, nutraceuticals, and sustainable food systems. It aims to provide both an updated scientific framework and a forward looking perspective on how bioactive compounds can shape the future of food innovation and health promotion.

Irene Dini
Guest Editor

molecules

MDPI

Editorial

Bioactive Compounds, Food Fermentation, and Sustainable Valorization: Advances, Gaps, and Future Directions

Irene Dini

Department of Pharmacy, University of Naples Federico II, 80131 Napoli, Italy; irdini@unina.it

The growing global emphasis on sustainability, public health, and food innovation is reshaping the scientific and industrial perceptions of plant-based bioactive compounds and agrofood by-products [1]. Once considered low-value residues, materials such as broccoli stalks [2], artichoke waste [3], mango leaves [4], olive pomace [5], grains [6], and non-commercial fruits [7] are now recognized as rich reservoirs of functional molecules with nutritional, technological, and health-promoting potential. The contributions collected in the Special Issue "Bioactive Compounds, Food Fermentation, and Sustainable Valorization: Advances, Gaps, and Future Directions" illustrate how these matrices can be transformed into high-value ingredients, supporting a circular bioeconomy while addressing modern dietary and health challenges. Plant by-products such as dietary fiber [8], polyphenols [9], flavonoids [10], phenolic acids [11], glucosinolates [12], carotenoids [13], and saponin [14] can be concentrated or structurally modified through innovative processing techniques, enhancing their functional properties and bioavailability [15]. Margarit et al. demonstrated that broccoli stalk powder obtained by freeze-drying can be incorporated into bread formulations, significantly increasing total dietary fiber, polyphenol content, and antioxidant activity [contribution 1]. Quizhpe et al. explored the use of enzymatically hydrolyzed artichoke and broccoli residues to fortify gluten-free breads, a category often criticized for poor nutritional quality. Their findings show substantial increases in soluble dietary fiber, phenolic compounds, and antioxidant capacity, accompanied by improved starch digestibility and reduced predicted glycemic index [contribution 2]. Lawrence et al. conducted a comprehensive prescreening of mango leaves from different cultivars, revealing significant variability in techno-functional properties and antioxidant profiles. Rich in mangiferin, quercetin, and kaempferol, mango leaves emerge as promising natural ingredients for functional foods and nutraceuticals [contribution 3]. Ziobro et al. evaluated the incorporation of a by-product mixture composed of apple pomace, red potato pulp, and sugar beet pulp (1:1:1) into corn-based extruded snacks. Their findings show that a 20% inclusion level provides the most balanced technological and nutritional outcome. Specifically, the formulation enhances dietary fiber, polyphenols, anthocyanins, tocopherols—particularly α-tocopherol—and key phytosterols (β-sitosterol, stigmasterol, campesterol), while reducing starch content and maintaining acceptable physical quality. This study further supports the concept that agro-industrial residues can be transformed into functional ingredients capable of improving both the health value and sustainability profile of widely consumed snack products [contribution 4]. Across these studies, two technological strategies stand out: enzymatic biotransformation, which releases bound phenolics and increases soluble fiber fractions, and freeze-drying, which preserves thermolabile compounds and enhances extractability. Together, these approaches demonstrate how processing can convert agricultural residues into potent functional ingredients. A second major theme concerns the biological activities of plant extracts and their potential applications in preventing or mitigating chronic diseases. Several contributions provide mechanistic insights into how bioactive compounds modulate oxidative stress, inflammation, metabolic pathways, and neuronal health. Gavia-García et al. reported that supplementation with *Sechium edule* var. *nigrum spinosum* in older adults with type 2 diabetes mellitus significantly increased SIRT1, SIRT3, SIRT5, and SIRT6 expression while improving systemic antioxidant status. These findings suggest a role for chayote in modulating mitochondrial function, metabolic homeostasis, and oxidative stress, key factors in diabetes management [contribution 5]. Bouaita et al. examined the influence of solvent polarity on the phytochemical profile and bioactivity of *Elaeagnus angustifolia* leaf extracts, demonstrating strong antioxidant activity and α-amylase inhibition, particularly in hydroethanolic fractions. These results support the plant's traditional use and highlight its potential as a natural antidiabetic agent [contribution 6]. Boutora et al. characterized the phytochemical composition of *Ajuga iva* and demonstrated its anti-inflammatory, analgesic, and antipyretic activities, supported by molecular docking analyses showing strong interactions with COX-2 [contribution 7]. This integrative approach reinforces the therapeutic relevance of traditional medicinal plants. You et al. investigated hydroxytyrosol-enriched olive pomace juice and showed that it activates the Nrf2, AMPK, SIRT1, PGC1α, and CREB/BDNF pathways, thereby protecting neuronal and microglial cells from oxidative stress and inflammation [contribution 8]. This study exemplifies how olive oil by-products can be repurposed into neuroprotective ingredients. Lee et al. standardized germinated oat extracts rich in avenanthramides and demonstrated their ability to reduce Aβ1-42-induced cytotoxicity and reactive oxygen species in SH-SY5Y cells [contribution 9]. These findings highlight the potential of germinated grains for preventing neurodegenerative diseases. Across these studies, a common mechanistic framework emerges: a reduction in oxidative stress (Nrf2 activation, ROS scavenging), modulation of inflammation (COX-2 inhibition, cytokine reduction), improvements in metabolic regulation (SIRT activation, α-amylase inhibition), and neuroprotection (BDNF upregulation, mitochondrial support). This convergence underscores the relevance of plant bioactives in addressing diabetes, neurodegeneration, and chronic inflammatory conditions. Several contributions demonstrate how plant extracts and by-products can be incorporated into food matrices to improve nutritional quality, sensory attributes, and functional properties. Incorporating broccoli stalk powder into wheat bread not only enhances fiber and antioxidant content but also aligns with consumer demand for sustainable, plant-forward products [contribution 1]. This approach exemplifies how bakery products can serve as vehicles for bioactive delivery. The enzymatically treated artichoke and broccoli extracts used by Quizhpe et al. significantly improved the nutritional profile of gluten-free breads, increasing soluble fiber and reducing glycemic response [contribution 2]. These findings are particularly relevant for individuals with celiac disease or gluten sensitivity, who often face limited access to nutritionally balanced products. Satora et al. investigated cloudy fruit juices fermented with *Lactiplantibacillus plantarum* and lactic acid-producing *Lachancea* yeasts, observing increases in organic acids, polyphenols, and fruity aromatic compounds. These beverages exhibit potential probiotic properties and represent a promising category of functional drinks [contribution 10]. Fermentation emerges as a key technology for increasing bioactive compound content, improving sensory profiles, generating new metabolites with health benefits, and stabilizing functional ingredients. A comprehensive review by Rymuszka and Gorczynska provides a modern reinterpretation of classical fermentations, including alcoholic, acetic, butyric, lactic, and propionic fermentation, highlighting their relevance in contemporary biotechnology [contribution 11]. Dini provided an extensive review of fermented foods,

 https://doi.org/10.3390/molecules31122115

probiotics, and postbiotics, emphasizing microbial metabolism, safety considerations, and the generation of bioactive metabolites such as bacteriocins, exopolysaccharides, and hydroxylated fatty acids [contribution 12]. The review highlights the need for standardized microbial characterization, improved regulatory frameworks, and deeper mechanistic understanding of strain-specific effects. Finally, studies by Birichi et al. [contribution 13], Sakuraoka et al. [contribution 14], and Bouaita et al. [contribution 6] reinforce the growing recognition that plant biodiversity, agricultural by-products, and staple crops represent a vast and largely untapped reservoir of functional molecules. Birinci and colleagues conducted a comprehensive evaluation of *Prunus armeniaca* cultivars, demonstrating significant variability in phenolic composition, antioxidant activity, and functional potential. Their findings highlight the importance of cultivar selection in the design of functional foods and underscore the value of fruit biodiversity as a source of bioactive compounds. This paper complements the studies in the Special Issue by reinforcing the concept that non-commercial or underutilized fruit varieties can serve as high-value functional ingredients [contribution 13]. Sakuraoka et al. identified and characterized active anti-inflammatory compounds in sweet potato roots, demonstrating their ability to modulate inflammatory pathways. This study provides compelling evidence that staple crops, often consumed for their macronutrient content, also harbor potent bioactive molecules with therapeutic potential [contribution 14]. Bouaita et al. (2026) provide a concise yet valuable contribution to the growing body of literature on *Elaeagnus angustifolia*, a species traditionally used in folk medicine across Central Asia and the Mediterranean. The study highlights the rich phytochemical profile of the leaves, emphasizing the presence of phenolic acids, flavonoids, and other antioxidant constituents that collectively underpin the plant's biological potential. The authors report marked antioxidant activity, which aligns with the high phenolic content and supports the traditional use of *Elaeagnus angustifolia* in managing inflammatory and oxidative stress-related conditions. Additionally, the study documents moderate antimicrobial and enzyme-modulating effects, suggesting possible applications in functional foods or nutraceutical formulations [contribution 6].

De Souza Aquino et al. (2026) highlight the functional potential of acerola fruit by-products within sustainable valorization strategies, showing that pulp and processing residues, particularly pomace, peels, and seeds, are enriched in ascorbic acid, hydroxycinnamic acids, flavonoids, carotenoids, and bioactive polysaccharides, often at levels exceeding those found in the edible fraction of food [contribution 15]. The contributions in this Special Issue demonstrate that plant by-products are valuable bioactive resources, not waste materials; processing technologies, including enzymatic hydrolysis, germination, freeze-drying, and fermentation, maximize the release and bioavailability of functional compounds; plant extracts can modulate key biological pathways (Nrf2, SIRT, AMPK, COX-2, NF-κB), supporting antioxidant, anti-inflammatory, antidiabetic, and neuroprotective effects; their incorporation into traditional foods (bread, gluten-free products, fermented beverages) enables the development of sustainable, health-promoting products, and classical fermentations can serve as advanced biotechnological platforms for producing functional ingredients and bio-based chemicals provided that selected microorganisms are used whose potential toxicity is known. Together, these studies offer a multidisciplinary, forward-looking perspective on the future of functional foods, in which sustainability, scientific innovation, and human health converge to transform plant matrices and agro-food by-products into powerful tools for nutrition and well-being.

Conflicts of Interest: The author declare no conflict of interest.

 https://doi.org/10.3390/molecules31122115

List of Contributions

1. Margarit, E.R.; Georgescu, A.A.; Popescu, E.C.; Tüğen, A.; Buruleanu, C.L. Application of Broccoli Stalk Powder in Bread Formulations. *Molecules* **2026**, *31*, 1414.
2. Quizhpe, J.; Peñalver, R.; Ayuso, P.; Nieto, G. Nutritionally Improved Gluten-Free Breads Fortified with Artichoke and Broccoli By-Products. *Molecules* **2026**, *31*, 152.
3. Lawrence, G.; Marchaux, I.; Pejcz, E.; Wojciechowicz-Budzisz, A.; Olędzki, R.; Zając, A.; Paroń, O.; Aurore, G.; Harasym, J. Prescreening of Mango (*Mangifera indica* L.) Leaves as a Potential Functional Food Ingredient: Techno-Functional and Antioxidative Characteristics. *Molecules* **2025**, *30*, 3381. https://doi.org/10.3390/molecules30163381.
4. Ziobro, R.; Gumul, D.; Sabat, R.; Wywrocka-Gurgul, A.; Zięba, T. Effect of Addition Levels of By-Product Mixture (Apple Pomace: Red Potato Pulp: Sugar Beet Pulp) on Phytochemical Profile, Antioxidant Activity and Physical Properties of Extruded Corn Snacks. *Molecules* **2026**, *31*, 1037. https://doi.org/10.3390/molecules31061037
5. Gavia-García, G.; Hernández-Álvarez, D.; Arista-Ugalde, T.L.; Aguiñiga-Sánchez, I.; Santiago-Osorio, E.; Cadena-Iñiguez, J.; Rosado-Pérez, J.; Mendoza-Núñez, V.M. *Sechium edule* var. *nigrum spinosum* (Chayote) Increases the mRNA Expression of Genes Encoding Sirtuins in Older Adults with Type 2 Diabetes Mellitus. *Molecules* **2026**, *31*, 1182. https://doi.org/10.3390/molecules31071182
6. Bouaita, R.; Djemil, R.; Bouhalit, S.; Lekmine, S.; Benslama, O.; Boutellaa, S.; Touzout, N.; Nieto, G.; Mustać, I.; Ondrašek, G.; et al. Phytochemistry and In Vitro Bioactivities of *Elaeagnus angustifolia* L. Leaves: Influence of Solvent Polarity, Antioxidant Properties, and α-Amylase Inhibition. *Molecules* **2026**, *31*, 861. https://doi.org/10.3390/molecules31050861.
7. Boutora, Y.; Boussekine, S.; Benslama, O.; Lekmine, S.; Mansouri, N.; Touzout, N.; Moussa, H.; Gacem, R.; Hfaiedh, N.; Nieto, G. Bioactive Potential and COX-2 Interaction of *Ajuga iva* (L.) Schreb. Hydroalcoholic Extract: Evidence from Experimental and Computational Studies. *Molecules* **2026**, *31*, 496. https://doi.org/10.3390/molecules31030496.
8. You, Y.-L.; Byun, H.-J.; Kang, N.; Lee, M.S.; Lee, J.-I.; Park, I.; Choi, H.-S. Standardized Hydroxytyrosol-Enriched Olive Pomace Juice Modulates Metabolic and Neurotrophic Signaling Pathways to Attenuate Neuroinflammation and Protect Neuronal Cells. *Molecules* **2026**, *31*, 336. https://doi.org/10.3390/molecules31020336.
9. Lee, Y.-Y.; Na, I.-S.; Kim, J.-E.; Song, J.-G.; Han, C.-E.; Kim, H.-W.; Shim, S.-M. Standardization of Germinated Oat Extracts and Their Neuroprotective Effects Against $A\beta_{1\text{-}42}$ Induced Cytotoxicity in SH-SY5Y Cells. *Molecules* **2025**, *30*, 3291. https://doi.org/10.3390/molecules30153291.
10. Satora, P.; Skotniczny, M.; Maziarek, M. Characteristics of Novel Fermented Cloudy Fruit Juices Produced Using *Lactiplantibacillus plantarum* and Lactic Acid-Producing *Lachancea* spp. Yeasts. *Molecules* **2025**, *30*, 3928. https://doi.org/10.3390/molecules30193928.
11. Rymuszka, A.; Gorczynska, W. Classical Food Fermentations as Modern Biotechnological Platforms: Alcoholic, Acetic, Butyric, Lactic and Propionic Pathways and Applications. *Molecules* **2026**, *31*, 333. https://doi.org/10.3390/molecules31020333.
12. Dini, I. Probiotics and Fermented Foods in Human Nutrition. *Molecules* **2026**, *31*, 1353. https://doi.org/10.3390/molecules31081353.
13. Birinci, C.; Kurek-Górecka, A.; Asadov, E.; Czuba, Z.P.; Kolaylı, S. Evaluation of Physicochemical Properties, Bioactive Composition, and Antioxidant Activity of *Prunus armeniaca* L. Cultivars for Functional Food and Nutraceutical Development. *Molecules* **2026**, *31*, 988. https://doi.org/10.3390/molecules31060988.
14. Sakuraoka, R.; Masutomi, H.; Ishihara, K. Identification of Active Anti-Inflammatory Compounds in Sweet Potato Storage Roots Extracted with Ethanol via LC-Q-TOF-MS. *Molecules* **2026**, *31*, 456. https://doi.org/10.3390/molecules31030456.
15. Aquino, J.d.S.; Araújo, A.N.V.d.; Araújo, J.M.D.d.; Santos, L.C.; Silva, J.C.C.; Batista, K.S.; Carvalho, L.R.R.A. Acerola and Its By-Products as Sources of Bioactive Compounds: Phytochemical Profile and Biological Effects in Experimental and Clinical Studies. *Molecules* **2026**, *31*, 1792. https://doi.org/10.3390/molecules31111792.

References

1. Kagueyam, S.S.; dos Santos Filho, J.R.; Contato, A.G.; de Souza, C.G.M.; Castoldi, R.; Corrêa, R.C.G.; Conte Junior, C.A.; Yamaguchi, N.U.; Bracht, A.; Peralta, R.M. Green Extraction of Bioactive Compounds from Plant-Based Agri-Food Residues: Advances Toward Sustainable Valorization. *Plants* **2025**, *14*, 3597. [CrossRef] [PubMed]
2. Gudiño, I.; Casquete, R.; Martín, A.; Wu, Y.; Benito, M.J. Comprehensive Analysis of Bioactive Compounds, Functional Properties, and Applications of Broccoli By-Products. *Foods* **2024**, *13*, 3918. [CrossRef] [PubMed]
3. Ayuso, P.; Quizhpe, J.; Rosell, M.d.l.Á.; Peñalver, R.; Nieto, G. Bioactive Compounds, Health Benefits and Food Applications of Artichoke (*Cynara scolymus* L.) and Artichoke By-Products: A Review. *Appl. Sci.* **2024**, *14*, 4940. [CrossRef]
4. Kučuk, N.; Primožič, M.; Kotnik, P.; Knez, Ž.; Leitgeb, M. Mango Peels as an Industrial By-Product: A Sustainable Source of Compounds with Antioxidant, Enzymatic, and Antimicrobial Activity. *Foods* **2024**, *13*, 553. [CrossRef] [PubMed]
5. Dantas Palmeira, J.; Araújo, D.; C. Mota, C.; Alves, R.C.; P. P. Oliveira, M.B.; Ferreira, H.M.N. Fermentation as a Strategy to Valorize Olive Pomace, a By-Product of the Olive Oil Industry. *Fermentation* **2023**, *9*, 442. [CrossRef]
6. Martínez-Villaluenga, C.; Peñas, E. Innovative Processing Technologies for Developing Functional Ingredients and Food Products with Health Benefits from Grains. *Foods* **2023**, *12*, 1356. [CrossRef] [PubMed]
7. Piñeiro, M.; Parera, V.; Ortiz, J.E.; Llalla-Cordova, O.; Manrique, S.; Castro, B.; Ighani, M.; Luna, L.C.; Feresin, G.E. Agro-Industrial Waste from *Pistacia vera*: Chemical Profile and Bioactive Properties. *Plants* **2025**, *14*, 1420. [CrossRef] [PubMed]
8. Plakantonaki, S.; Roussis, I.; Bilalis, D.; Priniotakis, G. Dietary Fiber from Plant-Based Food Wastes: A Comprehensive Approach to Cereal, Fruit, and Vegetable Waste Valorization. *Processes* **2023**, *11*, 1580. [CrossRef]
9. Saini, R.K.; Khan, M.I.; Kumar, V.; Shang, X.; Lee, J.-H.; Ko, E.-Y. Bioactive Compounds of Agro-Industrial By-Products: Current Trends, Recovery, and Possible Utilization. *Antioxidants* **2025**, *14*, 650. [CrossRef]
10. Martínez-Zamora, L.; Cano-Lamadrid, M.; Artés-Hernández, F.; Castillejo, N. Flavonoid Extracts from Lemon By-Products as a Functional Ingredient for New Foods: A Systematic Review. *Foods* **2023**, *12*, 3687. [CrossRef]
11. Nirmal, N.P.; Khanashyam, A.C.; Mundanat, A.S.; Shah, K.; Babu, K.S.; Thorakkattu, P.; Al-Asmari, F.; Pandiselvam, R. Valorization of Fruit Waste for Bioactive Compounds and Their Applications in the Food Industry. *Foods* **2023**, *12*, 556. [CrossRef]
12. Shinali, T.S.; Zhang, Y.; Altaf, M.; Nsabiyeze, A.; Han, Z.; Shi, S.; Shang, N. The Valorization of Wastes and Byproducts from Cruciferous Vegetables: A Review on the Potential Utilization of Cabbage, Cauliflower, and Broccoli Byproducts. *Foods* **2024**, *13*, 1163. [CrossRef]
13. Papapostolou, H.; Kachrimanidou, V.; Alexandri, M.; Plessas, S.; Papadaki, A.; Kopsahelis, N. Natural Carotenoids: Recent Advances on Separation from Microbial Biomass and Methods of Analysis. *Antioxidants* **2023**, *12*, 1030. [CrossRef] [PubMed]
14. Dini, I.; Tenore, G.C.; Dini, A. Oleanane Saponins in "Kancolla", a Sweet Variety of Chenopodium quinoa. *J. Nat. Prod.* **2002**, *65*, 1023–1026. [CrossRef] [PubMed]
15. Cavallo, P.; Dini, I.; Sepe, I.; Galasso, G.; Fedele, F.L.; Sicari, A.; Bolletti Censi, S.; Gaspari, A.; Ritieni, A.; Lorito, M.; et al. An Innovative Olive Pâté with Nutraceutical Properties. *Antioxidants* **2020**, *9*, 581. [CrossRef]

https://doi.org/10.3390/molecules31122115

molecules

MDPI

Article

Application of Broccoli Stalk Powder in Bread Formulations

Elena Roxana Margarit [1], Andreea Antonia Georgescu [1], Elena Corina Popescu [1], Aslıhan Tüğen [2] and Claudia Lavinia Buruleanu [1,*]

[1] Faculty of Environmental Engineering and Food Science, Valahia University of Targoviste, 13 Sinaia Alley, 130004 Targoviste, Romania; elena.margarit@valahia.ro (E.R.M.); andreea.georgescu@valahia.ro (A.A.G.); corina.popescu@valahia.ro (E.C.P.)

[2] Department of Food Engineering, Institute of Sciences, Afyon Kocatepe University, 03200 Afyonkarahisar, Türkiye; aslihan.tugen@usr.aku.edu.tr

* Correspondence: lavinia.buruleanu@valahia.ro

Abstract

Bread enriched with broccoli stalk powder is proposed as a newly formulated product with potential health benefits. Wheat flour in the bread recipe was enriched with powder obtained from freeze-drying broccoli stalks, a valuable by-product of vegetable processing. The effects of broccoli stalk powder (BSP) supplementation on the physicochemical and sensory properties of bread, as well as its bioactive profile, were evaluated. The results showed an increase in moisture content and acidity with increasing substitution levels from 0% (control bread—BC) to 7%, while some important parameters in terms of consumers' acceptability decreased (i.e., loaf volume and porosity). Elasticity exhibited moderate variations, with no major influence at lower substitution levels. A small-scale consumer test indicated good scores up to moderate substitution levels (3–5%). The antioxidant activity of broccoli stalk flour (62.13% ± 1.29%) positively influenced the antioxidant activity of bread with 3% BSP, which increased by approximately 4%. The total polyphenol content (TPC) of the bread with 5% BSP, together with its physicochemical and sensory characteristics, suggested that broccoli stalk powder is a promising functional ingredient for bakery applications.

Keywords: broccoli stalk powder; antioxidants; bread quality; sensory acceptance

1. Introduction

Global food systems are confronting major sustainability challenges arising from population growth, climate change, and resource scarcity. Sustainable food systems are defined as systems that balance environmental, economic, and social dimensions, meeting present needs without jeopardizing the capacity of future generations to meet their own needs [1]. Within this framework, the circular economy concept is increasingly displacing conventional linear "take–make–dispose" models. This model centers on closed-loop mechanisms that aim to retain resources within the system at the highest efficiency and for the longest possible duration, minimize waste generation, and reprocess by-products to reintegrate them into the economy [2]. The emergence of substantial quantities of by-products and waste during food production and processing positions the circular economy approach as critical for the food industry.

Moreover, the increasing consumer demand for safer and health-promoting foods encourages the use of naturally derived by-product components as alternatives to synthetic additives [3]. This trend highlights the strategic importance of vegetable by-products in addressing food insecurity and enhancing nutritional security.

The bakery sector is among the largest global food sectors and consumes considerable volumes of raw materials. Traditionally, the sector has relied predominantly on staple inputs such as wheat flour, fat, and water. However, increasing consumer awareness and the demand for healthier diets are steering the sector towards the development of more nutritious and functional products. In recent years, the enrichment of bread with functional compounds that impart health-promoting and preventive effects has attracted significant global attention [4]. Concurrently, the need to improve the sector's sustainability performance has necessitated the exploration of alternative and innovative raw material sources [5].

The bakery sector constitutes one of the most widely consumed food categories worldwide and therefore offers an ideal platform for delivering functional constituents to consumers. The low fiber content, high glycemic index, and limited bioactive compound profile of conventional baked goods create a clear imperative to enhance their nutritional quality. The incorporation of vegetable by-products into bakery products provides a dual benefit by improving nutritional quality while reducing food waste [1,4–6].

Transforming by-products into functional foods through appropriate processing techniques and hygiene protocols is critical for preserving the physical and organoleptic properties of the products. Nevertheless, during incorporation into food formulations, potential changes in color, texture, and overall consumer acceptability must be optimized [7]. In this context, technological approaches developed to address these issues facilitate the industrial-scale adoption of by-products and help overcome barriers associated with high operational costs.

The incorporation of vegetable processing by-products into bread as functional ingredients has been extensively investigated, not only from a sustainability perspective but also due to their potential to increase the nutritional density and bioactive richness of the final product. Within defined substitution levels, statistically and technologically meaningful increases can be achieved in the phenolic compounds, antioxidant capacity, dietary fiber, and certain minerals. However, as the substitution ratio increases, technology- and quality-related trade-offs become more pronounced, including color darkening, reduced product volume, increased textural firmness, and a concomitant decline in sensory acceptance [7,8]. Consequently, the practical challenge is to define, for each specific by-product, an optimal balance between the substitution range that maximizes bioactive gain and the sensory/technological quality level that remains acceptable.

Several studies have been conducted to develop functional bread. Enhancing the nutritional value of bread through enrichment with microalgae (spirulina and chlorella) [4], marine shellfishery products [9], broccoli sprouts [10], broccoli leaf powder [11], sea buckthorn berry [12], chickpea flour [13], kale [14], and lemon pomace fiber [15] has been widely reported.

Cruciferous vegetables including broccoli are known for their bioactive compounds with potent antibacterial and antioxidant properties. Bioactive substances from broccoli and broccoli by-products have been shown to exert cancer-preventive effects [16,17], lower glucose levels [18], and reduce inflammation and oxidative stress, which are key factors in cardiovascular disease [19]. Functional products enriched with dietary fiber and antioxidants contribute significantly to consumer health [20]. However, further animal studies are needed to validate the therapeutic potential of broccoli-derived bioactive compounds [16].

Broccoli (*Brassica oleracea* var. *italica*) is an agricultural crop belonging to the *Brassicaceae* family and originating from the Mediterranean region, with significant social and

https://doi.org/10.3390/molecules31091414

economic importance. It is now cultivated worldwide across a wide range of climates and regions, both in open fields and in greenhouse production systems [21], resulting in a considerable diversity among broccoli varieties [16]. In 2023, global broccoli production reached 26,472,040 thousand tons, representing a 1.65% increase from 2019 [22].

Marketable broccoli florets account for only a small fraction of the total above-ground plant biomass [23]. Thus, broccoli production generates considerable waste materials, such as leaves, stalks, and non-marketable vegetables [24,25]. Currently, more than 60% of global broccoli production is wasted during harvesting [26], resulting in environmental pollution [16,27,28]. The valorization of broccoli waste aligns with the Sustainable Development Goals (SDGs) initiative, contributing to SDG 12 (Responsible Consumption and Production) by reducing food loss, SDG 13 (Climate Action) by mitigating waste-related emissions, and SDG 2 (Zero Hunger) by enhancing global food security [29].

Research on broccoli has predominantly focused on its edible parts, whereas investigations into its by-products remain scarce [26]. However, various anatomical components of the broccoli plant have been documented as being incorporated into the formulation of novel functional food. Several foods, such as gluten-free cake [20], biscuits [25], gluten-free mini sponge cakes [30], crackers [31], pasta [32,33], noodles [34], spreadable cheese [35], salad dressing [36], and béchamel sauce [37] have been successfully formulated using broccoli co-products in recipes.

The valorization of broccoli by-product extracts has emerged as a promising approach, primarily attributed to the high dietary fiber (DF) content of the stalks [38]. The chemical composition of the broccoli stalk includes both soluble and insoluble fibers (pectins, cellulose, starch, hemicellulose, etc.), vitamins (including some B vitamins and ascorbic acid), minerals (K, Ca, Zn, Mg, P, Fe, Mn, etc.), flavonoids, and amino acids (including cysteine, a sulfur-containing amino acid) [23]. Compared to the floret tissue, the stem tissue contains a significantly lower total glucosinolate concentration. Nevertheless, the percentage of glucoraphanin to total glucosinolates in the stem indicates that it contains considerable amounts of health-promoting compounds [39]. Also, sulforaphane, a bioactive compound resulting from the hydrolysis of glucoraphanin, was reported in high amounts in the stem of some broccoli cultivars [40].

Genetic, geographical, and environmental factors may account for significant variations in the nutritional composition of the different botanical parts of broccoli, including the stems. The proximate composition of broccoli stalks was reported as follows: moisture 85 g/100 g FW, ash 10.34–12.37 g/100 g DW, protein 13.73–18.14 g/100 g DW, carbohydrate 75.42 g/100 g DW, total dietary fiber 77.28 g/100 g DW (8.30 g/100 g FW), and lipids 6.58 g/100 g DW [38]. To encourage the use of broccoli stalk, it is important to assess its nutritional value for human consumption [39] and also the mechanisms of action of its bioactive substances [16].

The conversion of broccoli by-products into functional ingredients for food applications exemplifies a circular economy strategy and provides a cost-effective means of valorization [25]. The transition from waste to a valuable resource faces challenges such as the scalability and cost-effectiveness of processing technologies and seasonal availability of broccoli by-products [29]. These challenges underscore the need for further research to fully realize the potential of broccoli by-product valorization.

Accordingly, this paper aims to examine the potential use of freeze-dried broccoli stalk in the bakery to produce functional bread with enhanced phenolic compounds and dietary fiber and to study the influence of its inclusion on the physico-chemical parameters and sensory attributes of bread, which are critical for consumers' acceptance.

 https://doi.org/10.3390/molecules31091414

2. Results

2.1. Physico-Chemical Parameters of Flours

The moisture contents of the broccoli stalk powder and wheat flour used in experimental were 6.23 ± 0.01% and 13.56 ± 0.05%, respectively. The hydration capacity, wet gluten, deformation index (DI), and elasticity index (I_e) were determined (Table 1).

Table 1. Characterization of wheat flour and broccoli stalk powder.

Sample	Moisture (%)	Hydration Capacity (%)	Wet Gluten (%)	DI (mm)	I_e (mm)	Acidity (°A)
Wheat flour	13.56 ± 0.05	58.66 ± 0.57	24.86 ± 0.11	4.06 ± 0.11	21 ± 0.09	2.56 ± 0.05
Broccoli powder	6.23 ± 0.01	290 ± 2.8	NA	NA	NA	5.6 ± 0.2

NA—not applicable.

2.2. Dough Properties During Mixing and Fermentation

Farinographic analysis provided the data needed for designing the enriched breads, namely water absorption of the wheat flour, defined as the amount of water required to achieve standard dough consistency, the development time, the duration of peak consistency, and dough softening after mixing. These key parameters are indicative of the gluten strength and mixing tolerance.

The correlation of farinographic and alveographic parameters (Table 2) consistently highlights the profile of a flour of medium rheological intensity, characterized by moderate gluten in terms of both strength and stability.

Table 2. Correlation of farinographic and alveographic parameters.

Farinographic Parameter	Value	Observations	Alveographic Parameter	Observations
Water absorption	56	Moderate hydration, easy-to-work dough	W = 188 × 10^{-4} J (average energy)	Moderate water retention capacity and gluten development
Dough developing time	2.8 min.	Rapid gluten development	L = 80 mm, P/L = 0.85	The gluten network forms quickly, corresponding to moderate extensibility
Dough stability	5.4 min.	Medium stability; low tolerance to kneading	P = 68 mm H_2O	The dough has moderate strength, according to medium stability; does not tolerate long mechanical stress
Dough softening	92 FU	High softening; dough loosens quickly	G = 19.9	The gluten network begins to relax and loses elasticity rapidly; correlated with the high degree of softening
Quality number	55	Average quality, suitable for standard baking	W = 188 ×10^{-4} J, P/L = 0.85	Energy and P/L balance confirm the moderate performance of the flour

W—baking strength; L—extensibility; P—tenacity (resistance to deformation); P/L—ratio between dough strength and extensibility; G—inflation index.

The Falling number of the wheat flour used in the experiment was 269 s. Although the value is at the lower limit of the optimal range for baking (250–350 s), it results that the alpha amylase activity in the flour had an optimum activity to support breadmaking, in accordance with the results provided by the farinogram and alveogram.

https://doi.org/10.3390/molecules31091414

2.3. Physico-Chemical Parameters of Bread

Figure 1 shows images of cross-sections of experimental bread obtained with and without broccoli stalk powder (BSP).

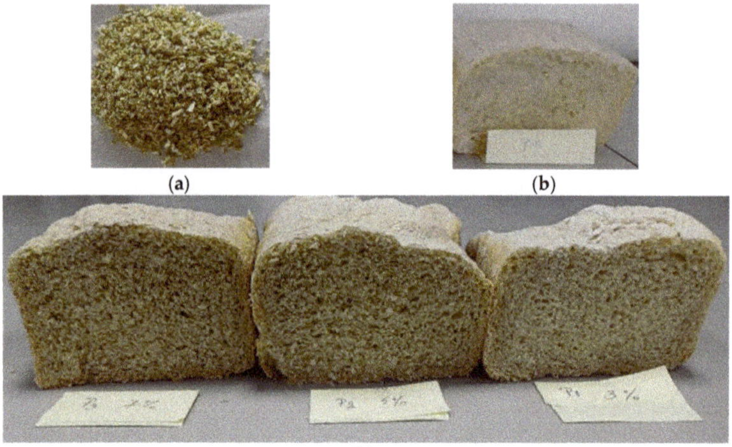

Figure 1. Aspect of the freeze-dried broccoli stalk powder (**a**) and experimental bread: control (**b**) and bread obtained through the enrichment of wheat flour with broccoli stalk powder (**c**). In figure (**c**), the supplementation of wheat flour with broccoli stalk powder decreased from left to right.

The moisture content of the bread increased with higher levels of broccoli powder added to wheat flour (Table 3), ranging from 5.72% (B3BSP) to 12.59% (B7BSP), when the control sample (BC) was included. The dietary fibers, non-starch polysaccharides, and other structural compounds present in broccoli stalk flour exhibit high water-binding capacity, which likely accounts for the observed increase in bread moisture. Enhanced water retention within the dough matrix contributes to a more hydrated crumb structure. Furthermore, the reduction in the gluten proportion resulting from the enrichment of wheat flour with BSP may alter the protein network of the dough, thereby facilitating additional water retention.

Table 3. Physico-chemical parameters of experimental bread.

Sample	Moisture (%)	Acidity (°A)	Porosity (%)	Loaf Volume (cm^{-3})	Elasticity (%)
BC	41.99 ± 0.02 ^{a}	2.7 ± 0.01 ^{a}	92.1 ± 0.05 ^{a}	320.66 ± 1.15 ^{a}	95.31 ± 0.02 ^{a}
B3BSP	44.39 ± 0.08 ^{b}	2.66 ± 0.05 ^{a}	72.46 ± 0.46 ^{b}	284 ± 4.0 ^{b}	92.78 ± 0.17 ^{b}
B5BSP	45.86 ± 0.14 ^{c}	3.1 ± 0.01 ^{b}	71.16 ± 0.11 ^{b}	265 ± 3.0 ^{c}	96.63 ± 0.23 ^{c}
B7BSP	47.28 ± 0.33 ^{d}	3.33 ± 0.05 ^{c}	72.69 ± 0.11 ^{c}	238.66 ± 1.52 ^{d}	97.44 ± 0.2 ^{d}

BC—control bread; B3BSP—bread supplemented with 3% (*w*/*w*) broccoli stalk powder; B5BSP—bread supplemented with 5% (*w*/*w*) broccoli stalk powder; B7BSP—bread supplemented with 7% (*w*/*w*) broccoli stalk powder; different lowercase letters within the column and for each group of samples represent significant differences according to Tukey's-b test ($p < 0.05$).

One-way ANOVA revealed a statistically significant effect of BSP on the moisture content of bread ($p < 0.001$). Post hoc analysis (Tukey'-b test) indicated significant differences ($p < 0.05$) between BC and bread with enrichment levels of BSP (3%, 5%, and 7%, respectively), confirming that BSP significantly increased water retention capacity in bread formulations. From a technological perspective, moisture is highly sensitive to recipe changes involving fiber-rich ingredients.

https://doi.org/10.3390/molecules31091414

As well as the moisture, the bread acidity increased with higher addition levels of broccoli stalk powder (Table 3), with this trend being manifested at 5% and 7% supplementation. Thus, the acidity of the B5BSP sample was 1.14 times higher than the control, while that of the B7BSP sample was 1.23 higher compared to the same referential. Tukey analysis indicated significant differences ($p < 0.01$) between BC and B5BSP and between BC and B7BSP. This increase may be associated with the biochemical composition of BSP, mainly the presence of organic acids (i.e., malic acid, citric acid), and its influence on the fermentation dynamics of the dough.

The porosity of the experimental bread decreased compared to the control sample (Table 3), affecting the crumb structure. This reduction is dramatic, by an average of 20% reported for the control, and starts from the lower level of BSP supplementation, probably due to the addition of non-gluten material and its negative influence on the gluten network development. According to ANOVA analysis, decreasing the bread porosity with increasing BSP addition levels is statistically significant ($p < 0.001$).

The loaf volume of bread decreased with increasing BSP levels (Table 3), from 320.66 ± 1.15 cm^3 (BC) to 238.66 ± 1.52 cm^3 (B7BSP). ANOVA analysis showed a significant decrease in bread volume with an increasing substitution level ($p < 0.001$). This parameter is strongly related to gluten network development and bread porosity, as wheat gluten proteins are responsible for forming the viscoelastic network able to retain the gases resulting from fermentation.

The loaf volume was identified as a sensitive technological parameter, affected by wheat flour addition, whereas bread elasticity remained comparatively stable, showing only moderate variations among samples (Table 3). Elasticity values suggest that the levels of BSP used did not significantly compromise the crumb resilience. It can be assumed that some interactions between dietary fibers and water counterbalanced for gluten reduction. ANOVA analysis showed no statistically significant effect ($p > 0.05$) of BSP supplementation on bread elasticity. These findings suggest that elasticity is less sensitive to broccoli stalk flour substitution than the bread volume and porosity.

2.4. *Effect of Broccoli Stalk Powder Addition on the Fiber Content*

The dietary fiber content of the experimental bread increased with the addition of BSP (Figure 2), due to the high content of insoluble fiber (cellulose, hemicellulose) and soluble fiber (pectin) in broccoli stalks. From a technological perspective, increasing BSP addition enhanced the water-retention capacity, as previously described.

Even a small addition (only 3% BSP) caused a visible increase in TDF, by 42% compared to the control. The supplementation of wheat flour with BSP at a 7% level resulted in a fiber content nearly double that of the control. The increase in the fiber content of bread with the addition of BSP is not linear, but it is consistent and significant. ANOVA analysis revealed significant differences between the means due to the addition of BSP to the wheat flour. The control (BC) differed significantly from all other samples ($p < 0.05$), whereas the addition of 3% and 5% BSP to the wheat flour did not result in significant changes in the fiber content of the breads. A post-hoc test indicated that sample B7BSP differed significantly ($p < 0.05$) from the other samples. The results highlight the potential of BSP as a functional ingredient for improving the nutritional value of bakery products.

The addition of broccoli fiber to bread may provide additional benefits compared to cereal fiber, such as potential prebiotic effect through the stimulation of beneficial microbiota, reduction of the glycemic index of bread, and increased satiety due to the water-retention capacity.

https://doi.org/10.3390/molecules31091414

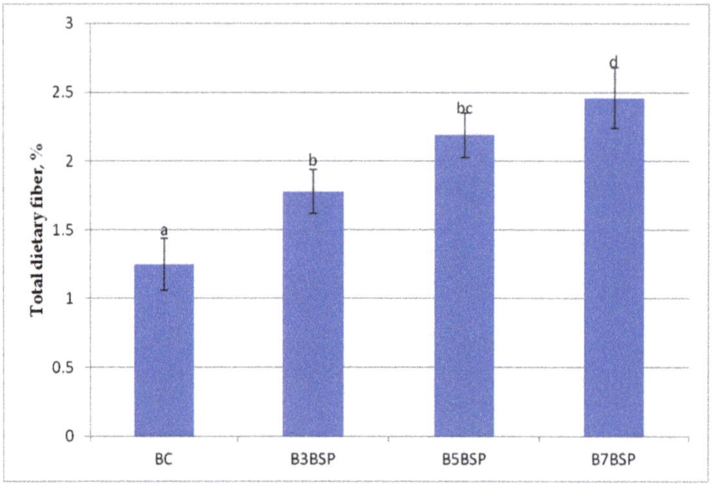

Figure 2. Total dietary fiber of experimental bread. Different lowercase letters within each sample represent significant differences according to Tukey's-b test ($p < 0.05$).

2.5. Effect of Broccoli Stalk Powder on Total Polyphenol Content (TPC) and Antioxidant Activity (AA) of Bread

Broccoli stalk powder was characterized by high antioxidant activity (62.13 ± 2.53%) and total polyphenol content (21.8 ± 1.19 mg GAE/g), with freeze drying used to protect the bioactive compounds. The incorporation of broccoli stalk powder significantly influenced both the antioxidant activity (AA) and total polyphenol content (TPC) of bread samples (Figure 3), consistent with the findings reported by Lafarga et al. [41].

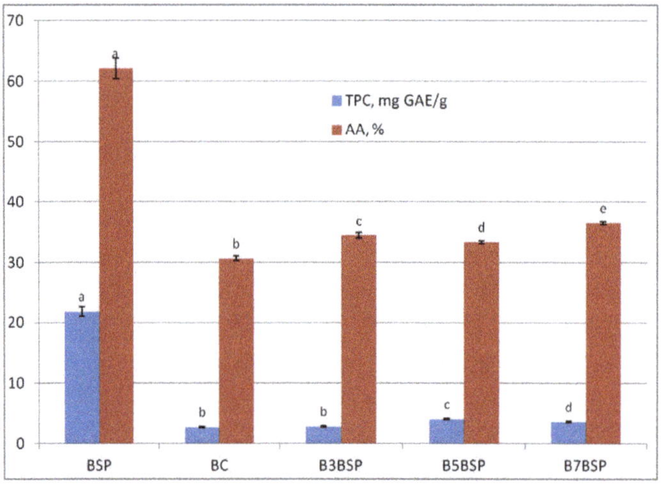

Figure 3. Antioxidant activity and total polyphenol content of broccoli stalk powder and bread samples. Different lowercase letters within each parameter (TPC and AA respectively) and for each sample represent significant differences according to Tukey's-b test ($p < 0.05$).

Increasing the substitution level of BSP led to increased values of AA and TPC, although this trend was not progressive for any of the parameters. Thus, the AA of

https://doi.org/10.3390/molecules31091414

bread with 7% BSP increased by approximately 20% compared with the control, while the TPC of bread with 5% BSP increased by 51.7% compared with the control. Positive correlations between AA and TPC have been reported, although the correlation coefficients varied relatively largely depending on the method used to determine AA, such as DPPH and FRAP [31]. No significant differences ($p < 0.05$) were determined between the total polyphenol content of BC and B3BSP. In terms of antioxidant activity, all samples were significantly different ($p < 0.05$).

The TPC content of the experimental sample increased at 3% and 5% supplementation with BSP, respectively, while a decreasing trend was observed at a high level (7%) of supplementation. This can be due to a different degree of inactivation of polyphenol oxidase at elevated temperatures [42].

2.6. Small-Scale Consumer Test and Colorimetric Analysis of Bread

The Small-scale Consumer Test was conducted in accordance with the Informed Consent and Ethics Statement for research involving human participants, as described below. The test indicated good acceptability of breads for 3% and 5% supplementation levels (Figure 4), in agreement with previous studies on the inclusion of broccoli powder in bread [10]. In contrast, a higher level of BSP (7%) negatively affected all visual and textural properties. The pronounced greenish color of this sample and its increased crumb density had low scores.

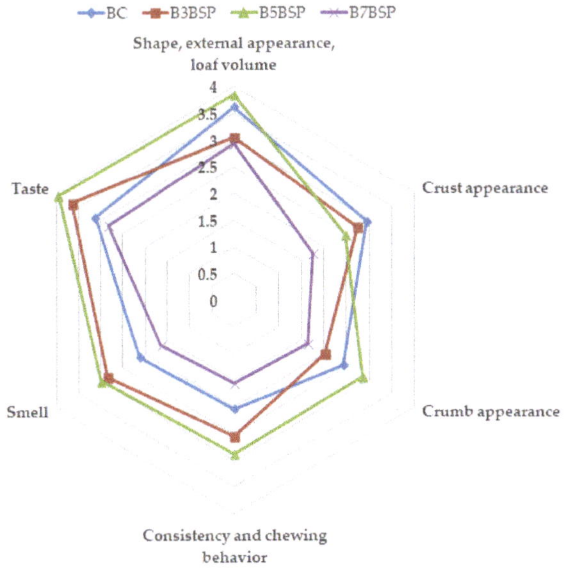

Figure 4. Sensory evaluation of control and BSP-enriched breads.

Lightness is an important characteristic in bread, particularly for assessing product quality and consumer acceptability. It reflects flour quality and influences consumer perception, as visual appearance is one of the primary criteria in choosing the assortment of bread. The higher lightness observed in fortified bread compared with the control may be attributed to the pale green color of BSF, which causes the dilution of the carotenoids in flour. Additionally, the fiber content of broccoli stalk flour may affect how the bread surface reflects light.

https://doi.org/10.3390/molecules31091414

The L^* values (Table 4) suggest a lighter crumb color in all samples with incorporated broccoli powder compared to the control, probably due to a reduced extent of Maillard reactions [41].

Table 4. Colorimetric parameters of bread crumb samples.

Sample	L^*	a^*	b^*
BC	52.39 ± 0.12 [a]	5.83 ± 0.14 [a]	11.15 ± 0.06 [a]
B3BSP	60.75 ± 0.12 [b]	−0.94 ± 0.08 [b]	12.89 ± 0.09 [b]
B5BSP	56.6 ± 0.3 [c]	−1.89 ± 0.01 [c]	14.10 ± 0.12 [c]
B7BSP	55.1 ± 0.09 [d]	−2.99 ± 0.1 [d]	18.2 ± 0.09 [d]

BC—control bread; B3BSP—bread supplemented with 3% (*w*/*w*) broccoli stalk powder; B5BSP—bread supplemented with 5% (*w*/*w*) broccoli stalk powder; B7BSP—bread supplemented with 7% (*w*/*w*) broccoli stalk powder; L^*—lightness; a^*—greenness; b^*—yellowness; different lowercase letters within the column and for each group of samples represent significant differences according to Tukey's-b test ($p < 0.05$).

All L^* values were higher than 50, indicating a relative light crumb color. Changes in crumb color characteristics were attributed to supplementation with BSP, although no direct relationship between the level of addition and L^* values was observed, as was also the case for a^* and b^* values. The observed differences in L^* values are likely associated with variations in the composition [43].

Negative a^* values were determined for all fortified breads, indicating a green color. The positive b^* values increased with increasing levels of addition, indicating a shift toward yellow coloration.

Delta E (ΔE) is a quantitative metric that expresses the difference between two colors, indicating how distinguishable they are to the human eye. In brief, ΔE represents the "distance" between two colors within a color space, typically the CIELAB system. The ΔE values for breads formulated with broccoli stalk powder were 10.89 for the 3% BSP formulation, 9.27 for the 5% BSP formulation, and 11.61 for the 7% BSP formulation.

2.7. Correlation Analysis of Chemical Parameters and Colorimetric Attributes

Pearson correlation analysis (Table 5) was applied to provide an integrated view on the chemical parameters characterizing the experimental bread.

Table 5. Pearson correlation analysis of key parameters.

	Moisture	Porosity	Elasticity	Volume	Acidity	AA	TPC	Lightness	TDF
Moisture	1	−0.844 **	0.547	−0.993 **	0.878 **	0.893 **	0.795 **	0.275	0.947 **
Porosity		1	−0.089	0.830 **	−0.515	−0.808 **	−0.645 *	−0.719 **	−0.803 **
Elasticity			1	−0.548	0.866 **	0.247	0.728 **	−0.617 *	0.552
Volume				1	−0.870 **	−0.913 **	−0.764 **	−0.269	−0.940 **
Acidity					1	0.663 *	0.825 **	−0.205	0.864 **
AA						1	0.459	0.459	0.834 **
TPC							1	0.013	0.782 **
Lightness								1	0.256
TDF									1

** correlation is significant at the 0.01 level; * correlation is significant at the 0.05 level.

Pearson correlation analysis revealed strong negative correlations between bread moisture and volume, bread volume and AA, and bread volume and TDF ($p < 0.01$). Strong positive correlations were determined between moisture and acidity, moisture and AA, and moisture and TDF ($p < 0.01$). A significant moderate positive correlation between TPC and moisture ($r = 0.795$, $p < 0.01$) was also established. No correlation was found

https://doi.org/10.3390/molecules31091414

between AA and TPC, possibly due to the influence of other chemical constituents on the antioxidant activity of bread. According to Zhao et al. [28], flavonoids are the primary contributors to the antioxidant activity in broccoli floret and broccoli by-products, including the stalk. In contrast, a strong correlation has been observed between TPC and DPPH-radical-scavenging activity in broccoli by-products [40].

Strong positive correlations were found between TDF and antioxidant activity (r = 0.834, $p < 0.01$), as well as TDF and TPC (r = 0.782, $p < 0.01$). These associations underline that broccoli stalk powder contributed both to the fiber enrichment of breads and to an increased content of bioactive compounds. The results are consistent with those reported by Quizhpe et al. [44].

Principal Component Analysis (Figure 5, Table 6) was applied to correlate the results of the physicochemical analysis with the profile of the four experimental breads. Principal component 1 (PC1) accounts for 67.425% of the variance, while Principal component 2 (PC2) accounted for 26.294% of the variance.

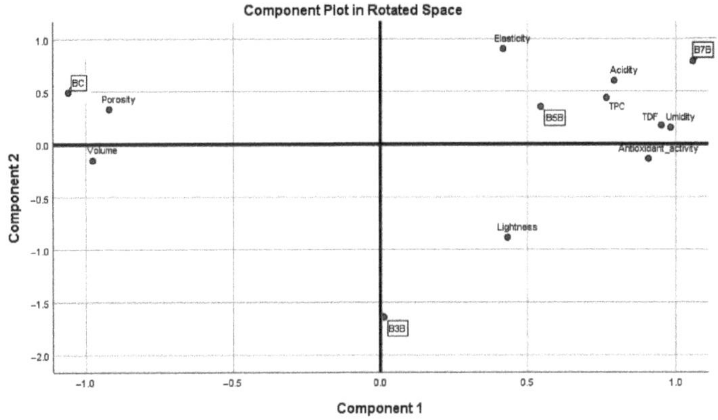

Figure 5. Principal Component Analysis.

Table 6. Component Score Coefficient Matrix.

Rotated Component Matrix [a]		
	Component	
	1	2
Moisture	0.982	0.156
Porosity	−0.921	0.332
Elasticity	0.418	0.905
Volume	−0.977	−0.153
Acidity	0.791	0.600
Antioxidant activity	0.908	−0.140
TPC	0.765	0.438
Lightness	0.433	−0.888
TDF	0.951	0.176

Extraction method: Principal Component Analysis; rotation method: Varimax with Kaiser Normalization [a] rotation converged in three iterations.

Moisture, acidity, antioxidant activity, total polyphenol content, and total dietary fiber are positioned in the positive part of PC1. This component is negatively associated with porosity and bread volume, indicating that PC1 characterizes breads with less porous and less voluminous structures.

 https://doi.org/10.3390/molecules31091414

Bread elasticity, as previously described in relation to the influence of BSP supplementation, had a high loading in the positive part of PC2. Component 2 was also primarily defined by lightness (negative loading), indicating a contrast between rheological and visual properties. Samples with higher elasticity tended to exhibit lower lightness values.

Overall, the PCA indicates that the main sources of variation arise from differences in compositional and structural characteristics (PC1) and from texture–appearance relationships (PC 2), highlighting the complexity and multidimensional nature of the breads.

Although bread porosity is important for consumer acceptance, this vector is located close to the control (BC). Figure 5 highlights the positive influence of wheat flour supplementation with 5% and 7% BSP respectively on the main physico-chemical and bioactive characteristics of the product. PC1 shows that composition and structure are the main drivers of variation, while PC2 indicates differences in texture and appearance. The control sample (BC) was structurally light/porous but firm, with lower moisture, TDF content, and antioxidant activity. The sample enriched with 3% BSP was located at the bottom center, characterized by a moderate chemical composition, low elasticity, and lighter color, suggesting that further optimization of the composition and/or texture is needed. In contrast, the samples enriched with 3% BSP and 5% BSP were positioned in the top-right region, exhibiting high elasticity and high values of the chemical parameters (moisture, TDF, and antioxidant activity). Thus, B5BSP and B7BSP combine high nutritional content with good texture.

3. Discussion

The flour used in the experiments presented a balanced rheological profile, with medium-strength gluten, capable of supporting standard technological processes. The farinogram–alveogram correlation confirmed that the analyzed flour is suitable for current bakery products, with relatively short processes and moderate hydration, where the P/L balance and rapid dough development are exploited. Accordingly, because the wheat flour had medium-strength gluten, it was assumed that the dough structure can absorb the addition of broccoli stalk powder without compromising its cohesion and extensibility. In moderate doses (e.g., 5–10%), broccoli fiber can improve the water-holding capacity, contributing to moister breads with improved softness. Dough supplemented with BSP required slightly longer and more careful kneading to properly activate the gluten [45]. Based on previous findings, the supplementation of wheat flour with BSP did not exceed 10%.

Previous studies have reported an increase in the moisture content of bread following the incorporation of powdered tomato co-products into their formulations [46]. Sayed-Ahmad et al. [47] observed an increase in the moisture content of bread following its fortification with chia seed flour, which was attributed to the higher content of water-soluble fiber in the chia-enriched bread.

The starch degradation, closely related to the water content of bread, is known as being responsible for the bread's staleness. Fiber incorporation into the dough through BSP supplementation enhances hydration of the crumb matrix, potentially contributing to the bread texture, staling retard, and improved freshness retention. However, the shelf life of the product could be threatened because high levels of moisture can negatively influence the microbiological stability, so that optimization of technological parameters, packaging, and storage processes will play a decisive role in both scopes.

High addition levels of BSP lead to modification of the carbohydrate composition of dough and consequently to changes in fermentation activity of the yeast. Thus, BSP brings fiber that can slow down fermentation, resulting in lower acidity in some cases. By another hand, the nutritional compounds can stimulate microbiota, with the effect being

https://doi.org/10.3390/molecules31091414

the high production of organic acids. Not in the last, the proteins from broccoli stalk flour can increase the buffering capacity of the dough. Supplementing wheat flour with BSP tends to increase the bread acidity, but the effect depends on the level of addition and its impact on yeast and lactic acid bacteria responsible for alcoholic fermentation and lactic acid fermentation of the dough respectively.

In terms of consumers' acceptability and shelf life of the product, high acidity levels may affect both the flavor profile of bread and its microbial stability. Although the effect of acidity appeared to be less pronounced than that on the loaf volume and porosity of bread, it remains technologically relevant. Finally, the moisture and acidity of bread behave increased proportionally with the level of fiber inclusion from broccoli stalk, suggesting that both hydration and biochemical effects are dose-dependent.

The gluten dilution due to the enrichment of wheat flour with broccoli stalk powder led to weakened gas-retention capacity of the dough [48]. Reducing the gas retention as a result of the disruption of the gluten network by BSP also led to a denser crumb structure of the bread, in opposition to its porosity. Korus et al. [14] observed a decrease in porosity of the bread enriched with powdered kale, but compared with the control, the differences were non-significant.

Interactions between phenolic compounds and gluten proteins have been reported to be responsible for various effects on the microstructure of bread and several of its sensory properties [49]. High concentrations of tea polyphenols decrease disulfide bonds, weaken the gluten network, and also reduce the loaf volume [50]. Similarly, a decrease in the loaf volume of vegetable-flour-loaded bread was determined by Mastromatteo et al. [51], Wang et al. [52], and González-Montemayor et al. [53]. The texture, an essential sensory attribute, was significantly influenced by the addition of celery powder, causing an increase in firmness and a denser core structure [52].

In agreement with the findings of this study, adding powdered broccoli co-products at a 2% (*w*/*w*) level to bread formulations resulted in a reduction in the specific volume compared to the control sample [41]. Similar results were reported by Lee [54], Anwar et al. [55], and Lafarga et al. [9]. This effect has been attributed to the dilution of starch and gluten following the addition of broccoli powder, as well as by a reduction in fully hydrated starch granules due to competition for water between the powder and the starch [41]. Furthermore, broccoli co-products are rich in glucosinolates, compounds with strong antimicrobial activity; their incorporation into bread formulations may therefore modulate yeast performance, potentially affecting the bread volume even at low inclusion levels [41].

Whole legume flours (green beans, peas, mesquite), rich in dietary fiber, contributed to increasing the fiber content of sourdough bread but moderately to the antioxidant activity of the bread. The additions produced a denser bread texture and higher water absorption of the dough [53]. Also, fibers isolated from broccoli stalks demonstrated a high water-absorption capacity, increasing the firmness of the bread crumb and reducing the elasticity of the dough. They contribute to a denser crumb structure. Fibers can dilute the gluten network, likely resulting in a decrease in specific volume and a more compact crumb structure. Fiber interaction with gluten may require a hydration adjustment and formulation optimization. Not in the end, dietary-fiber-rich fractions isolated from broccoli stalks may induce darker/greener hues [56].

The composition of broccoli dietary fiber (DF) is complex. The insoluble dietary fiber (IDF) in broccoli is predominantly composed of cellulose, hemicellulose, and lignin, whereas its soluble dietary fiber (SDF) comprises pectins, gums, mucilages, and β-glucans. The contents of IDF and SDF are also influenced by the broccoli cultivar [45]. However, the functional properties of individual components of DF remain insufficiently elucidated [27].

 https://doi.org/10.3390/molecules31091414

Consistent with total dietary fiber content, broccoli stems contain higher levels of both insoluble dietary fiber (IDF) and soluble dietary fiber (SDF) compared to leaves and inflorescences, with IDF representing the predominant fraction due to the physiological functions of the stalks [38,45]. Unlike SDF, IDF does not form gels upon contact with water, but is capable of retaining water within its structural matrix, thereby enhancing intestinal transit [38].

The total dietary fiber (TDF) of broccoli stalk powder was high, by 33.58 ± 1.31%, making it an excellent raw material for product enrichment. The processing method did not result in a significant change in the total fiber content of broccoli stems subjected to hot air drying; nevertheless, lower fiber content was observed in samples derived from lyophilized stems [45]. The cruciferous-based residues, including those derived from broccoli, can be efficiently valorized through innovative extraction and biotransformation techniques [57]. Extraction methods need to be optimized to obtain functional fibers. Accordingly, green extraction methods have been applied to obtain novel fiber-rich ingredients, with antioxidant and technological properties from broccoli stalks [56].

The incorporation of vegetable flours, including carrot, tomato, beetroot, and broccoli, into bread has been shown to enhance antioxidant levels [48]. Studies have revealed that extracts derived from broccoli stems possess higher antioxidant activity than those from florets [29]. Furthermore, although beneficial compounds present in cruciferous vegetables may be significantly reduced during thermal processing, some studies indicate that heat-sensitive ingredients can exhibit greater stability when incorporated into baked products [58]. The antioxidant properties of broccoli by-products are related to their comprehensive phytochemical profile. For example, they were found to be abundant in sulforaphane, an isothiocyanate renowned for its potent antioxidant properties [45], as the high content of value-added compounds, such as vitamins and phenolic compounds, gives broccoli and its various parts beneficial properties, including antioxidant activity [38]. It has been also suggested that the insoluble dietary fiber, which is present in high amounts in broccoli stalks, possesses the ability to associate with phenolic compounds through both covalent and non-covalent interactions, thereby serving as a structural matrix for the retention of these bioactive molecules [59].

The total phenolic concentration in stems, by 1.41 mg GAE/g DW, was 2.9-fold lower than TPC found in broccoli leaves [39]. The authors determined that broccoli stem tissue exhibited a DPPH-radical-scavenging activity of 16.4%. In another study, the TPC of broccoli stems was reported as 312 ± 41 mg/100 g DW [41], with a moisture content of 90.1 ± 1.4% for the stems used in breadmaking

Large variation among cultivars in terms of phenolic compounds and ROS-scavenging capacity has been reported [40]. Therefore, substantial genetic variation in broccoli tissue antioxidants should be considered by growers and breeders to enhance sustainability and productivity, particularly for by-product utilization [39]. Contrary to the abovementioned data reported by Liu et al. [39], higher TPC was found in broccoli stalks, reaching 9.39 mg GAE/g DW [38].

It has been observed that adding broccoli to bread affects oxidative stability during gastrointestinal digestion [60], while broccoli leaf powder significantly improves the antioxidant potential of gluten-free bread [11]. The incorporation of freeze-dried broccoli co-products into crackers has been reported to significantly increase the TPC and antioxidant capacity [14]. Gawlik-Dziki et al. [10] stated that incorporating broccoli sprouts into wheat bread at levels of 1–5% (w/w) led to an increase in antioxidant capacity and the significant enrichment with phenolic compounds; however the relationship between TPC and the percentage of addition was far from linear [10]. Similarly, Lee et al. [54] determined

https://doi.org/10.3390/molecules31091414

DPPH·-scavenging activity ranging from17.6% to 45.5% when broccoli was incorporated into breads at 2.5–10.0% (*w*/*w*).

The increases observed in antioxidant capacity and TPC in some experimental samples can be attributed to enzymatic action on the plant matrix [44]. This process degrades cell wall components, releasing bound polyphenols, which are subsequently converted into soluble and more bioavailable forms. However, although phenolic compounds are generally resistant to heat, their levels in the final product may decrease due to the breakdown of phenolics during processing [33]. Furthermore, the retention of bioactive compounds is strongly influenced by processing conditions, which may explain why the enriched breads did not exhibit significantly higher antioxidant activity or phenolic content compared to the control [61].

Incorporating vegetables into bread is generally well received by consumers and represents a convenient strategy to increase vegetable consumption [62] while delivering plant-based ingredients with demonstrated health effects [48]. Texture-related sensory changes in bread are primarily caused by interactions between phenolic compounds or other antioxidants present in extracts from flours obtained from agri-food industry by-products and gluten proteins. These antioxidants can reduce disulfide bonds, which play a crucial role in maintaining the structure of the gluten network [49].

In agreement with our results, Anwar et al. [55] reported that breads containing 1–2% (*w*/*w*) broccoli achieved the highest overall acceptability scores, while higher concentrations resulted in unacceptable textural and sensory properties. Additionally, bread formulations containing 2% (*w*/*w*) powdered broccoli stalks maintained their acceptability, appearance, and texture [41]. The overall sensory acceptability of gluten-free mini sponge cakes also decreased with increasing levels of broccoli leaf powder [30].

The color changes were consistent with findings reported in the literature, which show that incorporating vegetables or by-products from the vegetable industry into bakery formulations significantly affects product color [46]. Accordingly, the incorporation of powdered broccoli stalks [10] and broccoli sprouts [41] also resulted in lower a^* values in bread compared to the control. Similarly, such breads exhibited significantly more negative a^* values and higher b^* values than the control bread, indicating a greater contribution of green and yellow hues [11]. It was reported that breads containing broccoli exhibited a more pronounced green coloration, but the incorporation of broccoli did not affect the overall acceptability of the breads [41].

Color is a key quality parameter that significantly influences the consumer acceptance of food products [20,43]. Broccoli flour retains the carotenoids and chlorophylls of the vegetable, which confer an attractive green color when incorporated into food formulations [25]. However, β-carotene, which is most abundant in florets and leaves, is not present in stalks [38]. If broccoli is heat-processed before use, its behavior in dough may change, indirectly influencing the texture, crumb stability, and final color of the bread [42].

During the baking process, a range of compounds is formed as a result of chemical reactions, such as caramelization and Maillard reactions, leading to changes in the color of the final product. Consequently, variations in protein content and phenolic compounds may influence the lightness of the finished products [20,43]. A decrease in the L^* value of breads containing broccoli powder compared to the control was obtained by Lee [54]. However, the broccoli powder improved the overall nutritional value of the bread.

Finally, phenolic compounds increase functional value of the bread but may have negative effects on taste, texture, and appearance concentration, so that optimization is necessary to balance health and consumers' acceptability [49,55].

https://doi.org/10.3390/molecules31091414

In bread production with the addition of broccoli stalk powder, ΔE serves as a useful parameter for evaluating color changes relative to the control bread. All samples containing BSP exhibited high ΔE values, indicating pronounced chromatic differences. Broccoli stalk powder (BSP) contains pigments such as chlorophyll and phenolic compounds, which can contribute to greener or darker color tones, reduced lightness (darker crumb), and additional color changes during baking, including those associated with Maillard reactions. While ΔE objectively quantifies these differences, the bread does not merely appear "greener." For bread formulated with functional ingredients, higher levels of supplementation may be acceptable, provided the product is appropriately positioned and evaluated within the CIELAB color space.

According to Nikolov et al. [63], two thresholds can be mentioned separately: if $\Delta E < 1$, the color difference is imperceptible, while if $\Delta E > 3$, it is visible. In the present work, all ΔE values are greater than 5, so the color difference is clearly visible and significant compared to the control bread. There was no linear increase between the percentage of BSP addition and ΔE. Bread with 5% BSP had a lower ΔE than bread with 3% and 7% BSP, respectively, possibly as an effect of the non-uniform distribution of pigments and/or chemical reactions (Maillard/oxidation). A recent study clearly supports the threshold $\Delta E > 3$ as a perceptible difference in bakery products [44]. In the same study, ΔE ranged between 5.32 and 15.18 for reformulated breads containing broccoli extracts. Previous studies reported an increased green hue in broccoli-enriched crackers. The final product showed a ΔE value greater than 3, indicating a perceptible color difference compared with the control [31].

The incorporation of BSP resulted in high ΔE values, which were also accompanied by a reduction in the bread loaf volume. Although this relationship is not mathematically direct, it suggests a concurrent influence of BSP on both the crumb color and bread structure. The addition of 3% BSP resulted in a relatively high ΔE value (10.89), while the overall sensory score was comparable to that of the control sample, indicating no clear improvement in organoleptic quality. The incorporation of 7% BSP into wheat flour produced the highest ΔE value (11.61) but also the lowest sensory score among the tested formulations. The bread color was perceived as excessively intense (greenish/dark), which may have been associated with a perceived product defect. In contrast, the incorporation of 5% BSP resulted in a comparatively lower ΔE value (9.27) and the highest sensory score (3.34). The modified color of the bread was acceptable or even pleasant, possibly perceived as "natural/healthy", which suggests that this addition proportion, 5% BSP, is optimal. Spearman's rank correlation confirmed a perfect monotonic relationship ($\rho = -1.00$, $p < 0.01$), indicating that higher color differences consistently correspond to lower sensory acceptance.

Overall, the results suggest that BSP addition significantly influences both the physical and sensory properties of bread, highlighting a trade-off between nutritional enhancement and product acceptability. The 5% BSP level can be considered optimal, as it provides a balance among color changes, loaf volume, and sensory acceptance.

4. Materials and Methods

4.1. Material

Broccoli (*Brassica oleracea* L.) was purchased from a local market in Romania. The florets were removed, and the stalks were washed and cut into cubes of approximately 1 cm on each side. The prepared vegetable material was freeze-dried (Biobase BK-FD10P, Biobase Bioindustry, Jinan, China) and subsequently powdered using a laboratory mill. Care was taken during powdering to avoid heating the material.

 https://doi.org/10.3390/molecules31091414

4.2. Dough and Bread Making

All raw materials used in the bread formulation, including wheat flour, yeast, and salt, were purchased locally.

The flour used in the formulation of both control and experimental breads was wheat bread flour type 00 (average moisture content 13.56% wb, 0.6% ash content, 12% protein). Broccoli stalk powder (BSP) was incorporated into the wheat flour at levels of 3, 5, and 7%, as shown in Table 7.

Table 7. Experimental variants of bread.

Sample	Wheat Flour (g)	Water (mL)	Yeast (g)	Salt (g)	Broccoli Stalk Powder (g)
Control	500	320	20	7.5	-
Bread with 3% broccoli (B3BSP)	500	320	20	7.5	15
Bread with 5% broccoli (B5BSP)	500	320	20	7.5	25
Bread with 7% broccoli (B7BSP)	500	320	20	7.5	30

For each experimental batch, the dough was mixed for 7 min in a spiral mixer (MB-1300, Biovita, Cluj-Napoca, Romania). Mixing was performed in two stages: homogenization and actual mixing. The bread was produced using the direct method.

The dough batches were fermented in trays at 30–32 °C for 40 min, with a relative humidity of air by 75%.

The dough was baked in molds in an electric oven (Zilan, model ZLN2433, Istanbul, Turkey) at 180 °C for 25 min. After baking, the breads were removed from the molds and cooled rapidly.

4.3. Physico-Chemical Analysis

All chemicals used were purchased from Merck (Darmstadt, Germany).

Flour and bread moisture contents were determined by measuring the weight loss of the sample after drying at 130 $\pm$ 1 °C for one hour, until a constant mass was achieved.

The water-holding capacity of the wheat flour and BSP was determined by hydrating 1 g of sample for 30 min, followed by centrifugation for 5 min at 1000 rpm (EBA 21 Hettich Zentrifugen 1004, Tuttlingen, Germany).

The wet gluten content of the wheat flour was determined by washing the dough prepared from the flour sample with sodium chloride and drying off the obtained gluten. For the calculation, the quantity of retained water is divided by the amount of the sample.

The acidity of bread, defined as the total of acids and acid-reactive compounds present, was determined via titration with a base.

Fornet equipment (device reproducing Bread Volume Meter 13300, in-house developed system, Valahia University of Targoviste, Targoviste, Romania) was used for the determination of the loaf volume, as described in the standard AACC method 10-05.01 [64].

Porosity represents the total volume of pores within a given volume of bread crumb, expressed as a percentage. The total pore volume was determined from a known crumb volume, based on its mass and density. Portions of the crumb were cut from different points of the crumb section using a cylindrical perforator, and their volumes were calculated to obtain an average sample, as porosity of the bread varies across the section.

Porosity, together with pore structure (uniformity and size) and wall thickness, is an important indicator of product quality, which largely affects the digestibility of the bread.

https://doi.org/10.3390/molecules31091414

Elasticity values were determined by pressing a piece of bread crumb of a specified shape for a set period and measuring its return to the original position after the force was removed. The elasticity of the bread core is expressed as a percentage, calculated as the ratio between the height after pressing and recovery and the initial height of the crumb cylinder.

The Brabender Farinograph (Brabender GmbH & Co., KG, Duisburg, Germany) was used to measure the rheological properties of the dough in the time of the mixing, as described in the standard AACC method 54-21.02 [65]. The Chopin Alveograph (Alveolab, KPM Analytics, Westborough, MA, USA) was used to measure the quality of flour and the behavior of dough during fermentation by simulating the leavening process, as described in the standard AACC method 54-30.02 [66].

The Falling Number (FN) was determined by measuring the time (in seconds) needed for a piston to fall through a gel formed from a suspension of flour and heated water, according to the AACC method 56-81.03 [67].

Total dietary fiber in broccoli stalk powder and enriched breads was determined according to the AOAC 985.29 enzymatic gravimetric method, using the Total Dietary Fiber Assay kit from Megazyme (K-TDFR-100A/K-TDFR-200A 04/17, Megazyme International Ireland Ltd., Bray, Co., Wicklow, Ireland). The method involves enzymatic digestion followed by gravimetric determination of the residue.

Hydroalcoholic extracts of the samples were obtained via ultrasonication. Briefly, a solvent of 50% ethanol (absolute ethanol:double distilled water, 1:1, *v*/*v*) acidified with 0.1% (*v*/*v*) concentrated hydrochloric acid, was added to the material at a ratio of 0.5:10 (material: solvent). Extraction was performed via ultrasonication (PRO 70, 40 kHz, ASonic, Shenzhen, China) at room temperature for 20 min, with intermittent agitation. The supernatant was separated via centrifugation during two stages of extraction.

The total phenolic content (TPC) was determined using the Folin–Ciocalteu method. To 0.5 mL of extract, 10% Folin–Ciocalteu reagent was added with intermittent stirring, followed by 2 mL of 8% Na_2CO_3 solution. After 60 min at room temperature, the absorbance of the samples was measured against a blank sample at 765 nm (UV-VIS spectrophotometer model UV 1720, Shanghai Yoke Instrument Co., Ltd., Shanghai, China). The results were expressed as gallic acid equivalents (GAE).

The antioxidant activity of BSP and bread was determined using the DPPH method. To 0.75 mL of extract, 1.5 mL of DPPH solution was added and vortexed. The samples, along with a DPPH blank ($A_{blank\ DPPH}$) and an extract blank ($A_{blank\ extract}$), were kept at room temperature in the dark for 30 min. The absorbance of the samples (A_{sample}) was measured against ethanol as a control at 517 nm. The results were expressed as follows:

$$AA\% = \frac{\left[A_{blank\ DPPH} - \left(A_{sample} - A_{blank\ extract}\right)\right]}{A_{blank\ DPPH}} \times 100$$

The color of the samples was measured in the CIELAB system using an NR 110 Precision Colorimeter (3nh-ThreenhTechnology Co., Shenzhen, China), equipped with CQCS3 software version 3.4.3, a D65 light source, and a 4 mm aperture. The following color parameters were recorded: lightness (*L**, 0 = black, 100 = white), redness (*a**, <0 = green, >0 = red), and yellowness (*b**, <0 = blue, >0 = yellow).

The overall color difference (ΔE) for each formulation with BSP supplementation relative to the control bread was determined using the following equation:

$$\Delta E = \sqrt{(L_s - L_c)\hat{}2 + (a_s - a_c)\hat{}2 + (b_s - b_c)\hat{}2}$$

 https://doi.org/10.3390/molecules31091414

where the index "s" refers to the color components of the sample with BSP addition and the index "c" to the control sample.

4.4. Small-Scale Consumer Test

A small-scale consumer test was conducted on the bread samples. The panel consisted of 23 untrained assessors (aged 20–35 years), who evaluated overall acceptability using a five-point hedonic scale, ranging from 0 (dislike extremely) to 4 (like extremely).

4.5. Statistical Analysis

Experimental data are presented as means ± standard deviations ($n = 3$). Statistical analysis was performed using IBM SPSS Statistics 26. One-way Analysis of Variance (ANOVA) was conducted to evaluate the effect of the BSP addition level on the measured parameters ($p < 0.05$).

Pearson correlation analysis and Principal Component Analysis (PCA) were performed using the correlation matrix to gain deeper insights into the quality of the breads.

5. Conclusions

The incorporation of by-products into new food products allows for reduced production costs, lower waste-management expenditures, and the creation of new economic value streams. In line with the United Nations Sustainable Development Goals, converting by-products into functional ingredients contributes to the development of sustainable consumption and production patterns [7].

Broccoli stalk is a valuable by-product of the food industry, so its use can contribute to reducing food waste, valorizing unused nutritional components, and the development of functional products.

The moisture, acidity, porosity, and loaf volume of bread were significantly affected by the addition of broccoli stalk powder. The dietary fiber in broccoli stalk not only increases the total fiber content of the bread but also influences its technological, sensory, and nutritional properties. The incorporation of 5% freeze-dried broccoli stalk into wheat flour with a balanced rheological profile and gluten system resulted in a 50% increase in the total polyphenol content and an 8% increase in antioxidant activity compared to the control bread. Sensory acceptability decreased only at the highest level of broccoli stalk powder supplementation (namely, 7%) in the recipe.

Enrichment of the wheat flour with 5% broccoli stalk flour was identified, based on the evaluated parameters, as providing the optimal balance among functional enhancement, sensory acceptance, and the bioactive quality of bread. Further research is needed to investigate the rheological behavior of the dough, shelf-life stability of the bread, and the bioactive compounds present in the product. A complete characterization will support the recommendation of broccoli-stalk-flour-enriched bread for industrial applications. Special attention should be given to the dynamics of bread parameters during storage, as well as to the in vitro simulated digestion of the bioactive compounds.

Author Contributions: Conceptualization, E.R.M., A.T. and C.L.B.; methodology, E.R.M.; validation, C.L.B.; formal analysis, E.R.M., A.A.G., E.C.P. and A.T.; investigation, E.R.M., A.A.G., E.C.P. and A.T.; writing—original draft preparation, C.L.B. and A.T.; writing—review and editing, C.L.B.; supervision, C.L.B.; funding acquisition, C.L.B. All authors have read and agreed to the published version of the manuscript.

Funding: This work was supported by Valahia University of Targoviste.

 https://doi.org/10.3390/molecules31091414

Institutional Review Board Statement: All participants involved in the Small-scale Consumer Test were informed about the purpose and methodology of the study. The tested breads were safe for consumption and posed no health risks to participants. The data collected were used solely for research purposes. No sensitive personal information was recorded.

Informed Consent Statement: Informed consent was obtained from all participants voluntary involved in the Small-scale Consumer Test prior to inclusion.

Data Availability Statement: The original contributions presented in this study are included in the article. Further inquiries can be directed to the corresponding author.

Acknowledgments: The use of laboratory facilities of Valahia University of Targoviste is kindly acknowledged.

Conflicts of Interest: The authors declare no conflicts of interest.

References

1. Gallardo, M.A.; Martínez-Navarro, M.E.; Álvarez-Ortí, M.; Pardo, J.E. Utilization of Flours Derived from the Waste from the Frozen Vegetable Industry for Bakery Product Production. *Agriculture* **2024**, *14*, 1747. [CrossRef]
2. Martins, M.R.B.F. A Circular Economy Approach in the Agri-Food Chain: Using by-Products and Underexploited Raw Materials as Flours for Gluten-Free Bakery and as Coagulants for Wastewater Treatment. Ph.D. Thesis, Universidade de Trás-os-Montes e Alto Douro, Vila Real, Portugal, 2022.
3. Trigo, J.P.; Alexandre, E.M.C.; Saraiva, J.A.; Pintado, M.E. High Value-Added Compounds from Fruit and Vegetable by-Products–Characterization, Bioactivities, and Application in the Development of Novel Food Products. *Crit. Rev. Food Sci. Nutr.* **2020**, *60*, 1388–1416. [CrossRef]
4. Sukhikh, S.; Ivanova, S.; Dolganyuk, V.; Pilevinova, I.; Prosekov, A.; Ulrikh, E.; Noskova, S.; Michaud, P.; Babich, O. Evaluation of the Prospects for the Use of Microalgae in Functional Bread Production. *Appl. Sci.* **2022**, *12*, 12563. [CrossRef]
5. Duarte, S.; Sánchez-García, J.; Harasym, J.; Betoret, N. Enrichment of Rice Flour with Almond Bagasse Powder: The Impact on the Physicochemical and Functional Properties of Gluten-Free Bread. *Foods* **2025**, *14*, 2382. [CrossRef] [PubMed]
6. Hoxha, L.; Mestani, M.; Nuha, D.; Bytyçi, P.; Durmishi, B. Sustainable Applications of Cereal-Based Products Through the Utilisation of Fruit Waste. *Żywn. Nauka Technol. Jakość* **2024**, *31*, 174–198. [CrossRef]
7. Lau, K.Q.; Sabran, M.R.; Shafie, S.R. Utilization of Vegetable and Fruit By-Products as Functional Ingredient and Food. *Front. Nutr.* **2021**, *8*, 661693. [CrossRef]
8. Gómez, M.; Martinez, M.M. Fruit and Vegetable By-Products as Novel Ingredients to Improve the Nutritional Quality of Baked Goods. *Crit. Rev. Food Sci. Nutr.* **2018**, *58*, 2119–2135. [CrossRef]
9. Lafarga, T.; Gallagher, E.; Walsh, D.; Valverde, J.; Hayes, M. Chitosan-Containing Bread Made Using Marine Shellfishery Byproducts: Functional, Bioactive, and Quality Assessment of the End Product. *J. Agric. Food Chem.* **2013**, *61*, 8790–8796. [CrossRef]
10. Gawlik-Dziki, U.; Świeca, M.; Dziki, D.; Sęczyk, Ł.; Złotek, U.; Różyło, R.; Kaszuba, K.; Ryszawy, D.; Czyż, J. Anticancer and Antioxidant Activity of Bread Enriched with Broccoli Sprouts. *BioMed Res. Int.* **2014**, *2014*, 608053. [CrossRef] [PubMed]
11. Krupa-Kozak, U.; Drabińska, N.; Bączek, N.; Šimková, K.; Starowicz, M.; Jeliński, T. Application of Broccoli Leaf Powder in Gluten-Free Bread: An Innovative Approach to Improve Its Bioactive Potential and Technological Quality. *Foods* **2021**, *10*, 819. [CrossRef]
12. Ghendov-Mosanu, A.; Cristea, E.; Patras, A.; Sturza, R.; Padureanu, S.; Deseatnicova, O.; Turculet, N.; Boestean, O.; Niculaua, M. Potential Application of *Hippophae rhamnoides* in Wheat Bread Production. *Molecules* **2020**, *25*, 1272. [CrossRef]
13. Zafar, T.A.; Al-Hassawi, F.; Al-Khulaifi, F.; Al-Rayyes, G.; Waslien, C.; Huffman, F.G. Organoleptic and Glycemic Properties of Chickpea-Wheat Composite Breads. *J. Food Sci. Technol.* **2015**, *52*, 2256–2263. [CrossRef]
14. Korus, A.; Witczak, M.; Korus, J.; Juszczak, L. Dough Rheological Properties and Characteristics of Wheat Bread with the Addition of Lyophilized Kale (*Brassica oleracea* L. var. *sabellica*) Powder. *Appl. Sci.* **2022**, *13*, 29. [CrossRef]
15. Chang, R.-C.; Chia-Yen, L.; Shiau, S.-Y. Physico-Chemical and Sensory Properties of Bread Enriched with Lemon Pomace Fiber. *Czech J. Food Sci.* **2015**, *33*, 180–185. [CrossRef]
16. Yan, L.; Zhou, G.; Shahzad, K.; Zhang, H.; Yu, X.; Wang, Y.; Yang, N.; Wang, M.; Zhang, X. Research Progress on the Utilization Technology of Broccoli Stalk, Leaf Resources, and the Mechanism of Action of Its Bioactive Substances. *Front. Plant Sci.* **2023**, *14*, 1138700. [CrossRef]

https://doi.org/10.3390/molecules31091414

17. Kaiser, A.E.; Baniasadi, M.; Giansiracusa, D.; Giansiracusa, M.; Garcia, M.; Fryda, Z.; Wong, T.L.; Bishayee, A. Sulforaphane: A Broccoli Bioactive Phytocompound with Cancer Preventive Potential. *Cancers* **2021**, *13*, 4796. [CrossRef]
18. Crunkhorn, S. Broccoli Extract Lowers Glucose Levels. *Nat. Rev. Drug Discov.* **2017**, *16*, 530. [CrossRef]
19. Mthembu, S.X.H.; Mazibuko-Mbeje, S.E.; Moetlediwa, M.T.; Muvhulawa, N.; Silvestri, S.; Orlando, P.; Nkambule, B.B.; Muller, C.J.F.; Ndwandwe, D.; Basson, A.K.; et al. Sulforaphane: A Nutraceutical against Diabetes-Related Complications. *Pharmacol. Res.* **2023**, *196*, 106918. [CrossRef] [PubMed]
20. Mohammadi, M.; Nahidi, F.; Bolandi, M. Broccoli Seed Sprouts as a Functional Ingredient in Gluten-Free Cake Based on Rice Flour. *Appl. Food Res.* **2025**, *5*, 101198. [CrossRef]
21. Food and Agriculture Organization of the United Nations (FAO). Available online: https://www.fao.org/faostat/en/#home (accessed on 20 March 2026).
22. Food and Agriculture Organization of the United Nations (FAO). Crops and Livestock Products. Available online: https://www.fao.org/faostat/en/#data/QCL (accessed on 20 March 2026).
23. Granado-Castro, M.D.; Galindo-Riaño, M.D.; Gestoso-Rojas, J.; Sánchez-Ponce, L.; Casanueva-Marenco, M.J.; Díaz-de-Alba, M. Ecofriendly Application of Calabrese Broccoli Stalk Waste as a Biosorbent for the Removal of Pb(II) Ions from Aqueous Media. *Agronomy* **2024**, *14*, 554. [CrossRef]
24. Ares, A.M.; Nozal, M.J.; Bernal, J. Extraction, Chemical Characterization and Biological Activity Determination of Broccoli Health Promoting Compounds. *J. Chromatogr. A* **2013**, *1313*, 78–95. [CrossRef] [PubMed]
25. Fanesi, B.; Ismaiel, L.; Nartea, A.; Orhotohwo, O.L.; Kuhalskaya, A.; Pacetti, D.; Lucci, P.; Falcone, P.M. Bioactives and Technological Quality of Functional Biscuits Containing Flour and Liquid Extracts from Broccoli By-Products. *Antioxidants* **2023**, *12*, 2115. [CrossRef]
26. Petkowicz, C.L.O.; Williams, P.A. Pectins from Food Waste: Characterization and Functional Properties of a Pectin Extracted from Broccoli Stalk. *Food Hydrocoll.* **2020**, *107*, 105930. [CrossRef]
27. Shang, Y.; Zhang, W.; Dang, Y.; Gao, X. Physical Properties and Functional Characteristics of Broccoli-Soluble Dietary Fiber. *Food Biosci.* **2023**, *56*, 103272. [CrossRef]
28. Zhao, Y.; Zhang, Y.; Yang, H.; Xu, Z.; Li, Z.; Zhang, Z.; Zhang, W.; Deng, J. A Comparative Metabolomics Analysis of Phytochemcials and Antioxidant Activity between Broccoli Floret and By-Products (Leaves and Stalks). *Food Chem.* **2024**, *443*, 138517. [CrossRef] [PubMed]
29. Mallhi, I.Y.; Javed, F.; Hameed, S.; Tufail, T. Broccoli and Its By-Products: Turning Waste into Wellness. In *Importance of Plant Based Byproducts: Nutritional and Functional Properties*; Bader Ul Ain, H., Tufail, T., Saeed, F., Ansar Rasul Suleria, H., Eds.; Springer Nature: Cham, Switzerland, 2026; pp. 195–217.
30. Drabińska, N.; Ciska, E.; Szmatowicz, B.; Krupa-Kozak, U. Broccoli By-Products Improve the Nutraceutical Potential of Gluten-Free Mini Sponge Cakes. *Food Chem.* **2018**, *267*, 170–177. [CrossRef] [PubMed]
31. Lafarga, T.; Gallagher, E.; Bademunt, A.; Bobo, G.; Echeverria, G.; Viñas, I.; Aguiló-Aguayo, I. Physiochemical and Nutritional Characteristics, Bioaccessibility and Sensory Acceptance of Baked Crackers Containing Broccoli Co-products. *Int. J. Food Sci. Technol.* **2019**, *54*, 634–640. [CrossRef]
32. Drabińska, N.; Nogueira, M.; Szmatowicz, B. Valorisation of Broccoli By-Products: Technological, Sensory and Flavour Properties of Durum Pasta Fortified with Broccoli Leaf Powder. *Molecules* **2022**, *27*, 4672. [CrossRef]
33. Bokić, J.; Škrobot, D.; Tomić, J.; Šeregelj, V.; Abellán-Victorio, Á.; Moreno, D.A.; Ilić, N. Broccoli Sprouts as a Novel Food Ingredient: Nutritional, Functional and Sensory Aspects of Sprouts Enriched Pasta. *LWT* **2022**, *172*, 114203. [CrossRef]
34. Silva, E.; Sagis, L.M.C.; Van Der Linden, E.; Scholten, E. Effect of Matrix and Particle Type on Rheological, Textural and Structural Properties of Broccoli Pasta and Noodles. *J. Food Eng.* **2013**, *119*, 94–103. [CrossRef]
35. Lucera, A.; Costa, C.; Marinelli, V.; Saccotelli, M.A.; Del Nobile, M.A.; Conte, A. Fruit and Vegetable By-Products to Fortify Spreadable Cheese. *Antioxidants* **2018**, *7*, 61. [CrossRef] [PubMed]
36. Kamiloglu, S.; Ozdal, T.; Tomas, M.; Capanoglu, E. Oil Matrix Modulates the Bioaccessibility of Polyphenols: A Study of Salad Dressing Formulation with Industrial Broccoli By-products and Lemon Juice. *J. Sci. Food Agric.* **2022**, *102*, 5368–5377. [CrossRef] [PubMed]
37. Ferreira, S.S.; Passos, C.P.; Cardoso, S.M.; Wessel, D.F.; Coimbra, M.A. Microwave Assisted Dehydration of Broccoli By-Products and Simultaneous Extraction of Bioactive Compounds. *Food Chem.* **2018**, *246*, 386–393. [CrossRef] [PubMed]
38. Quizhpe, J.; Ayuso, P.; Rosell, M.D.L.Á.; Peñalver, R.; Nieto, G. Brassica Oleracea Var Italica and Their By-Products as Source of Bioactive Compounds and Food Applications in Bakery Products. *Foods* **2024**, *13*, 3513. [CrossRef]
39. Liu, M.; Zhang, L.; Ser, S.; Cumming, J.; Ku, K.-M. Comparative Phytonutrient Analysis of Broccoli By-Products: The Potentials for Broccoli By-Product Utilization. *Molecules* **2018**, *23*, 900. [CrossRef]

40. Hwang, J.-H.; Lim, S.-B. Antioxidant and Anticancer Activities of Broccoli By-Products from Different Cultivars and Maturity Stages at Harvest. *Prev. Nutr. Food Sci.* **2015**, *20*, 8–14. [CrossRef]
41. Lafarga, T.; Gallagher, E.; Bademunt, A.; Vinas, I.; Bobo, G.; Villaró, S.; Aguiló-Aguayo, I. Bioaccessibility, Physicochemical, Sensorial, and Nutritional Characteristics of Bread Containing Broccoli Co-Products. *J. Food Process. Preserv.* **2019**, *43*, e13861. [CrossRef]
42. Zhan, L.; Pang, L.; Ma, Y.; Zhang, C. Thermal Processing Affecting Phytochemical Contents and Total Antioxidant Capacity in Broccoli (*Brassica oleracea* L.). *J. Food Process. Preserv.* **2018**, *42*, e13548. [CrossRef]
43. Papagianni, E.; Kotsiou, K.; Biliaderis, C.G.; Lazaridou, A. Flaxseed and Sprouted Lentil Seeds as Functional Ingredients in the Development of Nutritionally Fortified "Clean-Label" Gluten-Free Breads. *Food Hydrocoll. Health* **2023**, *4*, 100165. [CrossRef]
44. Quizhpe, J.; Peñalver, R.; Ayuso, P.; Nieto, G. Nutritionally Improved Gluten-Free Breads Fortified with Soluble Fiber and Bioactive Compounds from Artichoke and Broccoli By-Products. *Molecules* **2026**, *31*, 152. [CrossRef] [PubMed]
45. Gudiño, I.; Casquete, R.; Martín, A.; Wu, Y.; Benito, M.J. Comprehensive Analysis of Bioactive Compounds, Functional Properties, and Applications of Broccoli By-Products. *Foods* **2024**, *13*, 3918. [CrossRef]
46. Majzoobi, M.; Ghavi, F.S.; Farahnaky, A.; Jamalian, J.; Mesbahi, G. Effect of Tomato Pomace Powder on the Physicochemical Properties of Flat Bread (Barbari Bread): Effect of Tomato Pomace on Properties of Flat Bread. *J. Food Process. Preserv.* **2011**, *35*, 247–256. [CrossRef]
47. Sayed-Ahmad, B.; Talou, T.; Straumite, E.; Sabovics, M.; Kruma, Z.; Saad, Z.; Hijazi, A.; Merah, O. Evaluation of Nutritional and Technological Attributes of Whole Wheat Based Bread Fortified with Chia Flour. *Foods* **2018**, *7*, 135. [CrossRef] [PubMed]
48. Amoah, I.; Taarji, N.; Johnson, P.-N.T.; Barrett, J.; Cairncross, C.; Rush, E. Plant-Based Food By-Products: Prospects for Valorisation in Functional Bread Development. *Sustainability* **2020**, *12*, 7785. [CrossRef]
49. Czajkowska–González, Y.A.; Alvarez–Parrilla, E.; Del Rocío Martínez–Ruiz, N.; Vázquez–Flores, A.A.; Gaytán–Martínez, M.; De La Rosa, L.A. Addition of Phenolic Compounds to Bread: Antioxidant Benefits and Impact on Food Structure and Sensory Characteristics. *Food Prod. Process. Nutr.* **2021**, *3*, 25. [CrossRef]
50. Qin, W.; Pi, J.; Zhang, G. The Interaction between Tea Polyphenols and Wheat Gluten in Dough Formation and Bread Making. *Food Funct.* **2022**, *13*, 12827–12835. [CrossRef]
51. Mastromatteo, M.; Danza, A.; Guida, M.; Del Nobile, M.A. Formulation Optimisation of Vegetable Flour-loaded Functional Bread Part I: Screening of Vegetable Flours and Structuring Agents. *Int. J. Food Sci. Technol.* **2012**, *47*, 1313–1320. [CrossRef]
52. Wang, N.; Xu, Y.; Chao, H.; Zhang, M.; Zhou, Y.; Wang, M. Effects of Celery Powder on Wheat Dough Properties and Textural, Antioxidant and Starch Digestibility Properties of Bread. *J. Food Sci. Technol.* **2020**, *57*, 1710–1718. [CrossRef]
53. González-Montemayor, A.M.; Solanilla-Duque, J.F.; Flores-Gallegos, A.C.; López-Badillo, C.M.; Ascacio-Valdés, J.A.; Rodríguez-Herrera, R. Green Bean, Pea and Mesquite Whole Pod Flours Nutritional and Functional Properties and Their Effect on Sourdough Bread. *Foods* **2021**, *10*, 2227. [CrossRef]
54. Lee, S.-H. Quality and Antioxidant Properties of White Breads Enhanced with Broccoli (*Brassica oleracea* L.) Powder. *Korean J. Food Cook. Sci.* **2015**, *31*, 614–622. [CrossRef]
55. Anwar, B.R.; Rakha, A.; Mahmood, M.A.; Batool, I.; Sohail, M.; Rashid, S. Enrichment of Wheat Flour Bread to Enhance Physicochemical and Sensory Attributes Using Broccoli Powder. *Pak. J. Food Sci.* **2017**, *27*, 39–45.
56. Núñez-Gómez, V.; González-Barrio, R.; Baenas, N.; Moreno, D.A.; Periago, M.J. Dietary-Fibre-Rich Fractions Isolated from Broccoli Stalks as a Potential Functional Ingredient with Phenolic Compounds and Glucosinolates. *Int. J. Mol. Sci.* **2022**, *23*, 13309. [CrossRef]
57. Shinali, T.S.; Zhang, Y.; Altaf, M.; Nsabiyeze, A.; Han, Z.; Shi, S.; Shang, N. The Valorization of Wastes and Byproducts from Cruciferous Vegetables: A Review on the Potential Utilization of Cabbage, Cauliflower, and Broccoli Byproducts. *Foods* **2024**, *13*, 1163. [CrossRef]
58. Lafarga, T.; Gallagher, E.; Aluko, R.E.; Auty, M.A.E.; Hayes, M. Addition of an Enzymatic Hydrolysate of Bovine Globulins to Bread and Determination of Hypotensive Effects in Spontaneously Hypertensive Rats. *J. Agric. Food Chem.* **2016**, *64*, 1741–1750. [CrossRef]
59. Zheng, S.; Zhang, Y.; Chen, Q.; Fu, X.; Huang, Q.; Zhang, B.; Dong, H.; Li, C. Exploring the Synergistic Benefits of Insoluble Dietary Fiber and Bound Phenolics: Unveiling the Role of Bound Phenolics in Enhancing Bioactivities of Insoluble Dietary Fiber. *Trends Food Sci. Technol.* **2024**, *149*, 104554. [CrossRef]
60. Ranawana, V.; Raikos, V.; Campbell, F.; Bestwick, C.; Nicol, P.; Milne, L.; Duthie, G. Breads Fortified with Freeze-Dried Vegetables: Quality and Nutritional Attributes. Part 1: Breads Containing Oil as an Ingredient. *Foods* **2016**, *5*, 19. [CrossRef]
61. Franke, K.; Djikeng, F.T.; Esatbeyoglu, T. Influence of Frying, Baking and Cooking on Food Bioactives. In *Retention of Bioactives in Food Processing*; Jafari, S.M., Capanoglu, E., Eds.; Food Bioactive Ingredients; Springer International Publishing: Cham, Switzerland, 2022; pp. 93–121.

62. Hobbs, D.A.; Ashouri, A.; George, T.W.; Lovegrove, J.A.; Methven, L. The Consumer Acceptance of Novel Vegetable-Enriched Bread Products as a Potential Vehicle to Increase Vegetable Consumption. *Food Res. Int.* **2014**, *58*, 15–22. [CrossRef]
63. Nikolov, A.; Grozeva, N.; Vasilev, M.; Orozova, D.; Zlatev, Z. Multi-Modal Characterization of Wheat Bread Enriched with Pigweed and Purslane Flour Using Colorimetry, Spectral Analysis, and 3D Imaging Techniques. *Analytica* **2025**, *6*, 31. [CrossRef]
64. *AACC Method 10-05.01*; Measurement of Volume by Rapeseed Displacement. AACC International: St. Paul, MN, USA, 2009.
65. *AACC Method 54-21.02*; Rheological Behaviour of Flour by Farinograph: Constant Flour Weight Procedure. AACC International: St. Paul, MN, USA, 1995.
66. *AACC Method 54-30.02*; Alveograph Method for Soft and Hard Wheat Flour. AACC International: St. Paul, MN, USA, 2009.
67. *AACC Method 56-81.03*; Determination of Falling Number. AACC International: St. Paul, MN, USA, 2010.

 https://doi.org/10.3390/molecules31091414

molecules

MDPI

Article

Nutritionally Improved Gluten-Free Breads Fortified with Soluble Fiber and Bioactive Compounds from Artichoke and Broccoli By-Products

Jhazmin Quizhpe, Rocío Peñalver, Pablo Ayuso and Gema Nieto *

Department of Food Technology, Food Science and Nutrition, Faculty of Veterinary Sciences, Regional Campus of International Excellence "Campus Mare Nostrum", University of Murcia, Espinardo, 30100 Murcia, Spain; jhazminedith.quizhper@um.es (J.Q.); rocio.penalver@um.es (R.P.); pablo.ayuson@um.es (P.A.)
* Correspondence: gnieto@um.es

Abstract

Background: Commercial gluten-free (GF) breads often exhibit low nutritional quality due to limited fiber and bioactive compounds. The enzymatic treatment of vegetable by-products, such as broccoli and artichoke, represents a sustainable strategy to release soluble dietary fiber and phenolic compounds, enhancing the functional value of GF products. Five GF bread formulations were developed: a control bread, breads containing broccoli or artichoke extracts (BB and BA), and breads with enzymatically treated extracts using Viscozyme® L and Celluclast® 1.5 L (BBE and BAE). A commercial GF bread (BC) served as a reference. Nutritional composition, dietary fiber fractions, phenolic content, antioxidant capacity, starch digestibility, physicochemical parameters, and sensory properties were evaluated. Results: Enzymatic treatments significantly improved the nutritional and functional properties of GF breads. Viscozyme L® produced the highest increases in antioxidant capacity and phenolic content (up to 30% higher in FRAP), while Celluclast® 1.5 L generated the highest rise in soluble dietary fiber (up to 2.75 g/100 g) and the best sensory acceptance. Moreover, Celluclast® 1.5 L significantly modified starch digestibility, reducing rapidly digestible starch by 14% and increasing slowly digestible starch by over 150%, suggesting a lower predicted glycemic response. Conclusions: Incorporating the enzyme-treated artichoke and broccoli by-products into GF breads effectively enhances soluble fiber, antioxidant potential, and sensory quality. Among treatments, Celluclast® 1.5 L applied to artichoke proved most effective overall, providing a balanced improvement in nutritional and functional attributes. These findings revealed the potential of Celluclast® 1.5 L-treated artichoke by-products as a source of natural bioactive compounds for developing clean-label, nutritionally enhanced GF breads.

Keywords: functional food; gluten-free bread; food by-products; enzymatic treatment; dietary fiber; digestible starch

1. Introduction

Celiac disease (CD) is a chronic autoimmune disorder triggered by the consumption of gluten, a protein found in cereals such as wheat, barley, and rye, in individuals who are genetically predisposed. This condition leads to inflammation and damage to the villi of the small intestine, resulting in malabsorption of nutrients and subsequent nutritional deficiencies [1]. Currently, a gluten-free (GF) diet is the only treatment that has been shown to produce positive results in these patients. In recent decades, the incidence of celiac disease has increased, and together with the strict dietary requirement, this has driven a growing demand for GF products on the market [2,3]. Despite the therapeutic efficacy of a GF diet, commercial GF products often contain low levels of dietary fiber, minerals such as zinc, magnesium, and iron, and vitamins (particularly B vitamins), while being high in calories, fats, and sugars [4]. These nutritional imbalances in commercial GF products can increase the risk of nutritional deficiencies and metabolic disorders in individuals who rely on them [5].

Several innovative approaches have recently emerged to enhance the nutritional value of GF products. One of the most relevant developments is the use of pseudocereals such as quinoa, amaranth, and buckwheat, as well as legume flours, which are rich in high-quality protein, dietary fiber, minerals, and bioactive compounds [6,7]. Similarly, there has been growing interest in the inclusion of flax seeds due to their high omega-3 fatty acid, lignan, and soluble fiber content, which contribute to cardiovascular health and glycemic control [8,9]. Additionally, extracts from plants, such as grapes, bananas, apples, and carrots, which are rich in phenolic compounds and antioxidants, are being investigated as functional ingredients to enhance the nutritional value and confer additional bioactive properties to GF products [10–12].

Food loss and waste represent a global challenge, with nearly one-third of food produced for human consumption discarded annually, resulting in inefficient resource use and considerable environmental impacts [13]. Within the framework of a circular economy, the valorization of vegetable by-products emerges as a sustainable strategy to mitigate these impacts while enabling the development of innovative products with enhanced nutritional value [14].

In this context, broccoli (*Brassica oleracea* var. *italica*) and artichoke (*Cynara scolymus* L.), edible inflorescence vegetables, are of great interest due to their high global production and numerous nutritional benefits. Broccoli production, whose combined output with cauliflower exceeded 26 million tons in 2023 [15], generates a considerable quantity of by-products, since the edible fraction accounts for only about 15% of the total weight, while the remaining parts—mainly stems and leaves—are commonly discarded post-harvest [16]. However, these non-edible fractions represent a valuable source of bioactive compounds, including glucosinolates and isothiocyanates, phenolic compounds, carotenoids, vitamin C, dietary fiber, and minerals [17]. These components have been widely associated with a range of health-promoting biological activities, such as anticarcinogenic, anti-inflammatory, antimicrobial, antioxidant, and antihypertensive effects [18]. Similarly, artichoke production reaches approximately 1.6 million tons annually [15], of which only the immature inflorescences are consumed, while the remaining 60–85%, primarily composed of outer bracts, leaves, and stems, is discarded as by-products [19]. In general, these by-product fractions are rich in nutrients and bioactive compounds of considerable interest, including phenolic acids derived from caffeoylquinic acid, flavonoids such as luteolin and apigenin, minerals, vitamins, and especially a high content of dietary fiber, particularly inulin and pectin [20,21]. Such compounds have been linked to antioxidant, anti-inflammatory, antimicrobial, anticancer, and prebiotic activities [22].

Although vegetables are rich in fiber, the insoluble fraction predominates, which is less fermentable and generates fewer health-promoting metabolites than the soluble fraction (SDF) [23]. Different techniques and methods, such as mechanical degradation, chemical treatments, thermal processes, and enzymatic methods, have been investigated to enhance the soluble fibers in foods [24]. Among these, enzymatic methods using cellulases, pectinases, and xylanases are particularly effective, operating under mild and environmentally friendly conditions [25,26].

https://doi.org/10.3390/molecules31010152

Considering the limited literature on exploring the use of enzyme-treated artichoke and broccoli by-products as natural functional ingredients in gluten-free breads, this study investigated their effect on the physico-chemical, nutritional, and sensory properties of gluten-free formulations. The enriched breads were compared with a control formulation and with commercial gluten-free bread to assess improvements in fiber content, antioxidant potential, and overall quality.

In this study, GF bread formulations were developed with the inclusion of broccoli and artichoke by-products after enzymatic treatment, which were compared with the Control and commercial GF bread. The effects of these vegetable by-products and enzymatic treatment on the nutritional quality and sensory attributes of the breads obtained were evaluated. This approach aims not only to enhance the functional properties of GF products for celiacs but also to contribute to the sustainable management of agro-industrial waste.

2. Results and Discussion

2.1. Chemical Composition and Physicochemical Characteristics

The chemical composition, antioxidant capacity, and total phenolic compounds (TPC) of the different GF bread samples, including formulated breads and the commercial bread, are detailed in Table 1.

Table 1. Proximal composition, antioxidant activity, and total phenolic content (TPC) of gluten-free breads.

	Control	BB	BBE	BA	BAE	BC
Proximal composition						
Energy content (kcal/100 g)	287.70 ± 5.66 a	287.56 ± 3.05 a	280.26 ± 4.90 a	209.22 ± 3.51 c	232.40 ± 0.66 b	235.70 ± 4.51 b
Fat (g/100 g)	12.33 ± 0.99 a	12.10 ± 0.44 a	11.90 ± 0.66 a	8.45 ± 1.34 b	8.08 ± 0.18 b	5.58 ± 0.15 b
Protein (g/100 g)	12.14 ± 0.58 a	12.88 ± 3.28 a	15.03 ± 0.71 a	10.61 ± 0.30 a	11.13 ± 0.47 a	2.90 ± 0.06 b
Moisture (g/100 g)	29.89 ± 1.44 c	27.64 ± 0.16 c	29.30 ± 0.31 c	34.22 ± 0.27 b	29.79 ± 0.10 c	42.37 ± 1.03 a
Ash (g/100 g)	2.38 ± 0.01 b	2.42 ± 0.00 b	2.58 ± 0.03 b	5.46 ± 0.68 a	4.82 ± 0.18 a	1.61 ± 0.10 b
IDF (g/100 g)	10.68 ± 1.29 b	12.07 ± 0.24 b	10.77 ± 0.56 b	17.05 ± 0.42 a	14.65 ± 0.05 a	3.07 ± 0.01 c
SDF (g/100 g)	0.54 ± 0.01 c	1.11 ± 0.30 bc	2.16 ± 0.69 ab	1.52 ± 0.57 abc	2.75 ± 0.27 a	1.00 ± 0.02 bc
TDF (g/100 g)	11.22 ± 1.28 b	13.18 ± 0.05 b	12.92 ± 0.13 b	18.58 ± 0.15 a	17.40 ± 0.22 a	4.07 ± 0.01 c
Carbohydrates (g/100 g)	32.05 ± 0.23 b	31.79 ± 3.50 b	28.27 ± 0.47 bc	22.70 ± 2.43 c	28.79 ± 0.23 bc	43.48 ± 0.85 a
Antioxidant activity and TPC						
FRAP (µmol TE/g)	13.73 ± 0.72 b	18.43 ± 2.75 ab	24.12 ± 0.97 a	19.27 ± 3.30 ab	18.92 ± 0.67 ab	5.72 ± 0.70 c
ABTS (µmol TE/g)	27.65 ± 1.36 cd	23.95 ± 1.33 d	32.88 ± 3.03 bc	37.05 ± 1.11 ab	40.48 ± 2.18 a	7.67 ± 1.11 e
DPPH (µmol TE/g)	8.38 ± 0.76 bc	11.29 ± 2.69 ab	14.64 ± 1.44 ab	15.17 ± 2.30 ab	19.35 ± 3.53 a	1.88 ± 1.18 c
TPC(mg GAE/g)	7.72 ± 0.14 b	7.05 ± 1.12 b	9.09 ± 0.29 b	12.28 ± 0.01 a	14.40 ± 0.64 a	2.98 ± 0.02 c

a–e: Different letters within the same row indicate statistically significant differences between samples ($p < 0.05$). Control: Bread without extract; BB: Bread with broccoli; BBE: Bread with enzymatically treated broccoli; BA: Bread with artichoke; BAE: Bread with enzymatically treated artichoke; BC: Commercial bread; IDF: Insoluble dietary fiber; SDF: Soluble dietary fiber; TDF: Total dietary fiber; FRAP: Ferric reducing antioxidant power assay; ABTS: 2,2′-azino-bis(3-ethylbenzothiazoline-6-sulfonic acid) radical cation scavenging assay; DPPH: 2,2-diphenyl-1-picrylhydrazyl radical scavenging assay; TPC: Total phenolic content.

Comparative analyses were performed using the Control as a reference to determine the effects of incorporating by-products into the formulation. On the other hand, the experimental GF breads were compared with a commercial product to establish nutritional differences relative to a market-available option. In terms of macronutrients, the broccoli samples (BB and BBE) showed no significant differences compared to the Control, maintaining similar profiles in energy, fat, protein, moisture, ash, and carbohydrates. The only exception was a slight increase in protein content in BBE, although it was not statistically significant. In contrast, samples with artichoke by-products showed a significant decrease in fat levels (about 8 g/100 g) compared to the Control (12.33 g/100 g). The observed decrease in fat content can be attributed to the fat-binding capacity of dietary fiber, which has been demonstrated to retain or immobilize lipids within the food matrix, thereby limiting their extractability

https://doi.org/10.3390/molecules31010152

during analysis [27,28]. In addition, both BA and BAE showed a significant increase in the ash fraction, 5.46 and 4.82 g/100 g, respectively, compared to 2.38 g/100 g for the Control. The increase in ash content observed after adding artichoke extract can be attributed to the naturally high mineral content of this vegetable [29]. Similar findings were reported by Dadali, who observed higher ash levels in cakes enriched with artichoke by-product powder [30].

With respect to dietary fiber (DF) content, the untreated breads (BB and BA) showed differences compared to the Control. BB exhibited similar insoluble (IDF, 10.68 g/100 g) and total fiber (TDF, 11.22 g/100 g) levels as the Control, whereas BA showed markedly higher IDF and TDF (17.05 and 18.58 g/100 g, respectively). Breads subjected to enzymatic treatment (BBE and BAE) displayed a significant increase in the soluble fiber fraction (SDF), with BBE reaching 2.16 g/100 g and BAE reaching the highest value of 2.75 g/100 g. However, IDF and TDF in the treated breads were slightly lower than in their respective untreated counterparts due to the conversion of insoluble fiber into soluble fiber, while remaining elevated compared to the Control. Similar results were observed in a previous study, where artichoke extracts generally had higher levels of dietary fiber than broccoli extracts, and when different enzymatic treatments were applied, their SDF levels increased significantly while reducing IDF levels [31]. In the case of artichoke, the commercial enzyme Celluclast® 1.5 L was used, which catalyzes the hydrolysis of β-1,4-glycosidic bonds in insoluble polysaccharides, resulting in the release of water-soluble polysaccharides and oligosaccharides [32]. On the other hand, the findings in the study of the enzymatic treatment of broccoli by-products are attributed to the use of Viscozyme® (an enzyme complex composed of arabinase, cellulase, β-glucanase, hemicellulase, and xylanase). This treatment is justified by the predominant presence of cellulose, xylans, and pectins in the wall of broccoli stems, which, when fragmented, release soluble oligosaccharides and, consequently, increase the SDF [33].

Additionally, the analyses revealed significant variations in antioxidant capacity (FRAP, ABTS, and DPPH) and total phenolic content (TPC) among the evaluated samples. All bread supplemented with vegetal extracts exhibited improvements compared to the Control. Treatment with both commercial enzymes caused significant changes ($p < 0.05$). In the GF bread with broccoli, enzymatic treatment resulted in increases of 30.9%, 37.3%, 29.7%, and 28.9% for FRAP, ABTS, DPPH, and TPC, respectively. In contrast, enzymatic treatment in artichoke bread produced more moderate changes, with increases of 9.3%, 27.6%, and 17.3% in ABTS, DPPH, and TPC, respectively. The differences observed between breads enriched with broccoli and artichoke extracts and the Control can be attributed to the higher content of polyphenols and other bioactive compounds in these plant matrices [34–39]. However, despite their abundance, the thermal processing involved in bread-making can partially degrade these compounds. As reported by Franke et al., the retention of bioactive compounds is highly dependent on processing conditions, which may explain why the enriched breads did not show markedly higher antioxidant activity or phenolic content compared to Control [40]. However, thermal processing during baking may reduce some heat-sensitive compounds, attenuating the beneficial effect of these extracts. On the other hand, the increases observed in antioxidant capacity and TPC in BBE and BAE can be attributed to the enzymatic action on the plant matrix. This action degrades cell wall components, releasing bound polyphenols, which are then transformed into soluble and bioavailable forms. Additionally, enzymatic hydrolysis generates oligosaccharides with inherent antioxidant activity, further contributing to the overall effect [31,41,42].

BC, predominantly composed of refined flours, starches, and processed vegetable fats, contains lower amounts of total dietary fiber (4.07 g/100 g), protein (2.90 g/100 g), and antioxidant capacity, together with a higher carbohydrate content (43.48 g/100 g) compared to the control formulation. In contrast, the Control bread, made with buckwheat flour,

 https://doi.org/10.3390/molecules31010152

vegetable proteins, and seeds, incorporates highly valuable ingredients that provide it with a clearly superior nutritional and functional profile [43–45].

Recent advances in enzyme-assisted technologies have significantly improved the value of agro-industrial by-products for functional food development. Innovative approaches such as ultrasound-assisted enzymatic hydrolysis and high-pressure–enzyme combinations have been shown to enhance enzyme accessibility to plant cell wall polysaccharides, increasing the release of soluble dietary fiber and bound phenolics [46–48]. In parallel, enzyme engineering has yielded cellulases and multi-enzyme complexes with superior thermal stability and catalytic efficiency, broadening their industrial applicability [49]. Despite these advances, their implementation can be limited by cost and complexity. Our findings demonstrate that commercially available enzyme preparations applied under mild conditions effectively enhance soluble fiber content and antioxidant capacity in gluten-free breads enriched with artichoke and broccoli by-products.

2.2. Starch and Sugar Content

Table 2 shows the results for the different fractions of digestible starch: slow (SDS), rapid (RDS), and total (TDS); and resistant starch (RS). The profile of the samples in terms of monosaccharides and disaccharides is also detailed.

Table 2. Starch and sugar content of gluten-free breads (g/100 g).

	Control	BB	BBE	BA	BAE	BC
Digestible and resistant starch						
RDS	44.20 ± 0.78 b	53.25 ± 0.76 a	41.11 ± 0.08 c	44.97 ± 0.57 b	38.00 ± 0.01 d	51.96 ± 0.30 a
SDS	7.17 ± 0.46 d	7.15 ± 0.03 d	21.72 ± 0.23 a	10.23 ± 0.33 c	18.23 ± 0.33 b	4.83 ± 0.28 e
TDS	51.37 ± 1.24 c	60.41 ± 0.74 a	62.83 ± 0.30 a	55.19 ± 0.23 b	56.22 ± 0.35 b	56.79 ± 0.58 b
RS	1.42 ± 0.04 b	1.22 ± 0.05 c	1.48 ± 0.03 ab	1.16 ± 0.01 c	1.58 ± 0.01 a	1.04 ± 0.01 d
Disaccharides and monosaccharides						
Sucrose	0.15 ± 0.01 c	0.10 ± 0.01 cd	0.05 ± 0.01 de	0.58 ± 0.04 a	0.40 ± 0.03 b	0.02 ± 0.01 e
Glucose	0.11 ± 0.01 d	0.17 ± 0.01 d	0.14 ± 0.01 d	0.33 ± 0.02 c	0.41 ± 0.03 b	0.80 ± 0.03 a
Fructose	0.04 ± 0.01 d	0.20 ± 0.01 b	0.15 ± 0.01 c	0.28 ± 0.02 a	0.18 ± 0.02 bc	0.04 ± 0.01 d
Maltose	0.07 ± 0.02 d	0.09 ± 0.02 d	0.16 ± 0.03 d	0.61 ± 0.06 c	0.81 ± 0.02 b	1.31 ± 0.02 a

a–e: Different letters within the same row indicate statistically significant differences between samples ($p < 0.05$). Control: Bread without extract; BB: Bread with broccoli; BBE: Bread with enzymatically treated broccoli; BA: Bread with artichoke; BAE: Bread with enzymatically treated artichoke; BC: Commercial bread; RDS: Rapidly digestible starch; SDS: Slowly digestible starch; TDS: Total digestible starch; RS: Resistant starch.

The different bread formulations showed significant variations in starch digestibility and simple sugar profile ($p < 0.05$). BC had a significantly higher RDS content than Control (51.96 and 44.20 g/100 g, respectively) and a lower SDS concentration (4.83 and 7.17 g/100 g). Additionally, it presented the lowest RS content among the samples (1.04 g/100 g). This profile corresponds to that described by Giuberti and Gallo, who reported that commercially available GF products tend to contain highly available starches and a low resistant fraction due to refined flours and aerated structure, resulting in a higher glycemic index [50]. In contrast, Control showed a more favorable profile, consistent with Di Cairano et al., who reported that less refined or underutilized flours, such as buckwheat flour used in this study, reduce rapid starch digestion and glycemic response [51].

Conversely, formulations containing broccoli and artichoke extracts, particularly those treated enzymatically (BBE and BAE), showed a significant reduction in RDS (41.11 and 38.00 g/100 g, respectively), and a marked increase in SDS (21.72 and 18.23 g/100 g). The RS content remained similar among the different formulations, ranging from 1.16 to 1.58 g/100 g, with BBE and BAE once again showing the highest values. These findings suggest that the incorporation

https://doi.org/10.3390/molecules31010152

of plant extracts, and especially their enzymatic treatment, significantly modifies the digestibility of starch in GF breads. A redistribution of the starch fractions towards a slower profile was observed, which could imply a lower glycemic index of the final product [50]. The lower proportion of RDS and the increase in SDS in breads enriched with plant extracts are consistent with the findings of Santamaría et al. and Melini et al., who reported similar effects when incorporating different flours or plant extracts into GF breads [52,53]. The observed effect was attributed to the presence of phenolic compounds and plant fibers, which interfere with starch and modify its gelatinization. Consequently, the broccoli and artichoke extracts used in this study, rich in indigestible polysaccharides and antioxidants, would have contributed to a bread matrix less accessible to digestive enzymes. Moreover, the more pronounced effects in the enzyme-treated samples can be explained by mechanisms similar to those reported by Kasprzak et al., where enzyme treatment of different starch sources modified their molecular structure, increasing the proportion of RS and reducing rapid digestibility [54]. All this suggests that the combination of plant extracts and enzymatic treatment may be an effective strategy for improving the digestion profile of starch in GF breads.

The sugar profile showed variations consistent with starch behavior. BC presented the highest glucose and maltose content (0.80 and 1.31 g/100 g, respectively), significantly higher than those of Control. This reflects faster hydrolysis and greater availability of simple sugars due to its composition of refined flours, added sugar, and glucose syrup. On the other hand, the incorporation of artichoke and broccoli extracts significantly modified the sugar profile compared to the Control. Bread containing broccoli exhibited a moderate increase in fructose and glucose levels, while those containing artichoke demonstrated higher concentrations, particularly in BAE, which exhibited a greater release of fructose and glucose. These effects can be explained by the release of intrinsic sugars from the plant matrices and by the partial hydrolysis of polysaccharides during enzymatic treatment [53,55]. Therefore, the observed increase in soluble sugars in these samples does not necessarily indicate greater starch digestibility, but rather the incorporation of simple carbohydrates from the functional ingredients.

2.3. Color and Acidity Profile

Table 3 presents the color and acidity parameters measured for the different GF breads analyzed in this study.

Table 3. Color and acidity parameters of gluten-free breads.

	Control	BB	BBE	BA	BAE	BC
*CIEL*a*b** parameters						
L^*	55.15 ± 1.58 b	52.93 ± 0.82 b	41.02 ± 1.61 d	47.67 ± 2.39 c	47.19 ± 0.74 c	79.45 ± 0.51 a
C	12.54 ± 0.87 c	16.92 ± 0.71 b	18.01 ± 1.31 ab	13.52 ± 2.20 c	17.27 ± 0.43 ab	20.36 ± 0.55 a
h	66.94 ± 2.08 c	71.69 ± 0.14 b	61.36 ± 0.14 d	71.03 ± 1.27 b	69.49 ± 1.12 bc	86.27 ± 0.14 a
a^*	5.32 ± 0.48 b	5.32 ± 0.19 b	8.63 ± 0.64 a	4.37 ± 0.53 c	6.05 ± 0.18 c	1.33 ± 0.05 a
b^*	11.41 ± 0.76 d	16.06 ± 0.69 b	15.81 ± 1.14 bc	12.80 ± 2.17 cd	16.17 ± 0.52 b	20.31 ± 0.54 a
ΔE^*	-	5.32 ± 0.72 c	15.18 ± 2.98 b	8.25 ± 2.78 c	9.37 ± 0.92 c	26.20 ± 1.89 a
Acidity profile						
pH	5.92 ± 0.15 a	6.01 ± 0.21 a	5.76 ± 0.01 ab	5.66 ± 0.01 ab	5.48 ± 0.02 b	4.99 ± 0.06 c
TTA (ml NaOH/10 g)	4.74 ± 0.06 c	5.92 ± 0.16 b	6.41 ± 0.14 ab	6.22 ± 0.04 b	6.82 ± 0.05 a	3.08 ± 0.25 d
L-Lactic Acid (g/kg)	0.25 ± 0.01 b	0.24 ± 0.01 b	0.25 ± 0.01 b	0.34 ± 0.03 a	0.39 ± 0.01 a	0.15 ± 0.01 c
Acetic Acid (g/kg)	1.98 ± 0.03 c	1.94 ± 0.01 c	2.04 ± 0.01 c	2.26 ± 0.09 b	2.47 ± 0.01 a	1.20 ± 0.02 d

a–d: Different letters within the same row indicate statistically significant differences between samples ($p < 0.05$). *CIEL*a*b**: International standard color space where L^* represents lightness, a^* the green–red axis, and b^* the blue–yellow axis; Control: Bread without extract; BB: Bread with broccoli; BBE: Bread with enzymatically treated broccoli; BA: Bread with artichoke; BAE: Bread with enzymatically treated artichoke; BC: Commercial bread; TTA: total titratable acidity; ΔE^*: Total color difference of each formulation compared to the control bread.

 https://doi.org/10.3390/molecules31010152

Regarding color measurements (*CIEL*a*b**), the addition of broccoli and artichoke extracts, as well as their enzymatically treated extracts, produced significant changes in the color of bread compared to the Control. The incorporation of these extracts led to a decrease in L^* and an increase in C and b^*, resulting in darker breads with greater color intensity and yellowish tones. Enzymatic treatment further enhanced these effects, significantly decreasing L^* and increasing C, a^*, and b^*, yielding breads that were darker, with more intense colors and pronounced reddish and yellowish tones. These changes can be attributed to the pigments present in the plant matrices, such as chlorophylls and carotenoids. Previous studies have shown that the use of pigmented plant by-products, such as broccoli or artichoke, in bakery formulations affects color parameters, often resulting in strong green or yellowish hues in the final product [56–59]. An important factor to consider in the development of new products is the total color difference (ΔE^*) relative to the Control, with values above 3 being perceptible to the human eye. In this study, all reformulated breads showed ΔE^* values ranging from 5.32 to 15.18, indicating noticeable color changes. Similar observations have been reported in studies using pigmented plant materials, such as green algae (*Chlorella vulgaris*), which also induced significant color modifications in breads [60]. BC was significantly lighter and more yellow than the Control, with higher saturation and hue, resulting in a noticeably brighter appearance. These differences can be attributed to its composition of refined, yellowish flours, whereas the Control, made with buckwheat flour and flax seeds, exhibits a darker, less saturated color.

In terms of acidity, BC exhibited a lower pH than Control, but its total titratable acidity (TTA) and levels of organic acids (L-lactic and acetic) were lower. This finding indicates that the acidity present in commercially available bread is not a consequence of yeast fermentation; rather, it is derived from other technological adjustments incorporated during its industrial formulation. Conversely, the incorporation of broccoli and artichoke extracts resulted in significant increases in TTA, which were more pronounced in enzymatically treated samples (6.41 and 6.82 mL NaOH/10 g, in BBE and BAE, respectively). This reflects a more intense fermentation process, increasing organic acids, particularly acetic acid. Consistent with the TTA results, the samples that exhibited higher concentrations were BBE and BAE, which presented 2.04 and 2.47 g acetic acid/kg of sample. These results can be attributed to enzymatic hydrolysis, which releases sugars and phenolic compounds from plant cell walls that serve as substrates for microbial fermentation during bread making. Recent studies have also shown that the fermentation of fiber- and protein-rich plant by-products with lactic acid bacteria significantly increases TTA and the concentration of organic acids [61]. These findings demonstrate that the synergistic action of plant extracts and enzymatic treatments effectively stimulates microbial activity, enhancing acid production in bread [62].

2.4. Sensory Analysis of Gluten-Free Breads

Figure 1 presents the results of the sensory evaluation and consumer acceptance of the GF bread formulations, assessed by a trained panel. In the sensory attribute analysis (Figure 1A), significant differences were observed in appearance, color, and texture ($p < 0.05$). The BC formulation obtained lower scores for appearance and color, whereas the other formulations enriched with broccoli and artichoke extracts exhibited sensory profiles comparable to the Control. To further explore the relationships among samples, a principal component analysis was conducted (Figure 1B). The biplot indicated that the Control bread was primarily associated with texture and appearance, while the BB and BBE formulations were positively associated with aroma and taste. In contrast, the BC formulation was clearly differentiated from the other treatments, confirming its less favorable sensory profile.

 https://doi.org/10.3390/molecules31010152

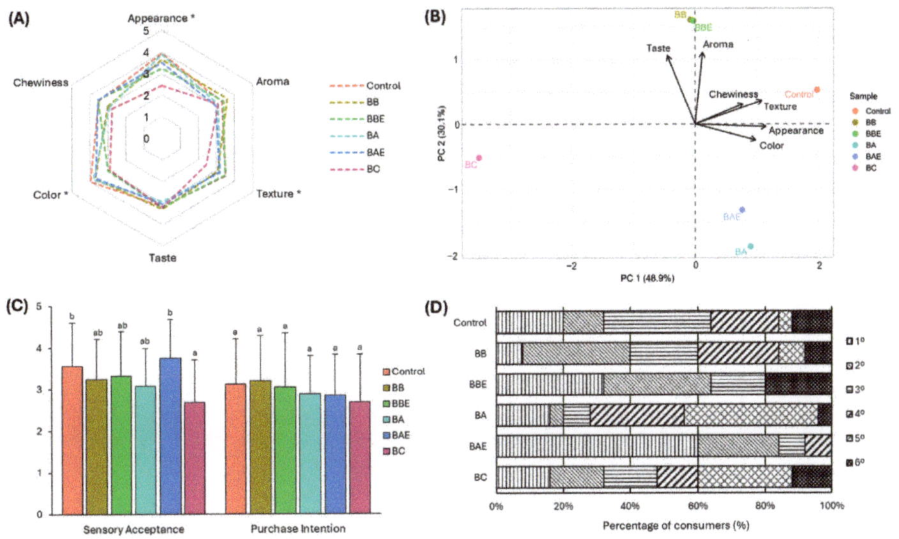

Figure 1. Sensory analysis of gluten-free breads formulations. Radar plots showing the sensory attributes of the different formulations (**A**). Principal component analysis (PCA) (**B**) biplot illustrating the distribution of gluten-free samples according to their sensory profiles and the contribution of each attribute to sample differentiation. Different letters (a, b) indicate statistically significant differences between samples (* $p < 0.05$). Sensory acceptance and purchase intention (**C**), and preference ordination (**D**) of gluten-free bread samples. Control: Bread without extract; BB: Bread with broccoli; BBE: Bread with enzymatically treated broccoli; BA: Bread with artichoke; BAE: Bread with enzymatically treated artichoke; BC: Commercial bread.

Regarding overall acceptance and purchase intention (Figure 1C), no significant differences were observed in the latter variable. However, significant differences in overall acceptance were detected ($p < 0.05$), with the Control and BAE formulations achieving the highest scores, while BC recorded the lowest values. Similar improvements in sensory acceptability have been reported for GF bakery products enriched with plant extracts, including *Moringa Oleifera*, *Linum usitatissimum*, *Citrus* × *sinensis*, and *Daucus carota* [63–66].

Finally, the preference ranking analysis (Figure 1D) revealed that the BBE and BAE formulations were the most highly valued by consumers, as they concentrated the highest percentages in the top positions. Conversely, the BC and BA formulations were predominantly ranked in the lowest positions, consistent with their lower acceptance levels.

2.5. Pearson Correlations

The results obtained for proximal composition, antioxidant activity, physicochemical properties, and sensory characteristics were subjected to a Pearson correlation analysis (Figure 2) to verify similarities and relationships between the attributes. The Pearson correlations of the different breads showed a positive correlation between fat and protein with purchase intention (r = 0.97 and r = 0.80, respectively). This association is to be expected, as fat contributes softness, juiciness, and better texture to the crumb, in addition to intensifying the flavor. On the other hand, protein contributes to better structure, volume, and elasticity [67]. This may explain the low results of BC compared to handmade breads, which had better nutritional values and better acceptance. These results are corroborated by the correlation between texture (r = 0.74) and chewiness (r = 0.88) and the

https://doi.org/10.3390/molecules31010152

sensory acceptance of sourdough breads, attributes in which BC obtained the lowest results compared to the other breads analyzed.

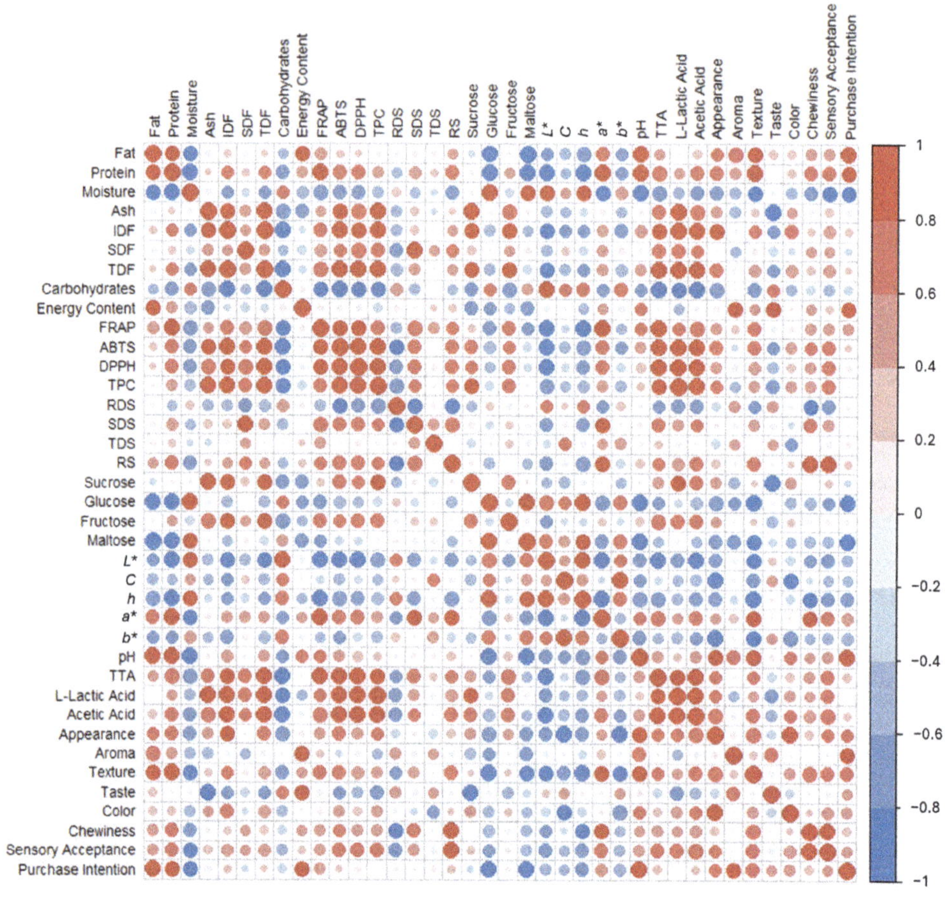

Figure 2. Pearson correlation heatmap between different measured parameters. Circle size represents the magnitude of the correlation coefficient, with larger circles indicating stronger correlations. IDF: Insoluble dietary fiber; SDF: Soluble dietary fiber; TDF: Total dietary fiber; TPC: Total phenolic content; RDS: Rapidly digestible starch; SDS: Slowly digestible starch; TDS: Total digestible starch; RS: Resistant starch; TTA: total titratable acidity.

On the other hand, strong positive correlations were found between IDF, TDF, and SDF and the results obtained in antioxidant activity (ABTS, FRAP, and DDPH) and TPC. These associations indicate that breads with higher fiber fractions also had higher levels of bioactive compounds, confirming that broccoli and artichoke extracts had an impact on both the fiber fortification of GF breads and a higher content of bioactive compounds. Moreover, this positive relationship can be attributed to the ability of dietary fiber, particularly the insoluble fraction, to interact with phenolic compounds through both covalent and non-covalent bonds, thereby acting as a binding matrix for these bioactive molecules [68]. Furthermore, breads with enzymatically treated extracts (BAE and BBE) showed higher antioxidant values than their corresponding analogs (BA and BB), suggesting that en-

https://doi.org/10.3390/molecules31010152

zymatically treated fiber may facilitate the release of bound phenols during extraction and analysis.

3. Materials and Methods

3.1. Broccoli and Artichoke Extracts Preparation

Agro-industrial by-products from broccoli (stems and leaves) and artichoke (stems, leaves, and external bracts) were supplied by Cricket (Lorca, Spain). All by-product powders were obtained following the method described by Ayuso et al. [31] The different broccoli and artichoke bio-residues were dried in a forced-air oven at 50 °C for 24 h. Once completely dried, the by-products were ground and sieved to produce fine powders with a particle size of 500 µm.

Viscozyme® L, a multi-enzyme complex (cellulase, arabanase, β-glucanase, and xylanase) with strong pectolytic activity from *Aspergillus aculeatus*, and Celluclast® 1.5 L, a cellulase obtained from *Trichoderma reesei*, were utilized for the enzymatic treatment of broccoli and artichoke by-products, respectively. Both enzymes were purchased from Novozymes (Bagsværd, Denmark).

For the enzymatic treatment, diluted broccoli and artichoke powders (1:5, *w*/*v*) were incubated with Viscozyme® L (pH 4.5, 45 °C) and Celluclast® 1.5 L (pH 6, 50 °C), respectively, at 0.45% (*v*/*w*) for 24 h under continuous magnetic stirring. The hydrolyzed mixtures were dried at 60 °C in a forced-air oven and subsequently milled again to achieve the same particle size as before enzymatic treatment.

3.2. Gluten-Free Breads Elaboration

The ingredients used in the GF bread formulation, as detailed in Table 4, were obtained from a local supermarket in Murcia, Spain. Ground flax seeds, sugar, and dry yeast were mixed with water and allowed to stand for 1 h at 25 °C to activate the yeast and form the flaxseed gel, thereby supporting proper fermentation and improving dough cohesion. The remaining bread ingredients were then added and mixed to obtain a fluid and homogeneous dough using an automatic bread machine (Silvercrest®, IAN 285058, Lidl Stiftung & Co. KG, Neckarsulm, Germany) set to the "gluten-free" program. This program comprises multiple stages of kneading, resting, and baking, with a total duration of 2 h.

Table 4. Gluten-free breads formulation.

Ingredients	Control	BB	BBE	BA	BAE
Buckwheat flour (g)	125	125	125	125	125
Maize starch (g)	75	75	75	75	75
Lentil protein (g)	20	20	20	20	20
Ground flax seeds (g)	35.2	35.2	35.2	35.2	35.2
Dry yeast (g)	7	7	7	7	7
Water (g)	175	175	175	175	175
Salt (g)	4.2	4.2	4.2	4.2	4.2
Sugar (g)	4.2	4.2	4.2	4.2	4.2
Olive oil (mL)	25	25	25	25	25
Broccoli extract (g)	-	12.5	-	-	-
Enzyme-treated broccoli extract (g)	-	-	12.5	-	-
Artichoke extract (g)	-	-	-	8.8	-
Enzyme-treated artichoke extract (g)	-	-	-	-	8.8

Control: Bread without extract; BB: Bread with broccoli; BBE: Bread with enzymatically treated broccoli; BA: Bread with artichoke; BAE: Bread with enzymatically treated artichoke.

https://doi.org/10.3390/molecules31010152

Five different GF formulations were developed: a control bread without any vegetable extract (Control); breads enriched with 12.5 g of dried broccoli extract, either untreated (BB) or enzymatically treated (BBE); and breads containing 8.8 g of dried artichoke extract, either untreated (BA) or enzymatically treated (BAE). After baking, the loaves were allowed to cool for 1 h, then sliced and vacuum-packed in polyethylene bags and stored at −20 °C until further analysis. The amounts of extract and broccoli were estimated based on the fiber content of all ingredients to ensure that the breads could be declared as "high fiber" products in accordance with Regulation (EC) No 1924/2006 [69].

For comparison, the five GF formulations were evaluated alongside a commercial GF bread (BC) purchased from a local supermarket. According to the product label, its composition includes: water, maize starch, rice flour, sugar, eggs (11%), vegetable margarine (palm fat, coconut fat, water, rapeseed oil, salt; emulsifier: mono- and diglycerides of fatty acids; natural flavor), glucose syrup, milk powder, vegetable fiber (psyllium), thickeners (hydroxypropyl methylcellulose, guar gum), yeast, emulsifier (mono- and diglycerides of fatty acids), salt, acidifying agent (tartaric acid), and natural vanilla flavor. The visual appearance of the GF breads is presented in Figure 3.

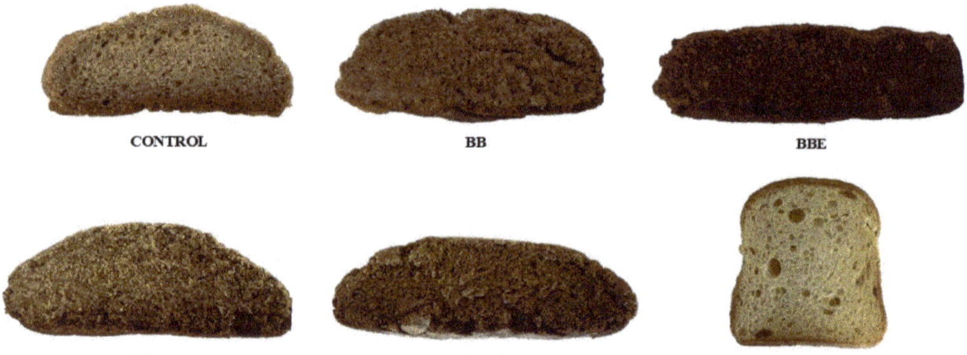

Figure 3. Visual appearance of the different gluten-free bread formulations. Control: Bread without extract; BB: Bread with broccoli; BBE: Bread with enzymatically treated broccoli; BA: Bread with artichoke; BAE: Bread with enzymatically treated artichoke; BC: Commercial bread.

3.3. Proximal Composition

The proximal composition of formulated and commercial GF breads was analyzed according to the official protocols set forth by the Association of Official Analytical Chemists (AOAC) employing the following methods: moisture (964.22), ash (923.03), protein (955.04), total fat (920.39), and dietary fiber fractions (insoluble, soluble, and total) (985.29) [70]. The carbohydrate content was calculated by difference, based on the values obtained for moisture, protein, fat, and ash [71]. The energy content was estimated using standard conversion factors, in accordance with the Food and Agriculture Organization of the United Nations (FAO) recommendations [72].

3.4. Starch Determination

Rapidly digestible starch (RDS), slowly digestible starch (SDS), total digestible starch (TDS), and resistant starch (RS) were determined using the enzymatic Digestible and Resistant Starch Assay Kit (Megazyme International, Bray, Ireland), following the procedure recommended by the manufacturer. This standardized and widely used method is based on the Englyst et al. approach and involves a kinetic enzymatic digestion coupled with colori-

https://doi.org/10.3390/molecules31010152

metric quantification of glucose released at 510 nm wavelength [73]. Enzymatic digestion was conducted at 37 °C for 240 min using pancreatic alpha-amylase and amyloglucosidase (PAA/AMG). Aliquots were collected at 20 min to quantify RDS, at 120 min for SDS, and at 240 min to determine TDS and RS. Results were expressed as g of starch per 100 g of bread.

3.5. *Physicochemical Parameters*

Color was assessed using a CR-400 portable colorimeter (Konica Minolta, Tokyo, Japan), calibrated with a standard white reference plate. The measured parameters included lightness (L^*), red-green component (a^*), yellow-blue component (b^*), chroma (C), and hue (h) following the *CIEL*a*b** color space system. Measurements were conducted in triplicate on 16 mm-thick slices from each bread sample. The total color difference (ΔE^*) of each formulation compared to the Control bread was calculated with the following formula:

$$\Delta E^* = \sqrt{(L_s^* - L_c^*)^2 + (a_s^* - a_c^*)^2 + (b_s^* - b_c^*)^2}$$

The pH was determined potentiometrically using a sensION+ PH31 pH meter (Hach-Lange, Barcelona, Spain) in a 1:10 (w/v) homogenized suspension of bread in distilled water. Total titratable acidity (TTA) was then assessed in the same suspension by titration with 0.1 N sodium hydroxide (NaOH) until the pH reached 8.5. Results were expressed as mL NaOH/10 g.

L-lactic acid and acetic acid concentrations were quantified using specific commercial enzymatic assay kits (BioSystems S.A., Barcelona, Spain), according to the manufacturer's instructions. Assays were performed on the BioSystems Y15 automatic analyzer (BioSystems S.A., Barcelona, Spain), measuring absorbance at 340 nm. Results were expressed as grams of L-lactic or acetic acid per kilogram of bread.

3.6. *Determination of Disaccharides and Monosaccharides*

Monosaccharides (glucose and fructose) and disaccharides (sucrose and maltose) concentrations were analyzed using enzymatic methods with three commercial kits (BioSystems S.A., Barcelona, Spain). Measurements were carried out spectrophotometrically at 340 nm using a BioSystems Y15 automatic analyzer (BioSystems S.A., Barcelona, Spain). Results were expressed as g of glucose, fructose, sucrose, or maltose per 100 g of bread.

3.7. *Antioxidant Capacity and Total Phenolic Content (TPC) Analysis*

Before analysis, bioactive compounds were extracted from the breads by homogenizing 2 g of sample with a 1:5 methanol/water (80:20, v/v) solution. The mixtures were incubated for 24 h at 4 °C in the dark, then centrifuged at 4500 rpm for 25 min at 4 °C. The supernatants were filtered through 0.45 µm nylon filters and stored at −20 °C until analysis.

The antioxidant capacity of the samples was evaluated using three complementary spectrophotometric assays. The ferric reducing antioxidant power (FRAP) assay was carried out by combining 100 µL of the extracted sample with 1 mL of FRAP reagent, which contained acetate buffer, $FeCl_3 \cdot 6H_2O$, and 2,4,6-tripyridyl-s-triazine (TPTZ) [74]. The absorbance was measured at 593 nm after 4 min in the dark. For the DPPH free radical scavenging 100 µL of the sample was mixed with 3.9 mL of a methanolic 2,2-diphenyl-1-picrylhydrazyl solution and incubated in dark conditions for 30 min [75]. The decrease in absorbance at 515 nm was recorded. The ABTS radical cation decolorization assay was conducted by incubating 100 µL of the sample with 2,2′-azino-bis (3-ethylbenzothiazoline-6-sulfonic acid) for 4 min, followed by measuring absorbance at 734 nm [76]. All antioxidant activity assays were performed in triplicate, and results expressed as µmol trolox equiv-

https://doi.org/10.3390/molecules31010152

alents (TE) per g of sample. In addition, the TPC of the bread samples was determined following the method described by Singleton et al., by measuring the absorbance at 750 nm after a 1-h incubation of the extracted sample mixed with the Folin–Ciocalteu reagent [77]. The analysis was carried out in triplicate, and results were expressed as mg gallic acid equivalents (GAE) per g of sample. All spectrophotometric measurements were performed on an Evolution 300 spectrophotometer (Thermo Scientific, Waltham, MA, USA).

3.8. Sensory Analysis

A hedonic sensory evaluation was conducted on the six GF bread samples using a structured 5-point scale to assess sensory attributes, including appearance, aroma, texture, taste, color, chewiness, overall acceptability, and purchase intention. The bread samples were cut into 3 cm^3 portions and coded with randomly selected three-digit numbers. The sensory panel consisted of 25 trained assessors (14 female, 11 male) aged between 18 and 60 years. Participants were required to be healthy adults without sensory impairments or food allergies related to the tested products. Exclusion criteria included smoking and medical conditions that could affect sensory perception. The evaluation was conducted in individual sensory booths according to ISO guidelines [78], with mineral water provided for rinsing their mouths between samples.

3.9. Statistical Analysis

All determinations were conducted in three independent experimental replicates ($n = 3$), and results were expressed as mean ± standard deviation (SD). Statistical analysis was performed using the SPSS software, version 28.0 (IBM Corporation, Armonk, NY, USA). Differences among GF breads were determined by one-way analysis of variance (ANOVA), setting a significance threshold of $p < 0.05$, followed by Tukey's post hoc test. The datasets from all experimental analyses were integrated and visualized through a heatmap of Pearson correlation coefficients, constructed using the 'pheatmap' package (version 2.5.0) in R software (version 4.5.0) (R Foundation for Statistical Computing, Vienna, Austria).

4. Conclusions

The present study expands current knowledge on the use of enzyme-treated vegetable by-products as natural functional ingredients in gluten-free bakery formulations. Specifically, the effects of enzymatically treated artichoke and broccoli extracts were assessed as sustainable alternatives to synthetic additives in gluten-free breads. The incorporation of the by-products represents a sustainable and effective strategy for developing gluten-free (GF) breads with enhanced nutritional, functional, and sensory properties compared with a commercial reference. Among the enzymatic treatments, Celluclast® 1.5 L applied to artichoke by-products (BAE) exhibited the most significant overall improvements. This enzyme increased soluble dietary fiber up to 2.75 g/100 g, while markedly modifying starch digestibility by reducing the rapidly digestible fraction by 14% and increasing the slowly digestible fraction by more than 150%, suggesting a lower predicted glycaemic response. Although Viscozyme® L mainly enhanced antioxidant capacity (30% in FRAP and 37% in ABTS), Celluclast® 1.5 L achieved a more balanced enhancement across nutritional, functional, and sensory attributes, with BAE showing the highest consumer acceptance.

In contrast, non-treated extracts exerted limited influence on antioxidant activity and product quality. The enzymatic treatment effectively improved the release of bioactive compounds and mitigated undesirable effects on crumb color and structure, demonstrating greater functional performance than the control formulation and the commercial reference bread.

https://doi.org/10.3390/molecules31010152

Correlations between compositional and sensory parameters indicated that increased soluble fiber and phenolic content were associated with improved sensory perception and antioxidant performance. Overall, the findings demonstrate the potential of enzyme-treated artichoke and broccoli by-products as sustainable, clean-label ingredients for the formulation of nutritionally improved GF breads, contributing to the valorization of agro-industrial residues and the advancement of circular food production systems. Further research is required to evaluate their impact on oxidative stability and storage performance under industrial conditions.

Author Contributions: Conceptualization, R.P., J.Q. and G.N.; methodology, R.P., and J.Q.; validation, J.Q., R.P., P.A. and G.N.; investigation, J.Q. and R.P.; writing—original draft preparation, J.Q.; writing—review and editing, J.Q., P.A., R.P. and G.N.; visualization, G.N.; supervision, G.N.; project administration, G.N.; funding acquisition, G.N. All authors have read and agreed to the published version of the manuscript.

Funding: This study forms part of the Agroalnext program and was supported by MCIU with funding from European Union Next GenerationEU (PRTR-C17.I1) and by Comunidad Autónoma de la Región de Murcia—Fundación Séneca. Jhazmin Quizhpe thanks the finantial support through Research Project PID2021-123628OB-C44 funded by MCIN/AEI/10.13039/501100011033 and by "ERDF A way of making Europe".

Institutional Review Board Statement: The study was conducted in accordance with the Declaration of Helsinki, and approved by the Ethics Committee of University of Murcia (protocol code 444/2024 and date of approval: 11 December 2024).

Informed Consent Statement: Informed consent was obtained from all subjects involved in the study. All data were anonymized before analysis.

Data Availability Statement: The original contributions presented in this study are included in the article. Further inquiries can be directed to the corresponding author.

Conflicts of Interest: The authors affirm that they have no financial interests or personal relationships that could be perceived as influencing the results or conclusions of this study.

References

1. Fasano, A.; Catassi, C. Celiac Disease. *N. Engl. J. Med.* **2012**, *367*, 2419–2426. [CrossRef] [PubMed]
2. Khemiri, S.; Khelifi, N.; Nunes, M.C.; Ferreira, A.; Gouveia, L.; Smaali, I.; Raymundo, A. Microalgae biomass as an additional ingredient of gluten-free bread: Dough rheology, texture quality and nutritional properties. *Algal Res.* **2020**, *50*, 101998. [CrossRef]
3. Lebwohl, B.; Rubio-Tapia, A. Presentation, and Diagnosis of Celiac Disease. *Gastroenterology* **2021**, *160*, 63–75. [CrossRef] [PubMed]
4. Vici, G.; Belli, L.; Biondi, M.; Polzonetti, V. Gluten free diet and nutrient deficiencies: A review. *Clin. Nutr.* **2016**, *35*, 1236–1241. [CrossRef]
5. Bianchi, P.I.; Aronico, N.; Santacroce, G.; Broglio, G.; Lenti, M.V.; Di Sabatino, A. Nutritional Consequences of Celiac Disease and Gluten-Free Diet. *Gastroenterol. Insights* **2024**, *15*, 878–894. [CrossRef]
6. Morales, D.; Miguel, M.; Garcés-Rimón, M. Pseudocereals: A novel source of biologically active peptides. *Crit. Rev. Food Sci. Nutr.* **2021**, *61*, 1537–1544. [CrossRef]
7. Peñalver, R.; Nieto, G. Developing a functional gluten-free sourdough bread by incorporating quinoa, amaranth, rice and spirulina. *LWT* **2024**, *201*, 116162. [CrossRef]
8. Kaur, R.; Ahluwalia, P.; Sachdev, P.A.; Kaur, A. Development of gluten-free cereal bar for gluten intolerant population by using quinoa as major ingredient. *J. Food Sci. Technol.* **2018**, *55*, 3584–3591. [CrossRef]
9. Parikh, M.; Netticadan, T.; Pierce, G.N. Flaxseed: Its bioactive components and their cardiovascular benefits. *Am. J. Physiol. Heart Circ. Physiol.* **2018**, *314*, H146–H159. [CrossRef]
10. Cantero, L.; Salmerón, J.; Miranda, J.; Larretxi, I.; Fernández-Gil, M.D.P.; Bustamante, M.Á.; Matias, S.; Navarro, V.; Simón, E.; Martínez, O. Performance of Apple Pomace for Gluten-Free Bread Manufacture: Effect on Physicochemical Characteristics and Nutritional Value. *Appl. Sci.* **2022**, *12*, 5934. [CrossRef]

 https://doi.org/10.3390/molecules31010152

11. Difonzo, G.; de Gennaro, G.; Pasqualone, A.; Caponio, F. Potential use of plant-based by-products and waste to improve the quality of gluten-free foods. *J. Sci. Food Agric.* **2022**, *102*, 2199–2211. [CrossRef]
12. Rodríguez, M.; Bianchi, F.; Simonato, B.; Rizzi, C.; Fontana, A.R.; Tironi, V.A. Exploration of grape pomace peels and amaranth flours as functional ingredients in the elaboration of breads: Phenolic composition, bioaccessibility, and antioxidant activity. *Food Funct.* **2024**, *15*, 608–624. [CrossRef] [PubMed]
13. Saadoun, J.H.; Bertani, G.; Levante, A.; Vezzosi, F.; Ricci, A.; Bernini, V.; Lazzi, C. Fermentation of Agri-Food Waste: A Promising Route for the Production of Aroma Compounds. *Foods* **2021**, *10*, 707. [CrossRef] [PubMed]
14. Eliopoulos, C.; Markou, G.; Langousi, I.; Arapoglou, D. Reintegration of Food Industry By-Products: Potential Applications. *Foods* **2022**, *11*, 3743. [CrossRef] [PubMed]
15. FAOSTAT. Food and Agriculture Organization of the United Nations Statistical Database. 2023. Available online: https://www.fao.org/faostat/ (accessed on 12 September 2025).
16. Gudiño, I.; Casquete, R.; Martín, A.; Wu, Y.; Benito, M.J. Comprehensive Analysis of Bioactive Compounds, Functional Properties, and Applications of Broccoli By-Products. *Foods* **2024**, *13*, 3918. [CrossRef]
17. Borja-Martínez, M.; Lozano-Sánchez, J.; Borrás-Linares, I.; A Pedreño, M.; Sabater-Jara, A.B. Revalorization of Broccoli By-Products for Cosmetic Uses Using Supercritical Fluid Extraction. *Antioxidants* **2020**, *9*, 1195. [CrossRef]
18. Quizhpe, J.; Ayuso, P.; Rosell, M.d.L.Á.; Peñalver, R.; Nieto, G. *Brassica oleracea* var *italica* and Their By-Products as Source of Bioactive Compounds and Food Applications in Bakery Products. *Foods* **2024**, *13*, 3513. [CrossRef]
19. Jiménez-Moreno, N.; Cimminelli, M.J.; Volpe, F.; Ansó, R.; Esparza, I.; Mármol, I.; Rodríguez-Yoldi, M.J.; Ancín-Azpilicueta, C. Phenolic Composition of Artichoke Waste and Its Antioxidant Capacity on Differentiated Caco-2 Cells. *Nutrients* **2019**, *11*, 1723. [CrossRef]
20. Ayuso, P.; Quizhpe, J.; Rosell, M.d.L.Á.; Peñalver, R.; Nieto, G. Antioxidant and Nutritional Potential of Artichoke (*Cynara scolymus* L.) By-Product Extracts in Fat-Replaced Beef Burgers with Hydrogel Emulsions from Olive Oil. *Appl. Sci.* **2024**, *14*, 10123. [CrossRef]
21. Ayuso, P.; Quizhpe, J.; Rosell, M.d.L.Á.; Peñalver, R.; Nieto, G. Bioactive Compounds, Health Benefits and Food Applications of Artichoke (*Cynara scolymus* L.) and Artichoke By-Products: A Review. *Appl. Sci.* **2024**, *14*, 4940. [CrossRef]
22. Colombo, R.; Moretto, G.; Pellicorio, V.; Papetti, A. Globe Artichoke (*Cynara scolymus* L.) By-Products in Food Applications: Functional and Biological Properties. *Foods* **2024**, *13*, 1427. [CrossRef]
23. Guan, Z.W.; Yu, E.Z.; Feng, Q. Soluble Dietary Fiber, One of the Most Important Nutrients for the Gut Microbiota. *Molecules* **2021**, *26*, 6802. [CrossRef]
24. Bader Ul Ain, H.; Saeed, F.; Ahmed, A.; Asif Khan, M.; Niaz, B.; Tufail, T. Improving the physicochemical properties of partially enhanced soluble dietary fiber through innovative techniques: A coherent review. *J. Food Process Preserv.* **2019**, *43*, e13917. [CrossRef]
25. Villanueva-Suárez, M.J.; Pérez-Cózar, M.L.; Redondo-Cuenca, A. Sequential extraction of polysaccharides from enzymatically hydrolyzed okara byproduct: Physicochemical properties and in vitro fermentability. *Food Chem.* **2013**, *141*, 1114–1119. [CrossRef] [PubMed]
26. Yang, Y.Y.; Ma, S.; Wang, X.X.; Zheng, X.L. Modification and Application of Dietary Fiber in Foods. *J. Chem.* **2017**, *2017*, 9340427. [CrossRef]
27. Luo, X.; Wang, Q.; Zheng, B.; Lin, L.; Chen, B.; Zheng, Y.; Xiao, J. Hydration properties and binding capacities of dietary fibers from bamboo shoot shell and its hypolipidemic effects in mice. *Food Chem. Toxicol.* **2017**, *109*, 1003–1009. [CrossRef]
28. Uebelhack, R.; Busch, R.; Alt, F.; Beah, Z.M.; Chong, P.W. Effects of Cactus Fiber on the Excretion of Dietary Fat in Healthy Subjects: A Double Blind, Randomized, Placebo-Controlled, Crossover Clinical Investigation. *Curr. Ther. Res.* **2014**, *76*, 39–44. [CrossRef]
29. Feiden, T.; Valduga, E.; Zeni, J.; Steffens, J. Bioactive Compounds from Artichoke and Application Potential. *Food Technol. Biotechnol.* **2023**, *61*, 312–327. [CrossRef]
30. Dadalı, C. Artichoke bracts as fat and wheat flour replacer in cake: Optimization of reduced fat and reduced wheat flour cake formulation. *J. Food Meas. Charact.* **2023**, *17*, 98–107. [CrossRef]
31. Ayuso, P.; Peñalver, R.; Quizhpe, J.; Rosell, M.d.l.Á.; Nieto, G. Broccoli, Artichoke, Carob and Apple By-Products as a Source of Soluble Fiber: How It Can Be Affected by Enzymatic Treatment with Pectinex® Ultra SP-L, Viscozyme® L and Celluclast® 1.5 L. *Foods* **2024**, *14*, 10. [CrossRef]
32. De la Peña-Armada, R.; Villanueva-Suárez, M.J.; Rupérez, P.; Mateos-Aparicio, I. High Hydrostatic Pressure Assisted by Celluclast® Releases Oligosaccharides from Apple By-Product. *Foods* **2020**, *9*, 1058. [CrossRef] [PubMed]
33. Schäfer, J.; Stanojlovic, L.; Trierweiler, B.; Bunzel, M. Storage related changes of cell wall based dietary fiber components of broccoli (*Brassica oleracea* var. *italica*) stems. *Food Res. Int.* **2017**, *93*, 43–51. [CrossRef] [PubMed]

https://doi.org/10.3390/molecules31010152

34. Favela-González, K.M.; Hernández-Almanza, A.Y.; De la Fuente-Salcido, N.M. The value of bioactive compounds of cruciferous vegetables (*Brassica*) as antimicrobials and antioxidants: A review. *J. Food Biochem.* **2020**, *44*, e13414. [CrossRef] [PubMed]
35. Pagano, I.; Piccinelli, A.L.; Celano, R.; Campone, L.; Gazzerro, P.; Russo, M.; Rastrelli, L. Pressurized hot water extraction of bioactive compounds from artichoke by-products. *Electrophoresis* **2018**, *39*, 1899–1907. [CrossRef]
36. Nieto, G. Incorporation of by-products of rosemary and thyme in the diet of ewes: Effect on the fatty acid profile of lamb. *Eur. Food Res. Technol.* **2013**, *236*, 379–389. [CrossRef]
37. Nieto, G.; Banón, S.; Garrido, M. Administration of distillate thyme leaves into the diet of Segureña ewes: Effect on lamb meat quality. *Animal* **2012**, *6*, 2048–2056. [CrossRef]
38. Banón, S.; Díaz, P.; Nieto, G.; Castillo, M.; Álvarez, D. Modelling the Yield and Texture of Comminuted Pork Products Using Color and Temperature. Effect of Fat/Lean Ratio and Starch. *Meat Sci.* **2008**, *80*, 649–655. [CrossRef]
39. Lekmine, S.; Boussekine, S.; Akkal, S.; Martín-García, A.I.; Boumegoura, A.; Kadi, K.; Djeghim, H.; Mekersi, N.; Bendjedid, S.; Bensouici, C.; et al. Investigation of Photoprotective, Anti-Inflammatory, Antioxidant Capacities and LC–ESI–MS Phenolic Profile of Astragalus gombiformis Pomel. *Foods* **2021**, *10*, 1937. [CrossRef]
40. Franke, K.; Djikeng, F.T.; Esatbeyoglu, T. Retention of bioactives in food processing. In *Influence of Frying, Baking and Cooking on Food Bioactives*; Esatbeyoglu, T., Ed.; Springer: Cham, Switzerland, 2022; pp. 93–121.
41. Barbosa, P.P.M.; Ruviaro, A.R.; Macedo, G.A. Conditions of enzyme-assisted extraction to increase the recovery of flavanone aglycones from pectin waste. *J. Food Sci. Technol.* **2020**, *58*, 4303–4312. [CrossRef]
42. Brienza, F.; Calani, L.; Bresciani, L.; Mena, P.; Rapacioli, S. Optimized Enzymatic Extraction of Phenolic Compounds from *Verbascum nigrum* L.: A Sustainable Approach for Enhanced Extraction of Bioactive Compounds. *Appl. Sci.* **2025**, *15*, 1405. [CrossRef]
43. Aguiar, E.V.; Santos, F.G.; Krupa-Kozak, U.; Capriles, V.D. Nutritional facts regarding commercially available gluten-free bread worldwide: Recent advances and future challenges. *Crit. Rev. Food Sci. Nutr.* **2023**, *63*, 693–705. [CrossRef] [PubMed]
44. Peñalver, R.; Ros, G.; Nieto, G. Development of Gluten-Free Functional Bread Adapted to the Nutritional Requirements of Celiac Patients. *Fermentation* **2023**, *9*, 631. [CrossRef]
45. Peñalver, R.; Ros, G.; Nieto, G. Development of Functional Gluten-Free Sourdough Bread with Pseudocereals and Enriched with Moringa oleifera. *Foods* **2023**, *12*, 3920. [CrossRef] [PubMed]
46. Romero-Garay, M.G.; Adaile-Pérez, V.M.; Montalvo-González, E.; Martínez-Montaño, E.; García-Magaña, M.D.L. Use of Poultry By-Products: Production of Functional Ingredients with High Bioactive Value through the Use of Ultrasound-Assisted Enzymatic Hydrolysis. *J. Food Meas. Charact.* **2024**, *19*, 1220–1233. [CrossRef]
47. Qian, J.; Chen, D.; Zhang, Y.; Gao, X.; Xu, L.; Guan, G.; Wang, F. Ultrasound-Assisted Enzymatic Protein Hydrolysis in Food Processing: Mechanism and Parameters. *Foods* **2023**, *12*, 4027. [CrossRef]
48. Stanek-Wandzel, N.; Krzyszowska, A.; Zarębska, M.; Gębura, K.; Wasilewski, T.; Hordyjewicz-Baran, Z.; Tomaka, M. Evaluation of Cellulase, Pectinase, and Hemicellulase Effectiveness in Extraction of Phenolic Compounds from Grape Pomace. *Int. J. Mol. Sci.* **2024**, *25*, 13538. [CrossRef]
49. Nemmaru, B.; Douglass, J.; Yarbrough, J.M.; DeChellis, A.; Shankar, S.; Thokkadam, A.; Wang, A.; Chundawat, S.P.S. Supercharged Cellulases Show Superior Thermal Stability and Enhanced Activity towards Pretreated Biomass and Cellulose. *Front. Energy Res.* **2024**, *12*, 1372916. [CrossRef]
50. Giuberti, G.; Gallo, A. Reducing the glycaemic index and increasing the slowly digestible starch content in gluten-free cereal-based foods: A review. *Int. J. Food Sci. Technol.* **2017**, *53*, 50–60. [CrossRef]
51. Di Cairano, M.; Condelli, N.; Caruso, M.C.; Cela, N.; Tolve, R.; Galgano, F. Use of Underexploited Flours for the Reduction of Glycaemic Index of Gluten-Free Biscuits: Physicochemical and Sensory Characterization. *Food Bioprocess. Technol.* **2021**, *14*, 1490–1502. [CrossRef]
52. Santamaria, M.; Ruiz, M.; Garzon, R.; Rosell, C.M. Comparison of vegetable powders as ingredients of flatbreads: Technological and nutritional properties. *Int. J. Food Sci. Technol.* **2024**, *59*, 7203–7212. [CrossRef]
53. Melini, V.; Melini, F.; Salvati, A.; Luziatelli, F.; Ruzzi, M. Effect of Artichoke Outer Bract Powder Addition on the Nutritional Profile of Gluten-Free Rusks. *Foods* **2025**, *14*, 2395. [CrossRef] [PubMed]
54. Kasprzak, M.M.; Lærke, H.N.; Larsen, F.H.; Knudsen, K.E.B.; Pedersen, S.; Jørgensen, A.S. Effect of Enzymatic Treatment of Different Starch Sources on the in Vitro Rate and Extent of Starch Digestion. *Int. J. Mol. Sci.* **2012**, *13*, 929–942. [CrossRef] [PubMed]
55. Pashaei, M.; Bahmanyar, F.; Tahmouzi, S.; Nasab, S.; Sadrabad, E.K.; Mollakhalili-Meybodi, N.; Mirmoghtadaie, L. The role of enzymes in gluten-free bakery products: A review of technological and nutritional perspectives. *Appl. Food Res.* **2025**, *5*, 100923. [CrossRef]
56. Canale, M.; Spina, A.; Summo, C.; Strano, M.C.; Bizzini, M.; Allegra, M.; Sanfilippo, R.; Amenta, M.; Pasqualone, A. Waste from Artichoke Processing Industry: Reuse in Bread-Making and Evaluation of the Physico-Chemical Characteristics of the Final Product. *Plants* **2022**, *11*, 3409. [CrossRef]

57. Cannas, M.; Conte, P.; Piga, A.; Del Caro, A. Artichoke By-Product Extracts as a Viable Alternative for Shelf-Life Extension of Breadsticks. *Foods* **2024**, *13*, 2639. [CrossRef]
58. Krupa-Kozak, U.; Drabińska, N.; Baczek, N.; Šimková, K.; Starowicz, M.; Jeliński, T. Application of Broccoli Leaf Powder in Gluten-Free Bread: An Innovative Approach to Improve Its Bioactive Potential and Technological Quality. *Foods* **2021**, *10*, 819. [CrossRef]
59. Krupa-Kozak, U.; Drabińska, N.; Rosell, C.M.; Fadda, C.; Anders, A.; Jeliński, T.; Ostaszyk, A. Broccoli leaf powder as an attractive by-product ingredient: Effect on batter behaviour, technological properties and sensory quality of gluten-free mini sponge cake. *Int. J. Food Sci. Technol.* **2019**, *54*, 1121–1129. [CrossRef]
60. Garzon, R.; Skendi, A.; Antonio Lazo-Velez, M.; Papageorgiou, M.; Rosell, C.M. Interaction of dough acidity and microalga level on bread quality and antioxidant properties. *Food Chem.* **2021**, *344*, 128710. [CrossRef]
61. Longo, A.; Amendolagine, G.; Miani, M.G.; Rizzello, C.G.; Verni, M. Effect of Air Classification and Enzymatic and Microbial Bioprocessing on Defatted Durum Wheat Germ: Characterization and Use as Bread Ingredient. *Foods* **2024**, *13*, 1953. [CrossRef]
62. Koirala, P.; Costantini, A.; Maina, H.N.; Rizzello, C.G.; Verni, M.; Beni, V.D.; Polo, A.; Katina, K.; Di Cagno, R.; Coda, R. Fermented Brewers' Spent Grain Containing Dextran and Oligosaccharides as Ingredient for Composite Wheat Bread and Its Impact on Gut Metabolome In Vitro. *Fermentation* **2022**, *8*, 487. [CrossRef]
63. Bourekoua, H.; Różyło, R.; Gawlik-Dziki, U.; Benatallah, L.; Zidoune, M.N.; Dziki, D. Evaluation of physical, sensorial, and antioxidant properties of gluten-free bread enriched with Moringa Oleifera leaf powder. *Eur. Food Res. Technol.* **2018**, *244*, 189–195. [CrossRef]
64. Krupa-Kozak, U.; Baczek, N.; Capriles, V.D.; Łopusiewicz, Ł. Novel Gluten-Free Bread with an Extract from Flaxseed By-Product: The Relationship between Water Replacement Level and Nutritional Value, Antioxidant Properties, and Sensory Quality. *Molecules* **2022**, *27*, 2690. [CrossRef] [PubMed]
65. Talens, C.; Álvarez-Sabatel, S.; Rios, Y.; Rodríguez, R. Effect of a new microwave-dried orange fibre ingredient vs. a commercial citrus fibre on texture and sensory properties of gluten-free muffins. *Innov. Food Sci. Emerg. Technol.* **2017**, *44*, 83–88. [CrossRef]
66. Majzoobi, M.; Poor, Z.V.; Jamalian, J.; Farahnaky, A. Improvement of the quality of gluten-free sponge cake using different levels and particle sizes of carrot pomace powder. *Int. J. Food Sci. Technol.* **2016**, *51*, 1369–1377. [CrossRef]
67. Mollakhalili-Meybodi, N.; Sheidaei, Z.; Khorshidian, N.; Nematollahi, A.; Khanniri, E. Sensory attributes of wheat bread: A review of influential factors. *J. Food Meas. Charact.* **2023**, *17*, 2172–2181. [CrossRef]
68. Zheng, S.; Zhang, Y.; Chen, Q.; Fu, X.; Huang, Q.; Zhang, B.; Dong, H.; Li, C. Exploring the synergistic benefits of insoluble dietary fiber and bound phenolics: Unveiling the role of bound phenolics in enhancing bioactivities of insoluble dietary fiber. *Trends Food Sci. Technol.* **2024**, *149*, 104554. [CrossRef]
69. European Parliament and Council of the European Union. Regulation (EC) No. 1924/2006 of the European Parliament and of the Council of 20 December 2006 on nutrition and health claims made on foods. *Off J. Eur. Union* **2006**, *L404*, 9–25.
70. AOAC International. *Official Methods of Analysis of AOAC International*, 19th ed.; AOAC International: Gaithersburg, MD, USA, 2012.
71. Nielsen, S.S. Introduction to Food Analysis. In *Food Analysis*, 6th ed.; Springer: Berlin/Heidelberg, Germany, 2024; pp. 3–14.
72. Charrondière, U.R.; Rittenschober, D.; Nowak, V.; Wijesinha-Bettoni, R.; Stadlmayr, B.; Haytowitz, D.; Persijn, D. *FAO/INFOODS Guidelines for Converting Units, Denominators and Expressions*; Version 1.0; FAO: Rome, Italy, 2012.
73. Englyst, H.N.; Kingman, S.M.; Cummings, J.H. Classification and measurement of nutritionally important starch fractions. *Eur. J. Clin. Nutr.* **1992**, *46*, S33–S50.
74. Benzie, I.F.F.; Strain, J.J. The Ferric Reducing Ability of Plasma (FRAP) as a Measure of "Antioxidant Power": The FRAP Assay. *Anal. Biochem.* **1996**, *239*, 70–76. [CrossRef]
75. Brand-Williams, W.; Cuvelier, M.E.; Berset, C. Use of a free radical method to evaluate antioxidant activity. *LWT Food Sci. Technol.* **1995**, *28*, 25–30. [CrossRef]
76. Re, R.; Pellegrini, N.; Proteggente, A.; Pannala, A.; Yang, M.; Rice-Evans, C. Antioxidant activity applying an improved ABTS radical cation decolorization assay. *Free Radic. Biol. Med.* **1999**, *26*, 1231–1237. [CrossRef]
77. Singleton, V.L.; Orthofer, R.; Lamuela-Raventós, R.M. Analysis of total phenols and other oxidation substrates and antioxidants by means of folin-ciocalteu reagent. *Methods Enzymol.* **1999**, *299*, 152–178. [CrossRef]
78. *ISO 8586*; Sensory Analysis: General Guidelines for the Selection, Training and Monitoring of Selected Assessors and Expert Sensory Assessors. International Organization for Standardization: Geneva, Switzerland, 2012.

https://doi.org/10.3390/molecules31010152

molecules

MDPI

Article

Prescreening of Mango (*Mangifera indica* L.) Leaves as a Potential Functional Food Ingredient: Techno-Functional and Antioxidative Characteristics

Génica Lawrence [1], Ingrid Marchaux [1], Ewa Pejcz [2,*], Agata Wojciechowicz-Budzisz [2], Remigiusz Olędzki [2], Adam Zając [3], Oliwia Paroń [2], Guylène Aurore [1,†] and Joanna Harasym [2,4,†]

1 Campus de Fouillole, Université des Antilles, COVACHIM-M2E (EA 3592), UFR SEN, F-97110 Pointe-à-Pitre, France; genica.lawrence@univ-antilles.fr (G.L.); imarchaux@gmail.com (I.M.); guylene.aurore@univ-antilles.fr (G.A.)

2 Department of Biotechnology and Food Analysis, Wroclaw University of Economics and Business, Komandorska 118/120, 53-345 Wroclaw, Poland; agata.wojciechowicz-budzisz@ue.wroc.pl (A.W.-B.); remigiusz.oledzki@ue.wroc.pl (R.O.); oliwia.paron@ue.wroc.pl (O.P.); joanna.harasym@ue.wroc.pl (J.H.)

3 Department of Bioorganic Chemistry, Wroclaw University of Economics and Business, Komandorska 118/120, 53-345 Wroclaw, Poland; adam.zajac@ue.wroc.pl

4 Adaptive Food Systems Accelerator-Science Centre, Wroclaw University of Economics and Business, Komadorska 118/120, 53-345 Wroclaw, Poland

* Correspondence: ewa.pejcz@ue.wroc.pl

† These authors contributed equally to this work.

Abstract: Mango (*Mangifera indica* L.) is cultivated in tropical and subtropical regions, with all parts of the tree—including leaves—used traditionally to treat diabetes, infections, pain, and other conditions. Mango leaves contain proteins, minerals, vitamins, and phenolic compounds, including mangiferin, quercetin, and kaempferol, whose content varies by cultivar. This study evaluated the functional and bioactive properties of dried mango leaves from five cultivars (Julie, DLO, Nam Dok Mai, Irwin, and Keïtt) to determine their potential for food and nutraceutical applications. Analyses included water- and oil-related parameters, swelling and solubility indices, foaming and emulsifying properties, and antioxidant activity (DPPH, ABTS, and FRAP in hydroalcoholic and water extracts), complemented by FT-IR/ATR spectroscopy. Significant differences between the five analyzed cultivars were observed. Irwin exhibited the highest antioxidant activity (2.65 $\pm$ 0.55 mg TE/g DM in DPPH assay), while Nam Dok Mai demonstrated superior foaming capacity (82.69 $\pm$ 7.79 mL). Strong correlations ($r > 0.9$) between reducing sugars and antioxidant capacity suggest cultivar selection based on sugar content could predict antioxidant potential. FT-IR confirmed the presence of polar phenolic and protein compounds. The results demonstrate that mango leaves offer cultivar-dependent functional and antioxidant attributes relevant to food systems. Their targeted valorization may support sustainable industrial applications and circular bioeconomy strategies, particularly in tropical regions where mango cultivation is widespread.

Keywords: mango leaves; *Mangifera indica* L.; functional properties; bioactive compounds; antioxidant activity; waste valorization

1. Introduction

Various plants and tree leaves can be considered for use in the food industry due to their nutritional quality, bioactive potential, and techno-functional properties. Mango

(*Mangifera indica* L.) leaves (MLs) are a good example in numerous studies that have demonstrated a wide range of their phytochemicals and associated health benefits, where polyphenols are the predominant compounds. Among them, mangiferin, a xanthone glycoside known for its strong antioxidant activity, is particularly abundant in MLs [1,2]. Evidence also suggests its potential antidiabetic [3], antimicrobial, antispasmodic, antigenotoxic [4], and anti-breast cancer effects [5]. MLs contain phenolic acids, benzophenones, ascorbic acid, carotenoids, tocopherols, and other antioxidants such as flavonoids [4]. Specific flavonoids like quercetin and kaempferol have been linked to reduced oxidative stress, a key factor in the development of many chronic diseases. Additionally, leaf extracts appear to modulate blood glucose levels and improve insulin sensitivity, indicating potential in the management of type 2 diabetes [3,6]. MLs are also a notable source of protein [3], minerals (nitrogen, potassium, phosphorus, iron, sodium, calcium, magnesium), and vitamins A, B, E, and C. These leaves are already used as an alternative livestock feed in developing countries.

Although MLs have long been used in traditional medicine in China, India, and Africa—typically as dried leaves or powder for herbal teas—scientific interest in their application has grown in recent years, particularly in the food, pharmaceutical, and cosmetic sectors. A more detailed characterization of their chemical composition and safety remains necessary before broader integration into modern medicine [4] and functional foods. Beyond their nutritional and antioxidant properties, only limited studies to date have explored the techno-functional properties of MLs, such as water-holding capacity, emulsifying ability, and foaming capacity, for potential food applications. Understanding these functional properties is essential for their appropriate use in food formulations, particularly considering the increasing demand for sustainable ingredients. Furthermore, the food industry's growing demand for natural antioxidants and functional ingredients, driven by consumer preference for clean-label products, creates opportunities for mango leaf valorization. The estimated 25–30% of mango tree biomass represented by leaves remains largely unutilized, representing a significant waste stream that could be converted into valuable functional ingredients.

In Guadeloupe, a French overseas region in the Caribbean, as many as 200 mango cultivars are grown. Previous studies have shown that mango fruit cultivars differ significantly in organoleptic, physicochemical, and nutritional properties. Preliminary results from our laboratory suggest that the composition of MLs—including physicochemical parameters, nutritional content, and antioxidant profile—also varies by cultivar.

This study was undertaken to comparatively evaluate the bioactive compound profiles and techno-functional properties of mango leaves from five selected cultivars grown in Guadeloupe. Although the antioxidant potential and phytochemical richness of mango leaves have been partially described in the literature, there is a noticeable lack of systematic studies addressing their functional attributes in relation to cultivar variability. It is hypothesized that the techno-functional properties and antioxidant potential of mango leaves are significantly influenced by varietal differences and that certain cultivars may offer superior profiles for targeted food applications. Furthermore, both aqueous and ethanolic extracts are expected to differ in their extractive efficiency and antioxidant activity, depending on the polarity and structure of the compounds involved. By bridging this knowledge gap, the present research not only expands the understanding of underutilized plant resources but also explores their potential valorization in functional food design and sustainable ingredient development. This work also aims to contribute to local agricultural diversification strategies and to foster circular bioeconomy initiatives in tropical regions such as Guadeloupe.

2. Results and Discussion

This study investigated the functional and bioactive properties of dried mango leaves from five cultivars (DLO, Nam Dok Mai, Irwin, Julie, and Keïtt) collected in Guadeloupe, France. After standardized drying and milling, the resulting powders were evaluated for their absorptional, emulsifying, foaming, and antioxidant properties. Representative images of the leaves and fruits are shown in Figure 1. Detailed procedures are described in Section 3.

Figure 1. The mango leaves of different cultivars (**A**), Irwin, Dlo, Julie, Nam Dok Mai, and Keïtt, and the different forms of mango fruits associated with the mango leaves studied (**B**).

2.1. Absorptional Characteristics of Mango Leaf Samples from Different Varieties

Understanding the water- and oil-binding properties of mango leaf powders is essential for determining their suitability in various food applications. These absorptional characteristics directly influence texture development, moisture retention, and processing behavior in food formulations. The evaluation encompassed both gravitational and mechanical water retention mechanisms, oil absorption capacity, and swelling behavior under thermal treatment conditions. Table 1 presents the absorptional characteristics of the five mango varieties, revealing distinct functional properties across water and oil-binding parameters.

Table 1. Absorptional characteristics of mango leaf samples from different varieties.

Sample	WHC [g H_2O/g DM]	WAC [g H_2O/g DM]	OAC [g oil/g DM]	HLI	WAI [g H_2O/g DM]	WSI [g H_2O/100 g DM]	SP [g H_2O/g DM]
Julie	5.08 ± 0.10 ^{b}	3.53 ± 0.15 ^{ab}	2.70 ± 0.17 ^{b}	1.307 ± 0.030 ^{a}	5.14 ± 0.18 ^{b}	13.89 ± 0.95 ^{a}	4.72 ± 0.18 ^{b}
DLO	5.19 ± 0.33 ^{b}	3.59 ± 0.08 ^{b}	2.69 ± 0.04 ^{b}	1.336 ± 0.040 ^{ab}	3.68 ± 0.34 ^{a}	28.69 ± 7.72 ^{c}	2.82 ± 0.56 ^{a}
Nam Dok Mai	5.53 ± 0.14 c	3.48 ± 0.12 ^{ab}	2.21 ± 0.08 ^{a}	1.576 ± 0.025 ^{c}	5.17 ± 0.29 ^{b}	19.41 ± 2.70 ^{b}	4.58 ± 0.33 ^{b}
Irwin	5.19 ± 1.33 ^{b}	3.76 ± 0.38 ^{b}	2.77 ± 0.13 ^{b}	1.354 ± 0.080 ^{b}	4.72 ± 0.07 ^{b}	15.68 ± 1.26 ^{ab}	4.10 ± 0.10 ^{b}
Keïtt	4.64 ± 0.05 a	3.73 ± 0.12 ^{b}	2.49 ± 0.08 ^{ab}	1.501 ± 0.088 ^{bc}	3.60 ± 0.19 ^{a}	30.83 ± 6.47 ^{c}	2.67 ± 0.39 ^{a}

WHC—water-holding capacity; WAC—water absorption capacity; OAC—oil absorption capacity; HLI—hydrophilic–lipophilic index; WAI—water absorption index; WSI—water solubility index; SP—swelling power; DM—dry matter. Lowercase letters mean significant differences in columns at $p = 0.05$.

The Nam Dok Mai variety demonstrated the highest water-holding capacity (WHC; 5.53 ± 0.14 g H_2O/g DM) and the highest hydrophilic–lipophilic index (HLI; 1.576 ± 0.025), while exhibiting the lowest oil absorption capacity (OAC; 2.21 ± 0.08 g oil/g DM). The Julie variety showed the highest swelling power (SP; 4.72 ± 0.18 g H_2O/g DM) and water absorption index (WAI; 5.14 ± 0.18 g H_2O/g DM), along with the lowest water solubility index (WSI; 13.89 ± 0.95 g H_2O/100 g DM) and the lowest HLI (1.307 ± 0.030). The Irwin variety exhibited the highest water absorption capacity (WAC; 3.76 ± 0.38 g H_2O/g DM) and OAC (2.77 ± 0.13 g oil/g DM). The Keïtt variety presented the lowest WHC (4.64 ± 0.05 g H_2O/g DM) but demonstrated the highest WSI (30.83 ± 6.47 g H_2O/100 g DM), as well as the lowest WAI (3.60 ± 0.19 g H_2O/g DM) and SP (2.67 ± 0.39 g H_2O/g DM). The DLO variety displayed intermediate WHC (5.19 ± 0.33 g H_2O/g DM) and moderate values across most absorption-related parameters.

Higher WAI values were observed (notably in Nam Dok Mai and Julie) than those reported by Kaur et al. [7], who obtained values around 3.0 g/g. This may suggest an increased water-binding capacity, potentially due to a greater number of hydrophilic groups or differences in the structural organization of polymers in the leaf matrix. Kaur et al. [7] also reported a WSI of 19.80%, while in the present study it ranged from 13.89 to 30.83 g H_2O/100 g DM, indicating a possibly higher proportion of soluble compounds (e.g., polyphenols, sugars) in certain cultivars such as Keïtt and DLO. Furthermore, their reported OAC (3.67 g/g) exceeded the values observed here (maximum of 2.77 g oil/g DM), which may be attributed to differences in drying conditions, protein content, hydrophobic compound concentration, or particle size resulting from milling processes.

The variations in WAI, WSI, and WHC among the mango leaf powders may be partly explained by differences in surface and structural properties. As noted by Dey et al. [8], mango-leaf-derived biosorbents exhibited relatively high specific surface areas (36–43 m^2/g) and abundant surface functional groups—such as hydroxyl, carboxyl, and phenolic moieties—which contribute to water retention and adsorption behavior. These features were also associated with enhanced performance in dye adsorption [9]. These findings support our results, where cultivars such as Nam Dok Mai and Julie exhibited the highest WHC and WAI, suggesting a more porous structure and a higher concentration of hydrophilic groups. Conversely, elevated WSI values observed in Keïtt and DLO may be linked to the increased solubility of low-molecular-weight components such as polyphenols and soluble fibers, which are known to be present in mango leaves and vary with genotype. These absorptional properties have direct implications for food applications: high WHC varieties like Nam Dok Mai (5.53 g H_2O/g DM) would be suitable for moisture-sensitive products, while high OAC varieties like Irwin (2.77 g oil/g DM) could function as natural emulsifiers in lipid-rich formulations.

Overall, the absorption-related parameters show distinct varietal differences that may influence the suitability of each cultivar for food applications requiring specific hydration, oil interaction, or swelling behavior.

2.2. *Foaming and Emulsification Characteristics of Mango Leaf Samples*

Surface-active properties are critical for applications requiring foam generation or emulsion stabilization in food systems. The ability to create and maintain foams and emulsions determines the potential utility of mango leaf powders in aerated products, sauces, and other complex food matrices. Table 2 demonstrates the foaming and emulsification properties of the mango leaf varieties, revealing distinct surface-active characteristics across foam generation, stability, and emulsion formation parameters.

Table 2. Foaming and emulsification characteristics of mango leaf samples from different varieties.

Sample	FC [mL]	FS [%]	EA [%]	ES [%]
Julie	65.29 ± 9.76^{a}	121.9 ± 16.6^{b}	1.33 ± 0.37^{ab}	3.32 ± 1.31^{b}
DLO	59.79 ± 2.25^{a}	112.8 ± 4.2^{a}	1.33 ± 0.37^{ab}	2.22 ± 1.49^{a}
Nam Dok Mai	82.69 ± 7.79^{b}	153.1 ± 14.4^{c}	1.66 ± 0.78^{b}	2.10 ± 1.40^{a}
Irwin	64.56 ± 1.55^{a}	125.3 ± 1.3^{b}	0.55 ± 0.00^{a}	3.28 ± 0.00^{b}
Keïtt	67.29 ± 2.32^{a}	129.4 ± 0.9^{b}	3.83 ± 0.78^{c}	2.47 ± 0.39^{a}

FC—foaming capacity; FS—foaming stability; EA—emulsifying activity; ES—emulsion stability. Lowercase letters mean significant differences in columns at $p = 0.05$.

Nam Dok Mai variety exhibited the highest foaming capacity (82.69 ± 7.79 mL) and the highest foam stability (153.1 ± 14.4%), indicating strong ability to generate and maintain foam structures. Keïtt variety showed the highest emulsifying activity (3.83 ± 0.78%), while presenting moderate foaming capacity (67.29 ± 2.32 mL) and relatively high foam stability (129.4 ± 0.9%).

Irwin variety presented the lowest emulsifying activity (0.55 ± 0.00%) but demonstrated the highest emulsion stability (3.28 ± 0.00%), along with moderate foaming capacity (64.56 ± 1.55 mL) and foam stability (125.3 ± 1.3%).

Julie variety exhibited moderate foaming capacity (65.29 ± 9.76 mL), intermediate foam stability (121.9 ± 16.6%), moderate emulsifying activity (1.33 ± 0.37%), and relatively high emulsion stability (3.32 ± 1.31%). DLO variety displayed the lowest foaming capacity (59.79 ± 2.25 mL) and the lowest foam stability (112.8 ± 4.2%), while showing moderate emulsifying activity (1.33 ± 0.37%) and lower emulsion stability (2.22 ± 1.49%). The FC and FS values obtained in this study are notably higher than those reported by Kaur et al. [7].

The foaming and emulsification characteristics reveal significant varietal differences that influence the suitability of these mango leaf varieties for applications requiring specific surface-active properties, foam formation, or emulsion stabilization in food systems.

2.3. *Bioactive Profiles of the Different Mango Leaf Varieties*

The antioxidant potential of mango leaves represents a key factor in their valorization as functional food ingredients. Quantifying bioactive compounds and their antioxidant capacity across different cultivars enables the identification of varieties with superior health-promoting properties. Table 3 shows the bioactive profiles of the different mango varieties, revealing distinct antioxidant capacities and reducing sugar contents across both ethanol and water extraction systems.

Table 3. Antioxidant activities and reducing sugar contents of mango leaf powders extracted in ethanol and water systems.

Sample	Reducing Sugars	DPPH ETOH	DPPH H_2O	ABTS ETOH	ABTS H_2O	FRAP ETOH	FRAP H_2O
	[GE mg/g DM]	[TE mg/g DM]				[$FeSO_4$ mg/g DM]	
Julie	20.61 ± 0.58 [a]	1.95 ± 0.25 [a]	2.25 ± 0.35 [a]	3.45 ± 0.45 [a]	1.65 ± 0.18 [a]	1.65 ± 0.15 [a]	1.95 ± 0.12 [a]
DLO	24.67 ± 1.57 [bc]	1.85 ± 0.35 [a]	2.45 ± 0.42 [ab]	3.85 ± 0.65 [b]	1.85 ± 0.28 [ab]	1.78 ± 0.18 [a]	2.05 ± 0.18 [ab]
Nam Dok Mai	25.92 ± 0.37 [c]	2.15 ± 0.45 [a]	2.89 ± 0.58 [b]	4.12 ± 0.78 [b]	2.05 ± 0.15 [b]	1.95 ± 0.12 [a]	2.18 ± 0.25 [b]
Irwin	28.04 ± 2.46 [d]	2.65 ± 0.55 [b]	3.15 ± 0.65 [c]	4.35 ± 0.85 [c]	2.15 ± 0.35 [c]	2.05 ± 0.22 [b]	2.35 ± 0.28 [c]
Keïtt	27.36 ± 0.77 [cd]	2.45 ± 0.35 [b]	2.95 ± 0.48 [bc]	4.15 ± 0.68 [b]	1.95 ± 0.25 [b]	1.85 ± 0.18 [a]	2.25 ± 0.22 [bc]

ETOH—ethanol extract; H_2O—water extract. Lowercase letters mean significant differences in columns at $p = 0.05$.

The Irwin variety demonstrated the highest antioxidant activity across most assays, exhibiting the highest reducing sugar content (28.04 ± 2.46 mg GE/g DM), the highest DPPH activity in both ethanol (2.65 ± 0.55 mg TE/g DM) and water extracts (3.15 ± 0.65 mg TE/g DM), the highest ABTS activity in ethanol extracts (4.35 ± 0.85 mg TE/g DM) and water extracts (2.15 ± 0.35 mg TE/g DM), and the highest FRAP activity in water extracts (2.35 ± 0.28 mg $FeSO_4$/g DM). The Keïtt variety showed strong antioxidant potential with high reducing sugar content (27.36 ± 0.77 mg GE/g DM) and substantial DPPH activity in water extracts (2.95 ± 0.48 mg TE/g DM), along with high ABTS activity in ethanol extracts (4.15 ± 0.68 mg TE/g DM). The Nam Dok Mai variety exhibited intermediate-to-high bioactive properties with moderate reducing sugar content (25.92 ± 0.37 mg GE/g DM) and good antioxidant activity across multiple assays, including DPPH water activity (2.89 ± 0.58 mg TE/g DM) and ABTS ethanol activity (4.12 ± 0.78 mg TE/g DM). DLO variety displayed moderate bioactive characteristics with reducing sugar content (24.67 ± 1.57 mg GE/g DM) and consistent antioxidant activities across both extraction systems. The Julie variety presented the lowest bioactive profile with the lowest reducing sugar content (20.61 ± 0.58 mg GE/g DM) and consistently lower antioxidant activities across all assays, including DPPH ethanol activity (1.95 ± 0.25 mg TE/g DM), ABTS water activity (1.65 ± 0.18 mg TE/g DM), and FRAP ethanol activity (1.65 ± 0.15 mg $FeSO_4$/g DM).

The conducted studies have shown that the Julie mango leaves have the lowest content of reducing sugars (20.61 ± 0.58 mg GE/g DM), while the Irwin mango leaves have the highest content of reducing sugars among the analyzed samples (28.04 ± 2.46 mg GE/g DM). The values of total sugar content obtained by us are consistent with results of studies conducted on the leaves of *Mangifera indica* L. grown in Phagwara, Punjab region, India. These studies have shown that the total carbohydrate content in mango leaves can be as high as 49.76% of the powder mass derived from this raw material [7].

Sucrose (apart from glucose and fructose) is indicated to be the main form of non-structural carbohydrates present in mango leaves [10]. It is possible that during powder production, partial sucrose degradation occurred, leading to increased glucose and fructose levels [11]. It has also been confirmed that mango leaves contain arabinose and xylose, both reducing sugars from the aldopentose group, which may react with the Folin–Ciocalteu reagent [12].

The bioactive profiles demonstrate significant varietal differences in antioxidant capacity and reducing sugar content, indicating that the selection of mango leaf varieties should consider the intended application's requirements for specific bioactive compounds and antioxidant properties. Therefore, due to the high sugar content, mango leaves (*Mangifera indica* L.) can be used in the production of bakery products, such as cookies and bread with desirable sensory qualities [13].

The observed antioxidant superiority of Irwin and Keïtt cultivars may reflect their genetic adaptation to environmental stressors. This suggests that leaf harvesting from stress-adapted cultivars could be strategically employed to maximize bioactive compound yields, supporting the development of standardized extraction protocols for commercial applications.

2.4. Pearson Correlation Analysis Between Functional and Bioactive Properties

The relationships between functional and bioactive properties enable rational cultivar selection and provide insights into the underlying mechanisms governing mango leaf functionality. Correlation statistical evaluation examined associations between all measured parameters to identify patterns that could inform practical decision making for ingredient selection (Table 4).

The Pearson correlation analysis revealed distinct correlation categories that provide comprehensive insights into the relationships between the functional properties and bioactive characteristics of the mango leaf varieties.

Antioxidant activity correlations demonstrated remarkably strong positive correlations ($r \geq 0.8$) across all assays, indicating consistent antioxidant profiles across varieties and reliable measurement regardless of solvent or method employed. The analysis revealed that sugar content serves as the primary driver of antioxidant capacity, with correlation coefficients ranging from 0.940 to 0.984 between reducing sugars and various antioxidant parameters. This key finding demonstrates that varieties with high reducing sugar content, specifically Irwin and Keïtt, consistently show superior antioxidant activity across all assays.

The observed high antioxidant activity values of the Irwin mango leaf powder may be due to the adaptation of this variety to living in conditions of higher temperature and greater sunlight than those occurring on the islands of the Lesser Antilles archipelago, where temperatures average 28–30 °C. The 'Irwin' mango was developed in Florida as a cross between 'Haden' and 'Lippens'. It has been reported that summers on the Florida peninsula are very warm and humid, with temperatures often exceeding 35 °C in the southern part of the peninsula [14]. This could make the 'Irwin' variety genetically better adapted to slightly higher temperatures and more sunlight (250–300 sunny days per year) than those prevailing on the island of Guadeloupe (180–240 sunny days per year) [15–17].

This may make the leaves of the 'Irwin' mango cultivar capable of producing higher amounts of antioxidants that counteract reactive oxygen species (ROS) formed as a result of the harmful effects of UV radiation on plant cells [18]. The results of the study confirm that exposure of young seed plants (developing from the embryo during germination), e.g., broad bean seedlings, to visible light or visible light combined with high-intensity UV-A and UV-C light causes a significant increase in the synthesis of polyphenolic compounds, including anthocyanins throughout the germination period. It was also confirmed that, as a result of intensive UV-A or UV-C radiation, there is a significant increase in the content of endogenous non-enzymatic antioxidants, such as ascorbate and glutathione (GSH) in cultivated broad bean seedlings [19]. Numerous studies have confirmed that increased synthesis of polyphenolic compounds, including flavonoids, reduces the damage caused by UV radiation by limiting the penetration of harmful ultraviolet (UV) radiation [20].

The 'Julie' mango variety comes from the islands of Trinidad and Jamaica and is very popular in cultivation in the Lesser Antilles [21]. It can be assumed that this variety shows a lower response to ultraviolet (UV) radiation from small-molecule (non-enzymatic) antioxidant systems. It has been confirmed that the leaves of many mango varieties are a rich source of highly polar antioxidants, such as mangiferin and benzophenone derivatives, such as iriflophenone 3-C-β-d-glucoside and iriflophenone 3-C-(2-O-p-hydroxybenzoyl)-β−d-glucoside, and phenolic acids, such as gallic acid [22].

Table 4. Pearson correlation coefficients between functional properties and antioxidant activities of mango leaf powders (*Mangifera indica* L.).

	WHC	WAC	OAC	HLI	WAI	WSI	SP	FC	FS	EA	ES	DPPH_H$_2$O	ABTS_H$_2$O	FRAP_H$_2$O
WHC	1.000	−0.629	−0.471	0.486	0.648	−0.515	0.626	0.378	0.537	−0.714	−0.204	0.495	0.528	0.527
WAC		1.000	0.213	−0.137	−0.271	0.447	−0.291	−0.043	−0.191	0.204	0.343	0.612	0.367	0.717
OAC			1.000	−0.942	0.348	0.120	0.368	−0.916	−0.898	−0.425	0.703	0.425	0.155	0.332
HLI				1.000	−0.449	−0.253	−0.465	0.847	0.877	0.545	−0.661	−0.384	−0.094	−0.281
WAI					1.000	−0.922	0.998	0.195	0.360	−0.556	0.399	0.527	0.588	0.580
WSI						1.000	−0.938	−0.357	−0.487	0.706	−0.694	−0.419	−0.484	−0.442
SP							1.000	0.229	0.393	−0.573	0.443	0.548	0.608	0.598
FC								1.000	0.990	0.137	−0.413	−0.217	0.003	−0.128
FS									1.000	0.173	−0.381	−0.165	0.055	−0.072
EA										1.000	−0.409	0.140	−0.086	0.117
ES											1.000	−0.075	−0.188	−0.001
DPPH_H$_2$O												1.000	0.949	0.989
ABTS_H$_2$O													1.000	0.930
FRAP_H$_2$O														1.000

WHC: water-holding capacity; WAC: water absorption capacity; OAC: oil absorption capacity; HLI: hydrophilic–lipophilic index; WAI: water absorption index; WSI: water solubility index; SP: swelling power; FC: foaming capacity; FS: foam stability; DPPH_H$_2$O, ABTS_H$_2$O, FRAP_H$_2$O: antioxidant activities in water extracts.

This may explain the higher total antioxidant activities we obtained in aqueous extracts of powdered leaves of particular mango varieties than in ethanol extracts. It has also been confirmed that *Mangifera indica* leaves can be a rich source of another highly polar antioxidant compound—ascorbic acid (vitamin C). The content of ascorbic acid in the leaves of *Mangifera indica* grown in the Dhaka region of Bangladesh can be as high as 3.25 mg/g of dry leaves [23]. Ascorbic acid is able to react strongly with singlet oxygen, superoxide, and hydroxyl radicals [19]. Therefore, mango leaf powder has high potential to become a raw material used in the production of supplements and food products that can counteract oxidative stress in the human body.

The conducted studies showed that the highest reducing activity was characteristic of the Irwin mango leaves in an aqueous solution (2.35 mg $FeSO_4$/g DM), and the lowest reducing activity was demonstrated by the Julie mango leaves in an ethanol solution (1.65 mg $FeSO_4$/g DM). Moreover, the reducing activity of all tested mango leaf varieties was higher in an aqueous solution than in an ethanol solution. The high reducing activity of the analyzed mango leaf powders (measured in the FRAP test) observed in our study is, to a large extent, confirmation of their high activity in neutralizing reactive oxygen species.

The observed high capacity of Irwin mango leaves to reduce iron ions Fe^{3+} to Fe^{2+} may be due to the presence of certain minerals, such as potassium, phosphorus, iron, sodium, calcium, and magnesium ions in the tested raw material [24]. The presence of vitamin C and vitamin E (tocopherols and tocotrienols) has also been confirmed in mango leaves, which have high reducing properties [25].

The high reducing activity of Irwin mango leaves may also result from the presence of protein fractions that act as the building blocks of cells and play a key role in the construction of the photosynthetic apparatus and in cell and tissue signaling [26]. Studies have confirmed that the powder made from mango leaves contains a high amount of crude protein at 171.4 g/kg DM [4]. The presence of amino acids such as cysteine (Cys) and methionine (Met) in these proteins causes them to have high reducing properties [27]. This makes mango leaves a valuable raw material for the production of human food or protein feed for animals [28], which shows not only nutritional (building) properties but also strong reducing properties.

Functional property correlations exhibited several significant relationships, particularly among water-related properties. The analysis revealed a nearly perfect correlation between water absorption index and swelling power ($r = +0.998$), indicating that these parameters measure essentially the same underlying property. The water absorption index showed a strong negative correlation with the water solubility index ($r = -0.922$), demonstrating that high absorption capacity corresponds to low solubility, while the water solubility index negatively correlated with swelling power ($r = -0.938$), showing that high solubility reduces swelling capacity. The oil–water balance analysis revealed a very strong negative correlation between oil absorption capacity and the hydrophilic–lipophilic index ($r = -0.942$), indicating that high oil absorption corresponds to lower hydrophilicity. Moisture characteristics showed that water-holding capacity positively relates to water absorption index ($r = +0.648$), while displaying a complex negative relationship with water absorption capacity ($r = -0.629$).

Foaming correlations demonstrated exceptional predictability, with foaming capacity showing a nearly perfect correlation with foam stability ($r = +0.990$), indicating that foaming capacity serves as an excellent predictor of foam stability. The hydrophilic–lipophilic index strongly enhanced both foaming capacity ($r = +0.847$) and foam stability ($r = +0.877$), while oil absorption capacity negatively affected foaming ability ($r = -0.916$), demonstrating that oil absorption reduces foaming capacity.

Cross-functional relationships revealed important interactions between different property categories. Water absorption capacity showed strong positive correlations with antioxidant parameters (r = +0.601 to +0.791), indicating that water absorption enhances antioxidant extraction efficiency. Reducing sugar content emerged as a strong driver of antioxidant capacity across multiple assays. The relationship between foaming properties and other characteristics showed that oil absorption capacity maintains strong negative relationships with foaming parameters, while the hydrophilic–lipophilic index demonstrates strong positive relationships with foaming properties.

Key insights from the correlation analysis include five major findings. First, antioxidant synergy is evident as all antioxidant assays are highly intercorrelated (r > 0.9), suggesting that varieties can be reliably ranked for antioxidant potential using any single method. Second, the perfect correlation between water absorption index and swelling power (r = 0.998) indicates these parameters measure the same underlying property. Third, foaming predictability is exceptional, as foaming capacity perfectly predicts stability (r = 0.990), with both properties driven by hydrophilic characteristics. Fourth, an oil–water trade-off exists, as the strong negative correlation between oil absorption and hydrophilicity suggests that varieties optimize for either oil or water interactions. Fifth, sugar-driven antioxidants are evident, as reducing sugar content serves as the strongest predictor of antioxidant capacity across all methods.

Variety implications for practical applications indicate specific variety selections based on desired properties. For applications requiring high antioxidant activity, Irwin or Keïtt varieties should be chosen due to their high reducing sugar content correlating with superior antioxidant activity. For water-based applications, Julie or Nam Dok Mai varieties are optimal choices due to their high water absorption index and swelling power for applications requiring swelling properties. For foaming applications, the Nam Dok Mai variety represents the best choice with the highest foaming capacity and foam stability combined with optimal hydrophilic–lipophilic index. For oil-based applications, Irwin or DLO varieties are recommended due to their higher oil absorption capacity suitable for oil-based systems.

Emulsifying activity relationships demonstrated contrasting behaviors with water-related properties. WHC showed a strong negative correlation with emulsifying activity (r = −0.714), indicating that varieties with high water retention capacity exhibit reduced ability to form emulsions. This relationship suggests that strong water-binding properties may interfere with the protein–lipid interactions necessary for emulsion formation. Conversely, WSI displayed a strong positive correlation with emulsifying activity (r = +0.706), demonstrating that higher concentrations of soluble compounds enhance emulsification capacity. This finding indicates that water-soluble components, likely including soluble proteins and low-molecular-weight emulsifiers, contribute significantly to emulsion formation.

Emulsion stability correlations revealed a different characteristic compared to emulsifying activity. OAC showed a strong positive correlation with emulsion stability (r = +0.703), suggesting that varieties with higher lipid-binding capacity can better maintain emulsion structure over time. This relationship indicates that oil-binding properties are crucial for preventing phase separation in formed emulsions. WSI demonstrated a strong negative correlation with emulsion stability (r = −0.694), showing that high concentrations of soluble compounds, while beneficial for emulsion formation, may compromise long-term stability through increased mobility and potential phase separation.

The hydrophilic–lipophilic index showed a strong negative correlation with emulsion stability (r = −0.661), indicating that highly hydrophilic varieties produce less stable emulsions. This relationship suggests that balanced hydrophilic–lipophilic properties, rather than extreme hydrophilicity, are optimal for maintaining emulsion integrity.

2.5. Mid-Infrared (IR) Spectra of Mango Leaves

In the mid-infrared (IR) spectra of mango (*Mangifera* L.) leaves, a broad and intense absorption band is observed at ca. 3325 cm^{-1} (Figure 2), corresponding to υ(OH) vibrations [29].

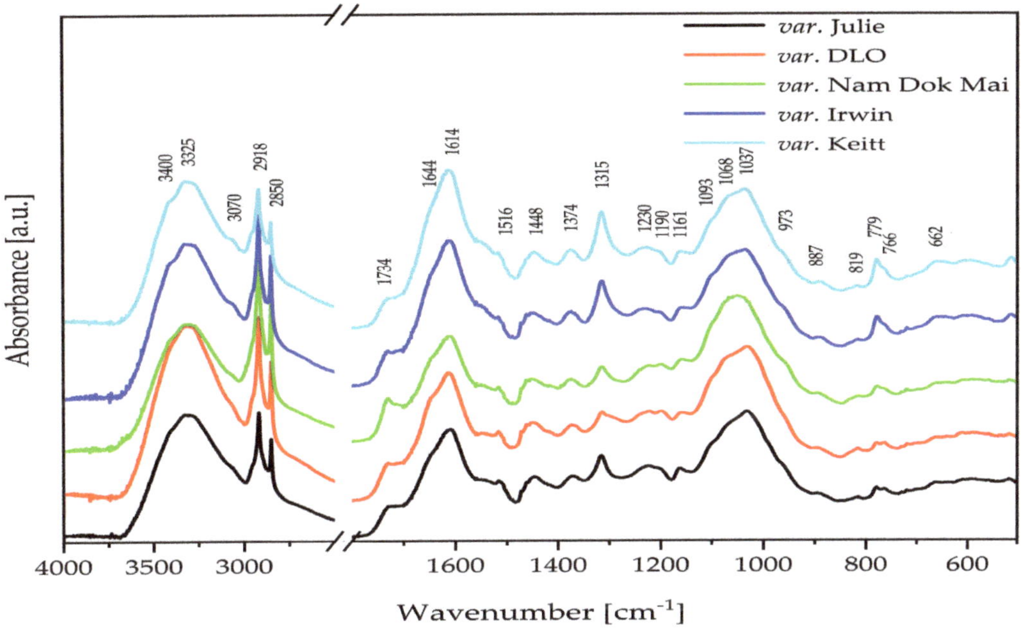

Figure 2. The FT-IR/ATR spectra of mango leaf samples from different varieties.

The above band includes contributions from both free hydroxyl groups' stretching vibrations υ(OH) and those involved in intra- and intermolecular hydrogen bonds (HBs). The band's pronounced broadening results from an extensive hydrogen-bonding network, typical of plant matrices rich in polar compounds. The spectra reveal several distinct absorption regions, reflecting the complex molecular composition of the samples, with the band near 3325 cm^{-1} being particularly dominant [29]. In addition to phenolic compounds—such as flavonoids (e.g., quercetin) and gallic acid—sugars also play a significant role in this spectral region due to their numerous hydroxyl groups capable of bonding hydrogen. Their presence further enhances the intensity and width of the υ(OH) band, confirming a high content of polar, hydrophilic compounds in mango leaves, which is further supported by the appearance of a very strong absorption band at 1614 cm^{-1}, which corresponds to the υ(C=C) vibrations in aromatic rings and conjugated C=O groups [29]. The band observed at 1516 cm^{-1} in the IR spectrum of mango leaves is primarily associated with δ(N-H) and υ(C–N) vibrations in amide groups (amide II), as well as υ(C=C) vibrations in the aromatic rings of flavonoids and other polyphenols. The presence of this band, together with the strong absorption at 1614 cm^{-1}, corresponding to υ(C=C) and conjugated C=O groups, indicates a significant content of both polyphenolic compounds and proteinaceous substances with polar, hydrophilic characteristics [29]. These contributions enhance the intensity and broadening of the υ(OH) band around 3325 cm^{-1}, confirming a high level of polar, hydrophilic matter in mango leaves. The vibration at 1644 can be attributed to the amide I (ν(C=O), δ(N-H)) bond from protein compounds [29–33].

The FT-IR spectrum of mango leaves, the bands at 2918 cm^{-1} and 2850 cm^{-1} (Figure 1), correspond to the stretching vibrations of methyl (ν_{as}(CH)) and methylene (ν_s(CH)) groups in the hydrocarbon chains of lipids and other organic compounds. These vibrations are characteristic of aliphatic alkyl groups commonly found in the cellular membrane structures and lipid components present in mango leaves. The FT-IR/ATR analysis suggests a low lipid content in the mango leaves. This is evidenced by sharp bands of medium/weak intensity at the following wavenumbers: 1734 (ν(C=O)), 1448 (δ(CH)/CH_2), 1374 (δ(CH)/CH_3), and 1230 and 1161 cm^{-1} (ν(C–O), δ(–CH_2–)) [29–34].

The strong bands in the 1093–973 cm^{-1} (Figure 1) range of the IR spectrum of mango leaves are primarily associated with ν(C–O) and C–O–C vibrations characteristic of polysaccharides such as cellulose. Additionally, vibrations related to phenolic structures of lignin, particularly ether and aromatic bonds, may also be present in this region. The presence of these bands indicates a significant contribution of structural cell wall components, including cellulose and lignin, which determine the mechanical properties of mango leaves [29–33,35].

The band at 1315 cm^{-1} in the IR spectra of mango leaves is primarily attributed to a combination of δ(O–H) deformation vibration and ν(C–O) vibration associated with lignin, hemicellulose, and polyphenolic compounds. The band observed near 779 cm^{-1} corresponds to aromatic ring vibrations of lignin and polyphenols, as well as bending modes of polysaccharide carbohydrate rings. Notably, the intensity of these bands varies among mango cultivars, with the Irwin and Keïtt varieties exhibiting the strongest bands, indicating a higher content of lignin, structural polysaccharides, and polyphenols that contribute not only to the mechanical strength but also to the antioxidant activity of the leaves [29–33,35].

3. Materials and Methods

3.1. Materials

The mango leaves (ML) of five cultivars were used in this study: DLO, Nam Dok Mai, Irwin, Julie, and Keïtt. MLs were collected from identified specimens at ASSOFWI farm, located in Bouchu district at Vieux-Habitants (16°05′17.8″ N, 61°44′08.2″ W), in Guadeloupe island (French West Indies), in February 2024. ASSOFWI farm is an independent association under the law of 1901, whose activities are dedicated to education for the public and research. It has a collection of 6 mango cultivars. On the same day of collection, MLs were carefully washed with tap water to remove any soil, spots, or debris and then washed a second time with distilled water and dried slightly with a tissue. The mango leaves were then placed in a food dryer (Tauro Essiccatori Biosec Pro 2200 W—Tauro Essiccatori Srl, Vicenza, Italy) set to 42 °C and 30% relative humidity for 72 h. After drying, the leaves were ground to a fine powder using an electric grinder (Taurus Aromatic 150 W, Group Taurus, Barcelona, Spain). The resulting powders were immediately vacuum-packed and stored at room temperature (23 ± 1 °C) under optimal preservation conditions until experimental use (maximum 48 h). Each cultivar of mango leaf powder (MLP) was clearly identified and separately vacuum-packed.

3.2. Absorptional Characteristics of Mango Leaf Samples from Different Varieties

The absorptional properties of mango leaf powders from various cultivars were evaluated through a series of standardized procedures assessing water-holding capacity (WHC), water absorption capacity (WAC), oil absorption capacity (OAC), hydrophilic/lipophilic index (HLI), water absorption index (WAI), water solubility index (WSI), and swelling power (SP). All analyses were conducted in triplicate.

Water-holding capacity (WHC) and water absorption capacity (WAC) were determined according to procedures adapted from [36], with modifications distinguishing gravitational and mechanical water retention. For WHC, 3 g of dried mango leaf powder was placed in a pre-weighed centrifuge tube containing 30 mL of distilled water. The dispersion was left to sediment undisturbed at room temperature for 24 h. After this period, the supernatant was carefully decanted, and the remaining hydrated sediment was weighed. WHC was expressed as the grams of water retained per gram of dry sample, under gravitational force only. For WAC, 3 g of dried powder was dispersed in 30 mL of distilled water, vortexed for 30 s, and centrifuged at 3000× *g* for 10 min. The supernatant was discarded, and the residue was dried at 50 °C until a constant weight was obtained. WAC was calculated as the total mass of water absorbed and retained (g/g) after centrifugation and drying [36].

OAC was measured by mixing 3 g of sample with 30 mL of rapeseed oil under identical conditions as for WAC. After vortexing and centrifugation, the unbound oil was carefully removed, and residual oil was evaporated by placing the tubes in a drying oven (Vindon Scientific, Rochdale, UK) at 50 °C for 25 min. The OAC was calculated as grams of oil retained per gram of dry sample [36]. The HLI was calculated as the ratio of WAC to OAC, providing a comparative measure of the sample's affinity for polar versus non-polar media.

For the determination of WAI, WSI, and SP, 3 g of mango leaf powder was suspended in 30 mL of distilled water in pre-weighed centrifuge tubes. The samples were heated in a water bath at 90 °C for 10 min (MLL147, AJL Electronics, Kraków, Poland) and then cooled to room temperature. Following centrifugation at 3000× *g* for 10 min (Thermo Fisher Scientific, Waltham, MA, USA), the supernatants were carefully decanted into pre-weighed Petri dishes and dried at 110 °C for 24 h in a laboratory oven (SML, Zalmed, Łomianki, Poland).

The sediment remaining in the centrifuge tubes was weighed to determine WAI, expressed as grams of water retained per gram of dry solids (g/g). WSI was calculated based on the mass of dried solubles in the supernatant and expressed as grams of solubles per 100 g of dry sample (g/100 g). Swelling power (SP) was determined as the ratio of the weight of the sediment to the dry matter content, expressed in g/g [37].

3.3. *Foaming and Emulsification Characteristics of Mango Leaf Samples*

The foaming and emulsifying properties of mango leaf powders were evaluated to assess their functionality in aerated and emulsion-based food systems. The following parameters were determined: foaming capacity (FC), foam stability (FS), emulsifying activity (EA), and emulsion stability (ES). All measurements were performed in triplicate.

To determine foaming properties, 1 g of mango leaf powder was dispersed in 50 mL of distilled water in a graduated 100 mL cylinder. The mixture was vigorously shaken by hand in a vertical motion for 5 min at room temperature. The volume of foam generated immediately after agitation was recorded and used to calculate FC (mL foam/g dry matter). The foam was then left undisturbed for 1 h, after which the foam volume was recorded again to determine FS, calculated as the percentage of the initial foam volume retained after 60 min [38].

EA and ES were measured following a modified centrifugation method. Briefly, 3 g of mango leaf powder was homogenized with 30 mL of distilled water and 30 mL of rapeseed oil for 1 min using a vortex mixer (Heidolph Reax, Schwabach, Germany). The emulsion was then centrifuged at 3000× *g* for 10 min (Thermo Fisher Scientific, Waltham, MA, USA), and the volume of the emulsified layer was measured. EA was expressed as the volume (mL) of the emulsified layer per total volume. For ES, the emulsion was heated to 70 °C for 30 min and then cooled to ambient temperature and left at rest for 1 h before remeasuring the emulsified layer. The stability was calculated as the percentage of emulsion volume retained after heat treatment and resting [39].

3.4. *Bioactive Profiles of the Different Mango Leaf Varieties*

3.4.1. Extraction Procedure

Bioactive compounds were extracted using two solvent systems: distilled water and a hydroalcoholic mixture (ethanol/water, 80:20 *v*/*v*), each acidified with 1% HCl. The alcohol-to-water ratio used (80:20 *v*/*v*) was optimal in terms of the effectiveness and efficiency of the extraction process. The 1% addition of hydrochloric acid (HCl) to the extraction mixture was intended to facilitate the process of decomposition of the cell walls of the raw material used. Exactly 0.5 g of dried mango leaf powder was weighed and suspended in 10 mL of extraction solvent. The mixtures were vortexed for 1 min (MX-S, Chemland, Stargard, Poland) and agitated for 2 h at ambient temperature using a laboratory rotary shaker (MX-RD PRO, Chemland, Stargard, Poland). After incubation, samples were centrifuged at 3500× *g* for 10 min at 4 °C (MPW-350, MPW MED. INSTRUMENTS, Warsaw, Poland). The obtained supernatants were stored at 8 °C and analyzed within 24 h [40].

3.4.2. Reducing Sugar Content

The reducing sugar content in the extracts was determined using a modified dinitrosalicylic acid (DNS) colorimetric method. A 0.5 mL aliquot of each extract was mixed with 0.25 mL of DNS reagent and heated in a boiling water bath for 5 min. After cooling to 50–60 °C, 3 mL of distilled water was added to dilute the reaction mixture. The absorbance was measured at 530 nm using a UV-Vis spectrophotometer (SEMCO S91E, EMCO, Sosnowiec, Poland). Glucose standards ranging from 100 to 800 µg/mL were used for calibration. The absorbance of a blank sample containing 0.5 mL of distilled water instead of the test extract was also measured. Results were expressed as milligrams of glucose equivalents (GEs) per gram of dry sample (mg GEs/g DM) [41].

3.4.3. DPPH Radical Scavenging Activity

Antioxidant activity was assessed using the DPPH free radical scavenging assay. For each measurement, 0.0345 mL of the extract was added to 1 mL of a 0.1 mM methanolic DPPH solution. The mixture was incubated for 20 min in the dark at room temperature, and absorbance was measured at 517 nm. The absorbance of a blank sample was also measured, which contained 1 mL of DPPH working solution and 0.0345 mL of distilled water instead of the tested extract. A calibration curve was prepared using Trolox as the standard (100–800 µmol/L). The results were expressed as milligrams of Trolox equivalents (TE) per gram of dry sample (mg TE/g DM) [42].

3.4.4. ABTS Radical Cation Decolorization Assay

The ABTS radical scavenging capacity was evaluated using a previously described method [40], with minor modifications. Briefly, 0.0204 mL of extract was added to 1 mL of a freshly prepared $ABTS^{\bullet+}$ solution, and the absorbance was measured at 734 nm precisely 10 s after mixing. The absorbance of a blank sample was measured, which contained 1 mL of ABTS working solution and 0.0204 mL of distilled water instead of the tested extract. Trolox was used as a standard (100–800 µmol/L), and antioxidant capacity was reported as mg TE/g DM [40].

3.4.5. Ferric Reducing Antioxidant Power (FRAP) Assay

The FRAP assay was used to determine the electron-donating ability of the mango leaf extracts. A 0.0345 mL aliquot of each extract was mixed with 0.998 mL of freshly prepared FRAP reagent. After incubation for 15 min at 36 °C, absorbance was measured at 593 nm. The absorbance of a blank sample was measured, which contained 0.998 mL of FRAP working solution and 0.0345 mL of distilled water instead of the test extract. The results were calculated using a ferrous sulfate standard curve (100–800 µmol/L) and expressed as milligrams of $FeSO_4$ equivalents per gram of dry sample (mg Fe^{2+}/g DM) [40].

3.5. Mid-Infrared (IR) Spectra of Mango Leaves

The dried mango leaf powders (MLPs) were used directly for this experiment. FT-IR spectra were acquired using a Nicolet 6700 spectrophotometer (Thermo Fisher Scientific, Waltham, MA, USA) equipped with a diamond crystal in attenuated total reflectance (ATR) mode. Each sample was scanned 128 times in the mid-infrared region (4000–500 cm^{-1}) at a resolution of 2 cm^{-1}. The wavenumbers and corresponding band assignments for all samples are summarized in Table S1 (Supplementary Materials), including a legend of the spectroscopic abbreviations used. All measurements were performed in triplicate. Spectral data were processed and compared using OriginPro 2024b software (OriginLab Corporation, Northampton, MA, USA).

3.6. Statistical Analysis

All analyses were performed in triplicate, and the results were expressed as means ± standard deviations. One-way analysis of variance (ANOVA) was conducted to evaluate the effect of mango leaf variety on the studied functional and bioactive properties. Post hoc comparisons between group means were performed using Tukey's honest significant difference (HSD) test at a significance level of $p < 0.05$. Pearson's correlation coefficients (r) were calculated to assess relationships between selected functional parameters and antioxidant activities. All statistical analyses were carried out using Statgraphics Centurion XVII.I (StatPoint Technologies, Inc., Warrenton, VA, USA).

4. Conclusions

This study establishes clear cultivar-dependent functional and bioactive profiles for mango leaves, with practical implications for targeted applications. The identification of reducing sugar content as a reliable predictor of antioxidant capacity ($r > 0.94$) provides a simple screening tool for cultivar selection. Irwin and Keïtt cultivars emerge as premium antioxidant sources, while Nam Dok Mai offers superior foaming properties for aerated food systems.

The strong correlations identified enable rational cultivar selection: high-sugar cultivars for antioxidant applications, high-hydrophilicity cultivars for water-based systems, and specific cultivars for specialized functional requirements. These findings support the development of standardized processing protocols and quality specifications for mango leaf-derived ingredients.

However, successful commercialization will require addressing safety evaluation, sensory optimization, and regulatory compliance. The demonstrated variability underscores the critical importance of cultivar identification and botanical standardization in product development.

Study Limitations and Future Directions

While this study provides valuable prescreening data, several limitations should be acknowledged. The analysis focused on dried leaves from a single geographic location and harvest time, which may not reflect seasonal or environmental variations. Future research should investigate the stability of these properties across different harvesting conditions and processing methods.

Additionally, safety evaluation, including anti-nutritional factors, sensory acceptability in food matrices, and the shelf-life stability of mango leaf ingredients, requires investigation. Scale-up feasibility and economic viability assessments are essential for commercial translation.

Supplementary Materials: The following supporting information can be downloaded at https://www.mdpi.com/article/10.3390/molecules30163381/s1. Table S1. Wavenumbers of the bands observed in the IR spectra of mango leaf samples from different varieties, together with the proposed assignments of the bands.

Author Contributions: Conceptualization, G.L. and J.H.; methodology, R.O., A.Z., and J.H.; software, J.H.; validation, R.O., A.Z., and J.H.; formal analysis, J.H.; investigation, I.M., A.Z., and O.P.; resources, G.L. and J.H.; data curation, J.H.; writing—original draft preparation, G.L., E.P., A.W.-B., R.O., A.Z., and O.P.; J.H.; writing—review and editing, E.P. and J.H.; visualization, G.L. and A.Z.; supervision, G.L., G.A., and J.H.; project administration, E.P., G.A., and J.H.; funding acquisition, E.P. and J.H. All authors have read and agreed to the published version of the manuscript.

Funding: This research received no external funding.

Institutional Review Board Statement: Not applicable.

Informed Consent Statement: Not applicable.

Data Availability Statement: The original contributions presented in this study are included in the article. Further inquiries can be directed to the corresponding authors.

Conflicts of Interest: The authors declare no conflicts of interest.

References

1. Kulkarni, V.M.; Rathod, V.K. Extraction of mangiferin from *Mangifera indica* leaves using three phase partitioning coupled with ultrasound. *Ind. Crops Prod.* **2014**, *52*, 292–297. [CrossRef]
2. Kulkarni, V.M.; Rathod, V.K. A novel method to augment extraction of mangiferin by application of microwave on three phase partitioning. *Biotechnol. Rep.* **2014**, *6*, 8–12. [CrossRef]
3. Saleem, M.; Tanvir, M.; Akhtar, M.F.; Iqbal, M.; Saleem, A. Antidiabetic potential of *Mangifera indica* L. cv. Anwar Ratol leaves: Medicinal application of food wastes. *Medicina* **2019**, *55*, 353. [CrossRef]
4. Kumar, M.; Saurabh, V.; Tomar, M.; Hasan, M.; Changan, S.; Sasi, M.; Maheshwari, C.; Prajapati, U.; Singh, S.; Prajapat, R.K.; et al. Mango (*Mangifera indica* L.) leaves: Nutritional composition, phytochemical profile, and health-promoting bioactivities. *Antioxidants* **2021**, *10*, 299. [CrossRef]
5. Yap, K.M.; Sekar, M.; Seow, L.J.; Gan, S.H.; Bonam, S.R.; Mat Rani, N.N.I.; Lum, P.T.; Subramaniyan, V.; Wu, Y.S.; Fuloria, N.K.; et al. *Mangifera indica* (mango): A promising medicinal plant for breast cancer therapy and understanding its potential mechanisms of action. *Breast Cancer Targets Ther.* **2021**, *13*, 471–503. [CrossRef] [PubMed]
6. Khan, J.T.; Richi, A.E.; Riju, S.A.; Jalal, T.; Orchi, R.J.; Singh, S.; Bhagat, P.; Abdel-Wahab, Y.H.A.; Ansari, P. Evaluation of antidiabetic potential of *Mangifera indica* leaf in streptozotocin-induced type 2 diabetic rats: Focus on glycemic control and cholesterol regulation. *Endocrines* **2024**, *5*, 137–152. [CrossRef]
7. Kaur, J.; Kaushik, D.; Kumar, M.; Babagil, G.E.; Amarowicz, R.; Proestos, C.; Oz, E.; Oz, F.; Esatbeyoglu, T.; Bordiga, M. Comprehensive analysis and characterization of *Mangifera indica* L. leaf powder. *Food Sci. Nutr.* **2025**, *13*, e70083. [CrossRef] [PubMed]
8. Dey, S.; Sreenivasulu, A.; Veerendra, G.T.N.; Phani Manoj, A.V.; Haripavan, N. Synthesis and characterization of mango leaves biosorbents for removal of iron and phosphorous from contaminated water. *Appl. Surf. Sci. Adv.* **2022**, *11*, 100292. [CrossRef]

9. Thiruketheeswaranathan, S. Comparative adsorption study of mango (*Mangifera indica*) leaves and neem (*Azadirachta indica*) leaves as an adsorbent for dye removal. *J. Sci.* **2022**, *13*, 49. [CrossRef]
10. Rossouw, G.C.; Orr, R.; Bennett, D.; Bally, I.S.E. The roles of non-structural carbohydrates in fruiting: A review focusing on mango (*Mangifera indica*). *Funct. Plant Biol.* **2024**, *51*, FP23195. [CrossRef]
11. Khaled, B.M.; Das, A.K.; Alam, S.M.S.; Saqib, N.; Rana, M.S.; Sweet, S.R.; Naznin, T.; Hossain, M.P.; Sardar, S.; Hossain, Z.; et al. Effect of different drying techniques on the physicochemical and nutritional properties of *Moringa oleifera* leaves powder and their application in bakery product. *Appl. Food Res.* **2024**, *4*, 100599. [CrossRef]
12. Laroque, D.; Inisan, C.; Berger, C.; Vouland, É.; Dufossé, L.; Guérard, F. Kinetic study on the Maillard reaction. Consideration of sugar reactivity. *Food Chem.* **2008**, *111*, 1032–1042. [CrossRef]
13. Pirca-Palomino, M.; Malange, Y.I.; Ramos-Escudero, F.; Muñoz, A.M.; Cancino-Chávez, K. Antioxidant properties, texture and sensory quality of sliced bread enriched with leaf powder from mango (*Mangifera indica*). *Pol. J. Food Nutr. Sci.* **2024**, *74*, 313–322. [CrossRef]
14. Mukherjee, S.K.; Litz, R.E. Introduction: Botany and importance. In *The Mango: Botany, Production and Uses*; CABI: Wallingford, UK, 2009; pp. 1–18.
15. Ueda, M.; Sasaki, K.; Utsunomiya, N.; Inaba, K.; Shimabayashi, Y. Changes in physical and chemical properties during maturation of mango fruit (*Mangifera indica* L. 'Irwin') cultured in a plastic greenhouse. *Food Sci. Technol. Res.* **2000**, *6*, 299–305. [CrossRef]
16. Feuillard, T.; Abillon, J.-M. Estimation of hourly insolation in Guadeloupe. *Sol. Wind Technol.* **1990**, *7*, 541–543. [CrossRef]
17. Bertin, A.; Frangi, J.P. Contribution to the study of the wind and solar radiation over Guadeloupe. *Energy Convers. Manag.* **2013**, *75*, 593–602. [CrossRef]
18. Close, D.C.; McArthur, C.; Hagerman, A.E.; Davies, N.W.; Beadle, C.L. Phenolic acclimation to ultraviolet-A irradiation in *Eucalyptus nitens* seedlings raised across a nutrient environment gradient. *Photosynthetica* **2007**, *45*, 36–42. [CrossRef]
19. Younis, M.E.B.; Hasaneen, M.N.A.; Abdel-Aziz, H.M. An enhancing effect of visible light and UV radiation on phenolic compounds and various antioxidants in broad bean seedlings. *Plant Signal. Behav.* **2010**, *5*, 1197–1203. [CrossRef]
20. Santos, I.; Fidalgo, F.; Almeida, J.M.; Salema, R. Biochemical and ultrastructural changes in leaves of potato plants grown under supplementary UV-B radiation. *Plant Sci.* **2004**, *167*, 925–935. [CrossRef]
21. Robin, G. *Julie Mango in the Eastern Caribbean: A Comprehensive Manual*; Caribbean Agricultural Research and Development Institute: St. Augustine, Trinidad and Tobago, 2001.
22. Guamán-Balcázar, M.C.; Montes, A.; Pereyra, C.; de la Ossa, E.M. Precipitation of mango leaves antioxidants by supercritical antisolvent process. *J. Supercrit. Fluids* **2017**, *128*, 218–226. [CrossRef]
23. Sadia, H.; Jeba, F.; Kamal, A.; Salam, A. Air pollution tolerance index of *Mangifera indica* plant species growing in the greater Dhaka region, Bangladesh. *J. Biodivers. Conserv. Bioresour. Manag.* **2019**, *5*, 1–12. [CrossRef]
24. Ribeiro, S.M.R.; Schieber, A. Bioactive compounds in mango (*Mangifera indica* L.). In *Bioactive Foods in Promoting Health*; Elsevier BV: Oxford, UK, 2010; pp. 507–523.
25. Parvez, G.M.M. Pharmacological activities of mango (*Mangifera indica*): A review. *J. Pharmacogn. Phytochem.* **2016**, *5*, 1–7.
26. Bohne, A.-V.; Schwenkert, S.; Grimm, B.; Nickelsen, J. Roles of tetratricopeptide repeat proteins in biogenesis of the photosynthetic apparatus. In *International Review of Cell and Molecular Biology*; Jeon, K.W., Ed.; Academic Press: Cambridge, MA, USA, 2016; Volume 324, pp. 187–227. [CrossRef]
27. Ralison, S.; Tounkara, F.; Karangwa, E.; Yong, S.; Guowei, L. In vitro antioxidant activities of protein hydrolysate from germinated black soybean (*Glycine max* L.). *Adv. J. Food Sci. Technol.* **2013**, *5*, 453–459. [CrossRef]
28. Fafiolu, A.; Odugunawa, O.; Bambose, A.; Oso, A.; Isah, O.; Olantanji, J.; Jegede, A. Feeding value of mango leaf (*Mangifera indica*) for growing rabbits. *J. Anim. Vet. Adv.* **2006**, *5*, 800–804.
29. Socrates, G. *Infrared and Raman Characteristic Group Frequencies: Tables and Charts*; Wiley: Chichester, UK, 2004; ISBN 978-0-47009-307-8.
30. Feng, Y.; Jin, C.; Lv, S.; Zhang, H.; Ren, F.; Wang, J. Molecular mechanisms and applications of polyphenol-protein complexes with antioxidant properties: A review. *Antioxidants* **2023**, *12*, 1577. [CrossRef]
31. Işık, M.; Dikici, E.; Altın, S.; Alp, C.; Kırboğa, K.K.; Köksal, E.; Beydemir, Ş. Phenolic content, antioxidant capacity, and therapeutic potential of mango (*Mangifera indica* L.) leaves. *Food Sci. Nutr.* **2025**, *13*, e70263. [CrossRef]
32. Sferrazzo, G.; Palmeri, R.; Restuccia, C.; Parafati, L.; Siracusa, L.; Spampinato, M.; Carota, G.; Distefano, A.; Di Rosa, M.; Tomasello, B.; et al. *Mangifera indica* L. leaves as a potential food source of phenolic compounds with biological activity. *Antioxidants* **2022**, *11*, 1313. [CrossRef]
33. Saxena, P.; Sharma, D.; Gautam, P.; Niranjan, A.; Rastogi, S. HPLC-DAD quantification of mangiferin, antioxidant potential and essential oil composition of the leaves of five varieties of *Mangifera indica* L. of North India. *Nat. Prod. Res.* **2024**, 1–12. [CrossRef]

34. Zając, A.; Dymińska, L.; Lorenc, J.; Ptak, M.; Hanuza, J. Depth profiling of horse fat tissue using mid-infrared and confocal Raman microscope. *Spectrosc. Lett.* **2018**, *51*, 81–88. [CrossRef]
35. Das, S.P.; Ravindran, R.; Deka, D.; Jawed, M.; Das, D.; Goyal, A. Bioethanol production from leafy biomass of mango (*Mangifera indica*) involving naturally isolated and recombinant enzymes. *Prep. Biochem. Biotechnol.* **2013**, *43*, 717–734. [CrossRef]
36. Villanueva, M.; De Lamo, B.; Harasym, J.; Ronda, F. Microwave radiation and protein addition modulate hydration, pasting and gel rheological characteristics of rice and potato starches. *Carbohydr. Polym.* **2018**, *201*, 374–381. [CrossRef]
37. Harasym, J.; Satta, E.; Kaim, U. Ultrasound treatment of buckwheat grains impacts important functional properties of resulting flour. *Molecules* **2020**, *25*, 3012. [CrossRef]
38. Kaushal, P.; Kumar, V.; Sharma, H.K. Comparative Study of Physicochemical, Functional, Antinutritional and Pasting Properties of Taro (*Colocasia esculenta*), Rice (*Oryza sativa*) Flour, Pigeonpea (*Cajanus cajan*) Flour and Their Blends. *LWT Food Sci. Technol.* **2012**, *48*, 59–68. [CrossRef]
39. Kiiru, S.; Ng'ang'a, J.; Konyole, S.; Roos, N.; Hetzer, B.; Marel, A.; Kinyuru, J. Physicochemical Characterisation and Impact of Gryllus bimaculatus Addition on Gluten-Free Flour Blends. *Int. J. Food Sci. Technol.* **2024**, *59*, 4620–4634. [CrossRef]
40. Olędzki, R.; Lutosławski, K.; Nowicka, P.; Wojdyło, A.; Harasym, J. Non-Commercial Grapevines Hybrids Fruits as a Novel Food of High Antioxidant Activity. *Foods* **2022**, *11*, 2216. [CrossRef]
41. Jain, A.; Jain, R.; Jain, S. Quantitative Analysis of Reducing Sugars by 3,5-Dinitrosalicylic Acid (DNSA Method). In *Basic Techniques in Biochemistry, Microbiology and Molecular Biology*; Humana: New York, NY, USA, 2020. [CrossRef]
42. Sridhar, K.; Charles, A.L. In vitro antioxidant activity of Kyoho grape extracts in DPPH and ABTS assays: Estimation methods for EC_{50} using advanced statistical programs. *Food Chem.* **2019**, *275*, 41–49. [CrossRef]

molecules

MDPI

Article

Effect of Addition Levels of By-Product Mixture (Apple Pomace: Red Potato Pulp: Sugar Beet Pulp) on Phytochemical Profile, Antioxidant Activity and Physical Properties of Extruded Corn Snacks

Rafał Ziobro [1], Dorota Gumul [1,*], Renata Sabat [1], Anna Wywrocka-Gurgul [1] and Tomasz Zięba [2,3]

1 Department of Carbohydrate Technology and Cereal Processing, Faculty of Food Technology, University of Agriculture in Krakow, 122 Balicka Street, 30-149 Krakow, Poland; rafal.ziobro@urk.edu.pl (R.Z.); renata.sabat@urk.edu.pl (R.S.); anna.wywrocka-gurgul@urk.edu.pl (A.W.-G.)
2 Institute of Sport, Tourism and Nutrition, Faculty of Biological Sciences, University of Zielona Góra, 65-417 Zielona Góra, Poland; tomasz.zieba@upwr.edu.pl
3 Department of Food Storage and Technology, The Faculty of Life Sciences and Technology, Wrocław University of Environmental and Life Sciences, Chełmońskiego 37, 51-630 Wrocław, Poland
* Correspondence: dorota.gumul@urk.edu.pl or rrgumul@cyf-kr.edu.pl; Tel.: +48-12-662-47-47

Abstract

Plant by-products such as apple pomace, potato pulp, and sugar beet pulp can be an excellent source of polyphenols, other phytochemicals and fiber, which is why they can be an excellent addition to snacks. Snacks, on the other hand, contain a lot of sugar and starch, which increases the risk of metabolic diseases and is unfavorable for diabetics, but after adding the above-mentioned by-products, their nutritional and health-promoting value increases. The aim of the study was to examine the effect of different addition levels of a mixture of by-products on the nutritional composition, phytochemical content, antioxidant activity, and quality of corn snacks. It was found that mixtures of by-products are an excellent addition to corn snacks in order to increase the health benefits of the product, as this additive increases the content of polyphenols, phenolic acids, anthocyanins, dietary fiber, tocopherols, especially α- and γ-tocopherol, as well as phytosterols, including β-sitosterol, stigmasterol, and campesterol. Furthermore, it increases the antioxidant potential of the snacks and the nutritional value of the final product, especially protein and ash content, reducing the starch content. Snacks containing a 20% by-product mixture may be recommended due to their improved nutritional and health-promoting properties and acceptable physical characteristics.

Keywords: by-products; snacks; phytochemical compounds; dietary fiber; nutritional composition; food quality

1. Introduction

Extrudates based on cornmeal are often used as snacks or ready-to-eat products (RTE) and contain high amounts of sugars and starch, which detrimentally affect their nutritional quality. They have a high glycemic index and can increase the risk of metabolic problems, such as cardiovascular disease, obesity, and type 2 diabetes [1,2]. According to the NOVA classification [3], snacks are usually classified as ultra-processed food, which is generally not preferred by dietitians. This problem could be solved by the addition of by-products with high content of dietary fiber (DF), which is often accompanied by polyphenols, and therefore could improve the nutritional quality and pro-health properties of cereal-based snacks. The above mentioned constituents (DF and polyphenols) play a chemo-preventive role in human nutrition, because they exhibit anti-radical anti-carcinogenic, anti-bacterial, anti-viral, anti-inflammatory, as well as hypoglycemic and hypotensive activity and diminish the risk of diseases such as diabetes, atherosclerosis and other cardiovascular diseases, genetic damage, degenerative bone changes, and neurodegenerative diseases, including Alzheimer's disease [4–7].

Plant by-products are an excellent source of these compounds, as they contain a wide spectrum of phytochemicals, e.g., phenolic compounds with highly valuable antioxidant properties and dietary fiber [2,8]. Phenolic compounds are present in the cell walls of plant material and can be an integral part of dietary fiber [9,10]. Therefore, by-products such as apple pomace, red potato pulp, and sugar beet pulp constitute a combination of many health-promoting compounds that can be successfully used in the production of health-promoting corn snacks [11–17].

Many authors [18–20] consider potatoes to be a functional food due to their polyphenol content. Among the polyphenols in potatoes, phenolic acids are the most abundant group found in potato tubers. The dominant phenolic acid is chlorogenic acid, followed by its derivatives, neo- and crypto-chlorogenic acids [11–13]. According to Rytel et al. [13], potatoes with colored flesh are richest in chlorogenic and neochlorogenic acids, while other acids are present in smaller amounts. Akyol et al. [21] and Mäder et al. [22] also noted that potatoes with different flesh colors are characterized by low levels of phenolic acids, such as caffeic, coumaric, ferulic, sinapic, and gallic acids. Another important group of polyphenolic compounds in potatoes is flavonoids, including catechin, epicatechin, kaempferol, and rutin [11]. Anthocyanins, a subgroup of flavonoids, are found only in red and purple potatoes and/or their skins, in concentrations ranging from 5.5 to 35 mg/100 g [12]. These anthocyanins are acylated with phenolic acids, mainly ferulic and p-coumaric acids. Purple potatoes contain petunidin and malvidin 3-rutinose-5-glucoside, acylated with p-coumaric and ferulic acids, while red potatoes are rich in pelargonidin and peonidin 3-rutinose-5-glucoside, also acylated with p-coumaric and ferulic acids [23].

Apple pomace is another good source of polyphenols and fiber. Polyphenols occur in soluble form or can be bound to cell wall structures [24]. The total polyphenol content measured using the Folin–Ciocalteu reagent in apples from 40 varieties was between 66.2 and 211.9 mg/100 g [25]. This corresponds to a total polyphenol content in dried apple pomace, which is estimated between 262 and 856 mg/100 g [26,27]. The increase in polyphenol content compared to fresh apples is related to water loss during the juice pressing and apple pomace drying processes. It should also be emphasized that some of the polyphenols are transferred to the juice during apple pressing (3–10%), and the largest source of these components is the skin, seeds, and stems of apples, which constitute fractions of apple pomace. The total polyphenol content in apple pomace is also influenced by enzymes released from vacuoles during juice pressing, such as polyphenol oxidases and peroxidases, which lead to the oxidation of polyphenols and thus to a reduction in their content [26]. It is estimated that the highest concentration of phenolic compounds could be found in seeds and petioles, 3593–23,607 and 6747–59,335 mg/kg d.m., respectively, and the lowest in pulp, 1655–5314 mg/kg d.m. The dominant polyphenol in the flesh fraction of apple pomace is chlorogenic acid (405–2299 mg/kg dry weight), while in seeds and stalks it is phloridzin (2899–19,600 and 3018–40,488 mg/kg dry weight, respectively). In addition to chlorogenic acid, apple pomace pulp is also a rich source of the following flavonol glycosides: quercetin-3-O-galactoside, quercetin-3-O-glucoside, quercetin-3-O-

 https://doi.org/10.3390/molecules31061037

xyloside, quercetin-3-O-arabinoside, and quercetin-3-O-rhamnoside. The content of these flavonols ranges from 27 to 579 mg/kg d.m. Flavonol glycosides are present in seeds at significantly lower levels than in other fractions of apple pomace, but the highest content of these compounds is found in petioles—84–3668 mg/kg d.m. Caffeic acid and procyanidins B2 occur at similar levels in all fractions of apple pomace, in amounts ranging from 8 to 555 mg/kg d.m. [14–16,24].

The main polysaccharides of the dietary fiber present in sugar beet pulp are alkali-soluble polysaccharides (ASP), consisting of arabinans, arabinogalactans, and galactans attached as side chains to pectin and partially bound to cellulose. Bioactive compounds from the polyphenol group are also partially bound to DF. These include hydroxy derivatives of cinnamic acids (e.g., ferulic acid) attached to arabinogalactans present in sugar beet [17], similarly to those found in wheat arabinoxylans [28,29]. In general, ferulic acid derivatives constitute a group of polyphenols that most often bind to fiber through covalent bonds (ester, carbon-carbon ether) [30]. In contrast, the connection between proanthocyanidins and fiber is usually formed through hydrogen bonds, hydrophobic effects, van der Waals forces, or other non-covalent interactions [31].

The above-mentioned plant by-products are an excellent source of a wide range of bioactive compounds belonging to the group of polyphenols and dietary fiber, which is why supplementation of cereal snacks with these compounds could be achieved by their addition. It should also be emphasized that extrusion is a technique that enables the utilization and full use of by-products, allowing the production of cereal snacks with enriching additives with proven health benefits [32]. The study attempted to obtain this type of snack using a mixture of the above-mentioned by-products to better utilize their unique composition. The aim of the study was to analyze the effect of different levels of a mixture of by-products on the nutritional value and content of polyphenolic compounds, tocopherols, and phytosterols in cereal snacks. In addition, the antioxidant activity of such snacks was estimated, and the functional and physical properties of the final products were characterized.

2. Results and Discussion

2.1. Characteristics of the Snack Enriching Mixture (Apple Pomace, Red Potato Pulp, Sugar Beet Pulp in a Ratio 1:1:1)

Considering the nutritional composition of the applied mixture of by-products consisting of apple pomace, sugar beet pulp, and red potato pulp in a ratio of 1:1:1, it was found that the level of dietary fiber exceeded all other compounds (Table 1). This is a logical consequence of the fact that the above-mentioned by-products are an excellent source of dietary fiber. It is especially true for apple pomace, where its content reaches 60% [33]. It should be emphasized that the determined dietary fiber consists of 2/3 insoluble fraction (IDF) and 1/3 soluble fraction (SDF). The ratio between the soluble and insoluble fractions of fiber, which is 1:2, is beneficial in terms of the health-promoting properties. Dietary fiber is characterized by hypocholesterolemic, hypoglycemic, and anticancer properties, and epidemiological studies clearly show a correlation between a decrease in dietary fiber consumption and an increase in the incidence of digestive system diseases, an increased risk of hypercholesterolemia and colon cancer [34,35]. Another very important component was total sugars, which amounted to 26 g/100 g d.m. (Table 1). This is the result of a much higher content of sugars in beet pulp, while the protein content fluctuated around 5.24 g/100 g d.m. (Table 1). It should be noted that protein from potato pulp is a complete protein with high biological value [36], which is very important when enriching grain feed mixtures with potato pulp, as grains are characterized by many limiting amino acids,

 https://doi.org/10.3390/molecules31061037

including lysine. The high ash content is due to the relatively high ash content in beet pulp and potato pulp, and the fat content fluctuates around 1.47 g/100 g d.m., which is mainly related to the fat from apple pomace (Table 1).

Table 1. Chemical composition of the applied mixture of by-products (apple pomace, red potato pulp, sugar beet pulp in a ratio 1:1:1; before processing).

Component	Content
protein [g/100 g d.m.]	5.24 ± 0.20
fat [g/100 g d.m.]	1.47 ± 0.13
ash [g/100 g d.m.]	2.33 ± 0.12
total sugars [g/100 g d.m.]	26.00 ± 1.1
insoluble dietary fiber (IDF) [g/100 g d.m.]	23.60 ± 0.57
soluble dietary fiber (SDF) [g/100 g d.m.]	6.33 ± 0.17
total dietary fiber (TDF) [g/100 g d.m.]	29.93 ± 0.74
TPC [mg of catechin/g d.m.]	9.23 ± 0.09
phenolic acids [mg of ferulic acid/g d.m.]	2.38 ± 0.02
flavonols [mg of quercetin/g d.m.]	0.87 ± 0.01
flavonoids [mg of rutin/g d.m.]	5.06 ± 0.01
anthocyanins [mg of cyanidin 3-glucoside/g d.m.]	2.12 ± 0.01
ABTS [µMTX/kg d.m.]	172 ± 11
DPPH [µMTX/kg d.m.]	57.03 ± 0.01
δ-tocopherol [mg/100 g dm]	n.d. [1]
α-tocopherol [mg/100 g dm]	1.04 ± 0.04
γ-tocopherol [mg/100 g dm]	0.33 ± 0.02
β-tocopherol [mg/100 g dm]	0.19 ± 0.01
cholesterol [mg/100 g dm]	0.16 ± 0.02
campesterol [mg/100 g dm]	0.89 ± 0.04
stigmasterol [mg/100 g dm]	1.34 ± 0.01
β-sitosterol (mg/100 g dm)	11.76 ± 0.01
δ-5-avenasterol [mg/100 g dm]	0.09 ± 0.01
δ-7-stigmasterol [mg/100 g dm]	n.d.
δ-7-avenasterol [mg/100 g dm]	n.d.

[1] n.d—not detected.

When considering the content of polyphenolic compounds, it should be observed that the level of polyphenols is significant, but flavonoids are the dominant group among them (Table 1), as the content of flavonoids is significant in each of these three by-products [11,15,23]. Next are phenolic acids and anthocyanins, which mainly come from red potato pulp [11,21,23]. Such a high polyphenol content results in very high antioxidant activity, as determined by two independent methods using the ABTS and DPPH free radicals (Table 1).

In the case of tocopherols in the discussed by-product mixture, three times more α-tocopherol was found than γ-tocopherol. This fact is related to the fact that apple pomace and red potato pulp are sources of α-tocopherol [37–39]. Among phytosterols, a significant amount of β-sitosterol was noted, followed by stigmasterol and campesterol, as all components of the by-product mixture are rich in these compounds [39,40].

In summary, it should be noted that the nutritional and health-promoting value of a mixture of apple pomace, sugar beet pulp, and potato pulp (1:1:1) is high. It could be described as a concentrate of dietary fiber and polyphenols, with high levels of total sugars, protein, and ash, which is why it can be successfully used as an ingredient to enrich corn snacks, which have low dietary value.

https://doi.org/10.3390/molecules31061037

2.2. Chemical Characteristics of Corn Snacks Produced with an Enriching Mixture of By-Products

2.2.1. Composition of Snacks with a Share of a Mixture of By-Products

The chemical composition of the snacks obtained with the addition of a mixture of by-products is noteworthy, as it reduces the content of an undesirable ingredient in these snacks, namely starch (Table 2). After adding 10% of the mixture, the amount of starch decreases by 6%, and after adding 30%, the starch content decreases by 23% in enriched snacks compared to the control (Table 2). A three- to six-fold increase in total sugars was also observed in snacks enriched with a mixture of by-products compared to snacks without it (Table 2). The observed increase in sugars, mainly from apple pomace, however, is compensated by a far greater increase in the soluble and insoluble fiber fraction, the source of which is the by-products present in the added mixture. The levels of both soluble and insoluble fiber in snacks containing a mixture of by-products were significantly elevated compared to the control. The change in the insoluble fiber fraction ranged from a 6- to 25-fold increase in this component in snacks containing a mixture of by-products compared to the control. In contrast, in snacks containing by-products, a 4- to 19-fold increase in the soluble fiber fraction was observed compared to the control. The total fiber (TDF) content ranged from 2.69 to 11.67 g/100 g of dry matter, which was 5 to 22.4 times higher in enriched snacks compared to the control of 0.52 g/100 g d.m. (Table 2), which is a logical consequence of the use of by-products that are a concentrate of dietary fiber.

Table 2. Chemical composition of snacks with a share of mixture of by-products (g/100 g d.m.).

Sample	Protein	Fat	Ash	IDF	SDF	TDF	Total Sugars	Starch
E0	4.56 ± 0.01^{a1}	1.11 ± 0.01^{a}	0.33 ± 0.01^{a}	0.30 ± 0.01^{a}	0.22 ± 0.05^{a}	0.52 ± 0.04^{a}	0.96 ± 0.04^{a}	83.31 ± 0.08^{d}
E10	4.42 ± 0.14^{a}	1.10 ± 0.07^{a}	0.57 ± 0.01^{b}	1.88 ± 0.03^{b}	0.81 ± 0.02^{b}	2.69 ± 0.01^{b}	2.82 ± 0.06^{b}	78.22 ± 0.14^{c}
E20	4.86 ± 0.04^{b}	1.01 ± 0.03^{a}	0.92 ± 0.01^{c}	5.58 ± 0.15^{c}	2.95 ± 0.02^{c}	8.52 ± 0.13^{c}	4.54 ± 0.04^{c}	72.04 ± 0.07^{b}
E30	5.35 ± 0.04^{c}	0.98 ± 0.01^{a}	1.24 ± 0.00^{d}	7.41 ± 0.07^{d}	4.26 ± 0.01^{d}	11.67 ± 0.06^{d}	5.75 ± 0.06^{d}	64.28 ± 0.14^{a}

[1] Values denoted by the same letters in columns do not differ significantly ($p < 0.05$).

Dietary fiber is a chemically heterogeneous complex consisting of insoluble and soluble fractions, each of which has a different physiological effect. Insoluble fiber is recommended for the prevention and treatment of colon diseases such as chronic constipation, irritable bowel syndrome, hemorrhoids, and diverticulosis. Soluble fiber, on the other hand, has hypocholesterolemic, hypoglycemic, and anticancer properties. The main sources of fiber in the diet are grains, but above all vegetables and fruits, especially by-products such as pomace and pulp [2,41,42]. The introduction of by-products rich in the earlier-mentioned polyphenolic compounds and dietary fiber is therefore highly justified to obtain snacks with high nutritional value. An increase in protein and ash content could also be noticed in snacks containing by-products compared to the control, in contrast to the constant fat content (Table 2). The increase in protein content by as much as 17% with a 30% addition of a mixture of by-products to snacks (Table 2) is mainly due to the use of potato pulp. In this regard, it should be emphasized that potato protein has high biological and nutritional value because it contains all essential amino acids (such as lysine, lecithin, phenylalanine, tyrosine) that are not synthesized by the human body. The biological value of potato protein is comparable to that of chicken egg protein (chemical score CS is in the range of 57–69%) [36] and other animal proteins and is significantly higher than the biological value of proteins from peas, wheat, and rice. In addition, potato protein is the best source of lysine in the plant world [36]. The increase in ash content (Table 2) in snacks is due to the use of a mixture of by-products, especially apple pomace and beet pulp [43,44], where the ash content is 1.88% and 7%, respectively.

 https://doi.org/10.3390/molecules31061037

A strong negative correlation was observed between the fat content in the analyzed snacks and individual dietary fiber fractions and total sugar content (r between −0.950 and −0.993). On the other hand, a strong positive correlation was found between starch and fat in the snacks (r = 0.958). A strong positive correlation was also noted between individual dietary fiber fractions and the content of anthocyanins, polyphenols, flavonoids and phenolic acids, as well as alpha-tocopherol in the snacks (r between 0.973 and 0.998; Table S1).

2.2.2. Polyphenol Compounds of Snacks with a Share of a Mixture of By-Products

It was found that the inclusion of by-products contributed to an increase (2 to 7 times) in total polyphenols (TPC) in snacks containing a mixture of by-products, compared to control snacks.

It can be concluded that the increase in total polyphenol content was gradual in relation to the proportion of by-products in the snacks (Table 3). In the studies by Gumul et al. [45] on the effect of adding defatted blackcurrant seeds to extruded corn products, it was also found that the amount of TPC increased with their rising level. The study by Reis et al. [46] analyzed the effect of apple pomace on extruded products. It was found that a 10% addition of pomace resulted in a 73% increase in total polyphenol content compared to unenriched snacks, and a 20% and 30% addition resulted in an increase of 37% and 82%, respectively, compared to the control [46]. In a study by Korkerd et al. [47], the effect of adding food processing by-products (defatted soybean meal, germinated brown rice flour, and mango peel fiber) to extruded snacks was investigated. It was noted that the by-products increased the polyphenol content in the snacks from 0.3 to 6.7–10.06 mg GAE/g of snack, depending on the extrusion conditions [47]. In studies by Oniszczuk et al. [48] on the effect of adding elderberry flowers and fruits to corn snacks, it was also proven that the polyphenol content in snacks increased proportionally with the addition of elderberry. Gumul et al. [49] came to similar conclusions when studying the effect of fortifying corn snacks with red and purple freeze-dried potatoes, as the polyphenol content in the snacks increased with the increasing proportion of potatoes in the mixture. A 5% addition of red potatoes to snack formulation resulted in a 25% increase in polyphenol content compared to control corn snacks. However, with 15% and 25% red potatoes in the enriched snacks, the polyphenol content increased by 99% and 188%, respectively, compared to the control [49]. Similar observations were made in studies by Drożdż et al. [50] on snacks containing currant and chokeberry pomace, where an increase in total polyphenols was observed in enriched snacks compared to corn extrudates. A 10% addition of chokeberry pomace resulted in a 44-fold increase in total polyphenol content compared to the control sample. Apart from that, a 15% and 20% addition of chokeberry to the crisps resulted in a 120-fold and 178-fold increase in total polyphenols, respectively, compared to the control. In a study by Bekele et al. [51], the effect of the composition of a chickpea and sorghum mixture on the polyphenol content in finished products—expanded snacks—was investigated. It was found that the content of phenolic compounds increased with the increase in the proportion of sorghum in the mixture. In a study by Yagci et al. [52], extruded snacks enriched with tomato pomace powder were examined. As expected, the phenolic content in the control sample was significantly lower than in snacks with added tomato pomace. With a 5% addition of tomato pomace, the total polyphenol content increased twofold compared to the control sample. In contrast, 10% and 15% tomato pomace content resulted in a 3.5- and 5-fold increase in total polyphenol content, respectively, compared to corn snacks [52]. In a study by Promsakha na Sakon Nakhon et al. [53] on the effect of adding pumpkin powder

 https://doi.org/10.3390/molecules31061037

to corn snacks, it was found that as the proportion of pumpkin flour increased, so did the polyphenol content.

The snacks produced with a mixture of by-products revealed high levels of flavonoids in comparison to the control (Table 3). Even the smallest—10% share of by-products contributed to a 5-fold increase in their amount as compared to non-enriched snacks. The application of 20 and 30% share of pomace/pulp mixture in snack formulation caused a 13- and 21-fold increase in flavonoid content compared to the control sample (Table 3). A similar increase was observed for phenolic acids and flavonols in extrudates with the above-mentioned additives, which were 2–7.5 times and 6–23 times higher than in the control, respectively (Table 3). In the case of anthocyanins, which appeared in snacks due to the presence of red potato pulp in the enrichment mixture, a significant amount of 1.05–10.2 mg/100 g d.m. was also observed (Table 3).

Table 3. Polyphenol compounds in snacks enriched with a mixture of by-products.

Sample	Anthocyanins [mg Cyanidin-3-Glycoside/100 g dm]	Total Phenolics TPC [mg Catechin/100 g d.m.]	Flavonoids [mg Rutin/100 g d.m.]	Content of Flavonols (mg Quercetin/100 g d.m.]	Content of Phenolic Acids [mg Ferulic Acid/100 g d.m.]
E0	n.d. [1]	53.24 ± 0.01 [a]	5.38 ± 0.07 [a]	0.92 ± 0 [a]	6.14 ± 0.72 [a]
E10	1.05 ± 0.01 [a2]	118.20 ± 1.32 [b]	28.24 ± 1.2 [b]	5.45 ± 0.74 [b]	15.81 ± 0.01 [b]
E20	7.42 ± 0.14 [b]	215.14 ± 1.17 [c]	68.53 ± 2.39 [c]	16.87 ± 0.2 [c]	34.24 ± 0.76 [c]
E30	10.2 ± 0.09 [c]	362.50 ± 1.98 [d]	117.57 ± 1.02 [d]	23.79 ± 0.09 [d]	47.83 ± 1.52 [d]

[1] n.d—not detected; [2] Values denoted by the same letters in columns do not differ significantly ($p < 0.05$).

In general, it was found that as the level of a mixture of by-products increased, so did the level of flavonoids, polyphenols, phenolic acids, flavonols, and anthocyanins, although this increase was not proportional to the amount of additive introduced, but many times greater. This can be explained by the various changes that occur during the extrusion process, including isomerization and the release of flavonoids and phenolic acids in particular from glycosidic and ester bonds, thanks to which they are more easily extracted during final assessment, hence their apparent quantity in the analyzed test material is high [49].

In a study by Oniszczuk et al. [48] on the effect of adding elderberry flowers and elderberries to corn snacks, it was found that the flavonoid content in the snacks increased proportionally with the addition of elderberries. Fortifying snacks with 5, 10, and 20% elderberry fruit resulted in an increase in flavonoid content of 30, 157, and 359%, respectively, compared to corn snacks. A 5% and 10% share of flowers in snacks resulted in a decrease in flavonoid content by 30% and 157%, respectively, compared to corn snacks. However, a 20% addition of flowers resulted in a more than threefold increase in flavonoid content compared to the control sample [48]. In studies by Gumul et al. [49] on the effect of adding red potatoes to corn snacks, an increase in flavonoid content was observed with a rising level of potato addition. A 5% share of red potatoes in snack formulation resulted in a 140% increase in flavonoid content compared to corn snacks (control). In contrast, 15% and 25% introduction of red potatoes in snacks resulted in a 387% and 484% increase in flavonoid content, respectively, compared to the control sample.

In the study by Drożdż et al. [50], the effect of adding currant and chokeberry pomace on the phenolic acid content in corn snacks was investigated. An increase in phenolic acid content was observed with increasing levels of fruit pomace addition. In the case of chlorogenic acid, a 10% addition of chokeberry pomace resulted in an almost 78-fold increase in its content compared to the control sample. Larger, 15% and 20% addition of chokeberry to the crisps resulted in an almost 102-fold and 154-fold increase in chlorogenic acid content, respectively, compared to the corn snacks. In the case of blackcurrant pomace, a 10% addition of the by-products to snack formulation resulted in an almost 39.5-fold increase in chlorogenic acid content compared to the control sample. Enriching the snacks with 15% and 20% blackcurrant resulted in an almost 43-fold and over 39-fold increase in chlorogenic acid content, respectively, compared to corn snacks [50]. In another study by Reis et al. [46] on the effect of incorporating apple pomace into extruded products, it was found that as the proportion of apple pomace in snacks increased, the flavonol content also increased. A 10% and 20% addition of apple pomace had the same effect. Both proportions contributed to a 3.25-fold increase in flavonol content compared to corn snacks. In contrast, a 30% proportion of pomace in snacks resulted in a 4-fold increase in flavonol content compared to the control sample [46]. Arribas et al. [54] also studied the effect of adding bean meal and carob meal to rice extrudates. They observed that as the amount of carob meal added increased, the flavonol content also increased.

The increase in polyphenols and their fractions was accompanied by an increase in antioxidant activity measured by ABTS and DPPH in snacks containing a mixture of by-products, compared to the control, by 20–80 times (ABTS) and 5–25 times (DPPH) (Figure 1).

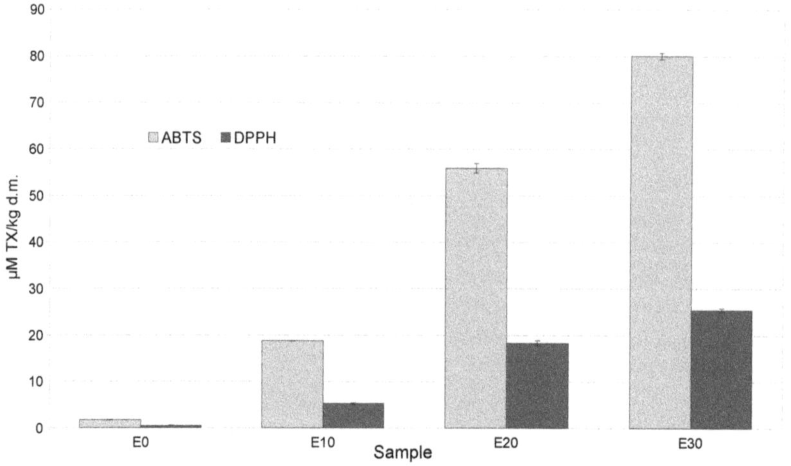

Figure 1. Antioxidant activities (ABTS, DPPH) of extrudates enriched with varying levels of a mixture of by-products.

This was confirmed by strong positive correlations between ABTS and anthocyanin content r = 0.992, total polyphenol content r = 0.983, flavonoid content r = 0.991, and phenolic acid content r = 0.999. In addition, strong positive correlations were found between DPPH and anthocyanin content at r = 0.996, total polyphenol content at r = 0.977, flavonoid content at r = 0.986, and phenolic acid content at r = 0.998. Furthermore, the antioxidant activity of the discussed snacks was influenced by tocopherols such as α-tocopherol, as the correlation between this isomer and ABTS and DPPH was r = 0.984 and r = 0.976, respectively (Table S1).

2.2.3. Tocopherols in Snacks with a Share of a Mixture of By-Products

In corn snacks used for control, δ and γ tocopherols were found, but α tocopherol and β tocopherols were not detected (Table 4). This is because cornmeal is the primary source

 https://doi.org/10.3390/molecules31061037

of γ tocopherol and δ tocopherol [55]. In the case of δ-tocopherol, whose content is low in both apple pomace and red potato pulp, and is not present at all in beet pulp [37,38,44], it was observed that enriching snacks with a mixture of by-products did not cause any changes in its content in snacks (Table 4).

Table 4. Tocopherols in extrudates enriched with varying levels of a mixture of by-products.

Sample	δ-Tocopherol	β-Tocopherol	γ-Tocopherol	α-Tocopherol
E0	0.17 ± 0.01 [a1]	n.d.	0.20 ± 0.01 [a]	n.d.
E10	0.2 ± 0.03 [a]	0.09 ± 0.04 [a]	0.25 ± 0.01 [b]	0.10 ± 0.03 [a]
E20	0.21 ± 0.02 [a]	0.15 ± 0.03 [a]	0.27 ± 0.02 [b]	0.18 ± 0.02 [b]
E30	0.18 ± 0.01 [a]	0.16 ± 0.02 [a]	0.38 ± 0.04 [c]	0.29 ± 0.03 [c]

[1] Values denoted by the same letters in columns do not differ significantly ($p < 0.05$), n.d.—not detected.

In contrast, the content of α-tocopherol, which was not detected in the control, increased with the increase in the content of by-products in the snacks (Table 4). This results from the very high content of α-tocopherol in both apple pomace and potato pulp [37–39], which make up the mixture added to enrich the snacks. Similarly, in the case of γ-tocopherol, an increase in its content was observed in snacks made with a mixture of by-products compared to the control. However, when comparing the increase in α-tocopherol content in snacks enriched with γ-tocopherol, it was clearly observed that the amount of α-tocopherol is significantly higher in these snacks (Table 4) due to the fact that the dominant tocopherol in apple pomace and red potato pulp is α-tocopherol [37–39]. In addition, β-tocopherol, which was absent from the control (Table 4), was found in snacks enriched with by-products. It should also be noted that during extrusion, deviations in the content of tocopherols and tocotrienols may occur as a result of changes in the chemical structure of these compounds due to their thermal degradation, depolymerization, and recombination of their molecular fragments under the influence of high pressure, temperature, and shear forces prevailing in the extruder [56]. Individual isomers of tocopherols and tocotrienols will react differently; it is assumed that α and β tocopherol are the least stable, while γ tocopherol is stable [37,38,56]. In addition, Zieliński et al. [57] noted a change in the content of tocopherols and tocotrienols in rye grain extrudates at a level of 91% when they used extrusion at a temperature of 120 degrees Celsius, while at the other temperatures of 160 and 200 degrees Celsius, the reduction was 75.68%. Tiwari and Cummins [58] had a similar opinion, stating that high extrusion temperatures and short times can cause tocopherols to stabilize, hence their lower losses at higher temperatures during this process. On the other hand, Shin et al. [59] noticed significant losses of tocopherols at extrusion temperatures of 110 to 140 degrees Celsius. Nevertheless, it should be emphasized that the mixtures used to produce the snacks enriched them mainly with α-tocopherol, γ-tocopherol and β-tocopherol. The amount of these compounds in the snacks was significantly lower than the percentage of these mixtures, which is due to the changes in these compounds during extrusion. By-product mixtures did not enrich the snacks in δ-tocopherol, as its amount in corn grits is very high, while in the by-product mixture it is low.

2.2.4. Phytosterols in Snacks with a Share of a Mixture of By-Products

Regarding phytosterols, it should be noted that snacks containing a mixture of by-products were characterized by a very high β-sitosterol content. The increase in β-sitosterol content after adding 10% of the mixture of by-products was twofold compared to the control, and with a 30% addition, it reached 11 times the amount of this phytosterol in relation to the control (Table 5).

https://doi.org/10.3390/molecules31061037

Table 5. Phytosterols in extrudates enriched with varying levels of a mixture of by-products.

Sample	Chole-Sterol	Campe-Sterol	Stigma-Sterol	β-Sitosterol	δ-5-Avena-Sterol	δ-7-Stigma-Sterol	δ-7-Avena-Sterol
E0	0.16 ± 0.01 [a1]	0.10 ± 0.01 [a]	0.30 ± 0.03 [a]	0.74 ± 0.03 [a]	0.03 ± 0.00 [a]	1.24 ± 0.01 [d]	0.21 ± 0.01 [a]
E10	0.17 ± 0.00 [a]	0.21 ± 0.01 [b]	0.29 ± 0.01 [a]	1.63 ± 0.11 [b]	0.14 ± 0.00 [b]	0.37 ± 0.00 [a]	0.17 ± 0.00 [a]
E20	0.25 ± 005 [b]	0.48 ± 0.01 [c]	0.42 ± 0.04 [b]	4.57 ± 0.07 [c]	0.42 ± 0.02 [c]	0.55 ± 0.03 [b]	0.21 ± 0.00 [a]
E30	0.35 ± 0.01 [c]	0.76 ± 0.13 [d]	1.17 ± 0.05 [c]	8.63 ± 0.05 [d]	0.66 ± 0.03 [d]	1.04 ± 0.00 [c]	0.25 ± 0.03 [a]

[1] Values denoted by the same letters in columns do not differ significantly ($p < 0.05$).

Such a large increase is because the mixture of by-products is rich in these types of compounds, especially apple pomace and red potato pulp, particularly in β-sitosterol. The amount of β-sitosterol in apple pomace reaches 1147 mg/g, and in red potatoes 1.82 mg/100 g [39,40,60]. Sugar beet pulp also contains β-sitosterol, stigmaceterol, and campesterol, which are predominant in plants [61]. An increase in campesterol and stigmasterol was also observed in snacks containing a mixture of by-products compared to the control snacks. The amount of other phytosterols, δ-5 and δ-7-avenasterols and δ-7-stigmasterol, varied in these products. The content of 5-avenasterol increased from 3 times in snacks with the addition of 10% of a mixture of by-products to 22 times in snacks with the addition of 30%. The amount of δ-7-stigmasterol in snacks decreased (about 70%) with the addition of 10% of a mixture of by-products. In the snacks with 20% and 30% addition of this mixture, the amount of δ-7-stigmasterol decreased by 55% and 11%, respectively, compared to the control. The exception was δ-7-avenasterol, the amount of which remained constant both in the control and in snacks produced with a mixture of by-products (Table 5). The amount of campesterol increased in snacks using by-products 2 to 7.6 times compared to the control, and stigmasterol from 25% to 3 times (Table 5). The amount of cholesterol at a 10% by-product content was identical to the control, but a 20% and 30% addition caused an increase in cholesterol by 56% and 118%, respectively, in snacks with by-products compared to the control (Table 5).

It should be noted that extrusion causes phytosterols degradation, which increases with the temperature of the process [62]. Therefore, a temperature of 160 °C was used here, along with an appropriate selection of by-products rich in phytosterols to counteract this problem. Phytosterols are important health-promoting substances with hypocholesterolemic properties, as well as anti-inflammatory, antioxidant, and anticancer effects [63,64]. The high amount of these compounds in snacks is an important achievement of this work.

2.3. Physical Properties of Snacks with a Share of a Mixture of By-Products

Table 6 represents the physical properties of corn snacks produced using extrusion technology. While the control sample seems to be well expanded, the use of an enriching mixture caused the detrimental effect even at 10% level of addition (Table 6). This is typical for fiber-rich raw materials, which are used to replace starch or flour, such as defatted blackcurrant seeds [65], orange pomace [66] and several other off-products of industrial origin [47], although depending on the raw material and processing parameters opposite trends could also be found—e.g., in the cases of carrot pomace [67]. Gumul et al. [65] explained the decrease in expansion and increase in density of maize extrudates after the addition of the enrichment formulation by the fact that these samples contained more protein, fiber and sugar and less starch compared to the control.

https://doi.org/10.3390/molecules31061037

Table 6. Physical properties of extrudates enriched with varying levels of a mixture of by-products (ER—expansion ratio, WEWAI—whole extrudate water absorption index, GEWAI—ground extrudate water absorption index, WSI—water solubility index, L*, a*, b*—color parameters).

Sample	Density [g/mL]	ER	WEWAI	GEWAI	WSI	L*	a*	b*	Cutting Force [N]
E0	0.05 ± 0.01^{b} [1]	0.31 ± 0.04^{b}	2.77 ± 0.24^{a}	4.35 ± 0.33^{a}	0.35 ± 0.03^{b}	51.6 ± 11.8	1.36 ± 0.31^{a}	11.6 ± 2.6^{a}	24.4 ± 14.3
E10	0.06 ± 0.01^{c}	0.24 ± 0.02^{a}	3.69 ± 0.95^{b}	5 ± 0.41^{b}	0.28 ± 0.04^{a}	58.3 ± 1.9	7.28 ± 0.96^{b}	20.2 ± 2.3^{b}	16.6 ± 6.1
E20	0.03 ± 0.01^{a}	0.22 ± 0.04^{a}	2.69 ± 0.65^{a}	5.7 ± 0.04^{c}	0.3 ± 0.05^{ab}	54.1 ± 1.8	10.24 ± 0.33^{c}	22.6 ± 0.6^{b}	11.6 ± 7.5
E30	0.05 ± 0.01^{b}	0.24 ± 0.05^{a}	3.2 ± 0.25^{b}	5.11 ± 0.06^{b}	0.78 ± 0.71^{c}	49.9 ± 4.5	12.58 ± 0.73^{d}	23.6 ± 1.9^{b}	12.4 ± 5.9

[1] Values denoted by the same letters in columns do not differ significantly ($p < 0.05$).

Water-binding properties of the snacks were improved in comparison to the control, as both WEWAI and GEWAI significantly increased. It could be discussed if this could be caused by the larger amount of insoluble dietary fiber, as a drop in solubility could be observed only for the smallest addition level, and the extrudates with 30% of enriching mixture were characterized by significantly higher solubility than the control, but the values of WSI were not very reliable (Table 6).

Among the color parameters, L* varied a lot in the case of the control sample, revealing its highly heterogeneous character (analysis of variance could not be applied here, due to lack of homoscedascity, Table 6). The addition of a mixture of by-products to extrudates resulted in more homogeneous samples becoming slightly darker while elevating the level of addition. In the case of the control, the a* and b* color parameters were more uniform than L*. A significant increase in redness against greenness (a*) and yellowness against blueness (b*) could be observed after using a mixture of by-products, and the change in color was correlated with the level of addition. It seems obvious that as the content of red potato pulp increases in the formulation, the red color is more pronounced. Other components are probably also involved in the shift towards higher yellowness of the samples, but in this case, the differences between the extrudates with the addition of by-products are not obvious (Table 6).

Significant variability could also be observed for the cutting force of the extrudates, which was mainly due to high non-uniformity of the samples (Table 6). The lowest average values were calculated for extrudates with a share of 20% enriching mixture, but the other enriched samples did not seem to differ statistically from it. On the other hand, the control sample exhibited the largest value of cutting force together with the largest value of standard deviation, meaning that the extrudate strand was exhibiting different hardness at different cutting points (Table 6). Slightly decreasing average values (softer products) observed for snacks with a mixture of by-products, as compared to the control sample, were probably due to the enrichment in fiber, which caused a formation of more uniform cellular structure but also diminished the hardness of the cell walls of the extrudates.

In summary, the use of the applied by-product mixture will contribute to the production of snacks with reduced starch content and significant protein, fat, ash, and dietary fiber content. In addition, snacks enriched with this by-product mixture will be rich in polyphenols, phytosterols (mainly β-sitosterol, stigmasterol and campesterol), and tocopherols such as α- and γ-tocopherol. The use of by-products in the production of snacks corresponds to the zero-waste technology, which is one of the goals of sustainable development. It can therefore be said that the appropriate selection of by-products and extrusion process parameters will contribute to the creation of health-promoting snacks. The results show how health-promoting compounds can be introduced into food by using by-products, which is a global trend in current scientific research [68,69].

https://doi.org/10.3390/molecules31061037

3. Materials and Methods

3.1. Materials

3.1.1. Preparation of By-Products

The by-products (apple pomace, red potato pulp and sugar beet pulp in a ratio 1:1:1) were freeze-dried for 40 h using a Gamma 1-16 LSC freeze dryer (Martin Christ Gefriertrocknungsanlagen GmbH, Osterode am Harz, Germany) at −20 °C and a pressure of 0.1 mbar. After the freeze-drying process, the samples were ground using a Grindomix GM-200 laboratory grinder (Retsch GmbH&Co. KG., Haan, Germany) and used for further analysis.

3.1.2. Preparation of Snacks

Production of extrudates: extrudates containing 10%, 20%, 30%, by-products (E10; E20; E30) together with a control sample (E0) were prepared in a single-screw laboratory extruder. The cornmeal and by-products (apple pomace, red potato pulp, and beet pulp in a 1:1:1 ratio) were pre-ground to the appropriate size and then mixed. The moisture level in the components of the premix used for extrusion was set at 14%, and the particle sizes ranged from 600 to 950 μm. Part of the cornmeal was replaced with a premix in the amounts of 10%, 20%, 30%, resulting in plant-based snacks (E10–E30%). The extrusion was carried out using a single-screw extruder (Brabender, Duisburg, Germany). The screw speed was set at about 100 rpm. The diameter of the screw was 60 mm, and the extrusion process temperatures in different zones were as follows: zone I 140 °C/zone II 160 °C/zone III 130 °C (compression degree = 1:3). A matrix with two nozzles, each with a diameter of 3 mm, was used (Figure 2).

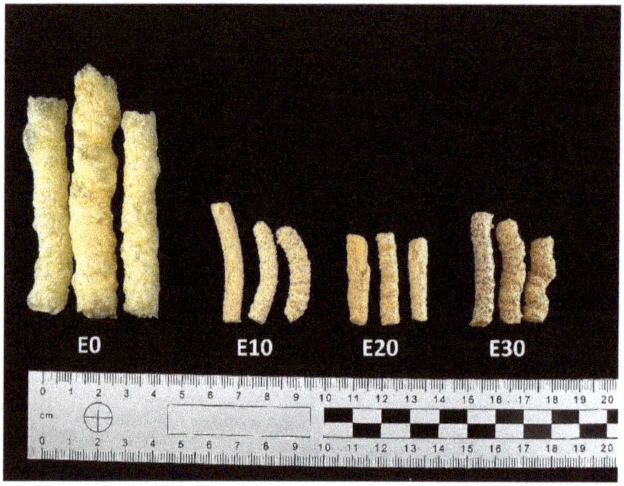

Figure 2. Snacks enriched with varying levels of a mixture of by-products.

3.2. Methods

The following analyses were performed on raw materials (mixture of by-products) and snacks:

3.2.1. Chemical Composition

The content of basic nutrients in the analyzed samples was determined by AOAC methods [70]. Protein (N × 6.25) was measured by the Kjeldahl method using the Kjeltec 2200 extraction device (Foss Analytical, Hillerød, Denmark). The total carbohydrate content

 https://doi.org/10.3390/molecules31061037

was determined using the AOAC method no. 974.06. Fat content was measured using the Soxhlet method (AOAC method no. 953.38) using the Soxtec Avanti 2055 device (Foss Analytical, Hillerød, Denmark). The ash content was determined using the AOAC method No. 930.05. The content of total, soluble and insoluble dietary fiber (TDF, SDF, IDF) was determined by the AACCI method 32-07 [71], and the starch content was determined according to the standard 122/1 of ICC [72]. All the above measurements were taken in at least two repetitions.

3.2.2. Analysis of Antioxidants and Anti-Radical Activities

Antioxidant compounds (polyphenols) and anti-radical activity in ethanol extracts were studied. A sample weighing 0.6 g was dissolved in 30 mL of 80% ethanol, shaken in the dark for 120 min (using an electric shaker, type WB22, Memmert, Schwabach, Germany), and then centrifuged at 4000 rpm for 15 min (MPW-350 centrifuge, MPW MED. Instruments, Warsaw, Poland). The supernatant was separated and stored at −20 °C for later analysis. The total polyphenol content (TPC) was determined by the spectrophotometric method with Folin–Ciocalteu reagent, according to Singleton et al. [73]. The contents of anthocyanins, phenolic acids and flavonols were determined using the methods of Mazza et al. [74], with modifications by Oomah et al. [75]. The methodology of El Hariri et al. [76] was applied to determine flavonoid content. Anti-radical potential was measured using the synthetic radical ABTS, according to Re et al. [77]. A diluted ethanol extract was combined with ABTS, shaken on a vortex mixer (WF2, Janke and Kunkel GmbH, Staufen im Breisgau, Germany), and its absorbance was measured using a spectrophotometer (Helios γ, 100–240 Thermo Spectronic, Runcorn, UK) at λ = 734 nm. The second reading was taken after 6 min at the same wavelength. The results were expressed in µmol TX/kg of dry matter (d.m.) Trolox equivalent (TEAC), where R2 = 0.999. Antioxidant activity in ethanol extracts was measured using 2,2-diphenyl-1-picrylhydrazyl (DPPH), according to Brand-Williams et al. [78]. Extracts (1 mL) were mixed with 4 mL of DPPH solution (0.012 g DPPH in 100 mL of ethanol). Absorbance was measured using a spectrophotometer at λ = 517 nm. Trolox (6-hydroxy2,5,7,8-tetramethylchromano-2-carboxylic acid) with R^2 = 0.983 was used as a standard. Results are expressed in µmol/kg d.m. Trolox equivalent (TEAC).

3.2.3. Determination of Tocopherols and Phytosterols

Determination of tocopherols and phytosterols was performed using gas chromatography. Tocopherols and phytosterols were determined by Hussain et al. [79], Oracz et al. [80] and Zhang et al. [81]. Samples were prepared by weighing 0.2 g of the sample (±0.0001 g) into a 20 mL vial. To this, 4 mL of freshly prepared saponification agent (3.9 mL of 2 M KOH in methanol and 0.5 mL of 10% ascorbic acid) was added. The vial was tightly closed, incubated at 85 °C for 40 min, and then cooled to room temperature. The contents were transferred to 30 mL centrifuge tubes containing 10 mL of hexane and 10 mL of saturated NaCl solution. The tubes were tightly closed, shaken for 10 min (175 rpm), and then centrifuged at 6000 rpm for 10 min. The top layer of hexane was collected, transferred to 20 mL vials, and evaporated under nitrogen. After drying, 1 mL of hexane was added, and the mixture was treated with ultrasound for 10 s. Samples were filtered through a nylon syringe filter (pore size 0.2–0.45 µm, ProSource Scientific, Calgary, AB, Canada). The analysis was performed using the Shimadzu GC 2010 Plus gas chromatograph with an FID detector (Shimadzu, Corp., Kyoto, Japan).

https://doi.org/10.3390/molecules31061037

3.2.4. Physical Parameters of Snacks

Density and coefficient of expansion were following Sokhey et al. [82]. Density determination was carried out by placing fragments of extrudate with a total weight of about 5 g in a cylinder with rapeseed and measuring the loss of their volume. The coefficient of expansion was calculated by dividing the average diameter of the extrudate by the diameter of the extruder nozzle (10 samples).

For determining the whole extrudate water absorption index (WEWAI), 1 g of whole extrudates was weighed and distilled water at room temperature was added. After wetting for 1 h, the extrudates were weighed, and the amount of water bound by them was obtained from the difference in mass. The result was then expressed per 1 g of product.

To measure the ground extrudate water absorption index (GEWAI) and solubility, 1 g of the product was ground, combined with 50 mL of distilled water and left at room temperature for 30 min. After this time, the sample was centrifuged for 5 min at 4000 rpm. Next, the supernatant was poured from above the sediment, and the amount of water absorbed by the sediment was determined for the calculation of GEWAI. The supernatant was poured into a Petri dish and dried at 50 °C for 24 h and the resulting mass was used to calculate the water solubility index [83].

The color of the samples was determined with a Konica Minolta CM-3500d spectrophotometer (Chiyoda, Japan) using a D65 illuminant and 10° observer angle. Five measurements were taken per sample and reported in the CIE Lab* system [84], with L* (lightness: 0 = black to 100 = white), a* (greenness to redness: negative to positive), and b* (blueness to yellowness: negative to positive).

3.2.5. Statistical Analysis

Experimental data were subjected to analysis of variance (ANOVA) with Duncan's multiple range test at a significance level of $p \leq 0.05$ using Statistica software (version 8.0, StatSoft Inc., Tulsa, OK, USA). All analyses were performed at least in duplicate. The results are expressed as mean values ± standard deviation (SD). Pearson correlation coefficients were additionally calculated in Microsoft Excel for Microsoft 365 (Version 2403) using selected data series.

4. Conclusions

Snacks containing a mixture of by-products (apple pomace, red potato pulp, sugar beet pulp) can have a very beneficial dietary effect due to their high content of polyphenols, phenolic acids, anthocyanins, dietary fiber, tocopherols, especially α-tocopherol, and phytosterols, in particular: β-sitosterol, stigmasterol, and campesterol. In the production of plant-based snacks, a mixture in a 1:1:1 ratio of apple pomace, potato pulp, and beet pulp was used. The snacks were characterized by powerful antioxidant activities (measured with ABTS and DPPH) and desirable physical and functional properties (up to 20% level of addition). In addition, the above-mentioned by-product mixture increased the protein and ash content and contributed to a significant reduction in starch content in snacks. It can be suggested that snacks with a 20% share of by-product mixtures are recommended on an industrial scale due to their high contents of protein, dietary fiber, ash, antioxidants, tocopherols (especially α-tocopherol), phytosterols (β-sitosterol, stigmasterol, and campesterol), and acceptable physical and functional properties.

Supplementary Materials: The following supporting information can be downloaded at: https://www.mdpi.com/article/10.3390/molecules31061037/s1.

https://doi.org/10.3390/molecules31061037

Author Contributions: Conceptualization, D.G. and R.Z.; methodology, D.G. and R.Z.; validation, D.G. and R.Z.; formal analysis, D.G., R.Z., A.W.-G., R.S. and T.Z.; investigation, D.G., R.Z., A.W.-G., R.S. and T.Z.; resources, D.G.; data curation, D.G. and R.Z.; writing—original draft preparation, D.G. and R.Z.; writing—review and editing, D.G. and R.Z.; visualization, D.G. and R.Z.; supervision, D.G. and R.Z.; project administration, D.G. and R.Z.; funding acquisition, D.G. All authors have read and agreed to the published version of the manuscript.

Funding: This research received no external funding.

Institutional Review Board Statement: Not applicable.

Informed Consent Statement: Not applicable.

Data Availability Statement: Data is contained within the article.

Conflicts of Interest: The authors declare no conflicts of interest.

Abbreviations

The following abbreviations are used in this manuscript:

ABTS	2,2′-azino-bis (3-ethylbenzothiazoline-6-sulfonic acid)
DPPH	2,2-diphenyl-1-picrylhydrazyl (DPPH)
ER	expansion ratio
WEWAI	whole extrudate water absorption index
GEWAI	ground extrudate water absorption index
WSI	water solubility index
TDF	total dietary fiber
SDF	soluble dietary fiber
IDF	insoluble dietary fiber

References

1. Clemens, R.; Kranz, S.; Mobley, A.R.; Nicklas, T.A.; Raimondi, M.P.; Rodriguez, J.C.; Slavin, J.L.; Warshaw, H. Filling America's Fiber Intake Gap: Summary of a Roundtable to Probe Realistic Solutions with a Focus on Grain-Based Foods. *J. Nutr.* **2012**, *142*, 1390S–1401S. [CrossRef]
2. Garcia-Amezquita, L.E.; Tejada-Ortigoza, V.; Serna-Saldivar, S.O.; Welti-Chanes, J. Dietary Fiber Concentrates from Fruit and Vegetable By-products: Processing, Modification, and Application as Functional Ingredients. *Food Bioprocess Technol.* **2018**, *11*, 1439–1463. [CrossRef]
3. Monteiro, C.A.; Cannon, G.; Moubarac, J.-C.; Levy, R.B.; Louzada, M.L.C.; Jaime, P.C. The UN Decade of Nutrition, the NOVA food classification and the trouble with ultra-processing. *Public Health Nutr.* **2017**, *21*, 5–17. [CrossRef] [PubMed]
4. Cetkovic, G.; Savatovic, S.; Canadanovic-Brunet, J.; Cetojevic-Simin, D.; Djilas, S.; Tumbas, V.; Skerget, M. Apple pomace: Antiradical activity and antiproliferative action in HeLa and HT-29 human tumor cell lines. *J. BUON* **2011**, *16*, 147–153.
5. Balasuriya, N.; Rupasinghe, H.V. Antihypertensive properties of flavonoid-rich apple peel extract. *Food Chem.* **2012**, *135*, 2320–2325. [CrossRef]
6. Makarova, E.; Górnaś, P.; Konrade, I.; Tirzite, D.; Cirule, H.; Gulbe, A.; Pugajeva, I.; Seglina, D.; Dambrova, M. Acute anti-hyperglycaemic effects of an unripe apple preparation containing phlorizin in healthy volunteers: A preliminary study. *J. Sci. Food Agric.* **2014**, *95*, 560–568. [CrossRef] [PubMed]
7. Rudrapal, M.; Rakshit, G.; Singh, R.P.; Garse, S.; Khan, J.; Chakraborty, S. Dietary Polyphenols: Review on Chemistry/Sources, Bioavailability/Metabolism, Antioxidant Effects, and Their Role in Disease Management. *Antioxidants* **2024**, *13*, 429. [CrossRef] [PubMed]
8. Messinese, E.; Pitirollo, O.; Grimaldi, M.; Milanese, D.; Sciancalepore, C.; Cavazza, A. By-Products as Sustainable Source of Bioactive Compounds for Potential Application in the Field of Food and New Materials for Packaging Development. *Food Bioprocess Technol.* **2023**, *17*, 606–627. [CrossRef]
9. Shirahigue, L.D.; Ceccato-Antonini, S.R. Agro-industrial wastes as sources of bioactive compounds for food and fermentation industries. *Ciência Rural* **2020**, *50*, e20190857. [CrossRef]

10. Saleh, M.; Amro, L.; Barakat, H.; Baker, R.; Reyash, A.A.; Amro, R.; Qasem, J. Fruit By-Product Processing and Bioactive Compounds. *J. Food Qual.* **2021**, *2021*, 5513358. [CrossRef]
11. Brown, C.; Culley, D.; Yang, C.-P.; Durst, R.; Wrolstad, R. Variation of Anthocyanin and Carotenoid Contents and Associated Antioxidant Values in Potato Breeding Lines. *J. Am. Soc. Hortic. Sci.* **2005**, *130*, 174–180. [CrossRef]
12. Brown, C.R. Breeding for Phytonutrient Enhancement of Potato. *Am. J. Potato Res.* **2008**, *85*, 298–307. [CrossRef]
13. Rytel, E.; Tajner-Czopek, A.; Kita, A.; Aniołowska, M.; Kucharska, A.; Sokół-Łętowska, A.; Hamouz, K. Content of polyphenols in coloured and yellow fleshed potatoes during dices processing. *Food Chem.* **2014**, *161*, 224–229. [CrossRef]
14. Górnaś, P.; Mišina, I.; Olšteine, A.; Krasnova, I.; Pugajeva, I.; Lācis, G.; Siger, A.; Michalak, M.; Soliven, A.; Segliņa, D. Phenolic compounds in different fruit parts of crab apple: Dihydrochalcones as promising quality markers of industrial apple pomace by-products. *Ind. Crops Prod.* **2015**, *74*, 607–612. [CrossRef]
15. Lv, Y. Triterpenes and Phenolic Compounds in Apple Fruit (*Malus domestica* Borkh.). Ph.D. Thesis, Swedish University of Agricultural Sciences, Uppsala, Sweden, 2016.
16. Sudha, M.L.; Dharmesh, S.M.; Pynam, H.; Bhimangouder, S.V.; Eipson, S.W.; Somasundaram, R.; Nanjarajurs, S.M. Antioxidant and cyto/DNA protective properties of apple pomace enriched bakery products. *J. Food Sci. Technol.* **2016**, *53*, 1909–1918. [CrossRef] [PubMed]
17. Rombouts, F.M.; Thibault, J.-F. Feruloylated pectic substances from sugar-beet pulp. *Carbohydr. Res.* **1986**, *154*, 177–187. [CrossRef]
18. Lemos, M.A.; Aliyu, M.M.; Hungerford, G. Influence of cooking on the levels of bioactive compounds in Purple Majesty potato observed via chemical and spectroscopic means. *Food Chem.* **2015**, *173*, 462–467. [CrossRef]
19. Navarre, D.A.; Brown, C.R.; Sathuvalli, V.R. Potato Vitamins, Minerals and Phytonutrients from a Plant Biology Perspective. *Am. J. Potato Res.* **2019**, *96*, 111–126. [CrossRef]
20. Xu, J.; Li, Y.; Kaur, L.; Singh, J.; Zeng, F. Functional Food Based on Potato. *Foods* **2023**, *12*, 2145. [CrossRef]
21. Akyol, H.; Riciputi, Y.; Capanoglu, E.; Caboni, M.; Verardo, V. Phenolic Compounds in the Potato and Its Byproducts: An Overview. *Int. J. Mol. Sci.* **2016**, *17*, 835. [CrossRef]
22. Mäder, J.; Rawel, H.; Kroh, L.W. Composition of Phenolic Compounds and Glycoalkaloids α-Solanine and α-Chaconine during Commercial Potato Processing. *J. Agric. Food Chem.* **2009**, *57*, 6292–6297. [CrossRef]
23. Lachman, J.; Hamouz, K.; Šulc, M.; Orsák, M.; Pivec, V.; Hejtmánková, A.; Dvořák, P.; Čepl, J. Cultivar differences of total anthocyanins and anthocyanidins in red and purple-fleshed potatoes and their relation to antioxidant activity. *Food Chem.* **2009**, *114*, 836–843. [CrossRef]
24. Opyd, P.; Jurgoński, A.; Juśkiewicz, J.; Milala, J.; Zduńczyk, Z.; Król, B. Nutritional and Health-Related Effects of a Diet Containing Apple Seed Meal in Rats: The Case of Amygdalin. *Nutrients* **2017**, *9*, 1091. [CrossRef] [PubMed]
25. Vrhovsek, U.; Rigo, A.; Tonon, D.; Mattivi, F. Quantitation of Polyphenols in Different Apple Varieties. *J. Agric. Food Chem.* **2004**, *52*, 6532–6538. [CrossRef]
26. Kołodziejczyk, K.; Kosmala, M.; Milala, J.; Sójka, M.; Uczciwek, M.; Król, B.; Markowski, J.; Renard, C.M.G.C. Characterisation of the chemical composition of scab-resistant apple pomaces. *J. Hortic. Sci. Biotechnol.* **2009**, *84*, 89–95. [CrossRef]
27. Sato, M.F.; Vieira, R.G.; Zardo, D.M.; Falcão, L.D.; Nogueira, A.; Wosiacki, G. Apple pomace from eleven cultivars: An approach to identify sources of bioactive compounds. *Acta Scientiarum. Agron.* **2010**, *32*, 29–35.
28. Tuyet Lam, T.B.; Iiyama, K.; Stone, B.A. Cinnamic acid bridges between cell wall polymers in wheat and phalaris internodes. *Phytochemistry* **1992**, *31*, 1179–1183. [CrossRef]
29. Lempereur, I.; Rouau, X.; Abecassis, J. Genetic and Agronomic Variation in Arabinoxylan and Ferulic Acid Contents of Durum Wheat (*Triticum durum* L.) Grain and Its Milling Fractions. *J. Cereal Sci.* **1997**, *25*, 103–110. [CrossRef]
30. Yuan, Y.; Xiang, J.; Zheng, B.; Sun, J.; Luo, D.; Li, P.; Fan, J. Diversity of phenolics including hydroxycinnamic acid amide derivatives, phenolic acids contribute to antioxidant properties of proso millet. *LWT* **2022**, *154*, 112611. [CrossRef]
31. Liu, X.; Le Bourvellec, C.; Renard, C.M.G.C. Interactions between cell wall polysaccharides and polyphenols: Effect of molecular internal structure. *Compr. Rev. Food Sci. Food Saf.* **2020**, *19*, 3574–3617. [CrossRef] [PubMed]
32. Riaz, M.N. *Extruders in Food Applications*; CRC Press: Boca Raton, FL, USA, 2000.
33. Gumul, D.; Ziobro, R.; Korus, J.; Kruczek, M. Apple pomace as a source of bioactive polyphenol compounds in gluten-free breads. *Antioxidants* **2021**, *10*, 807. [CrossRef]
34. Tinker, L.; Schneeman, B.; Davis, P.; Gallaher, D.; Waggoner, C. Consumption of prunes as a source of dietary fiber in men with mild hypercholesterolemia. *Am. J. Clin. Nutr.* **1991**, *53*, 1259–1265. [CrossRef] [PubMed]
35. Cassidy, A.; Bingham, S.; Cummings, J. Starch intake and colorectal cancer risk: An international comparison. *Br. J. Cancer* **1994**, *69*, 937–942. [CrossRef]
36. Pęksa, A.; Kita, A.; Kułakowska, K.; Aniołowska, M.; Hamouz, K.; Nemś, A. The quality of protein of coloured fleshed potatoes. *Food Chem.* **2013**, *141*, 2960–2966. [CrossRef] [PubMed]

 https://doi.org/10.3390/molecules31061037

37. Górnaś, P.; Soliven, A.; Seglina, D. Seed oils recovered from industrial fruit by-products are a rich source of tocopherols and tocotrienols: Rapid separation of α/β/γ/δ homologues by RP-HPLC/FLD. *Eur. J. Lipid Sci. Technol.* **2015**, *117*, 773–777. [CrossRef]
38. Akšić, M.F.; Lazarević, K.; Šegan, S.; Natić, M.; Tosti, T.; Ćirić, I.; Meland, M. Assessing the Fatty Acid, Carotenoid, and Tocopherol Compositions of Seeds from Apple Cultivars (*Malus domestica* Borkh.) Grown in Norway. *Foods* **2021**, *10*, 1956. [CrossRef]
39. Gumul, D.; Kruczek, M. Gluten-Free Snacks with Micronized and Freeze-Dried Red Potatoes: Nutritional and Pro-Health Values. *Molecules* **2025**, *30*, 1957. [CrossRef]
40. Han, J.-H.; Yang, Y.-X.; Feng, M.-Y. Contents of phytosterols in vegetables and fruits commonly consumed in China. *Biomed. Environ. Sci.* **2008**, *21*, 449–453. [CrossRef]
41. Hayyat, F.S.; Allayie, S.A.; Malik, J.A.; Dar, S.A. The Role of Dietary Fiber in Promoting Health: A Review of Choice and Outcomes. In *Food Process Engineering and Technology*; Springer Nature: Singapore, 2023; pp. 493–508.
42. Riccardi, B. Role of Dietary Fiber in Maintaining Well-Being, Critical Review. *Am. J. Biomed. Sci. Res.* **2024**, *25*, 425–434. [CrossRef]
43. Pieszka, M.; Gogol, P.; Pietras, M.; Pieszka, M. Valuable Components of Dried Pomaces of Chokeberry, Black Currant, Strawberry, Apple and Carrot as a Source of Natural Antioxidants and Nutraceuticals in the Animal Diet. *Ann. Anim. Sci.* **2015**, *15*, 475–491. [CrossRef]
44. Görgülü, A. Elevating sugar beet by-products into healthier, natural, and functional ingredients. *Explor. Foods Foodomics* **2025**, *3*, 101074. [CrossRef]
45. Gumul, D.; Ziobro, R.; Zięba, T.; Rój, E. The influence of addition of defatted blackcurrant seeds on pro-health constituents and texture of cereal extrudates: Cereal extrudates with defatted seeds. *J. Food Qual.* **2011**, *34*, 395–402. [CrossRef]
46. Reis, S.F.; Rai, D.K.; Abu-Ghannam, N. Apple pomace as a potential ingredient for the development of new functional foods. *Int. J. Food Sci. Technol.* **2014**, *49*, 1743–1750. [CrossRef]
47. Korkerd, S.; Wanlapa, S.; Puttanlek, C.; Uttapap, D.; Rungsardthong, V. Expansion and functional properties of extruded snacks enriched with nutrition sources from food processing by-products. *J. Food Sci. Technol.* **2015**, *53*, 561–570. [CrossRef]
48. Oniszczuk, A.; Olech, M.; Oniszczuk, T.; Wojtunik-Kulesza, K.; Wójtowicz, A. Extraction methods, LC-ESI-MS/MS analysis of phenolic compounds and antiradical properties of functional food enriched with elderberry flowers or fruits. *Arab. J. Chem.* **2019**, *12*, 4719–4730. [CrossRef]
49. Gumul, D.; Areczuk, A.; Ziobro, R.; Ivanišová, E.; Zięba, T. The influence of freeze-dried red and purple potatoes on content of bioactive polyphenolic compounds and quality properties of extruded maize snacks. *Qual. Assur. Saf. Crops Foods* **2018**, *10*, 51–60. [CrossRef]
50. Drożdż, W.; Boruczkowska, H.; Boruczkowski, T.; Tomaszewska-Ciosk, E.; Zdybel, E. Use of blackcurrant and chokeberry press residue in snack products. *Pol. J. Chem. Technol.* **2019**, *21*, 13–19. [CrossRef]
51. Bekele, E.K.; Nosworthy, M.G.; Tyler, R.T.; Henry, C.J. Antioxidant capacity and total phenolics content of direct-expanded chickpea–sorghum snacks. *J. Food Process. Preserv.* **2021**, *45*, e15439. [CrossRef]
52. Yagci, S.; Calıskan, R.; Gunes, Z.S.; Capanoglu, E.; Tomas, M. Impact of tomato pomace powder added to extruded snacks on the in vitro gastrointestinal behaviour and stability of bioactive compounds. *Food Chem.* **2022**, *368*, 130847. [CrossRef] [PubMed]
53. Promsakha na Sakon Nakhon, P.; Jangchud, K.; Jangchud, A.; Charunuch, C. Optimization of pumpkin and feed moisture content to produce healthy pumpkin-germinated brown rice extruded snacks. *Agric. Nat. Resour.* **2018**, *52*, 550–556. [CrossRef]
54. Arribas, C.; Cabellos, B.; Cuadrado, C.; Guillamón, E.; Pedrosa, M.M. Bioactive Compounds, Antioxidant Activity, and Sensory Analysis of Rice-Based Extruded Snacks-like Fortified with Bean and Carob Fruit Flours. *Foods* **2019**, *8*, 381. [CrossRef]
55. Jung, J.; Gim, S.Y.; Lee, C.; Kim, M.; Lee, J. Stability of tocopherol homologs in soybean, corn, canola, and olive oils under different moisture contents at 25 °C. *Eur. J. Lipid Sci. Technol.* **2016**, *119*, 1600157. [CrossRef]
56. Shahidi, F.; Ho, C.-T.; Van Chuyen, N. *Process-Induced Chemical Changes in Food*; Springer Science & Business Media: New York, NY, USA, 1998.
57. Zielinski, H.; Kozlowska, H.; Lewczuk, B. Bioactive compounds in the cereal grains before and after hydrothermal processing. *Innov. Food Sci. Emerg. Technol.* **2001**, *2*, 159–169. [CrossRef]
58. Tiwari, U.; Cummins, E. Nutritional importance and effect of processing on tocols in cereals. *Trends Food Sci. Technol.* **2009**, *20*, 511–520. [CrossRef]
59. Shin, T.-S.; Godber Samuel, J.; Martin Daniel, E.; Wells John, H. Hydrolytic stability and changes in e vitamers and oryzanol of extruded rice bran during storage. *J. Food Sci.* **1997**, *62*, 704–728. [CrossRef]
60. Lee, W.; Yeo, Y.; Oh, S.; Cho, K.-S.; Park, Y.-E.; Park, S.K.; Lee, S.M.; Cho, H.S.; Park, S.-Y. Compositional analyses of diverse phytochemicals and polar metabolites from different-colored potato (*Solanum tubersum* L.) tubers. *Food Sci. Biotechnol.* **2017**, *26*, 1379–1389. [CrossRef]
61. Schmitt, P.; Gaspar, T.; Hagège, D. Sterol composition of normal and habituated sugarbeet callus (*Beta vulgaris* L. Altissima). *In Vitro Cell. Dev. Biol. Plant* **1994**, *30*, 1–3. [CrossRef]

 https://doi.org/10.3390/molecules31061037

62. Pavani, M.; Singha, P.; Rajamanickam, D.T.; Singh, S.K. Impact of extrusion processing on bioactive compound enriched plant-based extrudates: A comprehensive study and optimization using RSM and ANN-GA. *Future Foods* **2024**, *9*, 100286. [CrossRef]
63. Corrêa, R.C.; Peralta, R.M.; Bracht, A.; Ferreira, I.C. The emerging use of mycosterols in food industry along with the current trend of extended use of bioactive phytosterols. *Trends Food Sci. Technol.* **2017**, *67*, 19–35. [CrossRef]
64. Pavani, M.; Singha, P.; Dash, D.R.; Asaithambi, N.; Singh, S.K. Novel encapsulation approaches for phytosterols and their importance in food products: A review. *J. Food Process Eng.* **2022**, *45*, e14041. [CrossRef]
65. Gumul, D.; Ziobro, R.; Zięba, T.; Rój, E. Physicochemical characteristics of cereal extrudates with different levels of defatted blackcurrant seeds. *J. Food Qual.* **2013**, *36*, 385–393. [CrossRef]
66. Navarro Cortez, R.O.; Gómez-Aldapa, C.A.; Aguilar-Palazuelos, E.; Delgado-Licon, E.; Castro Rosas, J.; Hernández-Ávila, J.; Solís-Soto, A.; Ochoa-Martínez, L.A.; Medrano-Roldán, H. Blue corn (*Zea mays* L.) with added orange (*Citrus sinensis*) fruit bagasse: Novel ingredients for extruded snacks. *CyTA J. Food* **2016**, *14*, 349–358. [CrossRef]
67. Kaisangsri, N.; Kowalski, R.J.; Wijesekara, I.; Kerdchoechuen, O.; Laohakunjit, N.; Ganjyal, G.M. Carrot pomace enhances the expansion and nutritional quality of corn starch extrudates. *LWT Food Sci. Technol.* **2016**, *68*, 391–399. [CrossRef]
68. Aktepe, N.; Baran, A.; Keskin, C. Cytotoxic and antimicrobial analysis of biosynthesized selenium nanoparticles from *Solanum tuberosum* peels. *Int. J. Agric. Environ. Food Sci.* **2025**, *9*, 559–569. [CrossRef]
69. Tanur Erkoyuncu, M. Modulation of phenolic profiles and antioxidant capacity by culture medium and tissue type in echinacea purpurea and echinacea pallida. *Int. J. Agric. Environ. Food Sci.* **2025**, *9*, 768–777. [CrossRef]
70. AOAC International. *Official Methods of Analysis*; AOAC International: Gaithersburg, MD, USA, 2006.
71. American Association of Cereal Chemists; Approved Methods Committee. *Approved Methods of the American Association of Cereal Chemists*; American Association of Cereal Chemists: St. Paul, MN, USA, 2000.
72. International Association for Cereal Science and Technology. *ICC Standard Methods of the International Association for Cereal Chemistry*; ICC: Vienna, Austria, 1976.
73. Singleton, V.L.; Orthofer, R.; Lamuela-Raventós, R.M. Analysis of total phenols and other oxidation substrates and antioxidants by means of folin-ciocalteu reagent. In *Oxidants and Antioxidants Part A*; Elsevier: Amsterdam, The Netherlands, 1999; pp. 152–178.
74. Mazza, G.; Fukumoto, L.; Delaquis, P.; Girard, B.; Ewert, B. Anthocyanins, Phenolics, and Color of Cabernet Franc, Merlot, and Pinot Noir Wines from British Columbia. *J. Agric. Food Chem.* **1999**, *47*, 4009–4017. [CrossRef] [PubMed]
75. Oomah, B.D.; Cardador-Martínez, A.; Loarca-Piña, G. Phenolics and antioxidative activities in common beans (*Phaseolus vulgaris* L.). *J. Sci. Food Agric.* **2005**, *85*, 935–942. [CrossRef]
76. El Hariri, B.; Sallé, G.; Andary, C. Involvement of flavonoids in the resistance of two poplar cultivars to mistletoe (*Viscum album* L.). *Protoplasma* **1991**, *162*, 20–26. [CrossRef]
77. Re, R.; Pellegrini, N.; Proteggente, A.; Pannala, A.; Yang, M.; Rice-Evans, C. Antioxidant activity applying an improved ABTS radical cation decolorization assay. *Free Radic. Biol. Med.* **1999**, *26*, 1231–1237. [CrossRef]
78. Brand-Williams, W.; Cuvelier, M.E.; Berset, C. Use of a free radical method to evaluate antioxidant activity. *LWT—Food Sci. Technol.* **1995**, *28*, 25–30. [CrossRef]
79. Hussain, N.; Jabeen, Z.; Li, Y.-L.; Chen, M.-X.; Li, Z.-L.; Guo, W.-L.; Shamsi, I.H.; Chen, X.-Y.; Jiang, L.-X. Detection of tocopherol in oilseed rape (*Brassica napus* L.) using gas chromatography with flame ionization detector. *J. Integr. Agric.* **2013**, *12*, 803–814. [CrossRef]
80. Oracz, J.; Nebesny, E.; Żyżelewicz, D. Effect of roasting conditions on the fat, tocopherol, and phytosterol content and antioxidant capacity of the lipid fraction from cocoa beans of different *Theobroma cacao* L. cultivars. *Eur. J. Lipid Sci. Technol.* **2014**, *116*, 1002–1014. [CrossRef]
81. Zhang, R.; Shen, W.; Wei, X.; Zhang, F.; Shen, C.; Wu, B.; Zhao, Z.; Liu, H.; Deng, X. Simultaneous determination of tocopherols and tocotrienols in vegetable oils by GC-MS. *Anal. Methods* **2016**, *8*, 7341–7346. [CrossRef]
82. Sokhey, A.S.; Kollengode, A.N.; Hanna, M.A. Screw configuration effects on corn starch expansion during extrusion. *J. Food Sci.* **1994**, *59*, 895–896. [CrossRef]
83. Stojceska, V.; Ainsworth, P.; Plunkett, A.; İbanoğlu, S. The recycling of brewer's processing by-product into ready-to-eat snacks using extrusion technology. *J. Cereal Sci.* **2008**, *47*, 469–479. [CrossRef]
84. International Commission on Illumination. *CIE 015: 2018 Colorimetry*; International Commission on Illumination: Vienna, Austria, 2018.

 https://doi.org/10.3390/molecules31061037

molecules

MDPI

Article

Sechium edule var. *nigrum spinosum* (Chayote) Increases the mRNA Expression of Genes Encoding Sirtuins in Older Adults with Type 2 Diabetes Mellitus

Graciela Gavia-García [1], David Hernández-Álvarez [1], Taide Laurita Arista-Ugalde [1], Itzen Aguiñiga-Sánchez [2], Edelmiro Santiago-Osorio [2], Jorge Cadena-Iñiguez [3], Juana Rosado-Pérez [1,*] and Víctor Manuel Mendoza-Núñez [1,*]

1 Research Unit on Gerontology, FES Zaragoza, National Autonomous University of Mexico, Mexico City 09230, Mexico; graciela_gavia_garcia@comunidad.unam.mx (G.G.-G.); hereda2912@gmail.com (D.H.-Á.); tdlarista@comunidad.unam.mx (T.L.A.-U.)

2 Hematopoiesis and Leukemia Laboratory, Research Unit on Cell Differentiation and Cancer, FES Zaragoza, National Autonomous University of Mexico, Mexico City 09230, Mexico; itzen.aguiniga@zaragoza.unam.mx (I.A.-S.); edelmiro@unam.mx (E.S.-O.)

3 Postgraduate College, Campus San Luis Potosi, Salinas de Hidalgo, San Luis Potosi 78600, Mexico; jocadena@colpos.mx

* Correspondence: juanarosadoperez@comunidad.unam.mx (J.R.-P.); mendovic@unam.mx (V.M.M.-N.); Tel.: +52-55-5623-0721 (J.R.-P. & V.M.M.-N.); Fax: +52-55-5773-6330 (J.R.-P. & V.M.M.-N.)

Abstract

Consumption of *Sechium edule* var. *nigrum spinosum* has antioxidant and hypoglycemic effects. Regarding the former, certain signaling pathways that influence these effects have already been proposed; however, the underlying molecular mechanisms of the hypoglycemic effects remain unknown. It has been recognized that the sirtuin-mediated signaling cascade responds to various stressors, such as oxidative stress, and regulates glucose metabolism. Therefore, it would be of great interest to determine whether there is a link between these two properties and whether it is mediated by sirtuins. Hence, the present study aimed to evaluate the effect of *Sechium edule* on the gene expression of the sirtuin family (SIRT1-SIRT6) in individuals with type 2 diabetes mellitus (T2DM). A quasi-experimental study was conducted with a convenience sample of 26 older adults diagnosed with T2DM, divided into a (i) placebo group (PG; $n = 14$) and (ii) experimental group (EG; $n = 12$). Clinical, biochemical, and anthropometric measurements were performed, and total oxidant/antioxidant capacity (TOS/TAS) and mRNA expression of genes encoding sirtuins were determined. All parameters were measured at baseline, three months, and six months after the intervention. In the EG, the expression levels of SIRT1, SIRT3, SIRT5, and SIRT6 increased by 52%, 69%, 62%, and 69%, respectively, six months after treatment. A 50% decrease in TOS and a 44% increase in TAS were also observed. Our findings suggest that the bioactive components of *Sechium edule* enhance sirtuin expression and exhibit antioxidant effects in older adults with T2DM.

Keywords: *Sechium edule*; type 2 diabetes mellitus; aging; sirtuins; mRNA

1. Introduction

Type 2 diabetes mellitus (T2DM) is a metabolic disorder that frequently occurs in older adults. Its pathophysiology has been linked to processes such as oxidative stress (OS) and alterations in markers associated with the hallmarks of aging, including chronic inflammation, telomere shortening, mitochondrial dysfunction, and genomic instability. Its high frequency and the debilitating nature of its complications pose challenges for healthcare systems worldwide [1–9].

Hence, there is a need to explore new therapeutic targets, with a focus on the pathophysiological mechanisms underlying T2DM, such as the activation of the family of enzymes structurally related to silent mating-type regulatory protein 2 (Sir2), better known as sirtuins (SIRTs) or longevity proteins. These enzymes are involved in DNA damage repair, stabilization of telomere shortening, regulation of the inflammatory process and glycemic metabolism, control of mitochondrial function, and protection against OS.

Among the nutritional interventions that improve glycemic response and have direct effects on sirtuins are those that focus on the use of polyphenols, bioactive compounds abundant in various fruits, such as berries, blueberries, raspberries, and grapefruit; and, as recently demonstrated in *Sechium edule* var. *nigrum spinosum* (which contains flavonoids with antioxidant activity and direct effects on the activation of certain sirtuins such as myricetin, naringenin, and quercetin), whose presence has been evidenced by our research group [10–12].

Previous studies have shown that consuming *Sechium edule* powder capsules has hypoglycemic, anti-inflammatory, hypotensive, and antioxidant effects in older adults with metabolic syndrome (MetS), with impacts at the molecular level through modulating the expression of mRNA of genes encoding enzymes with antioxidant functions and nuclear factor erythroid 2-related factor 2 (Nrf2) [13–18].

Sirtuins have been shown to deacetylate Nrf2, thereby promoting its activation and, consequently, the activation of antioxidant protection mechanisms, thereby reducing reactive oxygen species (ROS) levels [19]. Therefore, the objective of this study was to evaluate whether *Sechium edule* supplementation activates mRNA expression of genes encoding sirtuins in older adults with T2DM.

2. Results

Table 1 presents the anthropometric and clinical parameters, along with participants' ages. No statistically significant differences were found in these measurements.

Table 1. Anthropometric characteristics and blood pressure by study group.

Parameter	PG $n = 14$	EG $n = 12$	p-Value
Age (years)	67.1 ± 2.0	65.7 ± 4.8	
Weight (kg)			
Baseline	63.4 ± 2.5	69.1 ± 2.6	
Three months	63.7 ± 2.6	69.3 ± 2.7	0.16
Six months	64.4 ± 2.8	69.2 ± 2.9	0.25
BMI (kg/m^2)			
Baseline	26.0 ± 3.8	30.0 ± 1.6	
Three months	26.3 ± 3.6	30.1 ± 2.2	0.06
Six months	26.2 ± 3.6	30.1 ± 2.7	0.06
Waist circumference (cm)			
Baseline	95.7 ± 2.1	103.4 ± 2.1	
Three months	95.6 ± 2.3	102.6 ± 2.3	0.06
Six months	95.8 ± 2.2	101.8 ± 2.2	0.07

https://doi.org/10.3390/molecules31071182

Table 1. *Cont.*

Parameter	PG n = 14	EG n = 12	p-Value
SBP (mmHg)			
Baseline	127.7 ± 4.3	132.2 ± 4.3	
Three months	126.7 ± 6.3	126.5 ± 6.3	0.76
Six months	127.2 ± 6.3	129.4 ± 6.3	0.80
DBP (mmHg)			
Baseline	77.7 ± 2.5	82.7 ± 2.5	
Three months	79.7 ± 3.0	73.3 ± 3.0	0.15
Six months	80.0 ± 3.6	77.2 ± 3.6	0.59

Abbreviations: PG, placebo group; EG, experimental group; BMI, body mass index; SBP, systolic blood pressure; DBP, diastolic blood pressure. Data are expressed as means ± standard deviations. ANOVA of repeated measures test, significance level: 95%. Baseline vs. 3-month and baseline vs. 6-month inter-group p-values are shown.

Glucose, cholesterol, triglyceride, HDL, uric acid, urea, albumin, and %HbA1c levels did not show statistically significant changes in either group at 3 or 6 months post-treatment (Table 2).

Table 2. Biochemical parameters by study group.

Parameter	PG n = 14	EG n = 12	p-Value
Glucose (mg/dL)			
Baseline	182 ± 48	170 ± 42	
Three months	192 ± 75	164 ± 77	0.82
Six months	181 ± 38	170 ± 46	0.72
Cholesterol (mg/dL)			
Baseline	196 ± 49	183 ± 27	
Three months	200 ± 53	196 ± 18	0.86
Six months	161 ± 47	160 ± 45	0.99
Triglycerides (mg/dL)			
Baseline	164 ± 93	153 ± 86	
Three months	163 ± 54	151 ± 42	0.30
Six months	169 ± 67	139 ± 60	0.69
HDL-c (mg/dL)			
Baseline	47 ± 10	43 ± 10	
Three months	41 ± 8	41 ± 9	0.63
Six months	45 ± 10	44 ± 10	0.57
Uric acid (mg/dL)			
Baseline	4.84 ± 0.6	4.37 ± 0.6	
Three months	4.03 ± 1.2	4.63 ± 0.7	0.63
Six months	4.60 ± 1.5	4.87 ± 1.1	0.87
Urea (mg/dL)			
Baseline	32 ± 9.5	30 ± 7.8	
Three months	34 ± 9.0	30 ± 4.2	0.17
Six months	31 ± 9.0	27 ± 4.3	0.28

https://doi.org/10.3390/molecules31071182

Table 2. *Cont.*

Parameter	PG $n = 14$	EG $n = 12$	p-Value
Albumin (g/dL)			
Baseline	3.9 ± 0.19	3.67 ± 0.14	
Three months	4.0 ± 0.28	3.94 ± 0.15	0.21
Six months	4.1 ± 0.09	4.06 ± 0.21	0.91
HbA1c (%)			
Baseline	6.98 ± 0.80	8.72 ± 1.47	
Three months	7.02 ± 0.83	9.25 ± 1.30	0.45
Six months	6.86 ± 0.76	7.46 ± 1.93	0.97

Abbreviations: PG, placebo group; EG, experimental group; HDL-c; high-density lipoprotein cholesterol; HbA1c, glycosylated hemoglobin. Data are expressed as means ± standard deviations. ANOVA of repeated measures test, significance level: 95%. Baseline vs. 3-month and baseline vs. 6-month inter-group p-values are shown.

Regarding total oxidative capacity (Table 3), a statistically significant decrease was observed at six months post-treatment in the EG group (6.4 ± 2.9 (baseline) vs. 3.2 ± 2.1 (post)), which also showed a decrease in OSI (6.6 ± 3.3 (baseline) vs. 2.0 ± 1.4 (post)). Conversely, total antioxidant capacity increased (0.9 ± 0.3 (baseline) vs. 1.3 ± 0.2 (post)) in the same group.

Table 3. Oxidizing/antioxidant capacity and oxidative stress index by study group.

Parameter	PG $n = 14$	EG $n = 12$	p-Value
TOS (μmol H_2O_2 Equiv./L)			
Baseline	5.3 ± 2.3	6.4 ± 2.9	
Three months	4.9 ± 2.9	5.9 ± 3.0	0.63
Six months	4.7 ± 3.8	3.2 ± 2.1 *	0.03
TAS (mmol/L)			
Baseline	1.0 ± 0.3	0.9 ± 0.3	
Three months	1.1 ± 0.1	1.0 ± 0.2	0.86
Six months	1.0 ± 0.11	1.3 ± 0.2 *	0.04
OSI			
Baseline	6.0 ± 3.4	6.6 ± 3.3	
Three months	5.4 ± 3.6	4.9 ± 2.1	0.07
Six months	5.1 ± 4.4	2.0 ± 1.4 *	0.01

Abbreviations: PG, placebo group; EG, experimental group; TOS, total oxidation status; TAS, total antioxidant capacity; OSI, oxidative stress index. * Data are expressed as means ± standard deviations. ANOVA of repeated measures test, significance level: 95%. Baseline vs. 3-month and baseline vs. 6-month inter-group p-values are shown.

Figure 1 shows the relative expression of mRNA from genes encoding proteins that regulate responses to metabolic or energy stress. SIRT1 (Figure 1A), SIRT3 (Figure 1C), SIRT5 (Figure 1E), and SIRT6 (Figure 1F) showed statistically higher levels at 6 months post-treatment in the EG group compared to the PG group (52%, 69%, 62%, and 69% of EG levels, respectively).

 https://doi.org/10.3390/molecules31071182

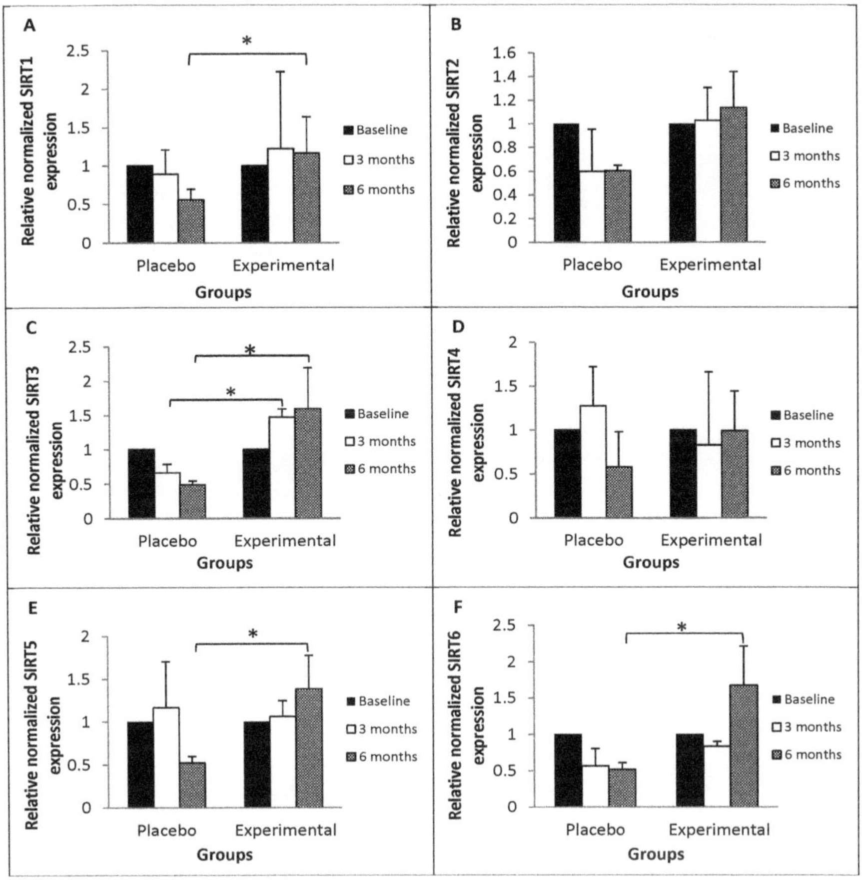

Figure 1. Relative mRNA expression of genes encoding proteins that coordinate the response to different types of metabolic or energy stress among the study subjects. Data are expressed as the mean ± standard deviation. Repeated measures ANOVA, significance level: 95%, $p < 0.05$. (**A**) SIRT1: sirtuin 1; (**B**) SIRT2: sirtuin 2; (**C**) SIRT3: sirtuin 3; (**D**) SIRT4: sirtuin 4; (**E**) SIRT5: sirtuin 5; (**F**) SIRT6: sirtuin 6. Relative mRNA expression levels were determined after normalization against β-actin. * Statistical significance between PG vs. EG at three- and six-month post-treatment intervals.

3. Discussion

Type 2 diabetes mellitus (T2DM) is a metabolic disorder characterized by chronic hyperglycemia secondary to defects in insulin secretion and/or action. Individuals with T2DM have a greater predisposition to developing pathologies such as cerebrovascular and cardiac diseases, as well as infections, among others [20,21]. It is estimated that by 2050, approximately 1.31 billion people will suffer from T2DM [22]; therefore, this pathology is and will continue to be a serious public health problem worldwide. Hence, there is a need to prevent its complications.

Sechium edule is a fruit that has been attributed to have hypotensive, anti-inflammatory, hypoglycemic, antioxidant, and even geroprotective properties when administered to older adults with metabolic syndrome (MetS) [17,21,23–25]. Therefore, it is possible that *Sechium edule* could be used as an adjunct in the treatment of T2DM in older adults [26,27].

https://doi.org/10.3390/molecules31071182

However, this investigation found no significant effect on clinical and biochemical markers in patients with T2DM. These findings can be explained by considering the specific differences between our population and those studied previously. It is important to note that the previous studies, although conducted in older adults, were performed in patients with MetS, who may present with hyperglycemia or glucose intolerance, but not with diabetes. Therefore, given the pathophysiology of T2DM, it is reasonable to assume that the effects of *Sechium edule* were insufficient to induce significant changes in this population. This proposal can be supported by the findings of a systematic review and meta-analysis showing that the hypoglycemic effect of *Sechium edule* is modest—an approximately 1% reduction in HbA1c levels [28], which is around 20 mg/dL, which may not be sufficient to generate a statistically significant change in patients with type 2 diabetes. However, studies suggest that any reduction in HbA1c decreases the risk of micro- and macrovascular complications and even death [29]. In this regard, we assume that, given the pathophysiology of T2DM, its natural history, and the development of metabolic memory, achieving significant changes in the identified markers is more complex. This explains why only some non-significant trends were observed in SBP, DBP, and HbA1c% levels.

On the other hand, it is interesting to observe the effect of *Sechium edule* on sirtuins and OS markers. Sirtuins are enzymes that belong to a family of seven members (SIRT1-SIRT7), with deacetylase (SIRT1-SIRT3, SIRT5-SIRT7) or ADP-ribosyltransferase (SIRT4 and SIRT6) activity, and are dependent on nicotinamide adenine dinucleotide (NAD+) and respond to various stressors, such as genotoxic and oxidative stressors. Accumulating evidence indicates that certain nutraceuticals present in a wide variety of fruits enhance sirtuin activity, leading to beneficial clinical outcomes in the treatment of cardiovascular diseases, arthritis, osteoporosis, dementia, and T2DM [30–33].

SIRT1 is a histone deacetylase that acts as a nutrient sensor. Its expression increases with caloric restriction and decreases with overfeeding. A decrease in SIRT1 expression leads to the recruitment or infiltration of macrophages into adipose tissue, resulting in histone hyperacetylation and, consequently, ectopic inflammatory expression. Conversely, its overexpression prevents this [34]. In human monocytes from patients with MetS, decreased SIRT1 expression levels have been associated with insulin resistance and atherosclerosis. Therefore, both glucotoxicity and lipotoxicity affect its expression [35]. In contrast, caloric restriction increases SIRT1 expression, which is associated with reduced inflammation and histological renal lesions in diabetic models. Consequently, it has been proposed as a promising therapy for preventing diabetic nephropathy with an anti-aging focus [36,37].

Along the same lines, it has been observed that certain nutraceuticals in *Sechium edule* enhance SIRT1 expression, thereby significantly improving mitochondrial function by mitigating OS and inflammation. Myricetin administration has been reported to promote mitochondrial biogenesis via SIRT1-mediated deacetylation of Peroxisome Proliferator-Activated Receptor Gamma Coactivator 1-alpha (PGC-1α) in experimental models [38].

It has also been observed that naringenin mitigates OS and inflammation and regulates mitochondrial function in human granulosa cells (KGN cells) by promoting SIRT1 expression [39]. Likewise, it has been reported that six months of resveratrol consumption increased SIRT1 levels by 50% in individuals with T2DM [30]. In the present study, we observed a 52% increase in SIRT1 mRNA expression in the EG group compared with the PG group, likely associated with the observed effects on oxidative markers [40].

It has been noted that SIRT1 activation increases the post-translational activity of Nrf2, a transcription factor that regulates antioxidant protection mechanisms [41]. Furthermore, *Sechium edule* has been shown to increase the mRNA levels of the genes encoding Nrf2 and antioxidant enzymes, such as superoxide dismutase (SOD) and catalase (CAT), in patients with

 https://doi.org/10.3390/molecules31071182

MetS [18]. This agrees with the results of the present study, where we found a 44% increase in TAS in the EG group at six months post-treatment. This could suggest that consumption of Sechium edule improves antioxidant capacity by increasing SIRT1 expression.

On the other hand, our results showed that *Sechium edule* consumption had no effect on SIRT2 and SIRT4 expression, probably because these enzymes maintain a delicate balance to maintain cellular homeostasis, so both extremes (overexpression or downregulation) can be dangerous. SIRT2 is inhibited in people with T2DM. This effect restores pancreatic β-cells' ability to enter the cell cycle and counteracts the decrease in the number of these cells [42]. For its part, SIRT4 overexpression leads to dyslipidemia, lipogenesis, and inhibition of insulin secretion [43]. At the same time, its absence increases insulin levels, leading to accelerated development of insulin resistance and OS in vivo models [44]. It can be seen that the effects of SIRT4 can vary, suggesting that its regulation is a controlled process in which the components of *Sechium edule* failed to exert any effect.

SIRT3, similar to SIRT1, regulates mitochondrial acetylation levels, protecting mitochondria from a wide range of damage, including oxidative injury [45,46]. Therefore, it is reasonable to assume that both sirtuins participate in similar protective mechanisms. It has been proposed that naringenin promotes mitochondrial biogenesis by reducing oxidative damage, thereby attenuating ischemia–reperfusion injury and cardiac damage via AMPK/SIRT3 signaling [47]. In cardiomyocytes from diabetic models, SIRT3 overexpression has been observed to attenuate hypertrophy and fibrosis and reduce ROS formation [48].

In the present study, we observed a 44% increase in TAS in the EG group, along with decreases in TOS and OSI (50% and 70%, respectively). This could indicate that *Sechium edule* supplementation promotes a balance between TOS/TAS levels, suggesting the maintenance of cellular homeostasis in these individuals and potentially leading to effective ROS clearance [49,50].

SIRT5, like SIRT3, promotes antioxidant defense mechanisms; its inactivation impairs NADPH production and reduces glutathione (GSH) synthesis. Therefore, the 62% increase in its expression in the EG group may be associated with the antioxidant effect observed in this group.

This behavior could be attributable to the presence of bioactive compounds in *Sechium edule*. One of them is quercetin, a flavonoid that promotes the desuccinylation of isocitrate dehydrogenase 2 (IDH2), a major source of NADPH, thereby supporting mitochondrial homeostasis, protecting against inflammation, and reducing oxidative damage [51]. In this regard, it is worth noting that supplementation with *Sechium edule* has been reported to reduce oxidative damage to lipids, proteins, and DNA [17].

Similarly, quercetin has been shown to activate not only SIRT5 but also SIRT6, thereby protecting against age-related metabolic diseases and regulating chromatin homeostasis for telomere maintenance [52,53]. Its decrease leads to telomere dysfunction, premature cellular senescence, and chromosome end fusions [54]. Our research group has shown that *consumption of Sechium edule* prevents telomere shortening [55]. Furthermore, like SIRT1, SIRT6 is essential for Nrf2 transcriptional activation under OS conditions. This reinforces the findings of the present study, which show an increase in TAS and decreases in TOS and OSI in the EG [56]. Therefore, the 69% increase in SIRT6 gene expression at six months post-treatment in the EG may also be related to greater antioxidant protection [57,58].

On the other hand, it is well established that NAD+ levels decrease during the aging process, leading to impaired nuclear and mitochondrial function [59]. Therefore, based on our results, we suggest that the antioxidant effect of *Sechium edule* supplementation may be associated with the restoration of NAD^+ levels, as evidenced by the evident increase in sirtuin transcriptional levels. This points to an increase in deacetylation [60,61], favoring

 https://doi.org/10.3390/molecules31071182

the maintenance of redox homeostasis, which is mediated by Nrf2. Similarly, sirtuins can directly deacetylate Nrf2 at lysine residues, promoting its nuclear translocation, its ability to bind the antioxidant response element (ARE) in DNA, and/or inhibiting its ubiquitination via Keap1 [62]. Furthermore, it has been noted that the flavonoids present in *Sechium edule* mimic the effects of caloric restriction (CRM) by modulating metabolic pathways, including AMPK (which detects cellular energy levels and increases NAD levels) [63]. Hence, a synergistic interaction could exist between the bioactive components of *Sechium edule*, involving the SIRT-AMPK-Nrf2 axis, which simultaneously integrates energy sensing and defenses against oxidative stress. This coincides with the observed results: increased expression of SIRT1, 3, 5, and 6 mRNA and increased total antioxidant/oxidative capacity.

One of the main limitations of this study is the very small, unrepresentative sample size and the variability in the data (very large standard deviations). Therefore, the reported results should be interpreted with caution, as they lack sufficient statistical power to generalize. For this reason, further clinical studies with representative samples and different clinical conditions are needed to confirm our findings.

4. Materials and Methods

4.1. Experimental Design

This study was approved by the Research and Biosafety Bioethics Committee of the Faculty of Higher Studies Zaragoza, UNAM (FESZ/DEPI/CE/023/22/; 21 October 2022), under trial registration number ISRCTN 43215432.

All procedures were conducted in accordance with the ethical principles of the Declaration of Helsinki of the World Medical Association. Informed consent was obtained from every participant. The fruits of *Sechium edule*, varietal group *nigrum spinosum*, were donated by the Interdisciplinary Research Group on *Sechium edule* A.C. (GISeM) of the RED-Chayote of the Agricultural Genetic Resources Subcommittee of the National Seed Inspection and Certification Service (NSICS), which focuses on the conservation, improvement, characterization, and enrichment of the genus *Sechium* in Mexico in the Municipality of Huatusco in the State of Veracruz; this is where the fruits used to make the capsules used in the study were harvested [64].

The fruit characterization was carried out according to the guidelines proposed by the International Union for the Protection of New Varieties of Plants (UPOV) and was validated, including morphological, phenotypic, and chromosomal characterization, by GISeM [16,64].

The biological material was collected at horticultural maturity (when the fruit is suitable for consumption), selected, washed, disinfected, and sectioned into slices, which were then dried at 40 °C and pulverized (epidermis, seeds, and spines). Previously, our research group identified the secondary metabolites present in each chayote capsule, which were determined using HPLC. These metabolites, in ascending order of concentration, were as follows: 0.71 μg of cucurbitacin I, 6.11 μg of cucurbitacin D, 89.9 μg of cucurbitacin B and 154.8 μg of cucurbitacin E; flavonoids: 0.014 μg of apigenin, 1.3 μg of quercetin, 2.38 μg of myricetin, 14.2 μg of phlorizin, 45.5 μg of rutin and 48.8 μg of naringenin; and phenolic acids: 0.11 μg of p-hydroxybenzoic, 1.4 μg of chlorogenic, 1.7 μg of p-coumaric, 3.3 μg of protocatechuic, 7.0 μg of ferulic, 8.7 μg of syringic, 9.3 μg of caffeic and 38.8 μg of gallic [17] (Supplementary Materials S1).

4.2. Intervention

The capsule formulation was designed in the pharmaceutical development laboratory of FES Zaragoza. The placebo was prepared using pharmaceutical-grade lactose monohydrate and talc (United States Pharmacopeia, USP) (Sigma, St. Louis, MO, USA). The

https://doi.org/10.3390/molecules31071182

optimal particle size of the *Sechium* powder was determined, and rheological studies were conducted to ensure capsule filling, weight, homogeneity, and stability. In accordance with the design, the treatments were manufactured and packaged by a pharmaceutical company specializing in nutraceutical products. The intervention consisted of consuming three capsules (placebo or active) per day (500 mg of powdered *Sechium edule*, one before each meal) for six months. For the selection of the *Sechium edule* dose, extrapolation between species (mouse and human) was performed using allometric scaling [65]. We also relied on data previously published by our research group, which used CD-1 strain mice as an experimental model and administered different doses of *Sechium edule* (8–5000 mg/kg) intraperitoneally; the LD_{50} was found to be greater than 5000 mg/kg. In this study, doses starting at 800 mg/kg did not cause significant changes in blood chemistry or toxicity in lymphoid organs, including the spleen and thymus, as well as the liver and kidneys. Furthermore, it significantly reduced glucose levels [66]. Subsequently, the dose was adjusted to a higher, intermediate level between 800 and 1600 mg/kg, that is, 1200 mg/kg. This is because intraperitoneal administration results in greater bioavailability compared to oral administration [67]. This dose was then extrapolated to humans, corresponding to 512.64 mg. Therefore, the dosage was adjusted to 500 mg per capsule. Subsequently, an exploratory study was conducted in which 12 older adults with MetS were administered three capsules of *Sechium edule* daily for six weeks, which demonstrated hepatoprotective, nephroprotective, and antioxidant effects after consumption of the fruit [28].

A quasi-experimental study was conducted in a convenience sample of 43 older adults, with a mean age of 66 ± 3 years. The participants were assigned to the (i) placebo group (PG; n = 21) or (ii) experimental group (EG; n = 22). Only some subjects agreed to donate venous blood for gene expression measurements during the study (PG, n = 14; EG, n = 12) (Figure 2). In the PG, seven patients dropped out of the study: five due to logistical issues (inability to attend measurements) and two due to a change in address. In the EG, ten people dropped out of the study: four due to logistical issues (inability to attend measurements) and six due to hypoglycemia. In both groups, all measurements were taken at baseline (before treatment) and at 3 and 6 months (post-treatment).

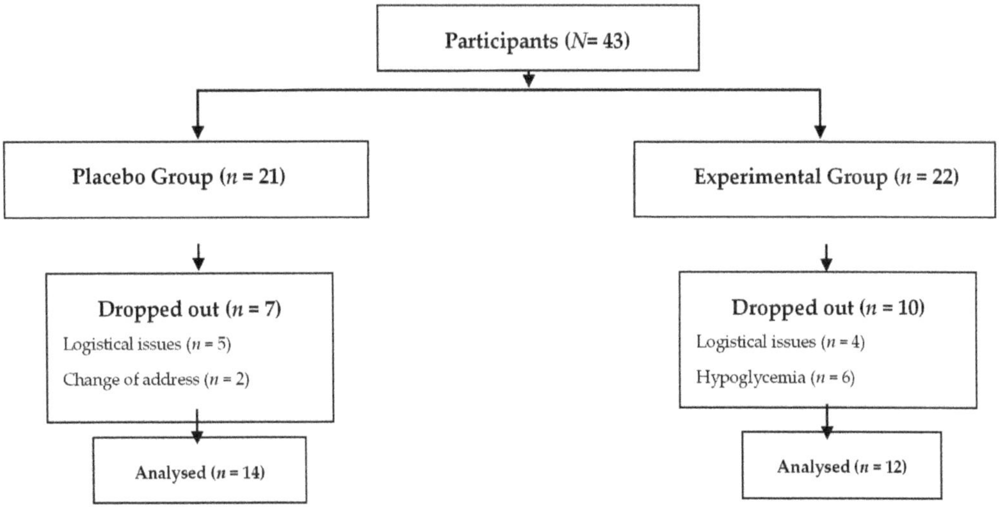

Figure 2. Flowchart of the participant monitoring process.

https://doi.org/10.3390/molecules31071182

4.3. Anthropometric and Blood Pressure Measurements

Body weight (kg) and waist circumference (cm) were recorded. Body weight was determined using a calibrated medical scale (SECA, Hamburg, Germany), while waist circumference was measured at the level of the umbilicus with a medical measuring tape (SECA, Hamburg, Germany). These measurements were performed by trained nursing staff [68]. Systolic (SBP) and diastolic (DBP) blood pressure were measured using a calibrated mercury sphygmomanometer. Patients were asked to rest for at least five minutes before the measurement, seated in a chair with a backrest, with their back straight, legs uncrossed, and feet flat on the floor. Finally, the Osler technique was used to identify pseudohypertension [69].

4.4. Biochemical Analysis

For blood sampling, participants were asked to fast for at least 8 h. Samples were obtained by venipuncture and collected in vacuum tubes without anticoagulant. For clinical chemistry determinations (glucose, cholesterol, triglycerides, high-density lipoproteins (HDL-c), uric acid, urea, and albumin), colorimetric techniques were used with a Selectra Junior automated clinical chemistry analyzer (Vital Scientific, Dieren, The Netherlands). An immunoturbidimetric assay with the same clinical chemistry analyzer determined the percentage of glycated hemoglobin. Total antioxidants and oxidant status (TOS/TAS) were determined from heparinized plasma, and, finally, samples were collected in tubes containing the anticoagulant EDTA for lymphocyte isolation.

4.5. Total Oxidation Status (TOS)

The TOS was determined using a commercial kit (Rel Assay Diagnostics, Gaziantep, Turkey). In this kit, the oxidants present in the sample can oxidize the ferrous ion-chelating complex to ferric ions. In an acidic medium, this ion forms a colored complex with the chromogen, which can be measured spectrophotometrically. Therefore, the color intensity is directly associated with the amount of oxidants present in the sample. This test uses hydrogen peroxide (H_2O_2) as a calibrator.

4.6. Total Antioxidant Status (TAS)

The TAS was quantified using a kit (Randox Laboratories Ltd., Antrim, UK) that uses metmyoglobin and H_2O_2, along with 2,2-azino-bis (3-ethylbenzthiazoline-6-sulfonic acid) (ABTS), to produce a blue-green stain from the ABTS+ cationic radical. The ABTS+ cation is relatively stable and was measured at 600 nm. Color intensity is inversely proportional to the amount of antioxidants present in the sample.

4.7. Oxidative Stress Index (OSI)

The OSI was determined as the ratio of TOS to TAS (TOS/TAS) [70].

4.8. Lymphocyte Isolation and RNA Extraction

Lymphocyte separation was performed using 5 mL of venous blood diluted in equal parts sterile phosphate-buffered saline (PBS) (Sigma, St. Louis, MO, USA) and 2% fetal bovine serum (FBS) (ThermoFisher Scientific, Waltham, MA, USA). Four milliliters of Ficoll-Paque (Gibco ThermoFisher Scientific, Waltham, MA, USA) was added. The mixture was centrifuged at 200× *g*, and the opaque interface was transferred to a sterile tube. RNA extraction from 2×10^6 lymphocytes was performed using the RNeasy Mini isolation kit (Qiagen, Hilden, Düsseldorf, Germany) according to the manufacturer's recommendations, and the samples were stored at 70 °C until use. RNA was quantified and its integrity

https://doi.org/10.3390/molecules31071182

was determined by loading 5 μg on a 1% agarose gel with ethidium bromide and Tris-acetate-EDTA buffer, and visualized using Kodak Molecular Imaging software (v.4.5.1). Simultaneously, its purity was calculated using the A260/A280 ratio. All RNA samples used were considered intact and of optimal purity.

4.9. Gene Expression Analysis

All reactions were performed using 10 ng of RNA and forward and reverse primers at a final concentration of 100 nM (IDT, Coralville, IA, USA) (Table 4). The primers were generated using a primer design tool (NCBI Primer-BLAST tool from NIH) software (v.2.5.0) [71].

Table 4. Details of primers used for real-time PCR assays.

Gene	Primer Name	Sequence
SIRT1	SIRT1-F	GGGCTGCGGTTCCTACTG
	SIRT1-R	TTATCTGGCTGCTGCGGAAA
SIRT2	SIRT2-F	CTCTCACCCTCTGGAGACCC
	SIRT2-R	ATGTCTGCTTCTCCACCAGC
SIRT3	SIRT3-F	GGTAGTTGAACGGGTCGAGG
	SIRT3-R	TAATAATCGTCCCTGCCGCC
SIRT4	SIRT4-F	CAATCAGACGGTCCCACTGT
	SIRT4-R	ATCCAACGGCCTTTTGCTGA
SIRT5	SIRT5-F	ACGTCGTGTGGTTTGGAGAA
	SIRT5-R	GGAAGTGCCCACCACTAGAC
SIRT6	SIRT6-F	GCAGTCTTCCAGTGTGGTGT
	SIRT6-R	TCCTCCATGGTCCAGACTCC
β-ACTIN	ACTIN-F	GAGCACAGAGCCTCGCC
	ACTIN-R	CGCGGCGATATCATCATCCA

SIRT1: (silent mating type information regulation 2 homolog 1 (*S. cerevisiae*), human; SIRT2: silent mating type information regulation 2 homolog 2 (*S. cerevisiae*) protein, human; SIRT3: silent mating type information regulation 2 homolog 3 (*S. cerevisiae*) protein, human; SIRT4: silent mating type information regulation 2 homolog 4, human; SIRT5: silent mating type information regulation 2 homolog 5, human; SIRT6: silent mating type information regulation 2 homolog 6 protein, human.

The QuantiFast SYBR Green RT-PCR kit (one-step RT-PCR) (Qiagen, Hilden, Düsseldorf, Germany) was also used for gene expression analysis, which allows for simultaneous execution of reverse transcription and PCR reactions. The reaction conditions are shown in Figure 3. The mean crossing threshold (Ct) of each gene was normalized to the mean Ct of the housekeeping gene β-actin.

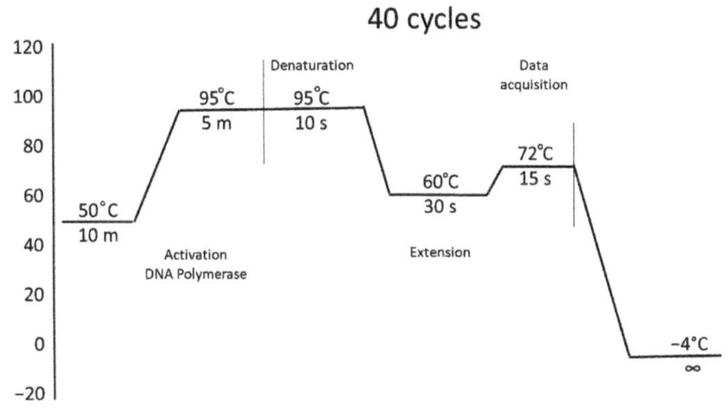

Figure 3. PCR stages. Temperatures, times, and cycles.

 https://doi.org/10.3390/molecules31071182

It should be noted that the gene selected for normalization by qPCR was based on a study, which compared different reference genes to evaluate transcription levels. In this study, they found that the β-actin and tyrosine 3-monooxygenase/tryptophan 5-monooxygenase activation protein zeta (YWHAZ) genes are the most stable genes in peripheral blood mononuclear cells of patients with type 2 diabetes mellitus (T2DM), which led us to select β-actin as the reference and control gene [72].

4.10. Statistical Analysis

The results are presented as the mean ± standard deviation and were analyzed using repeated measures ANOVA. Associations between sirtuin gene expression and the parameters analyzed were determined using Pearson's correlation coefficient in statistical SPSS (IBM, Armonk, NY, USA) software (v.25). Results were considered statistically significant when $p < 0.05$. All determinations were performed in duplicate.

5. Conclusions

Our findings suggest that consumption of *Sechium edule* for six months increases the transcriptional expression of SIRT1, 3, 5, and 6; likewise, it improves the antioxidant response capacity, as evidenced by the reduction in OSI. Therefore, supplementation with this fruit could be useful as a complementary treatment to delay the development of complications, given its antioxidant effects and role as a metabolic regulator in older adults with type 2 diabetes. However, further studies with representative population samples in different clinical contexts and conditions are needed.

Supplementary Materials: The following supporting information can be downloaded at: https://www.mdpi.com/article/10.3390/molecules31071182/s1. Supplementary Materials S1. Identification of secondary metabolites by High-Performance Liquid Chromatography (HPLC). References [66,73] are cited in the supplementary materials.

Author Contributions: V.M.M.-N. designed the study, wrote the manuscript, and analyzed the data. J.R.-P. performed the study, wrote the manuscript, and analyzed the data. D.H.-Á., T.L.A.-U., I.A.-S., E.S.-O., J.C.-I. and G.G.-G. performed the study and analyzed the data. All authors reviewed the final manuscript. All authors have read and agreed to the published version of the manuscript.

Funding: This work was supported by grants from the General Directorate of Academic Personnel Affairs, National Autonomous University of Mexico (DGAPA-UNAM, PAPIIT IN223923).

Institutional Review Board Statement: The study was conducted in accordance with the Declaration of Helsinki and approved by the Ethics Committee of the "National Autonomous University of Mexico (UNAM)—Zaragoza Campus".

Informed Consent Statement: Informed consent was obtained from all subjects involved in the study.

Data Availability Statement: The data that support the findings of this study are available from the corresponding author [V.M.M.-N.] upon reasonable request.

Acknowledgments: We appreciate the support from the National Council of Humanities, Science and Technology (CONAHCyT) for the scholarship granted to Graciela Gavia-García for her postdoctoral position.

Conflicts of Interest: The authors have no potential conflicts of interest.

https://doi.org/10.3390/molecules31071182

References

1. Mendoza-Núñez, V.M.; Martínez-Maldonado, M.L.; Vivaldo-Martínez, M. What is the onset age of human aging and old age? *Int. J. Gerontol.* **2016**, *10*, 56. [CrossRef]
2. WHO. Ageing and Health. Available online: https://www.who.int/news-room/fact-sheets/detail/ageing-and-health (accessed on 15 November 2025).
3. López-Otín, C.; Blasco, M.A.; Partridge, L.; Serrano, M.; Kroemer, G. Hallmarks of aging: An expanding universe. *Cell* **2023**, *186*, 243–278. [CrossRef] [PubMed]
4. López-Otín, C.; Kroemer, G. Hallmarks of aging: Integrating molecular and social determinants. *Geromedicine* **2025**, *1*, 202507. [CrossRef]
5. WHO. Diabetes. Available online: https://www.who.int/news-room/fact-sheets/detail/diabetes (accessed on 15 August 2025).
6. Yaribeygi, H.; Sathyapalan, T.; Atkin, S.L.; Sahebkar, A. Molecular mechanisms linking oxidative stress and Diabetes Mellitus. *Oxid. Med. Cell. Longev.* **2020**, *2020*, 8609213. [CrossRef]
7. Caturano, A.; D'Angelo, M.; Mormone, A.; Russo, V.; Mollica, M.P.; Salvatore, T.; Galiero, R.; Rinaldi, L.; Vetrano, E.; Marfella, R.; et al. Oxidative stress in Type 2 Diabetes: Impacts from pathogenesis to lifestyle modifications. *Curr. Issues Mol. Biol.* **2023**, *45*, 6651–6666. [CrossRef] [PubMed]
8. Yaribeygi, H.; Mohammadi, M.T.; Sahebkar, A. Crocin potentiates antioxidant defense system and improves oxidative damage in liver tissue in diabetic rats. *Biomed. Pharmacother.* **2018**, *98*, 333–337. [CrossRef]
9. Bellary, S.; Kyrou, I.; Brown, J.E. Type 2 diabetes mellitus in older adults: Clinical considerations and management. *Nat. Rev. Endocrinol.* **2021**, *17*, 534–548. [CrossRef]
10. Kitada, M.; Ogura, Y.; Monno, I.; Koya, D. Sirtuins and Type 2 Diabetes: Role in inflammation, oxidative stress, and mitochondrial function. *Front. Endocrinol.* **2019**, *10*, 187. [CrossRef]
11. Lingappa, N.; Mayrovitz, H.N. Role of sirtuins in Diabetes and age-related processes. *Cureus* **2022**, *14*, e28774. [CrossRef]
12. NLM. Sirtuins. Available online: https://www.ncbi.nlm.nih.gov/mesh/68037761 (accessed on 15 November 2025).
13. Abdelhaleem, I.A.; Brakat, A.M.; Adayel, H.M.; Asla, M.M.; Rizk, M.A.; Aboalfetoh, A.Y. The effects of resveratrol on glycemic control and cardiometabolic parameters in patients with T2DM: A systematic review and meta-analysis. *Med. Clin.* **2022**, *158*, 576–585. [CrossRef]
14. Cao, M.M.; Lu, X.; Liu, G.D.; Su, Y.; Li, Y.B.; Zhou, J. Resveratrol attenuates type 2 diabetes mellitus by mediating mitochondrial biogenesis and lipid metabolism via Sirtuin type 1. *Exp. Ther. Med.* **2018**, *15*, 576–584. [CrossRef] [PubMed]
15. Vieira, E.F.; Souza, S.; Moreira, M.M.; Cruz, R.; Silva, A.B.D.; Casal, S.; Delerue-Matos, C. Valorization of phenolic and carotenoid compounds of *Sechium edule* (Jacq. Swartz) leaves: Comparison between conventional, ultrasound- and microwave-assisted extraction approaches. *Molecules* **2022**, *27*, 7193. [CrossRef] [PubMed]
16. Avendaño, A.C.H.; Cadena, I.J.; Arévalo, G.M.L.; Campos, R.E.; Cisneros, S.V.M.; Aguirre, M.J.F. *Las Variedades del Chayote Mexicano, Recurso Ancestral con Potencial de Comercialización*; Grupo Interdisciplinario de Investigación en Sechium edule en México, A.C. (GISeM): Mexico City, Mexico, 2010.
17. Arista-Ugalde, T.L.; Santiago-Osorio, E.; Monroy-García, A.; Rosado-Pérez, J.; Aguiñiga-Sánchez, I.; Cadena-Iñiguez, J.; Gavia-García, G.; Mendoza-Núñez, V.M. Antioxidant and anti-inflammatory effect of the consumption of powdered concentrate of *Sechium edule* var. nigrum spinosum in Mexican older adults with metabolic syndrome. *Antioxidants* **2022**, *11*, 1076. [CrossRef] [PubMed]
18. Gavia-García, G.; Hernández-Álvarez, D.; Arista-Ugalde, T.L.; Aguiñiga-Sánchez, I.; Santiago-Osorio, E.; Mendoza-Núñez, V.M.; Rosado-Pérez, J. The supplementation of *Sechium edule* var. nigrum spinosum (chayote) promotes Nrf2-mediated antioxidant protection in older adults with metabolic syndrome. *Nutrients* **2023**, *15*, 4106. [CrossRef] [PubMed]
19. Patel, S.; Khan, H.; Majumdar, A. Crosstalk between Sirtuins and Nrf2: SIRT1 activators as emerging treatment for diabetic neuropathy. *Metab. Brain Dis.* **2022**, *37*, 2181–2195. [CrossRef]
20. WHO. Classification of Diabetes Mellitus. Available online: https://www.who.int/publications/i/item/classification-of-diabetes-mellitus (accessed on 16 August 2025).
21. Galicia-Garcia, U.; Benito-Vicente, A.; Jebari, S.; Larrea-Sebal, A.; Siddiqi, H.; Uribe, K.B.; Ostolaza, H.; Martín, C. Pathophysiology of Type 2 Diabetes Mellitus. *Int. J. Mol. Sci.* **2020**, *21*, 6275. [CrossRef]
22. GBD 2021 Diabetes Collaborators. Global, regional, and national burden of diabetes from 1990 to 2021, with projections of prevalence to 2050: A systematic analysis for the Global Burden of Disease Study 2021. *Lancet* **2023**, *402*, 203–234. [CrossRef]
23. Lombardo-Earl, G.; Roman-Ramos, R.; Zamilpa, A.; Herrera-Ruiz, M.; Rosas-Salgado, G.; Tortoriello, J.; Jiménez-Ferre, E. Extracts and fractions from edible roots of *Sechium edule* (Jacq.) Sw. with antihypertensive activity. *Evid. Based Complement. Alternat. Med.* **2014**, *2014*, 594326. [CrossRef]

 https://doi.org/10.3390/molecules31071182

24. Gordon, E.A.; Guppy, L.J.; Nelson, M. The antihypertensive effects of the Jamaican Cho-Cho (*Sechium edule*). *West Indian Med. J.* **2000**, *49*, 27–31.
25. Rosado-Pérez, J.; Aguiñiga-Sánchez, I.; Santiago-Osorio, E.; Mendoza-Núñez, V.M. Effect of *Sechium edule* var. nigrum spinosum (Chayote) on oxidative stress and pro-inflammatory markers in older adults with metabolic syndrome: An exploratory study. *Antioxidants* **2019**, *8*, 146. [CrossRef]
26. Agbabiaka, T.; Wider, B.; Watson, L.K.; Goodman, C. Concurrent use of prescription drugs and herbal medicinal products in older adults: A systematic review. *Drugs Aging* **2017**, *34*, s40017–s40266. [CrossRef] [PubMed]
27. Gavia-García, G.; Rosado-Pérez, J.; Aguiñiga-Sánchez, I.; Santiago-Osorio, E.; Mendoza-Núñez, V.M. Effect of *Sechium edule* var. nigrum spinosum (Chayote) on telomerase levels and antioxidant capacity in older adults with metabolic syndrome. *Antioxidants* **2020**, *9*, 634. [CrossRef] [PubMed]
28. Arista-Ugalde, T.L.; Delgado-Arroyo, S.; Gavia-García, G.; Hernández-Álvarez, D.; Aguiñiga-Sánchez, I.; Santiago-Osorio, E.; Rosado-Pérez, J.; Mendoza-Núñez, V.M. Hypoglycemic effects of *Sechium edule* (Chayote) in older adults: A systematic review and meta-analysis of clinical and preclinical trials. *Foods* **2025**, *14*, 2937. [CrossRef] [PubMed]
29. Stratton, I.M.; Adler, A.I.; Neil, H.A.W.; Matthews, D.R.; Manley, S.E.; Cull, C.A.; Hadden, D.; Turner, R.C.; Holman, R.R. Association of glycaemia with macrovascular and microvascular complications of type 2 diabetes (UKPDS 35): Prospective observational study. *BMJ* **2000**, *321*, 405. [CrossRef]
30. García-Martínez, B.I.; Ruiz-Ramos, M.; Pedraza-Chaverri, J.; Santiago-Osorio, E.; Mendoza-Núñez, V.M. Effect of resveratrol on markers of oxidative stress and sirtuin 1 in elderly adults with type 2 diabetes. *Int. J. Mol. Sci.* **2023**, *24*, 7422. [CrossRef]
31. Grabowska, W.; Sikora, E.; Bielak-Zmijewska, A. Sirtuins, a promising target in slowing down the ageing process. *Biogerontology* **2017**, *18*, 447–476. [CrossRef]
32. Morris, B.J. Seven sirtuins for seven deadly diseases of ageing. *Free Radic. Biol. Med.* **2013**, *56*, 133–171. [CrossRef]
33. Grabowska, W.; Suszek, M.; Wnuk, M.; Lewinska, A.; Wasiak, E.; Sikora, E.; Bielak-Zmijewska, A. Curcumin elevates sirtuin level but does not postpone in vitro senescence of human cells building the vasculature. *Oncotarget* **2016**, *7*, 19201–19213. [CrossRef]
34. Gillum, M.P.; Kotas, M.E.; Erion, D.M.; Kursawe, R.; Chatterjee, P.; Nead, K.T.; Muise, E.S.; Hsiao, J.J.; Frederick, D.W.; Yonemitsu, S.; et al. Sirt1 regulates adipose tissue inflammation. *Diabetes* **2011**, *60*, 3235–3245. [CrossRef]
35. de Kreutzenberg, S.V.; Ceolotto, G.; Papparella, I.; Bortoluzzi, A.; Semplicini, A.; Dalla, M.C.; Cobelli, C.; Fadini, G.P.; Avogaro, A. Downregulation of the longevity-associated protein sirtuin 1 in insulin resistance and metabolic syndrome: Potential biochemical mechanisms. *Diabetes* **2010**, *59*, 1006–1015. [CrossRef]
36. Li, Y.; Miao, Y.; Feng, Q.; Zhu, W.; Chen, Y.; Kang, Q.; Wang, Z.; Lu, F.; Zhang, Q. Mitochondrial dysfunction and onset of type 2 diabetes along with its complications: A multi-omics Mendelian randomization and colocalization study. *Front. Endocrinol.* **2024**, *15*, 1401531. [CrossRef]
37. Kitada, M.; Takeda, A.; Nagai, T.; Ito, H.; Kanasaki, K.; Koya, D. Dietary restriction ameliorates diabetic nephropathy through anti-inflammatory effects and regulation of the autophagy via restoration of Sirt1 in diabetic Wistar fatty (fa/fa) rats: A model of type 2 diabetes. *Exp. Diabetes Res.* **2011**, *2011*, 908185. [CrossRef] [PubMed]
38. Jung, H.Y.; Lee, D.; Ryu, H.G.; Choi, B.H.; Go, Y.; Lee, N.; Lee, D.; Son, H.G. Myricetin improves endurance capacity and mitochondrial density by activating SIRT1 and PGC-1α. *Sci. Rep.* **2017**, *7*, 6237. [CrossRef] [PubMed]
39. Yuan, B.; Mao, J.; Wang, J.; Luo, S.; Luo, B. Naringenin mitigates cadmium-induced cell death, oxidative stress, mitochondrial dysfunction, and inflammation in KGN cells by regulating the expression of sirtuin-1. *Drug Chem. Toxicol.* **2024**, *47*, 445–456. [CrossRef] [PubMed]
40. Rodgers, J.T.; Lerin, C.; Haas, W.; Gygi, S.P.; Spiegelman, B.M.; Puigserver, P. Nutrient control of glucose homeostasis through a complex of PGC-1 alpha and SIRT1. *Nature* **2005**, *434*, 113–118. [CrossRef]
41. Gureev, A.P.; Krutskikh, E.P. Regulation of the NRF2 transcription factor activity by SIRT1-induced deacetylation: A possible SIRT1–NRF2 feedback loop. *Russ. J. Genet.* **2025**, *61*, 133–139. [CrossRef]
42. Katz, L.S.; Scott, D.K.; Stewart, A.F. SIRT2 puts the brakes on human β cell proliferation: Therapeutic opportunities and next challenges. *J. Clin. Investig.* **2025**, *135*, e197142. [CrossRef]
43. Choubey, S.K.; Prabhu, D.; Nachiappan, M.; Biswal, J.; Jeyakanthan, J. Molecular modeling, dynamics studies and density functional theory approaches to identify potential inhibitors of SIRT4 protein from Homo sapiens: A novel target for the treatment of type 2 diabetes. *J. Biomol. Struct. Dyn.* **2017**, *35*, 3316–3329. [CrossRef]
44. Huynh, F.K.; Hu, X.; Lin, Z.; Johnson, J.D.; Hirschey, M.D. Loss of sirtuin 4 leads to elevated glucose- and leucine-stimulated insulin levels and accelerated age-induced insulin resistance in multiple murine genetic backgrounds. *J. Inherit. Metab. Dis.* **2018**, *41*, 59–72. [CrossRef]
45. Zhang, J.; Xiang, H.; Liu, J.; Chen, Y.; He, R.R.; Liu, B. Mitochondrial Sirtuin 3: New emerging biological function and therapeutic target. *Theranostics* **2020**, *10*, 8315–8342. [CrossRef]

46. Bellizzi, D.; Rose, G.; Cavalcante, P.; Covello, G.; Dato, S.; De Rango, F.; Greco, V.; Maggiolini, M.; Feraco, E.; Mari, V.; et al. A novel VNTR enhancer within the SIRT3 gene, a human homologue of SIR2, is associated with survival at oldest ages. *Genomics* **2005**, *85*, 258–263. [CrossRef]
47. Yu, L.M.; Dong, X.; Xue, X.D.; Zhang, J.; Li, Z.; Wu, H.J.; Yang, Z.L.; Yang, Y.; Wang, H.S. Naringenin improves mitochondrial function and reduces cardiac damage following ischemia-reperfusion injury: The role of the AMPK-SIRT3 signaling pathway. *Food Funct.* **2019**, *10*, 2752–2765. [CrossRef]
48. Li, L.; Zeng, H.; He, X.; Chen, J.X. Sirtuin 3 alleviates diabetic cardiomyopathy by regulating TIGAR and cardiomyocyte metabolism. *J. Am. Heart Assoc.* **2021**, *10*, e018913. [CrossRef] [PubMed]
49. Zhou, L.; Wang, F.; Sun, R.; Chen, X.; Zhang, M.; Xu, Q.; Wang, Y.; Wang, S.; Xiong, Y.; Guan, K.L.; et al. SIRT5 promotes IDH2 desuccinylation and G6PD deglutarylation to enhance cellular antioxidant defense. *EMBO Rep.* **2016**, *17*, 811–822. [CrossRef] [PubMed]
50. Zhou, B.; Yang, Y.; Pang, X.; Shi, J.; Jiang, T.; Zheng, X. Quercetin inhibits DNA damage responses to induce apoptosis via SIRT5/PI3K/AKT pathway in non-small cell lung cancer. *Biomed. Pharmacother.* **2023**, *165*, 115071. [CrossRef] [PubMed]
51. Chang, X.; Zhang, T.; Wang, J.; Liu, Y.; Yan, P.; Meng, Q.; Yin, Y.; Wang, S. SIRT5-related desuccinylation modification contributes to quercetin-induced protection against heart failure and high-glucose-prompted cardiomyocytes injured through regulation of mitochondrial quality surveillance. *Oxid. Med. Cell. Longev.* **2021**, *2021*, 5876841. [CrossRef]
52. You, W.; Zheng, W.; Weiss, S.; Chua, K.; Steegborn, C. Structural basis for the activation and inhibition of Sirtuin 6 by quercetin and its derivatives. *Sci. Rep.* **2019**, *9*, 19176. [CrossRef]
53. Zhang, H.; Zhang, J.; Zhang, H.X. Effect of quercetin on the protein-substrate interactions in SIRT6: Insight from MD simulations. *J. Mol. Graph. Model.* **2024**, *130*, 108778. [CrossRef]
54. Michishita, E.; McCord, R.A.; Berber, E.; Kioi, M.; Padilla-Nash, H.; Damian, M.; Cheung, P.; Kusumoto, R.; Kawahara, T.L.; Barrett, J.C.; et al. SIRT6 is a histone H3 lysine 9 deacetylase that modulates telomeric chromatin. *Nature* **2008**, *452*, 492–496. [CrossRef]
55. Gavia-García, G.; Rosado-Pérez, J.; Arista-Ugalde, T.L.; Aguiñiga-Sánchez, I.; Santiago-Osorio, E.; Mendoza-Núñez, V.M. The consumption of *Sechium edule* (chayote) has antioxidant effect and prevents telomere attrition in older adults with metabolic syndrome. *Redox Rep.* **2023**, *28*, 2207323. [CrossRef]
56. Rezazadeh, S.; Yang, D.; Tombline, G.; Simon, M.; Regan, S.P.; Seluanov, A.; Gorbunova, V. SIRT6 promotes transcription of a subset of NRF2 targets by mono-ADP-ribosylating BAF170. *Nucleic Acids Res.* **2019**, *47*, 7914–7928. [CrossRef] [PubMed]
57. Kuang, J.; Zhang, Y.; Liu, Q.; Shen, J.; Pu, S.; Cheng, S.; Chen, L.; Li, H.; Wu, T.; Li, R.; et al. Fat-specific Sirt6 ablation sensitizes mice to high-fat diet–induced obesity and insulin resistance by inhibiting lipolysis. *Diabetes* **2017**, *66*, 1159–1171. [CrossRef] [PubMed]
58. Kuang, J.; Chen, L.; Tang, Q.; Zhang, J.; Li, Y.; He, J. The role of Sirt6 in obesity and diabetes. *Front. Physiol.* **2018**, *9*, 135. [CrossRef] [PubMed]
59. Imai, S.; Guarente, L. NAD+ and sirtuins in aging and disease. *Trends Cell Biol.* **2014**, *24*, 464–471. [CrossRef]
60. Vedantham, S.; Thiagarajan, D.; Ananthakrishnan, R.; Wang, L.; Rosario, R.; Zou, Y.S.; Goldberg, I.; Yan, S.F.; Schmidt, A.M.; Ramasamy, R. Aldose reductase drives hyperacetylation of Egr-1 in hyperglycemia and consequent upregulation of proinflammatory and prothrombotic signals. *Diabetes* **2014**, *63*, 761–774. [CrossRef]
61. Kosanam, H.; Thai, K.; Zhang, Y.; Advani, A.; Connelly, K.A.; Diamandis, E.P.; Gilbert, R.E. Diabetes induces lysine acetylation of intermediary metabolism enzymes in the kidney. *Diabetes* **2014**, *63*, 2432–2439. [CrossRef]
62. Perez-Lao, E.J.; Fagerli, E.; Ferrier, F.; Young, J.I.; Perez-Pinzon, M.A. Regulatory dynamics of Nrf2 with sirtuins in the brain: Exploring cellular metabolism, synaptic plasticity, and defense mechanisms. *J. Neurochem.* **2025**, *169*, e70193. [CrossRef]
63. Murillo-Cancho, A.F.; Nievas-Soriano, B.J.; Lozano-Paniagua, D. Miméticos de la restricción energética: Un análisis comparativo de agentes naturales y sintéticos en la modulación de la autofagia. *Ars. Pharm.* **2025**, *66*, 385–398.
64. Cadena-Iñiguez, J.; Ruiz-Posadas, L.M.; Soto-Hernández, M.; Aguirre-Medina, J.F.; Avendaño-Arrazate, C.H.; Arévalo-Galarza, L. Infraspecific variation of *Sechium edule* (Jacq.) Sw. in the state of Veracruz, Mexico. *Genet. Resour. Crop. Evol.* **2008**, *55*, 835–847. [CrossRef]
65. Nair, A.B.; Jacob, S. A simple practice guide for dose conversion between animals and human. *J. Basic Clin. Pharm.* **2016**, *7*, 27–31. [CrossRef]
66. Aguiñiga-Sánchez, I.; Cadena-Íñiguez, J.; Santiago-Osorio, E.; Gómez-García, G.; Mendoza-Núñez, V.M.; Rosado-Pérez, J.; Ruíz-Ramos, M.; Cisneros-Solano, V.M.; Ledesma-Martínez, E.; Delgado-Bordonave, A.J.; et al. Chemical analyses and in vitro and in vivo toxicity of fruit methanol extract of *Sechium edule* var. *nigrum spinosum*. *Pharm. Biol.* **2017**, *55*, 1638–1645. [CrossRef]
67. Al Shoyaib, A.; Archie, S.R.; Karamyan, V.T. Intraperitoneal route of drug administration: Should it be used in experimental animal studies? *Pharm. Res.* **2019**, *37*, 12. [CrossRef]

 https://doi.org/10.3390/molecules31071182

68. Secretaría de Salud. *Toma de Medidas Clínicas y Antropométricas en el Adulto Mayor*; Subsecretaría de Prevención y Protección de La Salud: Mexico City, Mexico, 2002.
69. Secretaría de Salud. Norma Oficial Mexicana NOM-030-SSA-1999. In *Para la Prevención, Tratamiento y Control de la Hipertensión Arterial*; Secretaría de Salud: Mexico City, Mexico, 1999.
70. Sánchez-Rodríguez, M.A.; Mendoza-Núñez, V.M. Oxidative stress indexes for diagnosis of health or disease in humans. *Oxid. Med. Cell. Longev.* **2019**, *2019*, 4128152. [CrossRef]
71. Primer-BLAST-NCBI-NIH (Primer-BLAST). Available online: https://www.ncbi.nlm.nih.gov/tools/primer-blast/ (accessed on 1 October 2024).
72. Hazarika, A.; Nongkhlaw, B.; Mukhopadhyay, A. Identification of stable reference genes in peripheral blood mononuclear cells from type 2 diabetes mellitus patients. *Sci. Rep.* **2023**, *13*, 486. [CrossRef]
73. Arista-Ugalde, T.L. Efecto del Consumo de Frutos de *Sechium edule* Sobre Marcadores de Estrés Oxidativo, Inflamación Crónica y Daño Oxidativo al ADN en Adultos Mayores Con Síndrome Metabólico. Ph.D. Thesis, Universidad Nacional Autónoma de México, México city, Mexico, 2023.

 https://doi.org/10.3390/molecules31071182

molecules

MDPI

Article

Phytochemistry and In Vitro Bioactivities of *Elaeagnus angustifolia* L. Leaves: Influence of Solvent Polarity, Antioxidant Properties, and α-Amylase Inhibition

Rayene Bouaita [1,2], Randa Djemil [1,2], Samira Bouhalit [1,2], Sabrina Lekmine [1,2,*], Ouided Benslama [3], Saber Boutellaa [4], Nabil Touzout [5,6,*], Gema Nieto [7], Ivan Mustać [8], Gabrijel Ondrašek [8] and Muhammad Imtiaz Rashid [9,*]

1 Department of Molecular and Cellular Biology, Faculty of Nature and Life Sciences, University of Abbes Laghrour, Khenchela 40000, Algeria
2 Laboratory of Biotechnology, Water, Environment and Health (LBWEH), University of Abbes Laghrour, Khenchela 40000, Algeria
3 Department of Natural and Life Sciences, Faculty of Exact Sciences and Natural and Life Sciences, Larbi Ben M'Hidi University, Oum El Bouaghi 04000, Algeria
4 Laboratory of Natural Sciences and Materials, Abdelhafid Boussouf University Center, Mila 43000, Algeria
5 Laboratory Materials and Environment, University Yahia Fares of Medea, Urban Center, Medea 26000, Algeria
6 Department of Agronomy, Faculty of Sciences, Pole Urban Ouzera, University of Medea, Medea 26000, Algeria
7 Department of Food Technology, Nutrition and Food Science, Veterinary Faculty, University of Murcia, Regional Campus of International Excellence "Campus Mare Nostrum", Espinardo, 30100 Murcia, Spain
8 Department of Soil Amelioration, Faculty of Agriculture, University of Zagreb, Svetošimunska Cesta 25, 10000 Zagreb, Croatia
9 Center of Excellence in Environmental Studies, King Abdulaziz University, Jeddah 21589, Saudi Arabia
* Correspondence: lekmine.sabrina@univ-khenchela.dz (S.L.); touzout.nabil@univ-medea.dz (N.T.); mimmurad@kau.edu.sa or irmaliks@gmail.com (M.I.R.)

Abstract

This study investigated the effect of solvent polarity on extraction yield, phytochemical composition, antioxidant activity, and α-amylase inhibition of *Elaeagnus angustifolia* L. leaf extracts to evaluate their antidiabetic potential. Extraction yields varied with solvent polarity, with the hydroethanolic extract showing the highest (18.00%) and n-hexane the lowest (0.05%) yield. The n-butanol and ethyl acetate fractions contained the most phenolics (309.05 and 290.97 mg GAE/g), ethyl acetate was the richest in flavonoids (102.11 mg QE/g), and tannins were concentrated in dichloromethane (66.24 mg CE/g). HPLC revealed solvent-specific profiles: rutin and gallic acid dominated in n-butanol, quercetin in ethyl acetate, and 4-hydroxybenzoic and ferulic acids in dichloromethane, while chicoric acid appeared in hydroethanolic and n-hexane extracts. Antioxidant assays (DPPH, ABTS, and FRAP) showed strong activity in polar extracts, particularly hydroethanolic and ethyl acetate fractions. The n-hexane extract exhibited the highest α-amylase inhibition (IC_{50} = 36.70 μg/mL), surpassing acarbose (IC_{50} = 126.14 μg/mL), while other fractions were inactive (IC_{50} > 400 μg/mL). Molecular docking highlighted rutin, chlorogenic acid, and chicoric acid as potential enzyme binders. These findings demonstrate the chemical diversity and significant bioactivities of *E. angustifolia* leaves, supporting their potential as natural antidiabetic agents.

Keywords: *Elaeagnus angustifolia* L.; phenolic compounds; antioxidant activity; α-amylase inhibition; molecular docking

1. Introduction

An imbalance between reactive oxygen species (ROS) production and the body's antioxidant defenses results in oxidative stress, which causes cellular damage and plays a critical role in the pathogenesis of various chronic diseases, including diabetes mellitus [1]. The excess ROS can impair insulin signaling pathways and damage pancreatic β-cells, ultimately exacerbating hyperglycemia and diabetic complications [2]. Consequently, mitigating oxidative stress has become a key therapeutic target for management of diabetes [3,4]. In this context, antidiabetic agents that not only regulate blood glucose levels but also possess antioxidant properties are especially valuable, as they can address both glycemic control and oxidative damage, potentially reducing the risk of long-term complications [5–7].

Plants have historically been a rich source of bioactive compounds with medicinal properties, including antidiabetic effects [8]. Numerous phytochemicals such as polyphenols, flavonoids, and tannins modulate carbohydrate metabolism enzymes, enhance insulin secretion, improve insulin sensitivity, and scavenge free radicals [9]. The multifaceted biological activities of these natural compounds position them as promising candidates for alternative or complementary therapies in diabetes management. Moreover, the relative safety, accessibility, and longstanding traditional use of medicinal plants support their continued exploration for the development of novel antidiabetic agents.

The effective extraction of these bioactive compounds from plant matrices largely depends on the choice of solvents, as they differ in polarity and thus in their ability to solubilize specific classes of phytochemicals [10,11]. Selecting the appropriate solvent is vital to maximize the recovery of target molecules with desired biological activities while enhancing selectivity and purity [12]. Sequential solvent fractionation with solvents of increasing polarity allows for the separation of complex plant extracts into fractions enriched with distinct phytochemical groups [13]. This strategy aids in bioactivity-guided isolation and facilitates the identification of compounds responsible for the therapeutic effects.

On the other hand, high-performance liquid chromatography (HPLC) is an indispensable analytical technique employed to profile, identify, and quantify phenolic and other bioactive constituents within plant extracts [14,15]. HPLC provides high resolution, sensitivity, and reproducibility, enabling the detailed characterization of phytochemical composition [14,16,17]. This information is crucial for establishing correlations between specific compounds and observed biological activities, quality control, and standardization of herbal preparations, thereby enhancing their acceptance in pharmaceutical applications.

Focusing on *Elaeagnus angustifolia* L., a species belonging to the genus *Elaeagnus*, this plant has been traditionally used in various regions for its medicinal properties, particularly for its anti-inflammatory and antioxidant effects [18,19]. It is known to be rich in secondary metabolites, including flavonoids, phenolic acids, and sesquiterpenes, which contribute to its therapeutic potential [20,21].

In the present study, *E. angustifolia* leaves from the Baghaï region (Khenchela, Algeria) were collected and authenticated to investigate the influence of solvent polarity on extraction yield, phytochemical composition, and associated biological activities. A hydroalcoholic maceration was first performed, followed by sequential liquid–liquid partitioning using solvents of increasing polarity. The obtained fractions were subjected to high-performance liquid chromatography (HPLC) analysis to identify and quantify major bioactive phenolic compounds.

The biological potential of the extracts was evaluated through complementary in vitro assays. Antioxidant activity was assessed using DPPH, ABTS, and FRAP methods, enabling a comprehensive evaluation of radical scavenging and reducing capacities. In parallel,

 https://doi.org/10.3390/molecules31050861

antidiabetic activity was investigated via in vitro α-amylase inhibition and further supported by molecular docking analysis to elucidate the binding interactions between selected phytochemicals and the enzyme active site.

Although *E. angustifolia* has been studied in different geographical regions, no comprehensive phytochemical and biological evaluations have previously been reported for the Algerian population of this species. Considering that environmental, climatic, and edaphic factors significantly influence secondary metabolite biosynthesis, regional ecotypes may exhibit distinct phytochemical fingerprints and bioactivity profiles.

To the best of our knowledge, this study represents the first detailed HPLC-based characterization of phenolic compounds from Algerian *E. angustifolia* leaves, and the first to correlate solvent-dependent extraction patterns with both antioxidant and α-amylase inhibitory activities, supported by molecular docking analysis. This integrative strategy provides novel insights into the pharmacological potential of the Algerian ecotype and expands the existing knowledge on the species within the *Elaeagnaceae* family.

2. Results

2.1. The Extraction Yields

The extraction yields of *E. angustifolia* leaf powder varied significantly depending on the solvent used for maceration (Table 1). The hydroethanolic solvent produced the highest extraction yield (18%), demonstrating its superior efficiency in extracting bioactive compounds. This was followed by n-butanol, which yielded 4.32%, and ethyl acetate with a yield of 3.84%. The lower-polarity solvents, dichloromethane and n-hexane, resulted in markedly lower extraction yields of 0.32% and 0.05%, respectively. These results underscore the influence of solvent polarity on extraction efficiency, with higher-polarity solvents leading to higher extraction yields.

Table 1. Extraction yields of *E. angustifolia* leaf powder using different solvents.

Solvent	Extraction Yield (%)
Hydroethanolic	18.00
N-Butanol	4.32
Ethyl acetate	3.84
Dichloromethane	0.32
N-Hexane	0.05

2.2. Total Phenolic Content, Flavonoid Content, and Tannin Content

The quantification of total phenols, flavonoids, and tannins in *E. angustifolia* leaf extracts revealed significant variation depending on the extraction solvent used (Table 2). The n-butanol extract exhibited the highest total phenol content (309.05 ± 0.64 mg GAE/g), followed by the ethyl acetate extract (290.97 ± 1.21 mg GAE/g) and the hydroethanolic extract (244.09 ± 2.07 mg GAE/g), while the n-hexane extract had the lowest phenolic content (92.23 ± 0.65 mg GAE/g). Regarding flavonoids, the ethyl acetate extract showed the highest concentration (102.11 ± 0.79 mg QE/g), whereas the n-butanol extract contained the lowest (29.33 ± 0.58 mg QE/g). For tannins, the dichloromethane (DCM) extract demonstrated the highest content (66.24 ± 0.79 mg CE/g), with hydroethanolic and n-butanol extracts showing moderate levels (25.58 ± 0.36–0.67 mg CE/g). Statistical analysis using one-way ANOVA followed by Dunnett's post hoc test indicated significant differences among the extracts for all measured parameters ($p < 0.05$), confirming that solvent polarity greatly influences the extraction efficiency of phenolics, flavonoids, and tannins. Overall, polar solvents such as n-butanol, ethyl acetate, and hydroethanolic mixtures were more

 https://doi.org/10.3390/molecules31050861

effective in extracting total phenols and flavonoids compared to non-polar solvents like n-hexane, while DCM selectively extracted high levels of tannins. These results highlight that solvent choice critically affects the yield and composition of bioactive compounds in *E. angustifolia* leaves, which may in turn influence their biological activities.

Table 2. Total phenol, flavonoid, and tannin contents of *E. angustifolia* leaf powder extracts using different solvents.

Solvent	Total Phenol Content (mg GAE/g)	Flavonoid Content (mg QE/g)	Tannin Content (mg CE/g)
Hydroethanolic	244.09 ± 2.07^{c}	53.90 ± 0.41^{c}	25.58 ± 0.36^{a}
N-Hexane	92.23 ± 0.65^{a}	37.77 ± 0.30^{b}	22.26 ± 0.92^{a}
Dichloromethane (DCM)	136.44 ± 0.64^{b}	40.28 ± 0.42^{b}	66.24 ± 0.79^{c}
Ethyl Acetate	290.97 ± 1.21^{c}	102.11 ± 0.79^{d}	30.71 ± 0.25^{b}
N-Butanol	309.05 ± 0.64^{d}	29.33 ± 0.58^{a}	25.58 ± 0.67^{a}

Different superscript letters (a–d) within the same column indicate statistically significant differences ($p < 0.05$). Values are mean ± standard deviation (n = 3). GAE = gallic acid equivalents; QE = quercetin equivalents; CE = catechin equivalents.

2.3. HPLC Analysis

Figure 1 and Table 3 depict the phenolic profiles of *E. angustifolia* leaf extracts obtained using five solvents of varying polarity: hydroethanolic, n-hexane, dichloromethane, n-butanol, and ethyl acetate, as determined by high-performance liquid chromatography (HPLC). A comparative analysis of the extracts revealed clear differences in their phenolic compositions, which corresponded to solvent polarity and chemical affinity.

The n-butanol extract delivered the highest concentration of rutin (75.77 mg/g), substantially exceeding that of ethyl acetate extract (15.25 mg/g), whereas rutin was not detected in the hydroethanolic, n-hexane, or dichloromethane extracts. This suggests that n-butanol effectively solubilizes the relatively polar molecular structure of rutin. Similarly, gallic acid was most abundant in the n-butanol fraction (11.87 mg/g), moderately present in the ethyl acetate extract, and either absent or present in the other fractions, reinforcing the greater efficiency of polar solvents for extracting polar phenolic acids.

Cinnamic acid, on the other hand, exhibited low concentrations across most solvents but peaked in the ethyl acetate extract, indicating its preferential solubility in solvents with intermediate polarity. Chicoric acid was moderately concentrated in hydroethanolic and n-hexane extracts but was virtually absent elsewhere, reflecting a more complex solubility profile that may be influenced by structural features conducive to solubility in both polar and relatively non-polar environments.

Notably, quercetin was only detected in the n-hexane, dichloromethane, and ethyl acetate extracts, with ethyl acetate extracting the highest amount. The absence of quercetin in the hydroethanolic and n-butanol extracts suggests selectivity in solvent–phytochemical interactions, possibly due to the balance of polar and non-polar characteristics of quercetin.

The dichloromethane fraction uniquely contained significant amounts of 4-hydroxyben zoic acid and ferulic acid, compounds with moderate polarity that preferentially dissolve in solvents of intermediate polarity. Chlorogenic acid was exclusive to the n-butanol extract, consistent with its polar nature, whereas vanillic acid was found only in the ethyl acetate fraction, further emphasizing solvent selectivity

2.4. Antioxidant Activities

As shown in Table 4, the antioxidant activities of *E. angustifolia* leaf extracts were evaluated using ABTS, FRAP, and DPPH assays, revealing notable variation depending on the extraction solvent (Table 4).

 https://doi.org/10.3390/molecules31050861

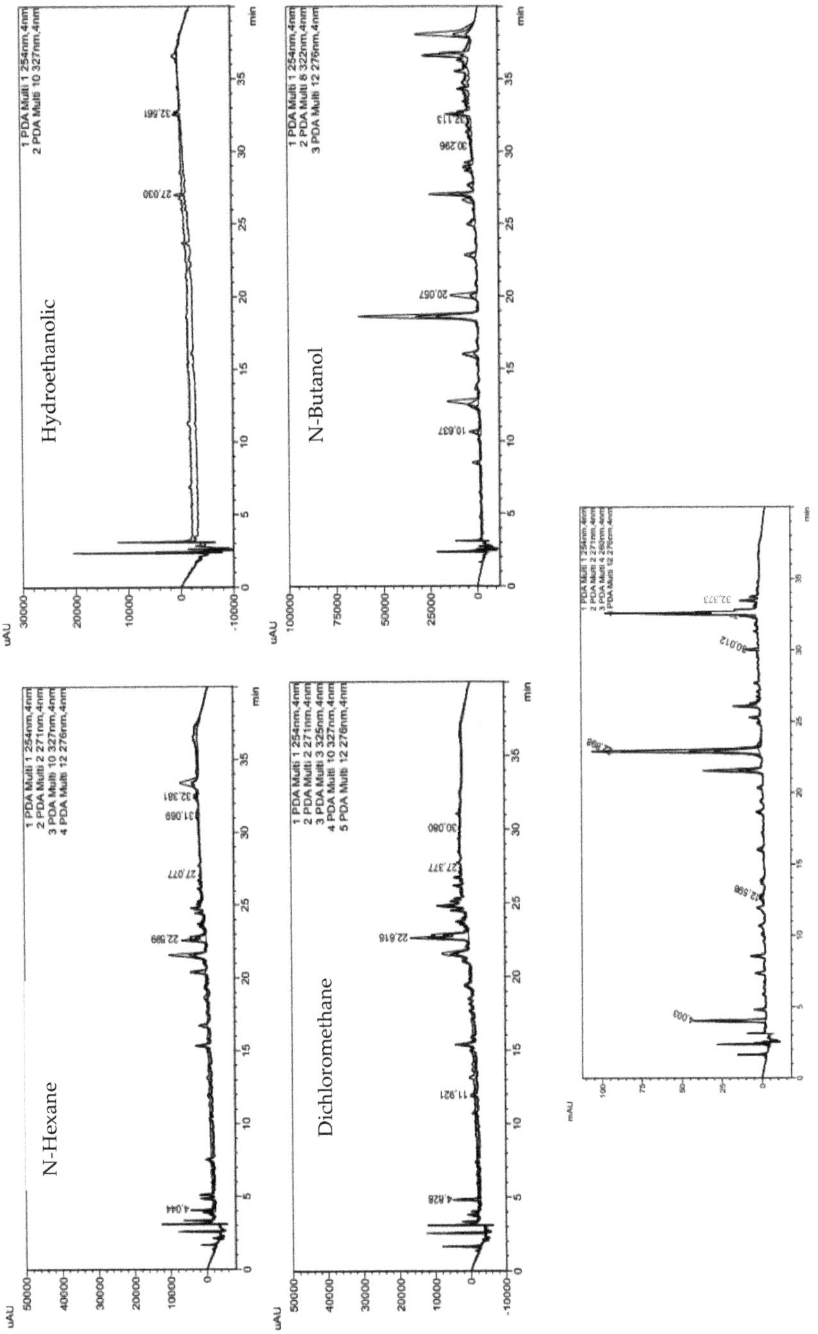

Figure 1. HPLC chromatographic profiles of *E. angustifolia* leaf extracts obtained with different solvents.

https://doi.org/10.3390/molecules31050861

Table 3. Identification and quantification of major phenolic compounds in *E. Angustifolia* leaf extracts obtained with different solvents using HPLC analysis.

	Hydroethanolic			N-Hexane			Dichloromethane			N-Butanol			Ethyl Acetate		
Compound	RT	Area	Conc.	RT	Area	Conc.	RT	Area	Conc.	RT	Area	Conc.	RT	Area	Conc.
Rutin	22.599	137,091	6.748 ± 0.21	ND	ND	ND	ND	ND	ND	22.616	309,807	15.250 ± 0.45	22.898	1,539,237	75.768 ± 1.32
Gallic Acid	4.044	34,380	1.048 ± 0.05	ND	ND	ND	ND	ND	ND	4.828	46,950	1.431 ± 0.07	4.003	389,470	11.873 ± 0.54
Cinnamic Acid	31.069	3617	0.041 ± 0.003	ND	ND	ND	30.296	2526	0.028 ± 0.002	5.620	1291	0.014 ± 0.001	30.012	52,781	0.592 ± 0.021
Chicoric Acid	27.077	3686	0.333 ± 0.012	27.030	10,046	0.906 ± 0.034	ND	ND	ND	27.377	1815	0.164 ± 0.006	ND	ND	ND
Quercetin	ND	ND	ND	32.561	9180	0.296 ± 0.011	32.113	3246	0.105 ± 0.004	ND	ND	ND	32.373	83,099	2.676 ± 0.09
4-Hydroxy benzoic Acid	ND	ND	ND	ND	ND	ND	10.637	98,936	1.508 ± 0.06	ND	ND	ND	ND	ND	ND
Ferulic Acid	ND	ND	ND	ND	ND	ND	20.057	230,241	3.834 ± 0.14	ND	ND	ND	ND	ND	ND
Chlorogenic Acid	ND	ND	ND	ND	ND	ND	ND	ND	ND	11.921	17,459	0.524 ± 0.018	ND	ND	ND
Vanillic Acid	ND	ND	ND	ND	ND	ND	ND	ND	ND	ND	ND	ND	12.598	32,291	0.684 ± 0.023

Concentration values of major phenolic compounds in *E. angustifolia* leaf extracts obtained with different solvents are presented as mean ± SD (n = 3). The symbol "ND" indicates that the compound was not detected in the corresponding extract. Standard deviations reflect variability among triplicate measurements.

https://doi.org/10.3390/molecules31050861

Table 4. Antioxidant activities of *E. angustifolia* leaf extracts evaluated by ABTS, FRAP, and DPPH assays.

Extract	ABTS (µg/mL)	FRAP (µg/mL)	DPPH (µg/mL)
Hydroethanolic	13.98 ± 0.20^{c}	35.88 ± 2.00^{b}	8.36 ± 0.09^{c}
N-hexane	41.15 ± 0.19^{f}	103.04 ± 1.83^{e}	35.55 ± 0.11^{h}
Dichloromethane	36.63 ± 3.31^{e}	138.65 ± 0.45^{f}	58.13 ± 0.13^{i}
Ethyl acetate	7.62 ± 0.17^{b}	41.31 ± 0.17^{c}	17.31 ± 0.29^{f}
N-butanolic	30.66 ± 0.81^{d}	81.32 ± 0.96^{d}	29.71 ± 0.12^{g}
BHT	1.29 ± 0.30^{a}	/	12.99 ± 0.15^{e}
BHA	1.81 ± 0.10^{a}	8.41 ± 0.67^{a}	6.14 ± 0.20^{b}
Ascorbic	/	9.01 ± 1.46^{a}	3 ± 1.12^{a}
α-Tocopherol	/	34.93 ± 2.38^{b}	13.02 ± 5.17^{d}

Different superscript letters (a–i) within the same column indicate statistically significant differences ($p < 0.05$).

The hydroethanolic extract showed moderate ABTS (13.98 ± 0.20 µg/mL) and DPPH (8.36 ± 0.09 µg/mL) activities and a FRAP value of 35.88 ± 2.00 µg/mL. The ethyl acetate extract exhibited the strongest ABTS activity (7.62 ± 0.17 µg/mL), while the dichloromethane extract demonstrated the highest FRAP (138.65 ± 0.45 µg/mL) and DPPH (58.13 ± 0.13 µg/mL) values. The n-butanol extract displayed moderate antioxidant activity across all assays, and the n-hexane extract showed comparatively weaker ABTS activity (41.15 ± 0.19 µg/mL) but strong FRAP and DPPH results, consistent with its composition. Among the standards, BHT, BHA, ascorbic acid, and α-tocopherol exhibited expected activities, with BHT and α-tocopherol showing strong DPPH scavenging, whereas BHA and ascorbic acid displayed generally lower activities. Statistical analysis using one-way ANOVA followed by Dunnett's post hoc test indicated significant differences among all extracts and standards for each assay ($p < 0.05$), confirming that both solvent polarity and extract composition critically influence antioxidant performance. Overall, the hydroethanolic and ethyl acetate extracts demonstrated antioxidant activities comparable to or approaching those of the standard compounds, highlighting their potential as effective natural antioxidants.

2.5. *In Vitro Antidiabetic Effect*

According to the results of the antidiabetic effects presented in Table 5, the inhibitory activity of the different extracts increased in a concentration-dependent manner, with clear differences observed between samples. The n-hexane extract exhibited the highest inhibitory activity, with inhibition percentages increasing from 29.03 ± 1.27% at 6.25 µg/mL to 82.04 ± 1.71% at 400 µg/mL. It showed an IC_{50} value of 36.70 ± 0.92 µg/mL, indicating strong inhibitory potency, since lower IC_{50} values correspond to higher biological activity. This value was significantly different ($p < 0.05$) from dichloromethane (278.4 ± 3.13 µg/mL) and the standard acarbose (126.14 ± 10.70 µg/mL), as indicated by different superscript letters (a–c), confirming statistically significant differences among samples.

Table 5. α-Amylase inhibitory activity of different *E. angustifolia* leaf extract fractions.

Samples	6.25 µg/mL	12.5 µg/mL	25 µg/mL	50 µg/mL	100 µg/mL	200 µg/mL	400 µg/mL	IC_{50} (µg/mL)
Hydroethanolic	ND	ND	ND	ND	ND	ND	ND	>400
n-Hexane	29.03 ± 1.27	34.44 ± 2.24	43.78 ± 0.16	58.06 ± 0.27	68.78 ± 3.53	79.74 ± 0.90	82.04 ± 1.71	36.70 ± 0.92^{a}
Dichloromethane	14.50 ± 1.81	18.74 ± 1.37	18.14 ± 2.88	24.75 ± 2.08	35.89 ± 0.55	40.66 ± 2.45	56.82 ± 0.27	278.4 ± 3.13^{c}
Ethyl acetate	ND	ND	ND	ND	ND	ND	ND	>400
n-Butanol	ND	ND	ND	ND	ND	ND	ND	>400
Acarbose	29.76 ± 0.17	33.36 ± 0.30	39.46 ± 0.11	43.68 ± 0.96	48.76 ± 0.17	57.36 ± 0.30	69.46 ± 0.11	126.14 ± 10.70^{b}

Different superscript letters (a–c) within the same column indicate statistically significant differences ($p < 0.05$).
Values are expressed as mean ± SD (n = 3). ND = not detected.

https://doi.org/10.3390/molecules31050861

The dichloromethane extract demonstrated moderate inhibition, reaching 56.82 ± 0.27% at 400 μg/mL, but its high IC_{50} value reflects weaker potency. The standard drug acarbose showed intermediate activity compared to the tested extracts. In contrast, the hydroethanolic, ethyl acetate, and n-butanol extracts exhibited no detectable inhibition (ND) at the tested concentrations, with IC_{50} values > 400 μg/mL, indicating negligible activity under the experimental conditions.

All values are expressed as mean ± SD (n = 3). Statistical analysis was carried out using one-way ANOVA followed by Tukey's post hoc test, and differences were considered significant at $p < 0.05$. Values with different superscript letters indicate statistically significant differences between groups.

2.6. *In Silico Molecular Docking*

To investigate the potential antidiabetic activity of the compounds identified in *E. angustifolia*, molecular docking simulations were performed targeting α-amylase (PDB ID: 3BAJ). The docking results were compared with those of acarbose, a known α-amylase inhibitor, which served as the co-crystallized reference ligand. All compounds identified in the HPLC profiles (Figure 1) of the five extracts were subjected to docking analysis. However, only those exhibiting the highest binding affinities were retained for detailed interpretation, and the results are represented in Table 6 and Figure 2.

Table 6. Molecular docking results of the most active compounds identified from HPLC profiles of *E. angustifolia*. extracts against α-amylase (PDB ID: 3BAJ).

Target	Compound	Binding Energy (kcal/mol)	Ki (μM)	Hydrogen Interactions (Å)	Hydrophobic Interactions
α-Amylase (3BAJ)	Acarbose (co-crystallized)	−9.54	0.10	Tyr62, His299, Asp300, Arg195, Glu233, Lys200, etc.	–
	Rutin	−8.66	0.45	Asp300, Glu240, His201, Glu233, Ala198	His305
	Chicoric acid	−7.82	1.90	Tyr62, His201, Lys200, Thr163, Ile148	Tyr151
	Chlorogenic acid	−7.32	4.40	Lys200, Glu240, Asp300, Asp197	Ile235, Tyr151
	Quercetin	−6.71	12.1	Lys200, His201	Ile235, Tyr151, Leu162, Ala198

The co-crystallized ligand acarbose displayed the strongest binding affinity with a binding energy of −9.54 kcal/mol. It formed multiple hydrogen bonds with key active-site residues such as Tyr62, His299, Asp300, Arg195, Glu233, Lys200, Glu240, Gln63, Thr163, and Trp59. These interactions validate the reliability of the docking protocol.

Among the tested compounds, rutin exhibited the most promising interaction with α-amylase, with a binding energy of −8.66 kcal/mol. It formed hydrogen bonds with Asp300, Glu240, His201, Glu233, and Ala198, along with a hydrophobic interaction involving His305.

Chicoric acid demonstrated a binding energy of −7.82 kcal/mol and formed hydrogen bonds with residues such as Tyr62, His201, Lys200, Thr163, and Ile148. A hydrophobic interaction with Tyr151 was also observed. Chlorogenic acid exhibited a binding energy of −7.32 kcal/mol and formed strong hydrogen bonds particularly with Asp300, Lys200, Glu240, and Asp197, in addition to hydrophobic contacts with Ile235 and Tyr151. Quercetin, although less active, still formed a stable complex with α-amylase, with a binding energy of −6.71 kcal/mol. Its binding involved hydrogen interactions with Lys200 and His201 and hydrophobic contacts with Ile235, Tyr151, Leu162, and Ala198.

https://doi.org/10.3390/molecules31050861

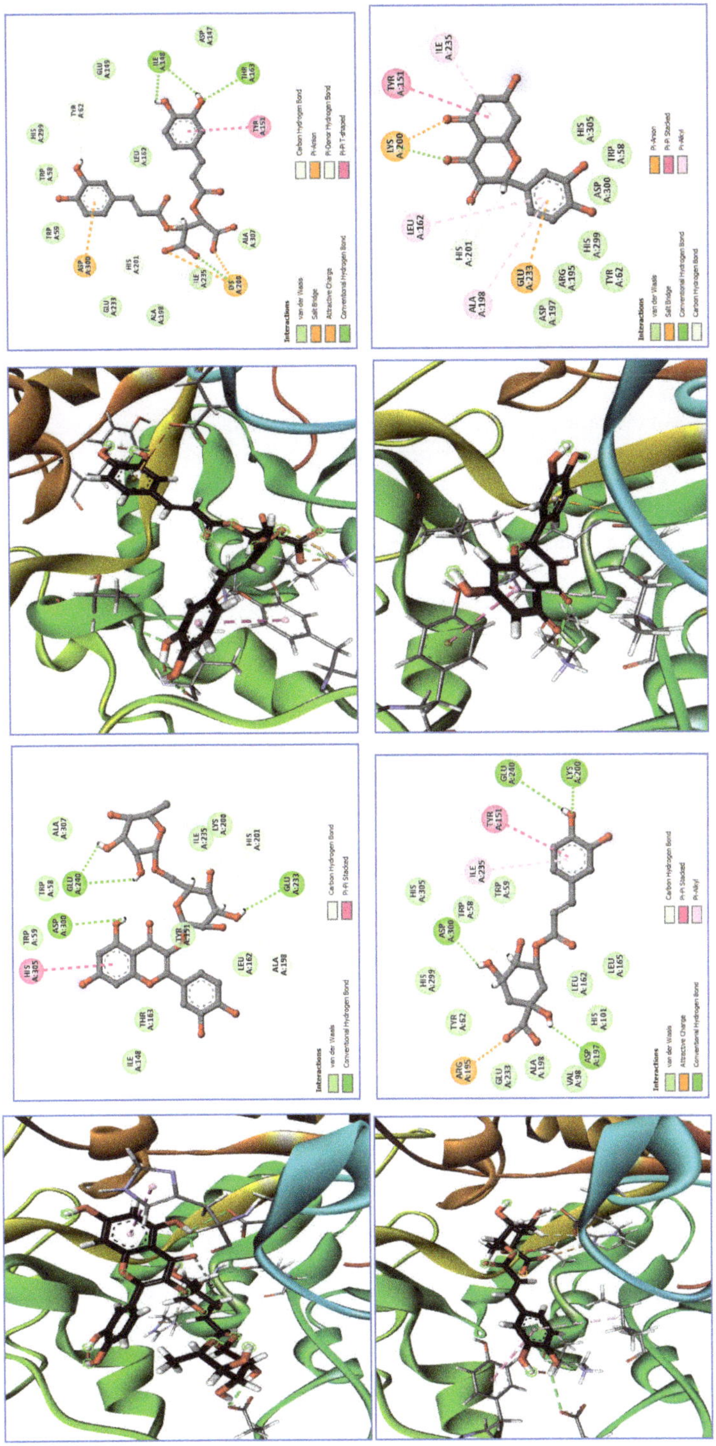

Figure 2. Two-dimensional and three-dimensional representations of the best docked compounds (rutin, chicoric acid, chlorogenic acid, and quercetin) within the active site of α-amylase (PDB ID: 3BAJ).

https://doi.org/10.3390/molecules31050861

3. Discussion

According to Tourabi et al., 2025 [22], the observed differences in extraction yields can be primarily attributed to the varying polarities of the solvents used during maceration. Solvent polarity is a critical factor influencing the solubility of phytochemicals, such as phenolics, flavonoids, and other bioactive compounds present in *E. angustifolia* leaves. Polar solvents like hydroethanol are more effective in extracting polar compounds because of their ability to disrupt plant cell walls and solubilize hydrophilic constituents [23]. In contrast, non-polar solvents such as n-hexane tend to dissolve fewer polar constituents, resulting in lower extraction yields [24]. Additionally, solvent penetration and interaction with plant matrices depend on the dielectric constant and hydrogen bonding capacity [25]. The variation in extraction efficiency aligns with these physicochemical principles, explaining why solvents with higher polarity generally yield more extractable materials. These aspects emphasize the importance of solvent selection in optimizing extraction protocols to maximize bioactive compound recovery.

A comparative analysis of the various solvent fractions used to extract bioactive compounds from *E. angustifolia* revealed distinct differences primarily driven by solvent polarity and chemical affinity. Polar solvents such as hydroethanolic, ethyl acetate, and n-butanol generally exhibit higher efficiency in extracting phenolic compounds, flavonoids, and tannins compared than less polar solvents as such as n-hexane and dichloromethane [26]. This aligns with the principle of "like dissolves like," as many phenolic and flavonoid molecules contain polar functional groups that facilitate the formation of hydrogen bonds with polar solvents, enhancing their solubility [26].

These observations correspond with findings from other studies, where polar organic solvents demonstrate superior extraction of polyphenolic compounds from *E. angustifolia* [27]. The efficiency differences among solvents can be explained by the chemical structures of the target compounds and their affinity for solvents of matching polarity [28]. Phenolic acids and flavonoids typically have multiple polar groups, favoring extraction by solvents capable of hydrogen bonding and polarity-based interactions, such as hydroethanolic and ethyl acetate mixtures. Conversely, non-polar solvents such as n-hexane are less effective because of their limited interactions with polar biomolecules [29,30].

Interestingly, the relatively high tannin content found in the dichloromethane fraction may be attributed to the structural complexity of tannins, which exist as both hydrolyzable and condensed forms with varying polarity. Some tannins possess moderately hydrophobic features, allowing partial solubility in solvents of intermediate polarity like dichloromethane [31–33]. This selective solubility illustrates that solvent polarity alone does not entirely determine extraction efficiency; the molecular size, degree of polymerization, and intramolecular hydrogen bonding within tannins also influence their affinity for specific solvents. Consequently, the dichloromethane fraction captures a unique profile of moderately polar tannins that are less accessible to highly polar or non-polar extraction media, highlighting the importance of tailored solvent selection to recover diverse bioactive constituents from *E. angustifolia* leaves.

Overall, the extraction profiles of *E. angustifolia* bioactive compounds were strongly influenced by solvent polarity and hydrogen bonding capabilities. Careful solvent selection based on the chemical nature of the phytochemicals is crucial for optimizing the recovery of phenolics, flavonoids, and tannins from *E. angustifolia* leaves.

Our findings add new information to the literature data on the *Elaeagnaceae* family. To the best of our knowledge, this is the first publication about the determination of the photochemical profile of the Algerian species *E angustifolia* via HPLC. The results proved the presence of bioactive compounds with interesting pharmacological activities. There-

https://doi.org/10.3390/molecules31050861

fore, this plant species could be identified as a potential source of compounds exhibiting antidiabetic and antioxidant properties.

The variations in the phenolic compound profiles observed across different *E. angustifolia* solvent extracts analyzed by HPLC are primarily due to the chemical structures of the compounds and their differential affinities for solvents with varying polarities. For instance, rutin, a flavonoid glycoside, contains multiple hydroxyl groups and sugar moieties, making it highly polar and capable of forming extensive hydrogen bonds [34]. This structural polarity explains the highest concentration of the compound in n-butanol solvent, which has a strong hydrogen bonding capacity, facilitating solubilization. Its absence in n-hexane or dichloromethane is due to their insufficient polarity to dissolve such hydrophilic compounds.

Similarly, gallic acid is a small phenolic acid with three hydroxyl groups that impart its pronounced polarity [35]. Their presence faciliates its effective extraction using polar solvents such as n-butanol and ethyl acetate, which can interact strongly with its polar phenolic groups. Its near absence in non-polar solvents aligns with its limited solubility in such media.

Cinnamic acid, which has a carboxyl group attached to an aromatic ring, exhibits moderate polarity. Therefore, it shows better extraction in solvents of intermediate polarity such as ethyl acetate, while being found only in low amounts in both highly polar and non-polar solvents [36]. Its amphiphilic nature allows some flexibility in solubility but generally favors solvents that balance polarity and hydrophobicity [37].

Furthermore, chicoric acid, a dicaffeoyltartaric acid derivative, possesses several polar hydroxyl and carboxyl groups as well as aromatic rings, giving it moderate-to-high polarity and hydrophobic characteristics [38]. Its extraction in both hydroethanolic (polar) and n-hexane (non-polar) solvents suggests structural features that allow partial solubility in both types of solvents, possibly because hydrophobic aromatic regions interact with less polar solvents and polar groups interact with polar solvents [39].

On the other hand, quercetin, a flavonoid aglycone without sugar attachments, is less polar than glycosides like rutin due to the absence of sugar moieties but still contains multiple hydroxyl groups [40]. It is preferentially extracted by medium-polarity solvents like ethyl acetate and dichloromethane, and to some extent n-hexane, reflecting its intermediate polarity and capability to interact with solvents that can balance polar and non-polar interactions [41]. The absence of highly polar solvents such as n-butanol may be due to the lower solubility of aglycones in very polar media.

Compounds such as 4-hydroxybenzoic acid and ferulic acid, detected mainly in the dichloromethane fraction, have moderate polarity and aromatic character. This favors extraction by solvents with intermediate polarity, which is consistent with their chemical structures that allow interaction via both hydrophobic and hydrogen bonding mechanisms [36].

Moreover, chlorogenic acid appeared only in the n-butanol extract, which can be explained by its highly polar ester linkage and multiple hydroxyl groups. These properties make it readily soluble in polar solvents [42].

Finally, vanillic acid, a methoxy-substituted hydroxybenzoic acid, was found exclusively in the ethyl acetate extract, indicating that its moderate polarity and hydrophobic character fit well with the polarity profile of this solvent.

Additionally, other factors can influence these extraction profiles, such as molecular size, possible formation of intramolecular hydrogen bonds that reduce polarity, and solvent viscosity, which affects diffusion rates. Moreover, the dielectric constant of the solvent and

https://doi.org/10.3390/molecules31050861

its ability to participate in specific interactions such as dipole–dipole and π-π stacking can preferentially stabilize certain compounds [43].

Thus, the selective extraction of phenolic compounds from *E. angustifolia* by different solvents is dictated by the intricate interplay of their chemical structures, polarity, hydrogen bonding capacity, and hydrophobic/hydrophilic balance. Understanding these relationships enables targeted extraction strategies to optimize yield of desired phytochemicals.

For the antioxidant activities of *E.angustifolia* leaf extracts, ABTS, FRAP, and DPPH assays revealed marked differences depending on the extraction solvent. The hydroethanolic and ethyl acetate extracts exhibited the strongest antioxidant potentials, which can be attributed to the presence of specific bioactive phenolic compounds, notably quercetin, ferulic acid, and 4-hydroxybenzoic acid [44]. These compounds are known for their potent radical scavenging and metal reducing properties, which likely contribute to the high performance in all three assays [45].

The n-butanol extract demonstrated moderate antioxidant capacity. This aligns with its relatively higher rutin and gallic acid content. These two compounds are widely recognized for their antioxidant potential, especially in aqueous or polar environments, which supports the observed activities [46]. The hydroethanolic extract also showed moderate effects, which can be correlated with the presence of rutin, gallic acid, and chicoric acid [47]. Although these phenolics were present at lower concentrations in the dichloromethane and n-hexane fractions, they still contributed synergistically to the antioxidant response. The differences observed among ABTS, DPPH, and FRAP results do indeed reflect the distinct reaction mechanisms and radical systems involved in each assay. The FRAP assay measures electron transfer capacity through Fe^{3+} reduction, the DPPH assay measures radical scavenging activity using a nitrogen-centered radical soluble in organic solvents, and the ABTS assay measures radical scavenging activity using a nitrogen-centered radical cation soluble in both aqueous and organic media [48–50]. These mechanistic differences, combined with differences in reaction pH, solvent, and sensitivity to steric effects, collectively explain the divergence among assay results for the same fraction [49,51]. The use of multiple complementary assays is therefore essential for a comprehensive characterization of antioxidant activity, as each assay captures a different dimension of the antioxidant capacity of phenolic-rich fractions [51].

Interestingly, the ethyl acetate extract, despite having the highest rutin concentration, displayed lower antioxidant activity. This might be explained by the limited diversity or synergistic interactions of phenolic compounds in this fraction, as well as possible degradation or reduced bioactivity in the solvent system. Additionally, the poor performance of certain standard antioxidants in some assays highlights the importance of assay specificity and the complex behavior of antioxidant molecules depending on their structure and polarity.

Overall, these findings suggest that the antioxidant potential of *E. angustifolia* extracts is closely linked to the total phenolic content and the specific composition and polarity of individual phenolic compounds. The variations observed between the solvent systems further underline the importance of solvent selection in optimizing the recovery of active antioxidants from plant materials.

For the antidiabetic effect, the observed α-amylase inhibitory activity of *E. angustifolia* leaf extract fractions, particularly the pronounced effect of the n-hexane and dichloromethane extracts, can be primarily attributed to the nature and chemical structures of the bioactive compounds selectively extracted by these solvents. The n-hexane fraction demonstrated the strongest inhibitory activity, which was significantly more potent

 https://doi.org/10.3390/molecules31050861

than that of the dichloromethane fraction and far exceeded the activity of the hydrophilic fractions and the acarbose control.

The pronounced α-amylase inhibitory activity observed in the n-hexane fraction indicates that non-polar constituents may contribute to the enzyme inhibition. However, in the absence of direct analytical characterization of this specific fraction (e.g., GC-MS profiling), this interpretation remains tentative. Non-polar solvents such as n-hexane are known to preferentially extract lipophilic molecules, including fatty acids and terpenoid derivatives previously reported in *E. angustifolia* [52–54]. These types of compounds may interact with hydrophobic regions of α-amylase or influence its conformational stability [55]. Nevertheless, the precise constituents responsible for the observed inhibition have not yet been identified, and further phytochemical investigation and bioactivity-guided fractionation are required to confirm the active molecules

Dichloromethane, which is moderately polar, extracts a mixture of lipophilic and some moderately polar compounds. The lower inhibitory potency compared to n-hexane could be related to the increased presence of less active polar constituents or dilution of the active principles. Moreover, compounds such as certain phenolics with mid-range polarity could partially contribute to enzyme inhibition but less effectively than the non-polar molecules abundant in the n-hexane fraction.

On the other hand, the negligible inhibition observed with the hydroethanolic, n-butanol, and ethyl acetate fractions correlates with their polarity profiles, favoring the extraction of more polar phenolic acids, flavonoid glycosides, and tannins. While many polar phenolics are known for their antioxidant activity, their interaction with α-amylase is often weaker or requires different structural features (e.g., number and position of hydroxyl groups) to exhibit significant enzyme inhibition. It is also possible that these polar compounds are unable to bind effectively or sterically hinder the active site of α-amylase.

Additional factors influencing inhibitory activity include molecular size and shape, functional groups capable of forming non-covalent interactions (hydrogen bonding and van der Waals forces), and the overall three-dimensional conformation of the bioactive molecules. For example, certain small-molecule terpenoids or phenolic aglycones present in non-polar extracts may better fit into the α-amylase active site or allosteric sites, enhancing inhibition [56].

Moreover, synergistic effects among phytochemicals within the n-hexane and dichloromethane fractions could amplify enzyme inhibition beyond the sum of the activities of individual constituents. Conversely, the inhibitory activity might be diminished in polar fractions due to antagonistic interactions or the presence of compounds that stabilize enzyme activity.

Finaly, solvent residues and extract matrix effects can influence bioassay outcomes by altering enzyme stability or assay conditions, although these should be minimized through appropriate extraction and purification procedures.

Thus, the strong α-amylase inhibitory effects of n-hexane and dichloromethane extracts from *E. angustifolia* highlight the importance of solvent selection in accordance with phytochemical polarity to target specific bioactivities. These findings underscore the effectiveness of non-polar bioactive components in inhibiting α-amylase, potentially contributing to the antidiabetic potential of these extracts. Further phytochemical characterization would be valuable for identifying and isolating the active constituents responsible for this activity.

E. angustifolia extract contains significant amounts of fatty acids (like n-Hexadecanoic acid, 9-Octadecenoic acid, and 1,2,3-propanetriyl ester), phytol, alkaloids, squalene, and

https://doi.org/10.3390/molecules31050861

isoquinoline derivatives [53]. This may explain the strong α-amylase inhibitory potential observed for this plant.

Our molecular docking analysis showed that rutin, chicoric acid, chlorogenic acid, and quercetin, isolated from *E. angustifolia* L., bind within the active site of α-amylase (PDB ID: 3BAJ), although none surpassed the binding affinity of acarbose (−9.54 kcal/mol).

Rutin, one of the top-scoring compounds from *E. angustifolia* L., exhibited a binding energy of −8.66 kcal/mol in our docking simulations against α-amylase (PDB ID: 3BAJ), which was similar to the stability and affinity of acarbose (−9.54 kcal/mol). It formed hydrogen bonds with Asp300, Glu240, His201, Glu233, and Ala198, and a hydrophobic interaction with His305. These interactions suggest well-oriented and energetically favorable binding within the enzyme's active site.

This in silico binding pattern echoes the findings of McMillan et al. [57], who identified rutin and acarbose as having among the best docking scores and highest stability when bound to pancreatic α-amylase and α-glucosidase. In their molecular dynamics simulations, rutin demonstrated conformational stability comparable to acarbose, and their in vitro assays showed that rutin's Ki values for both α-amylase and α-glucosidase inhibition were not significantly different ($p > 0.05$) from those of acarbose, suggesting nearly equivalent inhibitory potency.

The relevance of rutin as a potent α-amylase and α-glucosidase inhibitor is further reinforced by the findings of Lekmine et al. (2023) [58], who demonstrated that *Hyoscyamus albus* L. extract, rich in rutin among other phenolics, exhibited strong in vitro inhibition of these enzymes, with a notably greater effect on α-amylase (IC_{50} = 146.63 μg/mL). Their molecular docking simulations also identified rutin as one of the top contributors to the observed inhibitory effects, corroborating our own results.

Furthermore, Bouslamti et al. [59] reported rutin as the most active phenolic compound against DPP-IV in *Solanum elaeagnifolium*, with a glide docking score of −8.10 kcal/mol, corroborating its strong enzyme binding profile. Although their focus was on DPP-IV, the notable involvement of residues equivalent to those we identified (e.g., Tyr151) reinforces a shared binding mechanism across carbohydrate-related enzymes.

Additionally, Ibrahim et al. [15] showed that in sweet basil extracts enriched in rutin via elicitation protocols, rutin exhibited significant α-amylase inhibitory activity. Molecular docking into porcine pancreatic α-amylase revealed hydrogen bonding and hydrophobic interactions consistent with strong binding, with reported binding energies around −5.74 kcal/mol, reflecting similar mechanistic features to those observed in our model

Chicoric acid exhibited a noteworthy binding affinity for α-amylase, as reflected by a docking score of −7.82 kcal/mol. Its interaction profile revealed the formation of multiple hydrogen bonds with key residues in the enzyme active site, including Tyr62, His201, Lys200, Thr163, and Ile148. In particular, Ile148 formed two strong hydrogen bonds at distances of 2.27 Å and 2.52 Å, suggesting a stable and specific interaction with the catalytic groove of the enzyme. Additionally, a hydrophobic interaction with Tyr151 further reinforced the binding stability of chicoric acid in the active pocket. Despite its promising in silico interaction with α-amylase, to the best of our knowledge, no previous study has reported or investigated the anti-amylase activity of chicoric acid. This lack of data in the current scientific literature positions our findings as a novel contribution to the field. Given the strong binding of chicoric acid to several residues implicated in enzymatic catalysis and substrate recognition, chicoric acid has emerged as a potential natural α-amylase inhibitor. These results warrant further experimental validation, both in vitro and in vivo, to confirm the inhibitory potential of chicoric acid and explore its applicability in the management of hyperglycemia and related metabolic disorders.

https://doi.org/10.3390/molecules31050861

Chlorogenic acid, one of the major compounds identified in *E. angustifolia* demonstrated a binding energy of −7.32 kcal/mol in our docking experiments against α-amylase (PDB ID: 3BAJ). The interaction profile revealed that chlorogenic acid formed multiple hydrogen bonds and hydrophobic contacts with key catalytic residues within the active site of α-amylase, notably Asp197, Glu233, and Asp300. These interactions are consistent with the mechanisms underlying enzymatic inhibition, supporting the plausibility of chlorogenic acid as a competitive inhibitor.

This computational observation aligns with previous experimental findings that documented the α-amylase inhibitory activity of chlorogenic acid. In particular, Narita and Inouy [60] demonstrated that chlorogenic acid from green coffee beans significantly inhibited porcine pancreatic α-amylase in a dose-dependent manner, with IC_{50} values of 9.10 μg/mL and 9.24 μg/mL, respectively. These results suggest that chlorogenic acid can modulate postprandial glucose levels by limiting starch breakdown, a property that is highly desirable in the context of type 2 diabetes management. Furthermore, the ability of chlorogenic acid to inhibit α-amylase has been corroborated by other studies that emphasize its role in reducing carbohydrate digestion and glucose absorption. For instance, Santana-Gálvez et al. (2017) [61] highlighted the antidiabetic potential of chlorogenic acid-rich extracts, attributing part of their activity to α-amylase inhibition.

More recently, Zheng et al. (2020) [62] investigated porcine pancreatic α-amylase and identified chlorogenic acid (CHA) as a mixed-type inhibitor of the enzyme, with an IC_{50} of 0.498 ± 0.013 mg/mL. Their spectroscopic and circular dichroism analyses confirmed that CHA altered the enzyme's secondary structure via hydrogen bond interactions, and docking studies revealed binding energies around −7.8 to −7.2 kcal/mol at both active and secondary sites, values consistent with our calculated binding affinity. Similarly, Cardullo et al. (2021) [63] showed that CA can inhibit both α-amylase and α-glucosidase, thereby slowing intestinal glucose absorption. To enhance its activity, the authors synthesized eleven CA-derived amides, among which two compounds (amides 8 and 11) exhibited significantly stronger inhibitory effects than native CA, even surpassing acarbose in α-glucosidase inhibition. Kinetic studies revealed that these compounds act through mixed or competitive inhibition mechanisms depending on the enzyme. Molecular docking further supported the enhanced binding affinity, as these derivatives formed additional interactions with the enzyme active sites. Notably, benzothiazole-based amide 11 showed a synergistic inhibitory effect when combined with acarbose, suggesting its promising therapeutic potential.

Quercetin displayed a binding energy of −6.71 kcal/mol, forming hydrogen bonds with Lys200 and His201 and hydrophobic contacts with Ile235, Tyr151, Leu162, and Ala198. These interactions align well with the evidence presented by Günal-Köroğlu et al. (2024) [64], who reported the hypoglycemic activity of quercetin with IC_{50} values comparable to those of acarbose, highlighting hydrogen bond formation with key active-site residues via docking, MD simulations, and QSAR modeling to explain the mechanism of carbohydrate digestion inhibition and reduction in glucose absorption. Similarly, Raut et al. (2023) [65] identified quercetin as the most potent inhibitor in a polyphenolic library using molecular docking, MM/GBSA, and 100 ns MD simulations, confirming strong hydrogen bonding with the catalytic triad of human pancreatic α-amylase and a favorable free energy of −27.03 kcal/mol. Their in vitro data further validated the effectiveness of quercetin, with an IC_{50} of 57.37 ± 0.9 μg/mL. In a spectroscopic study by Huang et al. (2024) [66], quercetin was shown to be a mixed-type inhibitor of α-amylase. The authors observed static quenching of tryptophan and tyrosine residues, reflecting conformational alterations, and confirmed hydrogen bonds with Asp197 and Gln63, along with hydrophobic

 https://doi.org/10.3390/molecules31050861

interactions involving Ala198, Leu165, and Trp59. This finding supports the multifaceted binding observed between His201 and Lys200. Furthermore, Shen et al. (2023) [67] reported that quercetin acts via non-competitive inhibition, with an IC_{50} of 0.325 mg/mL. Their combined spectroscopic (UV-vis, FTIR, and CD), kinetic, docking and MD analyses showed hydrogen bonding with Asp197, Glu233, and Asp300, as well as stable enzyme–quercetin complexes, consistent with the critical active-site contacts. Altuner et al. (2022) [68] further corroborated the inhibitory potential of quercetin through in silico docking, LigPlot+ analysis, and ADMET profiling. They visualized the strong interactions of quercetin within the α-amylase binding site and highlighted its favorable pharmacokinetic and drug-likeness properties when compared to acarbose [68].

In silico molecular docking was performed on selected phenolic compounds identified in *E. angustifolia* to explore their potential interactions with α-amylase. Chicoric acid, rutin, and other phenolics exhibited favorable binding poses and interactions within the enzyme active site, suggesting possible mechanistic contributions to enzyme inhibition. It should be noted, however, that docking scores are predictive and do not provide experimental validation of binding affinity. Interestingly, the n-hexane extract showed the strongest α-amylase inhibitory activity experimentally, despite lacking the phenolic compounds included in the docking study. This discrepancy is likely due to the presence of lipophilic constituents (e.g., terpenoids, fatty acids) in the non-polar n-hexane fraction, which were not included in the docking panel. Furthermore, inhibition observed in complex extracts may result from synergistic or additive effects among multiple compounds. Therefore, while docking provides mechanistic insights for selected phenolics, it does not fully account for the bioactivity of the complete extracts, particularly non-polar fractions such as n-hexane.

4. Materials and Methods

4.1. Collection, Handling, and Extraction of Botanical Sample

E. angustifolia leaves were collected from the Baghaï region (Khenchela, Algeria) and taxonomically authenticated by Prof. Zereib Azzedine (Department of Agronomy, Abbas Laghrour University, Khenchela). A voucher specimen (EA-01) was deposited at the Herbarium of the Faculty of Natural Sciences, Abbas Laghrour University, Khenchela, Algeria, for future reference. The collection and identification of plant material were conducted in accordance with institutional guidelines, and the voucher specimen is available for consultation upon reasonable request.

Fresh leaves were air-dried at room temperature (25 ± 2 °C) under shade and protected from direct sunlight to prevent phytochemical degradation. The dried material was mechanically ground into a fine powder and stored in airtight containers under dry, light-protected conditions until extraction. For extraction, 100 g of powdered plant material was macerated in 80% ethanol (ethanol:water, 80:20 *v*/*v*) at a 1:10 (*w*/*v*) ratio. The extraction was performed under continuous magnetic stirring in darkness at room temperature for 96 h to minimize oxidative degradation. The mixture was filtered through Whatman No. 1 filter paper, and the solvent was removed under reduced pressure using a rotary evaporator at 40 °C.

The resulting crude hydroethanolic extract was suspended in distilled water and sequentially partitioned with solvents of increasing polarity: n-hexane, dichloromethane, ethyl acetate, and n-butanol. Each organic fraction was concentrated under reduced pressure at 40 °C and stored at 4 °C until further phytochemical and biological analyses.

https://doi.org/10.3390/molecules31050861

4.2. Determination of Total Phenolics, Flavonoids, and Condensed Tannins

The total phenolic, flavonoid, and condensed tannin contents of *E. angustifolia* extracts were quantified using microplate-based colorimetric methods. Total phenolics were assessed following the Folin–Ciocalteu procedure described by Singleton and Rossi (1965) [69], with absorbance recorded at 765 nm and results expressed as μg gallic acid equivalents per mg extract (μg GAE/mg). Total flavonoids were determined using a modified aluminum chloride assay according to Topçu et al. (2007) [52], measuring absorbance at 415 nm and expressing results as μg quercetin equivalents per mg extract (μg QE/mg). Condensed tannins were evaluated using the vanillin–HCl method adapted from Serif et al. (2023) [70], with absorbance read at 500 nm and values reported as μg catechin equivalents per mg extract (μg CE/mg).

4.3. High-Performance Liquid Chromatography (HPLC)

Phenolic acids in *E. angustifolia* extracts were analyzed by high-performance liquid chromatography (HPLC) with a diode array detector (Shimadzu, Kyoto, Japan; LC-20AT pump, SPD-M20A DAD). Separation was achieved on a reverse-phase Dionex Acclaim Polar Advantage C16 column (150 × 4.6 mm, 3 μm) (Thermo Fisher Scientific, Kyoto, Japan) using a gradient of methanol (A) and 1% formic acid in water (B): 0 min (100% A), 10 min (60% A), 17.5 min (100% B) for washing and re-equilibration. The flow rate was 1 mL/min, the column temperature was 30 °C, and detection occurred at 280 nm. Individual acids—including gallic, protocatechuic, p-hydroxybenzoic, vanillic, syringic, caftaric, caffeic, p-coumaric, ferulic, chicoric, rosmarinic, and cinnamic acids—were identified by matching retention times and UV spectra to authentic standards.

Quantification was performed using external calibration curves prepared with pure standard compounds at seven concentrations. The calibration curves showed excellent linearity with correlation coefficients (R^2) ≥ 0.995 for all phenolics. The limits of detection (LODs) and quantification (LOQs) were determined based on signal-to-noise ratios of 3:1 and 10:1, respectively, and ranged from 0.05 to 0.2 mg/kg for LODs and 0.15–0.6 mg/kg for LOQs, depending on the compound. Recovery studies were conducted by spiking known amounts of standards into the extracts, and recoveries ranged from 94 to 103%, demonstrating the accuracy and reliability of the method. The amounts of phenolic acids in the samples were then calculated from the calibration curves and are expressed in mg/kg dry weight.

4.4. Antioxidant Activity

The antioxidant potential of *E. angustifolia* extracts was evaluated using DPPH, ABTS, and reducing-power assays. For DPPH and ABTS scavenging, 40 μL of each extract at varying concentrations was incubated with the respective radicals (160 μL) in the dark, and absorbance was measured at 517 nm (DPPH) or 734 nm (ABTS) to calculate IC50 values, using BHA, BHT, Trolox, and ascorbic acid as standards [71,72]. Reducing power was determined by mixing 10 μL of extract with phosphate buffer and potassium ferricyanide, followed by incubation at 50 °C and reaction with trichloroacetic acid, water, and ferric chloride; absorbance at 700 nm was recorded, and A0.5 values were calculated [73]. All assays were performed in triplicate, and the results reflect the concentration required for 50% radical inhibition or an absorbance of 0.5, indicating antioxidant strength.

4.5. Alpha-Amylase Antidiabetic Activity Bioassay

The α-amylase inhibitory activity of *E. angustifolia* extracts was evaluated following the method described by Zengin et al. (2014) [74].

https://doi.org/10.3390/molecules31050861

Porcine pancreatic α-amylase (EC 3.2.1.1, Sigma-Aldrich, Saint Louis, MO, USA) was used as the enzyme source and prepared at a concentration of 10 U/mL in phosphate buffer (20 mM, pH 6.9, containing 6 mM NaCl).

Briefly, 25 μL of plant extract (1 mg/mL) was mixed with 50 μL of α-amylase solution and pre-incubated at 37 °C for 10 min to allow enzyme–inhibitor interaction. Subsequently, 50 μL of 0.1% (*w*/*v*) soluble starch solution (prepared in the same phosphate buffer) was added as substrate, and the mixture was incubated at 37 °C for 10 min.

The reaction was terminated by adding 25 μL of 1 M HCl. Then, 100 μL of iodine–potassium iodide reagent (I_2/KI solution) was added to detect residual starch. After 10 min of incubation at 37 °C, absorbance was measured at 630 nm using an EnSpire Multimode Plate Reader (PerkinElmer, Springfield, IL, USA).

Acarbose was used as a positive control under identical conditions.

The following controls were included:

Control (Ac): enzyme + substrate without extract

Sample (As): enzyme + substrate + extract

Blank (Ab): extract + substrate without enzyme

Enzyme blank (Ae): enzyme without substrate

The percentage of α-amylase inhibition was calculated using the equation

$$\alpha\text{-Amylase inhibition } (\%) = 1 - \frac{(A_c - A_e) - (A_s - A_b)}{A_c - A_e}$$

4.6. Statistical Analysis

The statistical analysis of the data was conducted utilizing GraphPad Prism DataEditor 6.0. Initially, one-way ANOVA was employed to analyze the data, followed by Dunnett's test to identify significant differences between the tests. The results were expressed as the mean ± standard error of the mean. A p-value of less than 0.05 was considered statistically significant, indicating meaningful differences between the experimental groups. This comprehensive analysis allowed for the accurate interpretation of the data and the identification of significant outcomes within the study.

4.7. In Silico Molecular Docking

Major phytochemicals identified in *E. angustifolia* extracts by HPLC were selected for molecular docking studies. Their 3D structures were retrieved from PubChem and converted to Mol2 format using Open Babel [75]. Ligand preparation was performed in AutoDock Tools [76]. by adding Gasteiger charges, merging non-polar hydrogens, and defining rotatable bonds before saving the structures as PDBQT files.

The crystal structure of human pancreatic α-amylase (PDB ID: 3BAJ) was obtained from the Protein Data Bank [76]. Specifically, the structure corresponds to human pancreatic α-amylase in complex with an inhibitor [77]. Water molecules and heteroatoms were removed, polar hydrogens were added, and Kollman charges were assigned using AutoDock Tools.

Docking simulations were carried out using AutoDock 4.2.6 [76]. employing the Lamarckian Genetic Algorithm. A grid box was centered on the enzyme active site to ensure adequate sampling of ligand conformations within the catalytic pocket. Each ligand was subjected to 50 independent docking runs, and the best binding pose was selected based on the lowest binding energy and clustering analysis.

Binding modes and molecular interactions, including hydrogen bonds and hydrophobic contacts, were visualized and analyzed using Discovery Studio Visualizer (BIOVIA, 2019).

https://doi.org/10.3390/molecules31050861

5. Conclusions

Collectively, these in vitro results establish a direct relationship between solvent polarity, phenolic compound profiles, and the respective antioxidant and antidiabetic activities of the extracts. This underscores the critical importance of optimizing extraction solvents to maximize the yield of therapeutically valuable phytochemicals. The study highlights, for the first time, the potent in vitro α-amylase inhibitory activity of non-polar leaf extracts from *E. angustifolia*, thereby supporting its potential as a promising source for the discovery of novel antidiabetic agents. However, these findings are preliminary and necessitate future bioassay-guided fractionation to isolate the pure active compound(s), followed by detailed in vivo pharmacological and toxicological studies to truly validate its therapeutic potential in managing diabetes and its complications.

Author Contributions: Data curation, R.B. and R.D.; Formal analysis, S.B. (Saber Boutellaa), G.O. and I.M.; Investigation, R.D.; Methodology, R.B.; Software, S.L., O.B. and M.I.R.; Validation, S.B. (Samira Bouhalit), N.T. and G.N. All authors have read and agreed to the published version of the manuscript.

Funding: This project was funded by the Deanship of Scientific Research (DSR) at King Abdulaziz University, Jeddah, Saudi Arabia under grant no. (**IPP: 373-188-2025**). The authors therefore acknowledge with thanks DSR for technical and financial support.

Institutional Review Board Statement: Not applicable.

Informed Consent Statement: Not applicable.

Data Availability Statement: The original contributions of this study are available in the article. For additional information, please contact the corresponding authors.

Acknowledgments: This project was funded by the Deanship of Scientific Research (DSR) at King Abdulaziz University, Jeddah, Saudi Arabia under grant no. (**IPP: 373-188-2025**). The authors therefore acknowledge with thanks DSR for technical and financial support.

Conflicts of Interest: The authors declare no conflicts of interest.

References

1. Rezvani, M. Oxidative Stress-Induced Gastrointestinal Diseases: Biology and Nanomedicines—A Review. *BioChem* **2024**, *4*, 189–216. [CrossRef]
2. Benslama, O.; Lekmine, S.; Moussa, H.; Tahraoui, H.; Ola, M.S.; Zhang, J.; Amrane, A. Silymarin as a Therapeutic Agent for Hepatocellular Carcinoma: A Multi-Approach Computational Study. *Metabolites* **2025**, *15*, 53. [CrossRef]
3. Afzal, S.; Abdul Manap, A.S.; Attiq, A.; Albokhadaim, I.; Kandeel, M.; Alhojaily, S.M. From Imbalance to Impairment: The Central Role of Reactive Oxygen Species in Oxidative Stress-Induced Disorders and Therapeutic Exploration. *Front. Pharmacol.* **2023**, *14*, 1269581. [CrossRef]
4. Guendouze, A.; Mokrani, E.H.; Benslama, O.; Lekmine, S. Exploring the Potential of Avenanthramides and Their Analogues as α-Glucosidase Inhibitors for Type 2 Diabetes Treatment Utilizing Virtual Screening, Molecular Dynamics, and Drug-Likeness Predictions. *Lett. Drug Des. Discov.* **2024**, *21*, 3858–3871. [CrossRef]
5. Tedjani, A.; Boual, Z.; El Hadj, M.D.O.; Lanez, T.; Belkhalfa, H.; El Alaoui-Talibi, Z.; El Modafar, C.; Abdelkafi, S.; Fendri, I.; Le Cerf, D. Antidiabetic Potential of Mucilage Fraction Extracted from Astragalus Gyzensis Seeds. *Eur. J. Biol. Res.* **2023**, *13*, 18–30.
6. Lekmine, S.; Benslama, O.; Ola, M.S.; Touzout, N.; Moussa, H.; Tahraoui, H.; Hafsa, H.; Zhang, J.; Amrane, A. Preliminary Data on Silybum Marianum Metabolites: Comprehensive Characterization, Antioxidant, Antidiabetic, Antimicrobial Activities, LC-MS/MS Profiling, and Predicted ADMET Analysis. *Metabolites* **2025**, *15*, 13. [CrossRef] [PubMed]
7. Mihoubi, N.; Ferhat, S.; Nedjhioui, M.; Zenati, B.; Lekmine, S.; Boudraa, R.; Ola, M.S.; Zhang, J.; Amrane, A.; Tahraoui, H. A Novel Halophilic Bacterium for Sustainable Pollution Control: From Pesticides to Industrial Effluents. *Water* **2025**, *17*, 888. [CrossRef]

 https://doi.org/10.3390/molecules31050861

8. Lekmine, S.; Benslama, O.; Bensalah, B.; Touzout, N.; Moussa, H.; Tahraoui, H.; Ola, M.S.; Hafsa, H.; Zhang, J.; Amrane, A. Bioactive Phenolics of *Hyoscyamus muticus* L. Subsp. Falezlez: A Molecular and Biochemical Approach to Antioxidant and Urease Inhibitory Activities. *Int. J. Mol. Sci.* **2025**, *26*, 370. [CrossRef]
9. Gondo, T.F. *Advancing Selectivity in Extraction and Analysis of Bioactive Compounds in Seaweed and Plant-Based Foods*; Lund University: Lund, Sweden, 2024.
10. Silva, A.M.; Lago, J.P.; Pinto, D.; Moreira, M.M.; Grosso, C.; Cruz Fernandes, V.; Delerue-Matos, C.; Rodrigues, F. Salicornia Ramosissima Bioactive Composition and Safety: Eco-Friendly Extractions Approach (Microwave-Assisted Extraction vs. Conventional Maceration). *Appl. Sci.* **2021**, *11*, 4744. [CrossRef]
11. Chen, R.-Z.; Tan, L.; Jin, C.-G.; Lu, J.; Tian, L.; Chang, Q.-Q.; Wang, K. Extraction, Isolation, Characterization and Antioxidant Activity of Polysaccharides from Astragalus Membranaceus. *Ind. Crops Prod.* **2015**, *77*, 434–443. [CrossRef]
12. Djellal, S.; Dahmoune, F.; Aoun, O.; Remini, H.; Belbahi, A.; Dairi, S.; Moussa, H.; Lakehal, M.; Kassouar, S.; Lakhdari, C. Optimization of Ultrasound-Assisted Extraction of Galactomannan from Carob Seeds "*Ceratonia siliqua* L." and Evaluation of Their Functional Properties and in Vitro Anti-Inflammatory Activity. *Sep. Sci. Technol.* **2024**, *59*, 41–58. [CrossRef]
13. Gherdaoui, D.; Yahoum, M.M.; Toumi, S.; Lekmine, S.; Lefnaoui, S.; Benslama, O.; Bouallouche, R.; Tahraoui, H.; Ola, M.S.; Ali, A. Elucidating Chiral Resolution of Aromatic Amino Acids Using Glycopeptide Selectors: A Combined Molecular Docking and Chromatographic Study. *Int. J. Mol. Sci.* **2024**, *25*, 9120. [CrossRef]
14. Shah, N.A.; Khan, M.R.; Sattar, S.; Ahmad, B.; Mirza, B. HPLC-DAD Analysis, Antioxidant Potential and Anti-Urease Activity of Asparagus Gracilis Collected from District Islamabad. *BMC Complement. Altern. Med.* **2014**, *14*, 347. [CrossRef]
15. Ibrahim, R.M.; Abdel-Baki, P.M.; El-Rashedy, A.A.; Mahdy, N.E. LC-MS/MS Profiling of Tipuana Tipu Flower, HPLC-DAD Quantification of Its Bioactive Components, and Interrelationships with Antioxidant, and Anti-Inflammatory Activity: In Vitro and in Silico Approaches. *BMC Complement. Med. Ther.* **2024**, *24*, 176. [CrossRef]
16. Moussa, H.; Hamid, S.; Mameri, A.; Lekmine, S.; Tahraoui, H.; Kebir, M.; Touzout, N.; Dahmoune, F.; Ola, M.S.; Zhang, J. From Green Chemistry to Healthy Environments: Silver Nanoparticles as a Dual Antioxidant and Antibacterial Agents for Advancing Biomedicine and Sustainable Wastewater Treatment. *Bioengineering* **2024**, *11*, 1205. [CrossRef]
17. Boukada, F.; Meddah, B. Flavonoids from aerial part of Algerian *Ajuga iva* (L.) Schreb.: The HPLC-UV analysis and antioxidant capacity. *Kragujev. J. Sci.* **2021**, *43*, 23–34. [CrossRef]
18. Sevindik, M.; Krupodorova, T.; Sevindik, E.; Koçer, O.; Uysal, I.; Ünal, O. *Elaeagnus angustifolia* L.: A Comprehensive Review of Its Biological Activities, Phenolic and Chemical Constituents, and Applications. *Appl. Fruit Sci.* **2025**, *67*, 70. [CrossRef]
19. Mohammed, F.S.; Pehlivan, M.; Sevindik, M. Antioxidant, Antibacterial and Antifungal Activities of Different Extracts of Silybum Marianum Collected from Duhok (Iraq). *Int. J. Second. Metab.* **2019**, *6*, 317–322. [CrossRef]
20. Francomano, F.; Caruso, A.; Barbarossa, A.; Fazio, A.; La Torre, C.; Ceramella, J.; Mallamaci, R.; Saturnino, C.; Iacopetta, D.; Sinicropi, M.S. β-Caryophyllene: A Sesquiterpene with Countless Biological Properties. *Appl. Sci.* **2019**, *9*, 5420. [CrossRef]
21. Silva, R.C.e.; Costa, J.S.d.; Figueiredo, R.O.d.; Setzer, W.N.; Silva, J.K.R.d.; Maia, J.G.S.; Figueiredo, P.L.B. Monoterpenes and Sesquiterpenes of Essential Oils from Psidium Species and Their Biological Properties. *Molecules* **2021**, *26*, 965. [CrossRef] [PubMed]
22. Tourabi, M.; Faiz, K.; Ezzouggari, R.; Louasté, B.; Merzouki, M.; Dauelbait, M.; Bourhia, M.; Almaary, K.S.; Siddique, F.; Lyoussi, B. Optimization of Extraction Process and Solvent Polarities to Enhance the Recovery of Phytochemical Compounds, Nutritional Content, and Biofunctional Properties of *Mentha longifolia* L. Extracts. *Bioresour. Bioprocess.* **2025**, *12*, 24. [CrossRef]
23. Wang, L.; Li, T.; Wu, C.; Fan, G.; Zhou, D.; Li, X. Unlocking the Potential of Plant Polyphenols: Advances in Extraction, Antibacterial Mechanisms, and Future Applications. *Food Sci. Biotechnol.* **2025**, *34*, 1235–1259. [CrossRef]
24. Yu, Z.-W.; Wang, J.; Lin, F.-H.; He, F.-M.; Li, H.-S.; Wang, S.-S.; Fang, M.-J.; Li, B.-C.; Qiu, Y.-K. A Comprehensive Study on Solvent Effect and Establishment of N-Hexane Solvent System Based Normal-Phase Liquid Chromatography× Reversed-Phase Liquid Chromatography for Isolation of Natural Products. *J. Chromatogr. A* **2024**, *1733*, 465278. [CrossRef] [PubMed]
25. Gil-Martín, E.; Forbes-Hernández, T.; Romero, A.; Cianciosi, D.; Giampieri, F.; Battino, M. Influence of the Extraction Method on the Recovery of Bioactive Phenolic Compounds from Food Industry By-Products. *Food Chem.* **2022**, *378*, 131918. [CrossRef] [PubMed]
26. Khatri, V.; Dey, P.K. Exploring the Dielectric Properties of Herbal Medicine and Modern Pharmaceuticals: An Integrative Review. *Front. Pharmacol.* **2025**, *15*, 1536397. [CrossRef] [PubMed]
27. Li, L.; Lv, J.; Wang, X.; Li, X.; Guo, D.; Wang, L.; Zhang, N.; Jia, Q. Green Extraction of Polyphenols from *Elaeagnus angustifolia* L. Using Natural Deep Eutectic Solvents and Evaluation of Bioactivity. *Molecules* **2024**, *29*, 2412. [CrossRef]
28. Barwick, V.J. Strategies for Solvent Selection—A Literature Review. *TrAC Trends Anal. Chem.* **1997**, *16*, 293–309. [CrossRef]
29. Palaiogiannis, D.; Chatzimitakos, T.; Athanasiadis, V.; Bozinou, E.; Makris, D.P.; Lalas, S.I. Successive Solvent Extraction of Polyphenols and Flavonoids from *Cistus creticus* L. Leaves. *Oxygen* **2023**, *3*, 274–286. [CrossRef]

30. Bitwell, C.; Indra, S.S.; Luke, C.; Kakoma, M.K. A Review of Modern and Conventional Extraction Techniques and Their Applications for Extracting Phytochemicals from Plants. *Sci. Afr.* **2023**, *19*, e01585. [CrossRef]
31. Bule, M.; Khan, F.; Nisar, M.F.; Niaz, K.; Nabavi, S.; Saeedi, M.; Sanches Silva, A. Tannins (Hydrolysable Tannins, Condensed Tannins, Phlorotannins, Flavono-Ellagitannins). In *Recent Advances in Natural Products Analysis*; Elsevier: Amsterdam, The Netherlands, 2020; pp. 132–146.
32. Molnar, M.; Jakovljević Kovač, M.; Pavić, V. A Comprehensive Analysis of Diversity, Structure, Biosynthesis and Extraction of Biologically Active Tannins from Various Plant-Based Materials Using Deep Eutectic Solvents. *Molecules* **2024**, *29*, 2615. [CrossRef]
33. Zeller, W.E. Activity, Purification, and Analysis of Condensed Tannins: Current State of Affairs and Future Endeavors. *Crop Sci.* **2019**, *59*, 886–904. [CrossRef]
34. Slámová, K.; Kapešová, J.; Valentová, K. "Sweet Flavonoids": Glycosidase-Catalyzed Modifications. *Int. J. Mol. Sci.* **2018**, *19*, 2126. [CrossRef] [PubMed]
35. Badhani, B.; Sharma, N.; Kakkar, R. Gallic Acid: A Versatile Antioxidant with Promising Therapeutic and Industrial Applications. *RSC Adv.* **2015**, *5*, 27540–27557. [CrossRef]
36. Lama-Muñoz, A.; Contreras, M.D.M. Extraction Systems and Analytical Techniques for Food Phenolic Compounds: A Review. *Foods* **2022**, *11*, 3671. [CrossRef]
37. Panda, S.K.; Mohanty, N.; Khuntia, P.; Mallad, A.; Jena, D. Influence of Solvent Polarity on Phytocompound Fingerprinting of Cinnamon Bark: Insights from HPTLC Analysis. *Food Anal. Methods* **2025**, *18*, 1942–1953. [CrossRef]
38. Yang, M.; Wu, C.; Zhang, T.; Shi, L.; Li, J.; Liang, H.; Lv, X.; Jing, F.; Qin, L.; Zhao, T. Chicoric Acid: Natural Occurrence, Chemical Synthesis, Biosynthesis, and Their Bioactive Effects. *Front. Chem.* **2022**, *10*, 888673. [CrossRef]
39. Novaes, F.J.M.; de Faria, D.C.; Ferraz, F.Z.; de Aquino Neto, F.R. Hansen Solubility Parameters Applied to the Extraction of Phytochemicals. *Plants* **2023**, *12*, 3008. [CrossRef]
40. Alizadeh, S.R.; Ebrahimzadeh, M.A. O-Glycoside Quercetin Derivatives: Biological Activities, Mechanisms of Action, and Structure–Activity Relationship for Drug Design, a Review. *Phytother. Res.* **2022**, *36*, 778–807. [CrossRef] [PubMed]
41. Zagrean-Tuza, C.; Mot, A.C.; Chmiel, T.; Bende, A.; Turcu, I. Sugar Matters: Sugar Moieties as Reactivity-Tuning Factors in Quercetin O-Glycosides. *Food Funct.* **2020**, *11*, 5293–5307. [CrossRef]
42. Lu, H.; Tian, Z.; Cui, Y.; Liu, Z.; Ma, X. Chlorogenic Acid: A Comprehensive Review of the Dietary Sources, Processing Effects, Bioavailability, Beneficial Properties, Mechanisms of Action, and Future Directions. *Compr. Rev. Food Sci. Food Saf.* **2020**, *19*, 3130–3158. [CrossRef] [PubMed]
43. Buncel, E.; Stairs, R.A. *Solvent Effects in Chemistry*; Wiley Online Library: Hoboken, NJ, USA, 2016.
44. Gunathilake, T.; Akanbi, T.O.; Suleria, H.A.R.; Nalder, T.D.; Francis, D.S.; Barrow, C.J. Seaweed Phenolics as Natural Antioxidants, Aquafeed Additives, Veterinary Treatments and Cross-Linkers for Microencapsulation. *Mar. Drugs* **2022**, *20*, 445. [CrossRef]
45. Luca, S.V.; Zengin, G.; Sinan, K.I.; Skalicka-Woźniak, K.; Trifan, A. Post-Distillation by-Products of Aromatic Plants from Lamiaceae Family as Rich Sources of Antioxidants and Enzyme Inhibitors. *Antioxidants* **2023**, *12*, 210. [CrossRef]
46. Jacobs, E.; Chambin, O.; Debeaufort, F.; Benbettaieb, N. Synergic versus Antagonist Effects of Rutin on Gallic Acid or Coumarin Incorporated into Chitosan Active Films: Impacts on Their Release Kinetics and Antioxidant Activity. *Antioxidants* **2023**, *12*, 1934. [CrossRef] [PubMed]
47. Shafi, W.; Mansoor, S.; Jan, S.; Singh, D.B.; Kazi, M.; Raish, M.; Alwadei, M.; Mir, J.I.; Ahmad, P. Variability in Catechin and Rutin Contents and Their Antioxidant Potential in Diverse Apple Genotypes. *Molecules* **2019**, *24*, 943. [CrossRef] [PubMed]
48. Yener, E.; Saroglu, O.; Sagdic, O.; Karadag, A. The Effects of Different Drying Methods on the in Vitro Bioaccessibility of Phenolics, Antioxidant Capacity, and Morphology of European Plums (*Prunes domestica* L.). *ACS Omega* **2024**, *9*, 12711–12724. [CrossRef]
49. Abcha, I.; Criado, P.; Salmieri, S.; Najjaa, H.; Isoda, H.; Neffati, M.; Lacroix, M. Edible Rhus Tripartita Fruit as Source of Health-Promoting Compounds: Characterization of Bioactive Components and Antioxidant Properties. *Eur. Food Res. Technol.* **2019**, *245*, 2641–2654. [CrossRef]
50. Abu, F.; Mat Taib, C.N.; Mohd Moklas, M.A.; Mohd Akhir, S. Antioxidant Properties of Crude Extract, Partition Extract, and Fermented Medium of Dendrobium Sabin Flower. *Evid.-Based Complement. Altern. Med.* **2017**, *2017*, 2907219.
51. Zhu, Z.; Zhong, B.; Yang, Z.; Zhao, W.; Shi, L.; Aziz, A.; Rauf, A.; Aljohani, A.S.M.; Alhumaydhi, F.A.; Suleria, H.A.R. LC-ESI-QTOF-MS/MS Characterization and Estimation of the Antioxidant Potential of Phenolic Compounds from Different Parts of the Lotus (Nelumbo Nucifera) Seed and Rhizome. *ACS Omega* **2022**, *7*, 14630–14642. [CrossRef]
52. Topçu, G.; Yücer, R.; Şenol, H. Bioactive Constituents of Anatolian Salvia Species. In *Salvia Biotechnology*; Springer: Berlin/Heidelberg, Germany, 2018; pp. 31–132.
53. Syed, A.L.; Shah, G.M.; Ashfaq, S.; Abrar, H.; e Baseerat, N.; Khan, N.; Hamza, M.; Hasan, M.A.; Ullah, K. GC-MS Analysis, Antimicrobial, Antioxidant, Phytochemical, and Phytotoxic Activities of *Elaeagnus angustifolia* L. Using Different Extracts. *Phytopharm. Res. J.* **2023**, *2*, 1–12.

 https://doi.org/10.3390/molecules31050861

54. Djeghim, H.; Bellil, I.; Benslama, O.; Lekmine, S.; Temim, E.; Boufendi, H.; Postigo, I.; Sánchez, P.; Khelifi, D. Effects of Genetic Diversity on the Allergenicity of Peanut (Arachis Hypogaea) Proteins: Identification of the Hypoallergenic Accessions Using BALB/c Mice Model and in Silico Analysis of Ara h 3 Allergen Cross-Reactivity. *J. Proteom.* **2024**, *306*, 105264. [CrossRef]
55. Zhu, J.; Chen, C.; Zhang, B.; Huang, Q. The Inhibitory Effects of Flavonoids on α-Amylase and α-Glucosidase. *Crit. Rev. Food Sci. Nutr.* **2020**, *60*, 695–708. [CrossRef]
56. Groussin, A.-L.; Antoniotti, S. Valuable Chemicals by the Enzymatic Modification of Molecules of Natural Origin: Terpenoids, Steroids, Phenolics and Related Compounds. *Bioresour. Technol.* **2012**, *115*, 237–243. [CrossRef]
57. McMillan, J.; Bester, M.J.; Apostolides, Z. In Silico Docking and ADMET Studies on Clinical Targets for Type 2 Diabetes Correlated to in Vitro Inhibition of Pancreatic Alpha-Amylase and Alpha-Glucosidase by Rutin, Caffeic Acid, p-Coumaric Acid, and Vanillin. *Silico Pharmacol.* **2025**, *13*, 42. [CrossRef]
58. Lekmine, S.; Benslama, O.; Kadi, K.; Martín-García, A.I.; Yilmaz, M.A.; Akkal, S.; Boumegoura, A.; Alhomida, A.S.; Ola, M.S.; Ali, A. LC/MS-MS Analysis of Phenolic Compounds in *Hyoscyamus albus* L. Extract: In Vitro Antidiabetic Activity, in Silico Molecular Docking, and in Vivo Investigation against STZ-Induced Diabetic Mice. *Pharmaceuticals* **2023**, *16*, 1015. [CrossRef]
59. Benslama, O.; Lekmine, S.; Mansouri, N. Phytochemical Constituents of Astragalus Monspessulanus and Integrative Analysis for Its Antioxidant, Photoprotective, and Antityrosinase Activities: Experimental and Computational Investigation. *Eur. J. Integr. Med.* **2023**, *60*, 102247. [CrossRef]
60. Narita, Y.; Inouye, K. Inhibition of Porcine Pancreas α-Amylase by Chlorogenic Acids from Green Coffee Beans and Cinnamic Acid Derivatives: A Focus on Kinetic. In *Coffee in Health and Disease Prevention*; Elsevier: Amsterdam, The Netherlands, 2015; pp. 757–763.
61. Santana-Gálvez, J.; Cisneros-Zevallos, L.; Jacobo-Velázquez, D.A. Chlorogenic Acid: Recent Advances on Its Dual Role as a Food Additive and a Nutraceutical against Metabolic Syndrome. *Molecules* **2017**, *22*, 358. [CrossRef] [PubMed]
62. Zheng, Y.; Yang, W.; Sun, W.; Chen, S.; Liu, D.; Kong, X.; Tian, J.; Ye, X. Inhibition of Porcine Pancreatic α-Amylase Activity by Chlorogenic Acid. *J. Funct. Foods* **2020**, *64*, 103587. [CrossRef]
63. Cardullo, N.; Floresta, G.; Rescifina, A.; Muccilli, V.; Tringali, C. Synthesis and in Vitro Evaluation of Chlorogenic Acid Amides as Potential Hypoglycemic Agents and Their Synergistic Effect with Acarbose. *Bioorg. Chem.* **2021**, *117*, 105458. [CrossRef]
64. Günal-Köroğlu, D.; Catalkaya, G.; Yusufoğlu, B.; Kezer, G.; Esatbeyoglu, T.; Abd El-Aty, A.M.; Capanoglu, E. Quercetin: Potential Antidiabetic Effects through Enzyme Inhibition and Starch Digestibility. *Food Saf. Health* **2025**, *3*, 9–22. [CrossRef]
65. Raut, B.K.; Upadhyaya, S.R.; Bashyal, J.; Parajuli, N. In Silico and in Vitro Analyses to Repurpose Quercetin as a Human Pancreatic α-Amylase Inhibitor. *ACS Omega* **2023**, *8*, 43617–43631. [CrossRef]
66. Huang, M.; Xiao, Q.; Li, Y.; Ahmad, M.; Tang, J.; Liao, Q.; Tan, C. Inhibition of α-Amylase Activity by Quercetin via Multi-Spectroscopic and Molecular Docking Approaches. *Food Biosci.* **2024**, *61*, 104951. [CrossRef]
67. Shen, Y.; Cheng, L.; Xu, M.; Wang, W.; Wan, Z.; Xiong, H.; Guo, W.; Cai, M.; Xu, F. SGLT2 Inhibitor Empagliflozin Downregulates MiRNA-34a-5p and Targets GREM2 to Inactivate Hepatic Stellate Cells and Ameliorate Non-Alcoholic Fatty Liver Disease-Associated Fibrosis. *Metabolism* **2023**, *146*, 155657. [CrossRef] [PubMed]
68. Altuner, E.M. In Silico Proof of the Effect of Quercetin and Umbelliferone as Alpha-Amylase Inhibitors, Which Can Be Used in the Treatment of Diabetes. *Kastamonu Univ. J. For. Fac.* **2022**, *22*, 202–216. [CrossRef]
69. Singleton, V.L.; Rossi, J.A. Colorimetry of Total Phenolics with Phosphomolybdic-Phosphotungstic Acid Reagents. *Am. J. Enol. Vitic.* **1965**, *16*, 144–158. [CrossRef]
70. Srief, M.; Bani, M.; Mokrani, E.H.; Mennai, I.; Hamdi, M.; Boumechhour, A.; Abou Mustapha, M.; Derdour, M.; Kerkatou, M.; El-Shazly, M. Evaluation of in Vitro and in Silico Anti-Alzheimer Potential of Nonpolar Extracts and Essential Oil from Mentha Piperita. *Foods* **2023**, *12*, 190. [CrossRef] [PubMed]
71. Sirivibulkovit, K.; Nouanthavong, S.; Sameenoi, Y. Based DPPH Assay for Antioxidant Activity Analysis. *Anal. Sci.* **2018**, *34*, 795–800. [CrossRef]
72. Oyaizu, M. Studies on Products of Browning Reaction Antioxidative Activities of Products of Browning Reaction Prepared from Glucosamine. *Jpn. J. Nutr. Diet.* **1986**, *44*, 307–315. [CrossRef]
73. Thaipong, K.; Boonprakob, U.; Crosby, K.; Cisneros-Zevallos, L.; Byrne, D.H. Comparison of ABTS, DPPH, FRAP, and ORAC Assays for Estimating Antioxidant Activity from Guava Fruit Extracts. *J. Food Compos. Anal.* **2006**, *19*, 669–675. [CrossRef]
74. Zengin, G.; Sarikurkcu, C.; Aktumsek, A.; Ceylan, R.; Ceylan, O. A Comprehensive Study on Phytochemical Characterization of Haplophyllum Myrtifolium Boiss. Endemic to Turkey and Its Inhibitory Potential against Key Enzymes Involved in Alzheimer, Skin Diseases and Type II Diabetes. *Ind. Crops Prod.* **2014**, *53*, 244–251. [CrossRef]
75. O'Boyle, N.M.; Banck, M.; James, C.A.; Morley, C.; Vandermeersch, T.; Hutchison, G.R. Open Babel: An Open Chemical Toolbox. *J. Cheminform.* **2011**, *3*, 33. [CrossRef]

 https://doi.org/10.3390/molecules31050861

76. Morris, G.M.; Huey, R.; Lindstrom, W.; Sanner, M.F.; Belew, R.K.; Goodsell, D.S.; Olson, A.J. AutoDock4 and AutoDockTools4: Automated Docking with Selective Receptor Flexibility. *J. Comput. Chem.* **2009**, *30*, 2785–2791. [CrossRef] [PubMed]
77. Brayer, G.D.; Luo, Y.; Withers, S.G. The Structure of Human Pancreatic A-amylase at 1.8 Å Resolution and Comparisons with Related Enzymes. *Protein Sci.* **1995**, *4*, 1730–1742. [CrossRef] [PubMed]

 https://doi.org/10.3390/molecules31050861

molecules

Article

Bioactive Potential and COX-2 Interaction of *Ajuga iva* (L.) Schreb. Hydroalcoholic Extract: Evidence from Experimental and Computational Studies

Yousra Boutora [1], Samira Boussekine [1], Ouided Benslama [2], Sabrina Lekmine [3], Nedjwa Mansouri [2,4], Nabil Touzout [5,6], Hamza Moussa [7,8], Rania Gacem [1], Najla Hfaiedh [9] and Gema Nieto [10,*]

1 Laboratory of Bioactive Molecules and Applications, Department of Applied Biology, Faculty of Exact Sciences and Natural and Life Sciences, Echahid Cheikh Larbi Tebessi University, Tebessa 12000, Algeria; yousra.boutoura@univ-tebessa.dz (Y.B.); boussekinesa@yahoo.fr (S.B.); rania.gace@univ-tebessa.dz (R.G.)
2 Department of Natural and Life Sciences, Faculty of Exact Sciences and Natural and Life Sciences, Larbi Ben M'Hidi University, Oum El Bouaghi 04000, Algeria; benslama.wided@hotmail.fr (O.B.)
3 Laboratory of Biotechnology, Water, Environment and Health (LBWEH), University of Abbes Laghrour, Khenchela 40000, Algeria
4 Higher National School of Forest, Khenchela 40000, Algeria
5 Department of Agronomy, Faculty of Sciences, Pole Urban Ouzera, University of Medea, Medea 26000, Algeria
6 Laboratory Materials and Environment, University Yahia Fares of Medea, Urban Center, Medea 26000, Algeria
7 Laboratoire de Gestion et Valorisation des Ressources Naturelles et Assurance Qualité (LGVRNAQ), Faculté des Sciences de la Nature et de la Vie et des Sciences de la Terre, Université de Bouira, Bouira 10000, Algeria; hamza.moussa49@gmail.com
8 Département des Sciences Biologiques, Faculté des Sciences de la Nature et de la Vie et des Sciences de la Terre, Université de Bouira, Bouira 10000, Algeria
9 Laboratory of Biotechnology and Biomonitoring of the Environment and Oasis Ecosystems (LBBEEO), Located at the Faculty of Sciences of Gafsa, University of Gafsa, Gafsa 2112, Tunisia; najlaharrathi@yahoo.fr
10 Department of Food Technology, Nutrition and Food Science, Veterinary Faculty, University of Murcia, Regional Campus of International Excellence "Campus Mare Nostrum", 30100 Espinardo, Murcia, Spain
* Correspondence: gnieto@um.es

Abstract

Ajuga iva (L.) Schreb. is traditionally used in North African ethnomedicine for the management of inflammation, pain, and fever. The present study aimed to characterize the phytochemical profile of the hydroalcoholic extract of its aerial parts and to evaluate its anti-inflammatory, analgesic, and antipyretic activities using established in vivo models. Preliminary phytochemical screening confirmed the presence of major classes of secondary metabolites, including polyphenols, flavonoids, tannins, and glycosidic compounds. Quantitative assays revealed appreciable levels of total phenolics (26.3 ± 1.2 mg GAE/g extract) and flavonoids (13.5 ± 0.9 mg QE/g extract). In vivo pharmacological evaluation demonstrated significant biological activities, with the highest tested dose (400 mg/kg) producing a marked inhibition of carrageenan-induced paw edema (44.9%), comparable to acetylsalicylic acid. At the same dose, the extract showed pronounced analgesic activity in the acetic acid-induced writhing test, with an inhibition rate of 64.2%, and a significant antipyretic effect in the brewer's yeast-induced fever model, as evidenced by a reduction in rectal temperature. In parallel, molecular docking was employed as an exploratory, hypothesis-generating in silico approach to investigate potential interactions between selected phenolic constituents identified in *A. iva* and cyclooxygenase-2 (COX-2). Several compounds, including rosmarinic acid, rutin, and apigenin-7-O-glucoside, displayed favorable predicted binding affinities and interactions with key residues of the COX-2 active site. It should be emphasized that molecular docking was used solely as a hypothesis-generating in silico tool and does not constitute direct biochemical evidence of COX-2 inhibition. Overall, these findings indicate that the hydroalcoholic extract of *Ajuga iva* exhibits notable anti-inflammatory, analgesic, and antipyretic activities in vivo. The in silico docking results provide supportive, predictive molecular insights that may help rationalize the observed bioactivities and encourage further biochemical and mechanistic investigations into this traditionally used medicinal plant.

Keywords: *Ajuga iva*; hydroalcoholic extract; anti-inflammatory; analgesic; antipyretic

1. Introduction

Medicinal plants continue to play a crucial role in the discovery of novel therapeutic agents, due to their richness in structurally diverse bioactive metabolites, including alkaloids, flavonoids, terpenoids, phenolics, polyphenols, and essential oils [1]. These naturally occurring compounds exhibit a wide range of pharmacological properties, such as antibacterial, antifungal, antioxidant, anti-inflammatory, and anticancer activities [2,3]. Among medicinal plants, *Ajuga iva* L. has been traditionally used for the treatment of various human disorders [4,5].

A. iva Schreb. is a perennial herbaceous species belonging to the Lamiaceae family and is native to the Mediterranean region [6,7]. In North African traditional medicine, it is commonly known as "Chandgoura" [8]. The aerial parts of the plant, including leaves and flowering tops, are traditionally prepared as decoctions or infusions to treat a variety of ailments, such as inflammatory disorders, digestive disturbances, diabetes, infections, pain, and rheumatism. Morphologically, *Ajuga iva* is a low-growing plant, reaching approximately 20 cm in height, with compact and dense leaves. Its inflorescences consist of solitary flowers developing in the leaf axils, typically flowering between May and June, while its fruits are reticulated nutlets, a characteristic feature of the species [8,9].

In North African ethnomedicine, particularly in Algeria, Tunisia, and Morocco, *A. iva* has been widely used to manage gastrointestinal disorders, hypertension, hyperglycemia, diabetes, alopecia, and pain-related conditions. Phytochemical investigations have revealed that the plant is rich in bioactive constituents, including flavonoids, phenolic acids, iridoids, tannins, phytoecdysteroids, terpenoids, organic acids, vitamins, minerals, glycosides, and fatty acid methyl esters [8]. This chemical diversity provides a strong scientific rationale for its traditional therapeutic applications. Inflammation is a complex biological response that plays a vital role in host defense; however, its dysregulation is closely associated with the development of chronic diseases, including autoimmune disorders and cancer. Conventional anti-inflammatory agents, such as glucocorticoids and non-steroidal anti-inflammatory drugs, exert their effects primarily through the inhibition of cyclooxygenase (COX) enzymes and other mediators. Despite their clinical efficacy, the long-term use of these drugs is often limited by serious adverse effects, including gastrointestinal, renal, and cardiovascular toxicity [10].

Modern discovery strategies increasingly integrate experimental and computational approaches. In silico molecular docking, in particular, allows rapid prediction of ligand–target interactions and provides valuable insight into potential mechanisms of action [11,12], especially regarding the inhibition of inflammation-related enzymes such as cyclooxygenase-2 (COX-2), a key therapeutic target. The combination of computational modeling with in vivo validation enhances the reliability of natural product-based drug discovery by linking chemical composition to pharmacological activity.

https://doi.org/10.3390/molecules31030496

Overall, this study adds value by integrating in vivo biological evaluations with LC–MS-based phytochemical profiling and molecular docking analysis, thereby providing complementary evidence supporting the pharmacological relevance of the regional *Ajuga iva* chemotype collected from the Tebessa region of Algeria. The observed anti-inflammatory, analgesic, and antipyretic activities are consistent with, and comparable to, previously reported findings. Furthermore, molecular docking results suggest potential interaction between the identified bioactive compounds and cyclooxygenase-2 (COX-2), offering a predictive mechanistic rationale for the observed biological effects. However, these in silico findings remain exploratory in nature and require further biochemical and molecular validation to confirm the involvement of COX-2 inhibition in the therapeutic potential of this traditionally valued medicinal plant.

2. Materials and Methods

2.1. Plant Material Collection and Preparation of Extract

The aerial parts of *Ajuga iva* were collected from the Tebessa region, Algeria. The botanical identity of the plant was confirmed by Dr. Hioun Soraya, a plant taxonomist at University of Tebessa, Algeria. Due to the unavailability of a local herbarium, a voucher specimen will be deposited at a recognized herbarium and assigned a voucher ID upon acceptance of the manuscript. The hydroalcoholic extract was prepared following standard procedures: 100 g of dried plant material were macerated in 1 L of 70% methanol for 24 h with occasional shaking. The mixture was filtered through Whatman No. 1 filter paper, and the solvent was evaporated under reduced pressure at 40 °C using a rotary evaporator. The obtained extract was then dried and stored at 4 °C until analysis. Extraction yield was calculated as the weight of the dried extract relative to the starting plant material, giving a yield of 14.08% *w*/*w*.

2.2. Phytochemical Screening

Various classical colorimetric tests were employed to detect the main groups of bioactive metabolites: ferric chloride test for tannins and polyphenols, Shinoda test for flavonoids, foam test for saponins, Liebermann–Burchard reaction for terpenes and sterols, Borntrager's test for glycosides, and Mayer's reagent for alkaloids [13].

2.3. Determination of Total Phenolic and Flavonoid Contents

The Folin–Ciocalteu procedure as used to measure total phenolic content (TPC). In brief, 200 μL of the extract was mixed with 600 μL of diluted Folin reagent (10%). After 3 min, 800 μL of Na_2CO_3 solution (7.5%) was added. The mixture was kept in the dark for 30 min, and absorbance value was recorded at 765 nm, the gallic acid equivalents (mg GAE/g extract) have been used to express the results [14].

The amount of total flavonoid (TFC) was measured using the aluminum chloride colorimetric technique. A mixture of 0.5 mL extract with 2 mL distilled water was prepared, followed by addition of 0.15 mL $NaNO_2$ (5%). After 5 min, 0.15 mL $AlCl_3$ (10%) was added. Following the addition 1 mL of NaOH (1 M) was introduced after six minutes later, and the absorbance was measured at 430 nm. Results were expressed as mg quercetin equivalents per gram (mg QE/g) [15].

2.4. LC-MS Analysis of Phenolic Compounds

The identification and quantification of phenolic compounds were performed using LC-MS techniques. Analyses were carried out on a Shimadzu UFLC XR system 5.91(Kyoto, Japan) equipped with a SIL-20AXR auto-sampler, (Shimadzu Corporation, Kyoto, Japan) CTO-20 AC column oven, LC-20ADXR binary pump, and a quadrupole 2020 detector. Sepa-

 https://doi.org/10.3390/molecules31030496

ration was achieved on an Inertsil ODS-4 C18 column (3 μm, 150 × 3.0 mm i.d.) maintained at 40 °C, with an injection volume of 20 μL and a flow rate of 0.5 mL/min. Mobile phases consisted of A (95% H_2O + 5% MeOH + 0.2% acetic acid) and B (50% ACN + 50% H_2O + 0.2% acetic acid), applied with a linear gradient: 0.01–14 min, 10–20% B; 14–27 min, 20–55% B; 27–37 min, 55–100% B; 37–45 min, 100% B; 45–50 min, 10% B. The desolvation line temperature was set at 275 °C, nebulizing gas flow at 1.50 L/min, drying gas at 15.00 L/min, and heat block temperature at 450 °C. LC-ESI(−) MS spectra $[M-H]^-$ were acquired using LabSolutions LCMS version 5.91 (Shimadzu Corporation, Kyoto, Japan), and compounds were identified by comparison of retention times and mass spectra with authentic standards. All analyses were performed in triplicate to ensure reproducibility [16].

Calibration curves were constructed from authentic standards across six concentration levels covering the analytical range. Peak areas were plotted against nominal concentrations, and weighted linear regression (1/x) was applied. Linearity was excellent (R^2 = 0.991–0.999), with back-calculated concentrations within ±15% of nominal values (±20% at the LLOQ). Method validation included accuracy, precision, sensitivity, and recovery. Intra-day reproducibility showed RSD% consistently below 5%, while inter-day reproducibility was below 7%. Limits of detection (LOD) ranged between 0.05 and 0.2 μg/mL, and limits of quantification (LOQ) between 0.15 and 0.5 μg/mL. Recovery rates for spiked samples were between 92% and 105%, confirming accuracy and robustness.

Carryover was assessed by injecting a blank matrix sample immediately after the highest calibration standard and high-concentration samples. Chromatograms of these blanks were examined for residual analyte peaks, with acceptance criteria requiring that any carryover signal not exceed 20% of the LLOQ response. No residual peaks were observed. Instrumental conditions were optimized to prevent saturation, including adjustment of injection volume, use of needle wash solutions, and gradient modifications to ensure complete elution. Representative chromatograms of high-concentration samples and subsequent blanks are provided in the Supplementary Materials to illustrate the absence of carryover.

Together, these validation parameters and carryover assessments demonstrate that the method is sensitive, accurate, and reproducible, and that chromatographic saturation does not compromise trace-level quantification.

2.5. Animal

Adult animals (Wistar rats and Swiss albino mice) were obtained from the Pasteur Institute, Algeria, a certified and accredited public institution for laboratory animal breeding and supply. They were acclimated for one week under standard laboratory conditions, including a controlled temperature of 22 ± 2 °C, a 12-h light/dark cycle, and free access to food and water. All experimental procedures were conducted in accordance with internationally accepted guidelines for the care and use of laboratory animals (ARRIVE guidelines and the principles of the 3Rs) and were approved by the Institutional Animal Care and Use Committee (IACUC) under registration number PI-AL-2025-07. Every effort was made to minimize animal suffering, and the minimum number of animals necessary to achieve statistically valid results was used. A post-hoc power analysis indicated that the chosen sample size (n = 5 for most experiments) provides >80% power to detect significant differences in anti-inflammatory and analgesic responses.

2.6. Anti-Inflammatory Activity

The anti-inflammatory potential was examined using the λ-carrageenan-induced paw edema model in rats following Elion Itou et al. (2017) [17], with slight adjustments. Rats

 https://doi.org/10.3390/molecules31030496

were randomly assigned into five groups ($n = 5$): a vehicle control group receiving distilled water (T−), three groups treated orally with the deferent doses of extract (EHA), dissolved in distilled water at doses (100, 200 and 400 mg/kg body weight) and a reference drug group (T+) receiving acetylsalicylic acid (75 mg/kg).

Paw volume was monitored before injection and every hour for five hours using the mercury displacement technique [18].

2.7. *Analgesic Activity*

According to Koster et al. (1959) [19], after a fasting period of 18 h five groups of mice were formed ($n = 5$). The vehicle control group (T−) were treated with saline solution, and the other groups were delivered the extract at doses (100, 200 and 400 mg/kg body weight), and a reference drug group (T+) receiving acetylsalicylic acid (75 mg/kg) by oral gavage using a stainless-steel gavage needle (18G, rounded tip). One hour later, 0.1 mL of 3% acetic acid was injected intraperitoneally, and the number of writhes was counted over 30 min.

2.8. *Antipyretic Activity*

Antipyretic activity was tested using the brewer's yeast-induced fever model following. Mice were fasted for 18 h, then injected subcutaneously with 10 mL/kg of yeast aqueous suspension (*Saccharomyces cerevisiae*, 15%) to induce hyperthermia. Four groups ($n = 5$) were used: vehicle control group receiving distilled water (T−), two groups receiving extract at doses (200 and 400 mg/kg body weight), and reference drug group (T+) receiving acetylsalicylic acid (75 mg/kg). Rectal temperature was measured hourly for five hours using a digital thermometer [20,21].

2.9. *In Silico Evaluation*

2.9.1. Protein Preparation for Docking

The crystal structure of COX-2 (PDB ID: 1CX2) was downloaded from the Protein Data Bank. All water molecules and heteroatoms were removed, while the co-crystallized ligand (S58) was kept to identify the active site pocket. Polar hydrogens and Kollman charges were added using AutoDock Tools.

2.9.2. Ligand Preparation

Phenolic compounds previously reported in *A. iva* were obtained from PubChem in SDF format. Their structures were minimized and converted into. PDBQT format through AutoDock Tools. The reference ligand S58 was prepared similarly for redocking.

2.9.3. Docking Procedure

Molecular docking studies were conducted with AutoDock version 4.2.6 [22]. The grid box was centered on the coordinates of the co-crystallized ligand to cover the COX-2 catalytic region. The Lamarckian Genetic Algorithm (LGA) was applied with default parameters, generating several poses for each compound, ranked by estimated binding free energy (kcal/mol) [16].

2.9.4. Docking Validation

To ensure docking accuracy, the ligand S58 was redocked into the active site. The resulting pose presented an RMSD of 0.557 Å relative to the experimental structure, indicating good reliability (RMSD < 2 Å).

https://doi.org/10.3390/molecules31030496

2.9.5. Visualization and Interaction Analysis

Analysis of hydrogen bonding, hydrophobic interaction, and electrostatic forces was performed using Discovery Studio Visualizer (version 21.1.0.20298). Binding energies and interaction patterns were examined to identify residues contributing to ligand stabilization.

2.10. Statistical Analysis

All graphical representations and statistical analyses were performed using GraphPad Prism version 10 (GraphPad Software 10, San Diego, CA, USA). Experimental data are expressed as mean ± SD. Statistical comparisons among groups were conducted using one-way analysis of variance (ANOVA), followed by Dunnett's post hoc test for multiple comparisons against the vehicle control group.

The selected group sizes (n = 5 per group) are consistent with those commonly employed in preliminary in vivo pharmacological studies using carrageenan-induced paw edema, acetic acid-induced writhing, and yeast-induced pyrexia models. Although no a priori formal power calculation was performed, statistically significant differences were observed between treated and vehicle groups under the present experimental conditions.

Differences were considered statistically significant at $p < 0.05$, with significance levels indicated as * $p < 0.05$, ** $p < 0.01$, and *** $p < 0.001$.

3. Results

3.1. Phytochemical Screening of the Hydroalcoholic Extract of A. iva

The qualitative phytochemical analysis of the hydroalcoholic extract of *A. iva* demonstrated the presence of several major classes of secondary metabolites (Table 1). The extract was particularly rich in tannins and polyphenols (+++), indicating a high content of phenolic constituents known for their antioxidant and therapeutic potential. Moderate levels (++) of flavonoids, glycosides, sterols, and terpenes were also observed, suggesting additional biological relevance. A weak reaction (+) for alkaloids was recorded, while saponins were absent (−). These findings confirm the chemical diversity of the extract and highlight its potential pharmacological importance.

Table 1. Qualitative phytochemical profile of the hydroalcoholic extract of *A. iva*.

Phytochemical Class	Observation
Tannins	+++
Polyphenols	+++
Flavonoids	++
Saponins	−
Alkaloids	+
Glycosides	++
Sterols	++
Terpenes	++

3.2. Overall Phenolic and Flavonoid Levels

The total amount of phenolics and flavonoids of the hydroalcoholic extract of *A iva* (L.) Schreb. were determined by colorimetric assays (Table 2). The total phenolic and flavonoid contents of the hydroalcoholic extract of *A. iva* (L.) Schreb. were quantified using standard colorimetric assays, and the results are summarized in Table 2. The extract exhibited a Total Phenolic Content (TPC) of 26.3 ± 1.2 mg GAE/g dry extract, based on a calibration curve obtained with gallic acid (R^2 = 0.9996). The Total Flavonoid Content (TFC) reached 13.5 ± 0.9 mg QE/g dry extract, calculated using a quercetin standard (R^2 = 0.9969). These values indicate that the hydroalcoholic extract is a rich source of phenolic and flavonoid compounds, which may contribute to its observed biological activities.

https://doi.org/10.3390/molecules31030496

Table 2. Total phenolic and flavonoid contents of the hydroalcoholic extract of *A iva* (L.) Schreb.

Parameter	Standard Used	Calibration Equation	R^2	Content (mg/g Dry Extract)	Unit Equivalent
Total Phenolic Content (TPC)	Gallic acid	y = 0.0095x + 0.012	0.9996	26.3 ± 1.2	mg GAE/g DE
Total Flavonoid Content (TFC)	Quercetin	y = 0.0072x + 0.010	0.9969	13.5 ± 0.9	mg QE/g DE

3.3. LC–MS/MS Identification of Phenolic and Flavonoid Compounds

The liquid chromatography–mass spectrometry (LC-MS) analysis of the hydroalcoholic extract of *A iva* (L.) Schreb. revealed the presence of a wide range of phenolic acids and flavonoids with significant pharmacological potential (Table 3). Among the identified compounds, kaempferol (0.3560 mg/g), apigenin (0.9980 mg/g), and rutin (0.0963 mg/g) were the most abundant, suggesting that these flavonoids are the major constituents contributing to the antioxidant and therapeutic effects of the extract. Other notable compounds included naringenin (0.0368 mg/g), the 7-O-glucosilated form of apigenin (0.0738 mg/g), and the 7-O-glucosylated form of luteolin (0.0123 mg/g), known for their anti-inflammatory and enzyme-inhibitory activities. Phenolic acids such as quinic (0.1810 mg/g), p-coumaric (0.0165 mg/g), and protocatechuic (0.0097 mg/g) acids were also detected, reinforcing the extract's richness in polyphenolic structures. Minor compounds like rosmarinic, syringic, and *trans*-cinnamic acids were found in lower concentrations but likely act synergistically with the dominant flavonoids. Overall, these findings demonstrate that *A. iva* is a valuable source of bioactive secondary metabolites with potential health-promoting properties.

Table 3. Phenolic and flavonoid molecule classes of *A iva* extract identified by the liquid chromatography–mass spectrometry (LC-MS).

No.	Compound Designation	RT (min)	$[M-H]^-$ (m/z)	Concentration (mg/g)
1	Quinic acid	1.935	191.00	0.1810
2	Protocatechuic acid	6.882	153.00	0.0097
3	Syringic acid	16.462	197.00	0.0010
4	p-Coumaric acid	21.028	163.00	0.0165
5	trans-Ferulic acid	23.322	193.00	0.0040
6	Rutin	24.750	609.00	0.0960
7	Luteolin-7-O-glucoside	25.387	447.00	0.0123
8	Apigenin-7-O-glucoside	27.657	431.00	0.0738
9	Rosmarinic acid	27.313	359.00	0.0020
10	Quercetin	32.480	301.00	0.0113
11	trans-Cinnamic acid	32.364	147.00	0.0007
12	Kaempferol	32.473	285.00	0.3560
13	Naringenin	34.351	271.00	0.0368
14	Apigenin	34.930	269.00	0.0998
15	Luteolin	35.441	285.00	0.0061
16	Acacetin	40.949	283.00	0.0293

3.4. Anti-Inflammatory Activity

At the 3rd hour after carrageenan injection, the vehicle control group (T−), treated with distilled water, showed the highest edema volume (0.69 ± 0.08 mL), confirming successful induction of inflammation. In contrast, the reference drug group (T+), treated with acetylsalicylic acid (75 mg/kg), exhibited a marked decrease in paw edema (0.36 ± 0.07 mL). Administration of *A. iva* extract at doses of 100, 200, and 400 mg/kg produced a progressive, dose-dependent reduction in paw edema (0.52 ± 0.05 mL, 0.43 ± 0.05 mL, and 0.38 ± 0.06 mL, respectively) (Figure 1). Statistical analysis (ANOVA, $p < 0.0001$) followed by Dunnett's test showed that the 200 and 400 mg/kg doses produced a very highly significant reduction compared with the vehicle control (T−), and the 100 mg/kg dose showed a significant effect ($p < 0.01$).

 https://doi.org/10.3390/molecules31030496

3.5. Analgesic Activity

Pain-relieving effect of the hydroalcoholic extract of *A. iva* was assessed using the acetic acid-induced writhing test in mice. The vehicle control group (T−), treated with distilled water, exhibited the highest number of writhes (120 ± 10.86), confirming the successful induction of nociception. In contrast, the reference drug group (T+), administered acetylsalicylic acid (75 mg/kg), exhibited a notable decrease to 27 ± 1.35 writhes. Treatment with *A. iva* extract at 100, 200, and 400 mg/kg produced a writhing inhibition showing a dose-dependent increase, yielding 61 ± 3.31, 43 ± 2.82, and 32 ± 4.04 writhes, respectively (Figure 2). All experimental groups showed a statistically highly significant effect relative to the vehicle control, with all comparisons reaching **** ($p < 0.0001$).

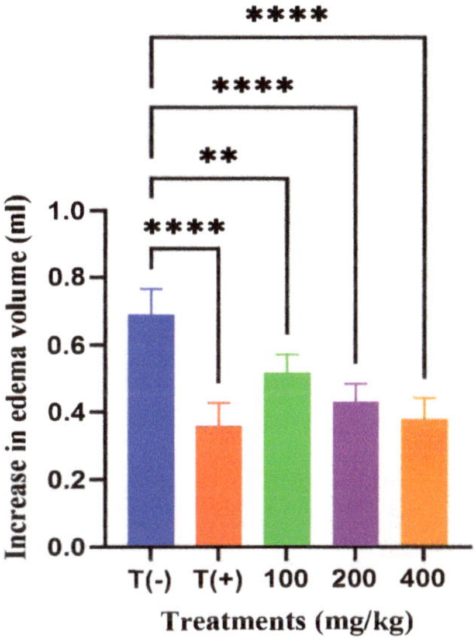

Figure 1. Effect of *A. iva* hydroalcoholic extract on carrageenan-induced paw edema in rats at the 3rd hour ** $p < 0.01$, **** $p < 0.0001$ vs. vehicle control (T−).

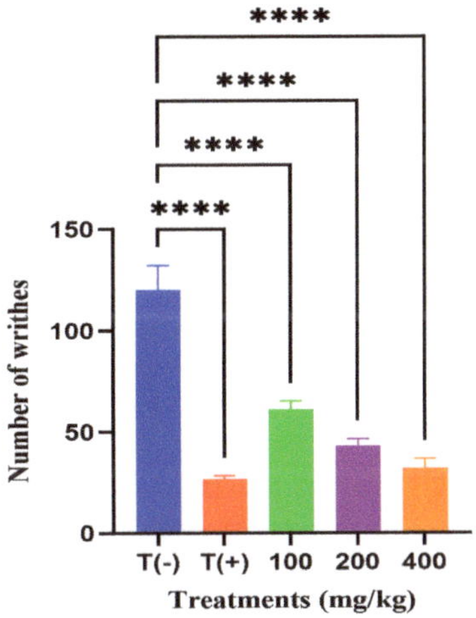

Figure 2. Effect of *A iva* hydroalcoholic extract on analgesic activity as measured by the number of writhes in mice **** $p < 0.0001$ vs. vehicle control (T−).

https://doi.org/10.3390/molecules31030496

3.6. Antipyretic Activity

Fever-reducing effect of the hydroalcoholic extract of *A. iva* was assessed (Figure 3). The vehicle control group (T−), treated with distilled water, exhibited the highest mean rectal temperature (38.57 °C), confirming successful induction of fever. In contrast, the reference drug group (T+), treated with acetylsalicylic acid (75 mg/kg),), showed a significant reduction in body temperature to 37.72 °C (ANOVA, $p < 0.0001$). Administration of *A. iva* extract at 200 mg/kg produced a comparable reduction (37.74 °C), and 400 mg/kg reduced body temperature to 37.28 °C. All experimental groups showed a statistically highly significant effect relative to the vehicle control, with all comparisons reaching **** ($p < 0.0001$).

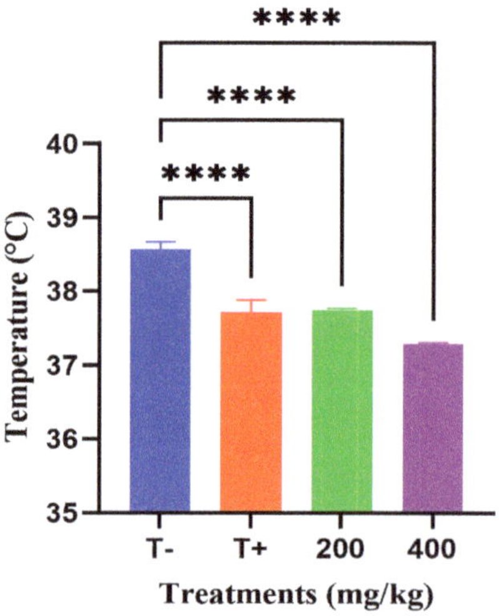

Figure 3. Effect of *A. iva* hydroalcoholic extract in the brewer's yeast induced pyrexia model in rats. **** $p < 0.0001$ vs. vehicle control (T−).

3.7. In Silico Evaluation

The docking study aimed to assess the potential inhibitory binding of *A. iva* phenolic compounds with the cyclooxygenase-2 (COX-2) enzyme (PDB ID: 1CX2), which plays a key role in the inflammatory response. All the identified compounds from the hydroalcoholic extract were analyzed, and their binding affinities were found to range from -8.8 kcal·mol^{-1} for rosmarinic acid to -5.3 kcal·mol^{-1} for the smallest molecule, quinic acid.

The best ten compounds, based on their docking scores, are presented in Table 4. These compounds exhibited predicted binding energies ranging from -8.8 to -6.2 kcal·mol^{-1}, indicating a favorable theoretical accommodation within the COX-2 active site at a structure-based, predictive level. These docking scores should be interpreted as indicative of potential binding modes rather than direct evidence of inhibitory activity. Two-dimensional representations of the best-ranked docked complexes are provided in Figures S1 and S2, highlighting the main predicted ligand–residue interactions within the COX-2 binding pocket.

Among the tested compounds, rosmarinic acid displayed the most favorable interaction (–8.8 kcal/mol), forming multiple hydrogen bonds with key residues such as Gln192, Ser353, Tyr355, and Tyr385, and additional hydrophobic contacts with Val523, Ser353, and Gly526, important residues known to stabilize inhibitors in the COX-2 catalytic site.

Rutin and apigenin-7-O-glucoside also exhibited strong binding energies (–8.7 and –8.5 kcal/mol, respectively). According to molecular docking predictions, rutin was predicted

https://doi.org/10.3390/molecules31030496

to form an extensive hydrogen-bonding network involving Val523, Ser353, Phe518, Gln192, and His90. Similarly, apigenin-7-O-glucoside was predicted to establish multiple hydrogen-bonding and hydrophobic contacts with Tyr385, Gln192, Leu352, and Ser353. These interactions illustrate the plausible binding orientations of the ligands within the COX-2 binding pocket, as visualized in Figure S1 Compounds such as acacetin (–8.3 kcal/mol) and luteolin-7-O-glucoside (–7.6 kcal/mol) also displayed favorable docking scores, engaging with catalytic residues His90, Tyr355, Arg120, and Ser530, which are essential for enzyme inhibition.

Table 4. The most favorable docking outcomes of *A. iva* phenolic constituents were docked with cyclooxygenase-2 (Protein Data Bank: 1CX2).

	Affinity Energy (Kcal·mol^{-1})	H-Bonds (Separation Å)	Hydrophobic Contacts	Additional Contacts
Native ligand (S58)	−9.4	Arg120 (2.32), Tyr355 (2.44), His90 (2.58), Gln192 (3.06)	Leu231, Ala527, Val349, Gly526, Tyr385, Leu384, Trp387, Leu352, Phe518, Val526	His90
Rosmarinic acid	−8.8	Gln192 (2.47), Ser353 (2.33) Tyr355 (2.65), Tyr385 (2.36), Leu384 (2.55)	Val523, Ser353, Gly526	Arg120
Rutin	−8.7	Val523 (2.22), Ser353 (3.07), Leu352 (2.26), Leu352 (2.23), Phe518 (2.45), Phe518 (2.04), Gln192 (2.93), Gln192 (2.73), Gln192 (2.51), His90 (2.08)	Phe518, Pro514	-
Apigenin-7-o-glucoside	−8.5	Tyr385 (2.62, 2.54), Gln192, Leu352 (3.10, 2.32, 2.85), Ser353 (2.64, 2.48), His90 (2.57), Tyr355 (2.33)	Leu352, Gly526, Ala527, Ser530, Val349, Leu359, Leu531	-
Acacetin	−8.3	Gln192 (2.82), Ser353 (2.65), His90 (2.49), Tyr355 (2.06), Leu384 (3.25, 2.39)	Ser353, Ala527, Val523, Leu352, Gly526, Leu526, Leu384, Tyr385, Trp387	-
Luteolin-7-o-glucoside	−7.6	Ser530 (2.60), Gln192 (2.52), His90 (2.71), Arg120 (2.35), Val116 (2.42)	Ala527, Val349, Leu352, Val523, Ser353	-
Naringenin	−7.5	Tyr355 (2.24), Val523 (2.40)	Leu352, Gly526, Ser353, Val355	Met522, Arg120
Kampherol	−7.1	His90 (2.65), Val523 (2.47), Ser530 (2.82)	Ala527, Val349, Leu531, Val523, Ser353	-
Apigenine	−6.7	Gln192 (2.59), His90 (2.75), Tyr355 (2.81)	Ser353, Ala527, Val523, Leu352, Gly526	Tyr385
Luteolin	−6.5	Leu384 (2.54), His90 (3.08)	Leu352, Gly526, Ser353, Val523, Val349, Ala527	-
Quercetin	−6.2	Ser353 (3.23), Met522 (2.82)	Gly526, Ala527, Leu527, Leu531, Val349, Leu352	Tyr385, Arg120

Moderate interactions were observed for naringenin (–7.5 kcal/mol), kaempferol (–7.1 kcal/mol), apigenin (–6.7 kcal/mol), luteolin (–6.5 kcal/mol), and quercetin (–6.2 kcal/mol), forming stable hydrogen bonds and hydrophobic interactions within the COX-2 binding pocket (Figure S2).

4. Discussion

Phytochemical screening of the hydroalcoholic extract of *Ajuga iva* revealed the presence of multiple classes of secondary metabolites, including polyphenols, flavonoids, tannins, alkaloids, sterols, and terpenes. These findings are consistent with previous reports on *A. iva*

 https://doi.org/10.3390/molecules31030496

(Saidi et al. 2023., Makni 2013), confirming the chemical richness of this species and supporting its pharmacological relevance [23,24]. Natural plant extracts such as *A. iva* typically contain diverse bioactive compounds may act in a complementary or synergistic manner, leading to complex pharmacological effects rather than single-compound activity [25–27].

LC–MS/MS analysis further demonstrated that the extract is particularly rich in flavonoids and phenolic acids, compound classes widely associated with antioxidant, anti-inflammatory, and antipyretic activities [28]. Importantly, all pharmacological evaluations in the present study were performed using the crude hydroalcoholic extract. Therefore, the observed biological effects should be interpreted at the extract level and are more plausibly explained by the combined and potentially synergistic action of multiple phytochemical classes rather than by isolated individual compounds [29,30]. The phytochemical composition of *A. iva* further supports its pharmacological relevance, as these compounds have been previously implicated in anti-inflammatory, analgesic, antimicrobial antioxidant activities, with applications extending to the food and cosmetic industries and involvement in nociceptive pathways [31–33].

Quantitative analysis showed that the hydroalcoholic extract of *A. iva* contained substantial amounts of total polyphenols (26.3 ± 1.2 mg GAE/g dry extract) and total flavonoids (13.5 ± 0.9 mg QE/g dry extract). These values are higher than those reported by Mohamed et al. [24] for a methanolic extract of *A. iva* (25.69 ± 2.6 mg GAE/g and 3.00 ± 0.3 mg QE/g extract), lower than the total phenolic content of an ethanolic extract reported by Bouyahya et al. [34] (49.75 ± 2.08 mg GAE/g), and comparable to the values reported by Medjeldi et al. [35] (28.3 ± 1.12 mg GAE/g). Total flavonoid contents reported for other Ajuga species, such as A. chamaepitys (24.72 ± 0.24 mg QE/g dry extract) [36] further highlight interspecific and extraction-related variability.

Inflammation, pain, and fever are closely interconnected physiological defense responses that protect tissues against damage and infection; however, when dysregulated, they contribute to pathological conditions, including arthritis, diabetes, neurodegenerative disorders, and autoimmune diseases [37,38]. Consequently, plant-derived bioactive molecules with comparable therapeutic efficacy but fewer adverse effects have attracted increasing scientific and clinical interest [39,40].

In the present study, the hydroalcoholic extract of *A. iva* exhibited significant anti-inflammatory, analgesic, and antipyretic activities, all of which followed a clear dose-dependent pattern. In the carrageenan-induced paw edema model, the extract significantly inhibited inflammation in a dose-dependent manner, suggesting modulation of inflammatory mediators involved in both early and late phases of the inflammatory response [41,42]. This pattern consistent dose–response relationship strengthens the pharmacological interpretation of the results.

Similarly, the extract produced a marked, dose-dependent reduction in acetic acid-induced writhing, indicative of peripheral analgesic activity potentially related to the modulation of inflammatory mediators such as prostaglandins and bradykinins [43]. The antipyretic effect observed in the brewer's yeast-induced fever model also followed a dose-dependent trend, with significant reduction in elevated body temperature, suggesting possible interference with fever-mediating pathways involving prostaglandin E_2 and pro-inflammatory cytokines [43].

Recent computational studies highlighting the multitarget therapeutic potential of natural compounds, such as silymarin, further support the relevance of plant-derived bioactive metabolites in pharmacological research [44–48]. Cyclooxygenase-2 (COX-2) is widely recognized as a central molecular target in anti-inflammatory drug discovery due to its inducible expression during inflammation conditions and its role in catalyzing the con-

 https://doi.org/10.3390/molecules31030496

version of arachidonic acid into pro-inflammatory prostaglandins that mediate pain, fever, and inflammation [49,50]. Structurally, COX-2 possesses a larger and more flexible active site than COX-1, allowing selective accommodation of bulkier ligands and enabling the development of COX-2-preferential inhibitors with improved safety profiles [51,52]. Within this context, molecular docking represents a valuable computational tool for exploring potential ligand–target recognition patterns at the structural level.

The novelty of the present work lies in the integration of LC–MS/MS-based phytochemical profiling with structure-based molecular docking applied to a region-specific *Ajuga iva* chemotype collected from the Tebessa region of Algeria. This combined approach provides compound-level mechanistic hypotheses that complement the extract-level in vivo pharmacological findings.

Molecular docking was employed as a hypothesis-generating, structure-based approach to explore potential interactions between phenolic constituents identified in *Ajuga iva* and the COX-2 active site. Several compounds displayed favorable predicted binding energies (–8.8 to –6.2 kcal·mol^{-1}), for the top-ranked ligands. while the co-crystallized reference ligand (S58) exhibited a predicted binding energy of –9.4 kcal·mol^{-1}. Among the screened compounds, rosmarinic acid, rutin, and apigenin-7-O-glucoside showed the most favorable docking scores, suggesting enhanced accommodation within the COX-2 binding pocket.

Analysis of the predicted binding modes revealed recurrent interaction with amino acid residues commonly involved in COX-2 ligand recognition, including Arg120 and Tyr355 at the entrance of the catalytic channel. were frequently engaged in hydrogen-bonding or electrostatic interactions across several docked complexes. These residues have been widely reported in structural and computational studies as recurrent interaction sites involved in ligand positioning within the COX channel [51]. In the present docking models, rosmarinic acid was predicted to form an electrostatic interaction with Arg120, while other ligands, including luteolin-7-O-glucoside, naringenin, and quercetin, were also predicted to interact with this residue across several docked complexes.

Additional predicted interactions involved residues such as Gln192, Ser353, and Tyr385. For example, rosmarinic acid was predicted to form hydrogen-bond interactions with Ser353 and Tyr385 in the docking models. His90, located near the entrance of the active site, was also predicted to interact with several ligands, including acacetin, apigenin-7-O-glucoside, and kaempferol. A network of hydrophobic contacts involving Leu352, Val349, Ala527, Phe518, Gly526, and Trp387 was observed, suggesting possible van der Waals and π–π interactions.

These predicted interactions are generally in agreement with residues previously reported in computational studies of COX-2 ligand recognition, including Arg120, Tyr355, Gln192, Ser530, and surrounding hydrophobic residues [52]. The overlap between residues identified in the present docking analysis and those reported in the literature supports the structural plausibility of the predicted binding modes.

Importantly, molecular docking does not constitute experimental evidence of enzymatic inhibition. In this work, docking was used solely as a structure-based, hypothesis-generating approach to explore potential interactions between phenolic constituents of *Ajuga iva* and COX-2. No direct COX-2 enzymatic inhibition or prostaglandin quantification assays were performed, and the docking results should therefore be interpreted with caution. Therefore, the computational results should be interpreted with caution and considered as a structural rationale to guide future experimental investigations.

Finally, it should be emphasized that all pharmacological evaluations were conducted using the crude hydroalcoholic extract of *Ajuga iva*, and the observed biological effects cannot be directly attributed to individual phytochemicals. The overall pharmacological activity

https://doi.org/10.3390/molecules31030496

is more plausibly explained by the combined and potentially synergistic effects of multiple constituents rather than by a single dominant molecule.

A limitation of the present study is the relatively small sample size used in the in vivo experiments, which may reduce statistical power and limit the detection of subtle effects. Future studies employing larger cohorts and power-based experimental designs will be necessary to further substantiate these findings.

5. Conclusions

This study demonstrated that the hydroalcoholic extract of the aerial parts of *Ajuga iva* exhibits significant analgesic, antipyretic, and anti-inflammatory activities. These effects are likely associated with its rich phytochemical composition, particularly its high content of polyphenols and flavonoids. Molecular docking was used as an exploratory, hypothesis-generating in silico approach, suggesting potential interactions between selected bioactive metabolites and the COX-2 active site; however, these computational findings do not constitute direct evidence of enzymatic inhibition. Notably, the present study highlights a distinct regional chemotype of *Ajuga iva,* which may contribute to its unique pharmacological profile and provides added value compared with previously reported populations. Overall, the results confirm the pharmacological potential of *Ajuga iva* at the extract level and support its traditional use in the management of inflammation, pain, and fever, while underscoring the need for further studies to isolate active constituents and elucidate their precise mechanisms of action.

Supplementary Materials: The following supporting information can be downloaded at: https://www.mdpi.com/article/10.3390/molecules31030496/s1, Figure S1: Two-dimensional (2D) chemical structures of the top bioactive metabolites identified from *Ajuga iva* extract (rosmarinic acid, rutin, apigenin-7-o-2 glucoside, acacetin, and luteolin-7-o-glucoside), selected based on their predicted binding affinities toward the cyclooxygenase-2 (COX-2) enzyme; Figure S2: Two-dimensional (2D) chemical structures of the top bioactive metabolites identified from *A. iva* extract (naringenin, kampherol, apigenine, luteolin, and quercetin), selected based on their predicted binding affinities toward the cyclooxygenase-2 (COX-2) enzyme.

Author Contributions: Y.B. and S.B. contributed to the conceptualization, methodology design, and data curation. O.B. and N.M. performed phytochemical extraction, LC-MS/MS analysis, and formal data interpretation. S.L. supervised experimental protocols, harmonized the manuscript, and contributed to critical revision. N.T. carried out computational modeling and molecular docking simulations, while H.M. conducted bioassays, antioxidant and cholinesterase activity testing, and cytotoxicity evaluation. R.G. assisted with data validation, preparation of figures and tables, and manuscript editing. N.H. provided pharmacological contextualization and critical review of neurodegenerative relevance. G.N. coordinated project administration, supervised the overall study, and approved the final manuscript. All authors have read and agreed to the published version of the manuscript.

Funding: This research received no external funding.

Institutional Review Board Statement: Not applicable.

Informed Consent Statement: Not applicable.

Data Availability Statement: Data supporting the findings of this study are available from the corresponding author upon reasonable request.

Conflicts of Interest: The authors declare no conflict of interest.

https://doi.org/10.3390/molecules31030496

References

1. El-Saadony, M.T.; Saad, A.M.; Mohammed, D.M.; Korma, S.A.; Alshahrani, M.Y.; Ahmed, A.E.; Ibrahim, E.H.; Salem, H.M.; Alkafaas, S.S.; Saif, A.M. Medicinal Plants: Bioactive Compounds, Biological Activities, Combating Multidrug-Resistant Microorganisms, and Human Health Benefits-a Comprehensive Review. *Front. Immunol.* **2025**, *16*, 1491777. [CrossRef] [PubMed]
2. Boussekine, S.; Lekmine, S.; Gasmi, S.; Benkhedir, A.; Saker, H.; Lidoughi, A. The protective effect of selenium on diabetic nephropathy in wistar rats. *J. Microbiol. Biotechnol. Food Sci.* **2022**, *12*, e5960. [CrossRef]
3. Guendouze, A.; Mokrani, E.H.; Benslama, O.; Lekmine, S. Exploring the Potential of Avenanthramides and Their Analogues as α-Glucosidase Inhibitors for Type 2 Diabetes Treatment Utilizing Virtual Screening, Molecular Dynamics, and Drug-Likeness Predictions. *Lett. Drug Des. Discov.* **2024**, *21*, 3858–3871. [CrossRef]
4. Touzout, N.; Mihoub, A.; Ahmad, I.; Lekmine, S.; Tahraoui, H.; Krache, F.; Ali, E.; Hussain, S.; Danish, S.; Amrane, A. Deciphering the Mitigating Role of Silicon on Tomato Seedlings under Lambda Cyhalothrin and Difenoconazole Coexposure. *Sci. Rep.* **2025**, *15*, 26512. [CrossRef]
5. Benslama, O.; Lekmine, S.; Moussa, H.; Tahraoui, H.; Ola, M.S.; Zhang, J.; Amrane, A. Silymarin as a Therapeutic Agent for Hepatocellular Carcinoma: A Multi-Approach Computational Study. *Metabolites* **2025**, *15*, 53. [CrossRef]
6. Bennaghmouch, L.; Hajjaji, N.; Gmira, N. Flavonoïdes *d'Ajuga iva* (L.) Schreb (Labiée). *Rev. Marocaine Des Sci. Agron. Vétérinaires* **2002**, *22*, 25–30.
7. Istifli, E.S.; Sarıkürkcü, C. Assessment of Apigenin-7-Glucoside and Luteolin-7-Glucoside as Multi-Targeted Agents against Alzheimer's Disease: A Molecular Docking Study. *Int. J. Plant Based Pharm.* **2021**, *1*, 56–64. [CrossRef]
8. Ghedira, K.; Chemli, R.; Richard, B.; Zeches, M.; Men-Olivier, L.L. Part of a survey on the traditional pharmacopoeia of Tunisia: A study of the aerial parts of Ajuga iva (L.) Schreb. 1991.
9. Dzoyem, J.P.; Kuete, V.; Eloff, J.N. Biochemical Parameters in Toxicological Studies in Africa: Significance, Principle of Methods, Data Interpretation, and Use in Plant Screenings. In *Toxicological Survey of African Medicinal Plants*; Elsevier: Amsterdam, The Netherlands, 2014; pp. 659–715.
10. Wang, J.; Luo, J.; Rotili, D.; Mai, A.; Steegborn, C.; Xu, S.; Jin, Z.G. SIRT6 Protects against Lipopolysaccharide-Induced Inflammation in Human Pulmonary Lung Microvascular Endothelial Cells. *Inflammation* **2024**, *47*, 323–332. [CrossRef]
11. Raut, B.K.; Upadhyaya, S.R.; Bashyal, J.; Parajuli, N. In Silico and in Vitro Analyses to Repurpose Quercetin as a Human Pancreatic α-Amylase Inhibitor. *ACS Omega* **2023**, *8*, 43617–43631. [CrossRef]
12. Djeghim, H.; Bellil, I.; Benslama, O.; Lekmine, S.; Temim, E.; Boufendi, H.; Postigo, I.; Sánchez, P.; Khelifi, D. Effects of Genetic Diversity on the Allergenicity of Peanut (*Arachis hypogaea*) Proteins: Identification of the Hypoallergenic Accessions Using BALB/c Mice Model and in Silico Analysis of Ara h 3 Allergen Cross-Reactivity. *J. Proteom.* **2024**, *306*, 105264. [CrossRef] [PubMed]
13. Harborne, A.J. *Phytochemical Methods a Guide to Modern Techniques of Plant Analysis*; Springer Science & Business Media: Berlin/Heidelberg, Germany, 1998.
14. Singleton, V.L.; Rossi, J.A. Colorimetry of Total Phenolics with Phosphomolybdic-Phosphotungstic Acid Reagents. *Am. J. Enol. Vitic.* **1965**, *16*, 144–158. [CrossRef]
15. Zhishen, J.; Mengcheng, T.; Jianming, W. The Determination of Flavonoid Contents in Mulberry and Their Scavenging Effects on Superoxide Radicals. *Food Chem.* **1999**, *64*, 555–559. [CrossRef]
16. Lekmine, S.; Benslama, O.; Ola, M.S.; Touzout, N.; Moussa, H.; Tahraoui, H.; Hafsa, H.; Zhang, J.; Amrane, A. Preliminary Data on Silybum Marianum Metabolites: Comprehensive Characterization, Antioxidant, Antidiabetic, Antimicrobial Activities, LC-MS/MS Profiling, and Predicted ADMET Analysis. *Metabolites* **2025**, *15*, 13. [CrossRef]
17. Elion Itou, R.D.G.; Etou Ossibi, A.W.; Epa, C.; Bokia, C.B.; Ouamba, J.M.; Abena, A.A. Anti-Inflammatory and Analgesic Effects of Leaves of *Chromolaena odorata* L. (King and Robinson). *Afr. J. Pharm. Pharmacol.* **2017**, *11*, 217–223. [CrossRef]
18. Van Arman, C.G.; Begany, A.J.; Miller, L.M.; Pless, H.H. Some Details of the Inflammations Caused by Yeast and Carrageenin (with Appendix on Kinetics of the Reaction). *J. Pharmacol. Exp. Ther.* **1965**, *150*, 328–334. [CrossRef]
19. Koster, R.; Anderson, M.; De Beer, E.J. Acetic Acid-Induced Analgesic Screening. *Fed. Proc.* **1959**, *18*, 412–417.
20. De Dosso, S.; Siebenhuener, A.R.; Winder, T.; Meisel, A.; Fritsch, R.; Astaras, C.; Szturz, P.; Borner, M. Treatment Landscape of Metastatic Pancreatic Cancer. *Cancer Treat. Rev.* **2021**, *96*, 102180. [CrossRef]
21. Tam, M.T.; Dosso, J.A.; Robillard, J.M. The Impact of a Global Pandemic on People Living with Dementia and Their Care Partners: Analysis of 417 Lived Experience Reports. *J. Alzheimer's Dis.* **2021**, *80*, 865–875. [CrossRef]
22. Morris, G.M.; Huey, R.; Lindstrom, W.; Sanner, M.F.; Belew, R.K.; Goodsell, D.S.; Olson, A.J. AutoDock4 and AutoDockTools4: Automated Docking with Selective Receptor Flexibility. *J. Comput. Chem.* **2009**, *30*, 2785–2791. [CrossRef]

 https://doi.org/10.3390/molecules31030496

23. Saidi, S.; Remok, F.; Handaq, N.; Drioiche, A.; Gourich, A.A.; El Menyiy, N.; Amalich, S.; Elouardi, M.; Touijer, H.; Bouhrim, M. Phytochemical Profile, Antioxidant, Antimicrobial, and Antidiabetic Activities of *Ajuga iva* (L.). *Life* **2023**, *13*, 1165. [CrossRef] [PubMed]
24. Makni, M.; Haddar, A.; Kriaa, W.; Zeghal, N. Antioxidant, Free Radical Scavenging, and Antimicrobial Activities of Ajuga Iva Leaf Extracts. *Int. J. Food Prop.* **2013**, *16*, 756–765. [CrossRef]
25. Chaachouay, N. Synergy, Additive Effects, and Antagonism of Drugs with Plant Bioactive Compounds. *Drugs Drug Candidates* **2025**, *4*, 4. [CrossRef]
26. Kadi, K.; Mrah, R.; Hamli, S.; Lekmine, S.; Dib, D.; Addad, D.; Boukeria, S.; Gueboudji, Z.; Hafsaoui, I. Evaluation of the Anticoagulant Activity of Margins from Olives Extraction in the Khenchela Region. *J. Fundam. Appl. Sci.* **2020**, *12*, 634–649. [CrossRef]
27. Redjili, S.; Ghodbane, H.; Tahraoui, H.; Abdelouahed, L.; Chebli, D.; Ola, M.S.; Assadi, A.A.; Kebir, M.; Zhang, J.; Amrane, A. Green Innovation: Multifunctional Zinc Oxide Nanoparticles Synthesized Using Quercus Robur for Photocatalytic Performance, Environmental, and Antimicrobial Applications. *Catalysts* **2025**, *15*, 256. [CrossRef]
28. Khatteli, A.; Benabderrahim, M.A.; Triki, T.; Guasmi, F. Aroma Volatiles, Phenolic Profile and Hypoglycaemic Activity of *Ajuga iva* L. *Food Biosci.* **2020**, *36*, 100578. [CrossRef]
29. Alcázar Magaña, A.; Kamimura, N.; Soumyanath, A.; Stevens, J.F.; Maier, C.S. Caffeoylquinic Acids: Chemistry, Biosynthesis, Occurrence, Analytical Challenges, and Bioactivity. *Plant J.* **2021**, *107*, 1299–1319. [CrossRef]
30. Benali, T.; Bakrim, S.; Ghchime, R.; Benkhaira, N.; El Omari, N.; Balahbib, A.; Taha, D.; Zengin, G.; Hasan, M.M.; Bibi, S. Pharmacological Insights into the Multifaceted Biological Properties of Quinic Acid. *Biotechnol. Genet. Eng. Rev.* **2024**, *40*, 3408–3437. [CrossRef]
31. Mohammed, A.E.; Abdul-Hameed, Z.H.; Alotaibi, M.O.; Bawakid, N.O.; Sobahi, T.R.; Abdel-Lateff, A.; Alarif, W.M. Chemical Diversity and Bioactivities of Monoterpene Indole Alkaloids (MIAs) from Six Apocynaceae Genera. *Molecules* **2021**, *26*, 488. [CrossRef]
32. Ammar, H.; Touihri, I.; Kholif, A.E.; M'Rabet, Y.; Jaouadi, R.; Chahine, M.; Marti, M.E.d.H.; Vargas-Bello-Pérez, E.; Hosni, K. Chemical Composition, Antioxidant, and Antimicrobial Activities of Leaves of Ajuga Iva. *Molecules* **2022**, *27*, 7102. [CrossRef]
33. Lahrizi, L.; Errachidi, F.; Nekhla, H.; El Ghadraoui, L. *Ajuga iva* L.: An Overview of Phytochemical Profile and Biological Functionalities. *Chem. Rev. Lett.* **2024**, *7*, 31–44.
34. Bouyahya, A.; Abrini, J.; El-Baabou, A.; Bakri, Y.; Dakka, N. Determination of Phenol Content and Antibacterial Activity of Five Medicinal Plants Ethanolic Extracts from North-West of Morocco. *J. Plant Pathol. Microbiol.* **2016**, *7*, 2. [CrossRef]
35. Medjeldi, S.; Bouslama, L.; Benabdallah, A.; Essid, R.; Haou, S.; Elkahoui, S. Biological Activities, and Phytocompounds of Northwest Algeria Ajuga Iva (L) Extracts: Partial Identification of the Antibacterial Fraction. *Microb. Pathog.* **2018**, *121*, 173–178. [CrossRef] [PubMed]
36. Llorent-Martínez, E.J.; Zengin, G.; Ortega-Barrales, P.; Zakariyyah Aumeeruddy, M.; Locatelli, M.; Mollica, A.; Mahomoodally, M.F. Characterization of the Phytochemical Profiles and Biological Activities of *Ajuga chamaepitys* Subsp. Chia Var. Chia and Ajuga Bombycina by High-Performance Liquid Chromatography–Electrospray Ionization–Tandem Mass Spectrometry (HPLC–ESI–MSn). *Anal. Lett.* **2019**, *52*, 852–868. [CrossRef]
37. Cumming, E.; Peters, C. Immune Response to Infection. *Anaesth. Intensive Care Med.* **2024**, *25*, 439–443. [CrossRef]
38. Moussa, H.; Hamid, S.; Mameri, A.; Lekmine, S.; Tahraoui, H.; Kebir, M.; Touzout, N.; Dahmoune, F.; Ola, M.S.; Zhang, J. From Green Chemistry to Healthy Environments: Silver Nanoparticles as a Dual Antioxidant and Antibacterial Agents for Advancing Biomedicine and Sustainable Wastewater Treatment. *Bioengineering* **2024**, *11*, 1205. [CrossRef]
39. Lekmine, S.; Benslama, O.; Bensalah, B.; Touzout, N.; Moussa, H.; Tahraoui, H.; Ola, M.S.; Hafsa, H.; Zhang, J.; Amrane, A. Bioactive Phenolics of *Hyoscyamus muticus* L. Subsp. Falezlez: A Molecular and Biochemical Approach to Antioxidant and Urease Inhibitory Activities. *Int. J. Mol. Sci.* **2025**, *26*, 370. [CrossRef]
40. Mihoubi, N.; Ferhat, S.; Nedjhioui, M.; Zenati, B.; Lekmine, S.; Boudraa, R.; Ola, M.S.; Zhang, J.; Amrane, A.; Tahraoui, H. A Novel Halophilic Bacterium for Sustainable Pollution Control: From Pesticides to Industrial Effluents. *Water* **2025**, *17*, 888. [CrossRef]
41. Martínez-Rizo, A.B.; Fosado-Rodríguez, R.; Torres-Romero, J.C.; Lara-Riegos, J.C.; Ramírez-Camacho, M.A.; Herrera, A.L.A.; de la Torre, F.E.V.; Góngora, E.C.; Arana-Argáez, V.E. Models in Vivo and in Vitro for the Study of Acute and Chronic Inflammatory Activity: A Comprehensive Review. *Int. Immunopharmacol.* **2024**, *135*, 112292. [CrossRef]
42. Alblihed, M.A. Astragalin Attenuates Oxidative Stress and Acute Inflammatory Responses in Carrageenan-Induced Paw Edema in Mice. *Mol. Biol. Rep.* **2020**, *47*, 6611–6620. [CrossRef]
43. Phillips, J.C.; Hardy, D.J.; Maia, J.D.C.; Stone, J.E.; Ribeiro, J.V.; Bernardi, R.C.; Buch, R.; Fiorin, G.; Hénin, J.; Jiang, W. Scalable Molecular Dynamics on CPU and GPU Architectures with NAMD. *J. Chem. Phys.* **2020**, *153*, 044130. [CrossRef] [PubMed]

 https://doi.org/10.3390/molecules31030496

44. Hafsa, H.; Nasrallah, N.; Zeghbib, S.; Kebir, M.; Tahraoui, H.; Assadi, A.A.; Khezami, L.; Alghamdi, A.; Dahdouh, N.; Lekmine, S. Enhanced Photocatalytic Degradation of Pharmaceutical Pollutants Using Copper-Doped TiO2: Optimization, Machine Learning Integration, and Ecological Safety Assessment. *Arab. J. Sci. Eng.* **2025**, 1–25. [CrossRef]
45. Nieto, G.; Bañon, S.; Garrido, M.D. Incorporation of thyme leaves in the diet of pregnant and lactating ewes: Effect on the fatty acid profile of lamb. *Small Rumin. Res.* **2012**, *105*, 140–147. [CrossRef]
46. Nieto, G.; Banón, S.; Garrido, M. Administration of distillate thyme leaves into the diet of Segureña ewes: Effect on lamb meat quality. *Animal* **2012**, *6*, 2048–2056. [CrossRef]
47. Díaz, P.; Nieto, G.; Bañón, S.; Garrido, M.D. Determination of shelf life of sous vide salmon (Salmo salar) based on sensory attributes. *J. Food Sci.* **2009**, *74*, 371–376. [CrossRef]
48. Mohsin, N.U.A.; Aslam, S.; Ahmad, M.; Irfan, M.; Al-Hussain, S.A.; Zaki, M.E.A. Cyclooxygenase-2 (COX-2) as a Target of Anticancer Agents: A Review of Novel Synthesized Scaffolds Having Anticancer and COX-2 Inhibitory Potentialities. *Pharmaceuticals* **2022**, *15*, 1471. [CrossRef]
49. Nørregaard, R.; Kwon, T.-H.; Frøkiær, J. Physiology and Pathophysiology of Cyclooxygenase-2 and Prostaglandin E2 in the Kidney. *Kidney Res. Clin. Pr.* **2015**, *34*, 194–200. [CrossRef]
50. Faki, Y.; Er, A. Different Chemical Structures and Physiological/Pathological Roles of Cyclooxygenases. *Rambam Maimonides Med. J.* **2021**, *12*, e0003. [CrossRef]
51. Arora, M.; Choudhary, S.; Singh, P.K.; Sapra, B.; Silakari, O. Structural Investigation on the Selective COX-2 Inhibitors Mediated Cardiotoxicity: A Review. *Life Sci.* **2020**, *251*, 117631. [CrossRef]
52. Baek, S.-H.; Hwang, S.; Park, T.; Kwon, Y.-J.; Cho, M.; Park, D. Evaluation of Selective COX-2 Inhibition and in Silico Study of Kuwanon Derivatives Isolated from Morus Alba. *Int. J. Mol. Sci.* **2021**, *22*, 3659. [CrossRef]

https://doi.org/10.3390/molecules31030496

molecules

MDPI

Article

Standardized Hydroxytyrosol-Enriched Olive Pomace Juice Modulates Metabolic and Neurotrophic Signaling Pathways to Attenuate Neuroinflammation and Protect Neuronal Cells

Ye-Lim You [1], Ha-Jun Byun [1], Namgil Kang [2], Min Soo Lee [2], Jeong-In Lee [2], Ilbum Park [3] and Hyeon-Son Choi [1,*]

1 Department of Food Nutrition, Sangmyung University, Hongjimun 2-gil 20, Jongno-gu, Seoul 03016, Republic of Korea; 202333023@sangmyung.kr (Y.-L.Y.); byunhajun@gmail.com (H.-J.B.)
2 JY Globalfoods Co., Ltd., #304, 1-Dong, SKV1 Center, 48, Achasan-ro 17-gil, Seongdong-gu, Seoul 04799, Republic of Korea; team17@unibio.kr (N.K.); lms@jyglobalfoods.com (M.S.L.)
3 Kwangdong Pharm Co., Ltd., 52, Gwacheon-daero 7da-gil, Gwacheon-si 13840, Republic of Korea; pib975@ekdp.com
* Correspondence: hsc1970@smu.ac.kr; Tel.: +82-2-2287-6155; Fax: +82-2-2287-0104

Abstract

Olive pomace (OP), a by-product of olive oil production, is a sustainable resource rich in bioactive compounds with potential applications in cosmetics and pharmaceuticals. This study investigates the protective effects of olive pomace juice (OPJ) against H_2O_2-induced neuronal damage and LPS-induced inflammatory responses in HT22 and BV2 cells, respectively. OPJ suppressed H_2O_2-induced cell death and exerted anti-apoptotic effects by reducing the BAX/BCL2 ratio and caspase-3 cleavage. OPJ also mitigated neurodegenerative hallmarks by decreasing amyloid fibrils formation and inhibiting β-secretase and acetylcholinesterase (AChE) activity. Mechanistically, OPJ enhanced antioxidant response by upregulating Nrf2 and its downstream molecule HO-1, along with increasing mRNA levels of antioxidant enzymes, including catalase, SOD1, and GPx. OPJ further activated AMPKα–SIRT1–PGC1α signaling and CREB–BDNF–TrkB signaling, suggesting modulation of key antioxidant, anti-apoptotic, and neurotrophic pathways. In BV2 cells, OPJ downregulated pro-inflammatory cytokines (IL-6 and IL-1β) and decreased iNOS and COX-2 expression through suppression of NF-κB and MAPK signaling pathways. HPLC analysis identified hydroxytyrosol (10.92%) as the major active compound in OPJ, which compared with tyrosol (2.18%), and hydroxytyrosol exhibited greater neuroprotective and anti-inflammatory effects than tyrosol. This study highlights the potential of OPJ and its major compound, hydroxytyrosol, as functional agents for mitigating neurodegeneration-related cellular response, supporting its application in the food and pharmaceutical industries.

Keywords: olive pomace juice (OPJ); hydroxytyrosol; neuroprotection; neuroinflammation; Nrf2; AMPK; CREB

1. Introduction

The olive tree (*Olea europaea* L.) and its fruit have been highly valued for centuries for their culinary significance and diverse applications in food, health, and cosmetics [1,2]. Olives are prized for their rich flavor, nutritional value, and versatility, packed with bioactive compounds primarily found in olive oil, olive leaves, and the fruit itself [3].

These compounds offer a wide range of health benefits, including antioxidant and anti-inflammatory effects, and contribute to cardiovascular health, skin rejuvenation, and brain function [4,5]. Beyond these applications, olive derivatives also play an important role in industrial processes, including biofuel production and animal feed, highlighting their remarkable versatility [6,7]. One by-product of olive oil production, olive pomace, consists of the solid residue left after oil extraction, including skins, pulp, seeds, and stems [2]. Traditionally considered waste, olive pomace is now gaining attention due to its nutritional value, bioactive compounds, and functional properties, opening up potential applications across various industries [1]. In the broader context of sustainability, the food industry—despite its wide-ranging applications—remains one of the largest contributors to global waste, creating significant environmental, social, and economic challenges [8]. Addressing these challenges requires effective waste management strategies that emphasize resource efficiency. Upcycling, which transforms by-products into higher value goods, is a promising approach [9], making the olive industry an excellent example of contributing to a more sustainable food system [10].

Neurodegenerative diseases are global health challenges of growing concern, standing out due to their widespread impact and increasing prevalence [11,12]. These conditions such as Alzheimer's, Parkinson's diseases, are marked by the accumulation of toxic proteins, oxidative stress, inflammation, and mitochondrial dysfunction [13–15]. The rising incidence is largely due to the aging population, the most significant risk factor [16]. While current pharmacological treatments can help slow symptom progression, they are often associated with side effects such as nausea, diarrhea, and loss of appetite [5], leading to a growing demand for natural products or food-based therapies that offer long-term benefits with minimal side effects.

Recent research has highlighted the neuroprotective potential of olives and their bioactive compounds against the molecular and cellular mechanisms that underlie neurodegenerative diseases [1,5,17–19]. However, although the neuroprotective effects of major olive polyphenols have been widely reported, these effects have not been investigated in an integrated manner that includes the AMP-activated protein kinase alpha (AMPKα)–sirtuin 1 (SIRT1)–peroxisome proliferator-activated receptor gamma coactivator 1 alpha (PGC1α) signaling axis. Notably, the AMPKα–SIRT1–PGC1α signaling axis promotes mitochondrial biogenesis and enhances antioxidant defenses, thereby countering oxidative stress, a major contributor to neuronal degeneration [20]. AMPKα represents the catalytically active α-subunit of AMP-activated protein kinase and serves as a central regulator of cellular energy sensing and redox homeostasis. Similarly, PGC1α is a major α-isoform-specific transcriptional coactivator that governs mitochondrial biogenesis and oxidative metabolism downstream of AMPK signaling. These molecules were therefore selected as key signaling targets to evaluate OPJ-mediated regulation of energy homeostasis and redox balance in neuronal cells. Similarly, the cAMP response element-binding protein (CREB)–brain-derived neurotrophic factor (BDNF)–tropomyosin receptor kinase B (TrkB) pathway plays a crucial role in neuronal survival and synaptic plasticity, providing resilience against progressive neuronal loss [8]. Despite their relevance, involvement of these pathways in neuroprotective effects of olive pomace has not yet been explored, representing a key mechanistic gap that this study aims to address.

Although the neuroprotective effects of individual olive polyphenols, such as hydroxytyrosol, tyrosol, and oleuropein, have been extensively reported—primarily in relation to antioxidant and anti-inflammatory mechanisms—these studies have largely focused on isolated compounds and single signaling pathways. Consequently, whether olive pomace

 https://doi.org/10.3390/molecules31020336

juice, as a complex polyphenolic matrix, is associated with integrated modulation of multiple stress-responsive signaling networks relevant to neuronal resilience remains unclear.

Based on this mechanistic gap, the present study tests the hypothesis that olive pomace juice is associated with changes in key signaling pathways involved in oxidative stress regulation, metabolic homeostasis, neurotrophic support, and inflammatory responses, including the AMPKα–SIRT1–PGC1α, CREB–BDNF–TrkB, and NF-κB pathways, in mouse hippocampal neuronal (HT22) and microglial (BV2) cell models. By focusing on extract-level associations rather than attributing effects to individual compounds, this study provides novel integrative insight into how olive pomace-derived polyphenols may collectively influence neuronal stress responses. In addition, the findings support the potential value of olive pomace as a sustainable, food-derived resource for neuroprotective research.

2. Results and Discussion

2.1. High-Performance Liquid Chromatography (HPLC) Analysis of Hydroxytyrosol and Tyrosol in OPJ

To characterize the major phenolic constituents of olive pomace juice (OPJ), hydroxytyrosol and tyrosol were quantified using high-performance liquid chromatography (HPLC). Representative HPLC chromatograms of OPJ showed well-resolved peaks corresponding to hydroxytyrosol and tyrosol, which were identified by comparison with authentic standards based on retention times and UV spectral characteristics (Figure S1A–C). Quantitative analysis revealed that hydroxytyrosol was the predominant phenolic constituent in OPJ, accounting for 10.92% of the total detected compounds, whereas tyrosol was present at a substantially lower proportion (2.18%). In addition, a considerable fraction of the total polyphenol content (21.46%), as determined by the Folin–Ciocalteu colorimetric assay, consisted of unidentified polyphenolic compounds. Notably, oleuropein, which is commonly reported as a major phenolic component in olive-derived products, was not observed under the specific HPLC conditions employed in this study, which were optimized for the targeted quantification of hydroxytyrosol and tyrosol rather than comprehensive phenolic profiling. Based on their relative abundance among the quantified phenolic compounds, hydroxytyrosol and tyrosol were selected as representative phenolic constituents for subsequent cellular experiments. The concentrations of OPJ used in subsequent cell-based experiments were expressed as μg/mL of the OPJ extract, reflecting the total extract mass containing the quantified phenolic constituents, rather than the concentrations of individual polyphenols.

An overview of the experimental workflow is summarized in Figure 1.

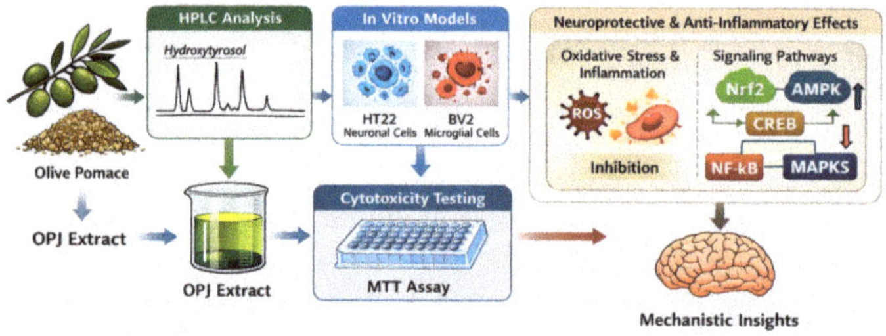

Figure 1. Experimental workflow for olive pomace juice (OPJ) study.

2.2. OPJ Protects HT22 Neurons Against H_2O_2-Induced Oxidative Damage

To evaluate whether olive pomace juice (OPJ) confers direct neuroprotective effects against oxidative stress-induced neuronal injury, HT22 hippocampal neuronal cells were used as an in vitro model. Cell viability was assessed using an MTT assay under hydrogen

 https://doi.org/10.3390/molecules31020336

peroxide (H_2O_2)-induced oxidative stress conditions. In this study, the normal untreated group without H_2O_2 exposure was defined as the normal control group (NOR), while cells treated with H_2O_2 alone served as the oxidative stress control group (CON).

The cytotoxicity of OPJ was first evaluated in HT22 neuronal cells under basal conditions without H_2O_2-induced oxidative stress to determine appropriate treatment concentrations. OPJ slightly increased HT22 cell viability at low concentrations ranging from 3.12 to 12.5 μg/mL. OPJ treatment at 25 and 50 μg/mL did not significantly affect cell viability compared with the control (CON) group. In contrast, OPJ at concentrations of 100 μg/mL and above significantly reduced HT22 cell viability in a dose-dependent manner, with a 23% reduction observed at 400 μg/mL (Figure 2A). Based on these results, non-cytotoxic concentrations of OPJ (12.5, 25, and 50 μg/mL) were selected for all subsequent experiments.

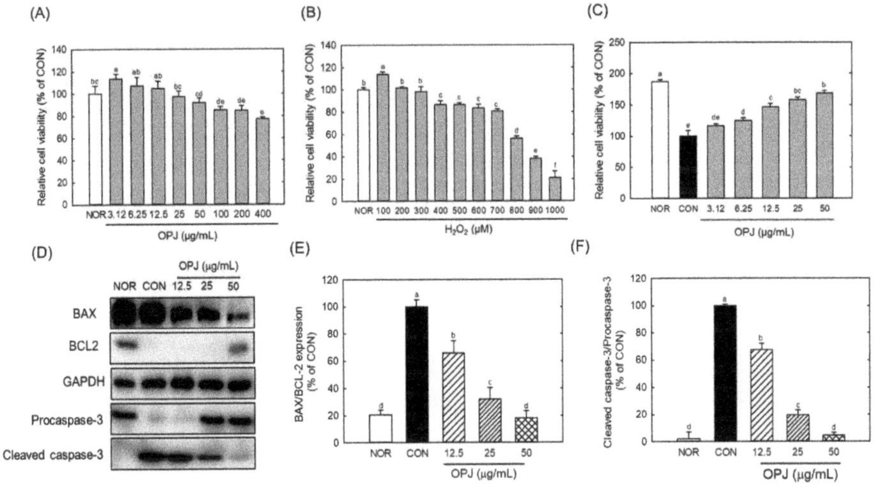

Figure 2. Effect of OPJ on viability and anti-apoptotic signaling in HT22 cells exposed to oxidative stress. HT22 hippocampal neuronal cells were incubated for 24 h with OPJ (3.125–400 μg/mL) (**A**), hydrogen peroxide (H_2O_2; 100–1000 μM) alone (**B**), or OPJ under H_2O_2-induced oxidative stress (800 μM) (**C**), and cell viability was assessed using an MTT assay. Protein abundances of BAX, BCL2, total caspase-3, cleaved caspase-3, and GAPDH were determined by Western blot analysis (**D–F**). Relative protein levels of BAX/BCL2 (**E**) and cleaved caspase-3/total caspase-3 (**F**) were quantified using ImageJ (Version 1.53t). All values represent mean ± SEM. Means with different letters indicate significant differences ($p < 0.05$). NOR, normal untreated group; CON, H_2O_2-treated control group (800 μM).

Hydrogen peroxide (H_2O_2) treatment induced a dose-dependent reduction in HT22 cell viability. Notably, exposure to 800 μM H_2O_2 decreased cell viability by 44.6% relative to the CON group, indicating substantial oxidative cytotoxicity (Figure 2B). Under oxidative stress conditions induced by 800 μM H_2O_2, OPJ treatment significantly enhanced HT22 cell viability in a dose-dependent manner. OPJ concentrations up to 50 μg/mL increased cell viability by approximately 68% compared with the H_2O_2-treated group (Figure 2C), demonstrating a pronounced neuroprotective effect against oxidative injury.

Sustained oxidative stress induced by H_2O_2, a potent reactive oxygen species (ROS) generator, is a major contributor to neuronal injury through disruption of cellular redox homeostasis and mitochondrial dysfunction. In HT22 cells, the marked reduction in cell viability following H_2O_2 exposure reflects neuronal vulnerability to oxidative insults,

 https://doi.org/10.3390/molecules31020336

which are closely associated with early molecular events in neurodegenerative processes, including dysregulation of β-secretase (BACE1) and acetylcholinesterase (AChE). The restoration of cell viability by OPJ under oxidative stress conditions suggests that OPJ effectively mitigates ROS-induced neuronal damage and preserves cellular integrity.

These findings establish HT22 neurons as a primary model for evaluating the neuroprotective effects of OPJ and provide a foundation for subsequent analyses of apoptosis regulation and neurodegeneration-related biomarkers described in the following sections.

2.3. OPJ Inhibits Oxidative Stress-Induced Apoptosis by Modulating BAX/BCL2 Signaling in HT22 Cells

To determine whether the neuroprotective effects of OPJ against oxidative stress involve regulation of apoptotic signaling, the expression of key apoptosis-related proteins was analyzed in H_2O_2-exposed HT22 neuronal cells.

Western blot analysis revealed that H_2O_2-induced oxidative stress markedly altered the expression of multiple apoptosis-related proteins, including BAX, BCL2, and caspase-3, in HT22 neuronal cells (Figure 2D). Specifically, H_2O_2 treatment significantly reduced the levels of BCL2, an anti-apoptotic factor. In contrast, OPJ treatment markedly increased its protein abundance at concentrations of 50 μg/mL, compared to the control group (CON) (Figure 2D). OPJ demonstrated an anti-apoptotic effect by reducing the ratio of Bcl-2–associated X protein (BAX) to B-cell lymphoma 2 (BCL2), an indicator of apoptosis, by 34.4%, 68.2%, and 81.8% at concentrations of 12.5, 25, and 50 μg/mL, respectively (Figure 2E). Additionally, caspase-3, a terminal effector of apoptosis, was significantly regulated by OPJ treatment. The inactive form of caspase-3 (procaspase-3) increased with OPJ treatment, while the activated form (cleaved caspase-3) decreased. OPJ reduced the ratio of cleaved caspase-3/caspase-3, another indicator of apoptosis, by 32.1%, 80.4%, and 95.3% at 12.5, 25, and 50 μg/mL, respectively (Figure 2F). These results suggest that OPJ effectively protects neuronal cells from H_2O_2-induced damage by modulating apoptosis-related proteins and pathways. Hydrogen peroxide (H_2O_2)-induced oxidative stress is a well-established trigger of neuronal apoptosis, in which ROS-mediated damage to proteins, lipids, and DNA disrupts cellular integrity [21]. In neurons, such oxidative insults shift the balance toward pro-apoptotic signaling, creating an environment that favors programmed cell death [21]. Our findings indicate that OPJ counteracts these oxidative stress-driven mechanisms, exerting neuroprotective effects by modulating the apoptotic machinery rather than merely preventing ROS accumulation. Consistent with these observations, OPJ attenuated the increase of pro-apoptotic mediators, including BAX induction and caspase-3 cleavage, while enhancing the stability of anti-apoptotic proteins such as BCL2 and procaspase-3. This modulation restores the balance between survival and death pathways, suggesting that OPJ not only protects structural integrity but also actively reprograms the apoptotic threshold in neuronal cells. Beyond the regulation of apoptosis-related proteins, OPJ may help preserve mitochondrial function and maintain cellular energy homeostasis under oxidative stress conditions [22,23]. By recalibrating the BAX/BCL2 ratio and inhibiting caspase-3 induction, OPJ enables neurons to withstand transient oxidative insults that would otherwise be lethal. These findings emphasize the potential of OPJ as a neuroprotective agent and provide a foundation for further studies exploring its mechanisms and therapeutic applications in oxidative stress-related neurodegenerative diseases.

 https://doi.org/10.3390/molecules31020336

2.4. OPJ Augments Antioxidant Defense via Nrf2-HO-1 Signaling Pathway

To determine whether OPJ enhances endogenous antioxidant defense mechanisms under oxidative stress, Nrf2 signaling and the expression of downstream antioxidant enzymes were examined in H_2O_2-exposed HT22 neuronal cells.

OPJ effectively enhanced the abundance of nuclear factor erythroid 2–related factor 2 (Nrf2), a key regulator of antioxidant responses, and its downstream target, HO-1, under H_2O_2-induced oxidative stress conditions (Figure 3A). Nrf2 protein levels increased by 1.6-fold and 2.4-fold at OPJ concentrations of 25 and 50 μg/mL, respectively, while HO-1 levels rose by 1.9-fold and 3.2-fold at the same concentrations (Figure 3B). Furthermore, oxidative stress caused by H_2O_2 significantly downregulated the mRNA levels of antioxidant enzymes, including catalase, superoxide dismutase 1 (SOD1), and glutathione peroxidase (Gpx). OPJ treatment counteracted this effect by upregulating the expression of these genes. Catalase mRNA expression increased by 12.5% and 42.4% at 25 and 50 μg/mL of OPJ, respectively (Figure 3C). Similarly, SOD1 mRNA levels showed a modest increase of 10.7% at the lower concentration (not statistically significant) and a significant increase of 17.1% at the higher concentration, while Gpx mRNA levels increased by 18.9% and 23.1% at the corresponding concentrations (Figure 3D,E). These findings suggest that OPJ mitigates oxidative stress-induced neuronal cell death by upregulating the Nrf2 signaling pathway and enhancing the expression of antioxidant enzymes. Nrf2 is a transcription factor that regulates a wide range of antioxidant and cytoprotective genes [21]. Under basal conditions, Nrf2 is bound by Kelch-like ECH-associated protein 1 (KEAP1), which facilitates its ubiquitination and subsequent degradation [24]. Oxidative stress induces conformational changes in KEAP1, releasing Nrf2 to translocate into the nucleus. There, Nrf2 binds to antioxidant response elements (AREs) and induces the expression of enzymes such as superoxide dismutase (SOD), glutathione peroxidase (GPx), and heme oxygenase-1 (HO-1), as observed in this study (Figure 4), confirming the role of Nrf2 in OPJ-mediated antioxidant defense. In addition to its antioxidant role, Nrf2 modulates neuroinflammatory responses by limiting pro-inflammatory cytokines and suppressing the activation of microglia and astrocytes [25]. Moreover, Nrf2 supports neuronal survival by facilitating repair of damaged DNA and proteins [26]. This dual function of Nrf2—attenuating oxidative stress and suppressing neuroinflammation—highlights its significance as a neuroprotective target and provides mechanistic insight into OPJ's protective effects. Several natural compounds have been identified as Nrf2 activators with therapeutic potential in neurodegenerative diseases. For instance, sulforaphane, found in cruciferous vegetables such as broccoli, has demonstrated neuroprotective effects through Nrf2 activation [27]. Similarly, olive-derived polyphenols have been reported to exert neuroprotective actions via Nrf2 signaling [28]. These protective effects of OPJ are consistent with previous studies on natural Nrf2 activators, suggesting that OPJ may confer neuroprotection by simultaneously enhancing antioxidant defenses and limiting neuroinflammation, reinforcing its potential as a functional food-derived neuroprotective agent.

https://doi.org/10.3390/molecules31020336

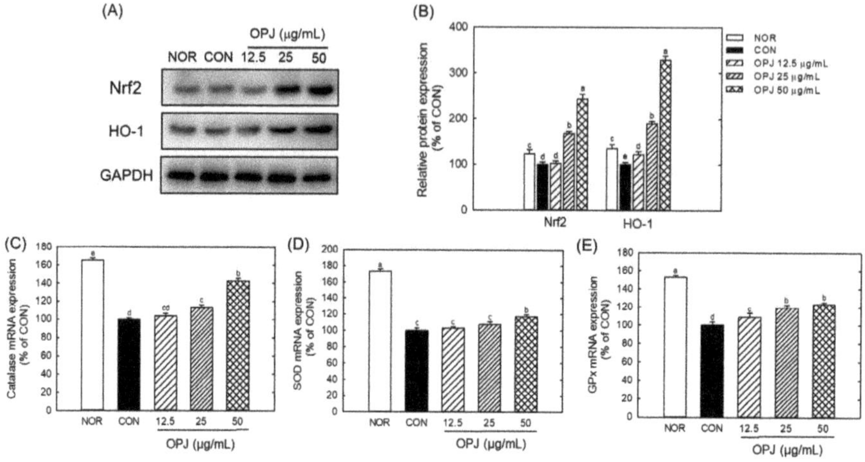

Figure 3. Effect of OPJ on antioxidant response in hydrogen peroxide-exposed HT22 cells. Nrf2 and HO-1 protein abundances were determined by Western blot and their relative levels were quantified by Image J (Version 1.53t) (**A**,**B**). mRNA expressions of catalase (**C**), superoxide dismutase 1 (SOD1) (**D**), and glutathione peroxidase (Gpx) (**E**) were analyzed by real-time PCR. All values represent mean ± SEM. Different letters indicate statistical significance among groups (p < 0.05). NOR, normal (untreated) group; CON, H_2O_2 (800 μM) treated group.

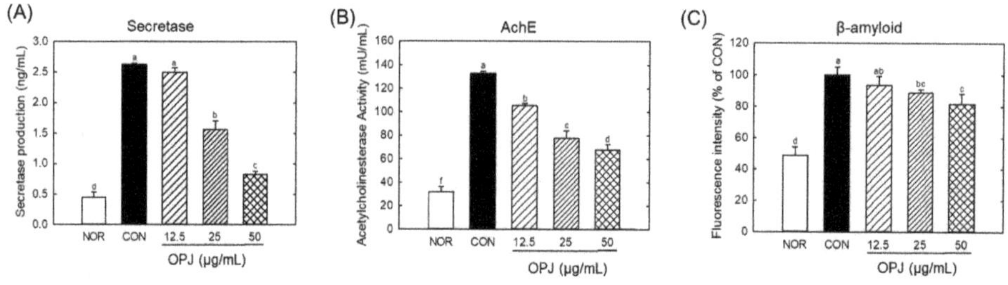

Figure 4. Effect of OPJ on neurogenerative biomarkers in hydrogen peroxide-exposed HT22 cells. β-secretase production (**A**) and acetylcholineesterase (AChE) activity (**B**) were determined by ELISA kit. β-amyloid fibril level (**C**) was examined by Thioflavin staining. All values represent mean ± SEM. Different letters indicate statistical significance among groups (p < 0.05). NOR, normal (untreated) group; CON, H_2O_2 (800 μM) treated group.

2.5. OPJ Modulates Neurodegenerative Pathological Markers

To assess whether OPJ influences molecular markers associated with early neurodegenerative pathology, β-secretase activity, acetylcholinesterase (AChE) activity, and amyloid fibril formation were evaluated in H_2O_2-exposed HT22 neuronal cells.

The level of β-secretase, an enzyme involved in amyloid β production, was significantly elevated under H_2O_2-induced oxidative stress. OPJ effectively suppressed β-secretase production, reducing its levels by 40.4% and 68.7% at 25 and 50 μg/mL, respectively (Figure 4A). A similar regulatory effect was observed for acetylcholinesterase (AChE), an enzyme responsible for breaking down acetylcholine, a critical neurotransmitter. OPJ significantly decreased the H_2O_2-induced increase in AChE activity by 20.8%, 41.3%, and 48.6% at concentrations of 12.5, 25, and 50 μg/mL, respectively (Figure 4B). Furthermore, OPJ reduced the formation of amyloid fibrils, as detected by thioflavin fluorescence. The fluorescence intensity, indicative of amyloid fibril accumulation, decreased

 https://doi.org/10.3390/molecules31020336

by 11.9% and 19.2% at 25 and 50 μg/mL of OPJ, respectively (Figure 4C). These findings collectively indicate that OPJ effectively suppresses neurodegenerative responses induced by H_2O_2-mediated oxidative stress in neuronal cells.

Prolonged oxidative stress is a key driver of early neurodegenerative changes associated with Alzheimer's disease (AD). In particular, markers such as Tau phosphorylation, β-secretase (BACE1) upregulation, and AChE dysregulation are closely linked to amyloid-β production, cholinergic dysfunction, and cytoskeletal instability, respectively [29,30]. In the present study, H_2O_2 exposure markedly increased BACE1, AChE activity, and amyloid fibril accumulation in HT22 cells, reflecting key molecular events of AD-like pathology. OPJ treatment effectively attenuated these H_2O_2-induced neurodegenerative markers, indicating a neuroprotective effect. The observed reduction in BACE1 and AChE indicates that OPJ mitigates ROS-mediated molecular alterations that drive early AD pathology. Additionally, the decrease in amyloid fibril formation implies that OPJ can prevent protein aggregation under oxidative stress. These effects are consistent with previous studies showing that polyphenol-rich natural products, such as resveratrol, curcumin and hydroxytyrosol, protect neuronal cells by modulating BACE1 expression and AChE activity in oxidative stress models [31,32]. These findings suggest that OPJ confers multi-level neuroprotection: it preserves neuronal survival, modulates apoptotic pathways, and suppresses early molecular events associated with neurodegeneration. By concurrently targeting ROS accumulation, apoptosis, and key neurodegenerative biomarkers, OPJ demonstrates a protective effect against oxidative stress-induced neuronal injury, highlighting its potential as a functional food-derived intervention for neurodegenerative disorders.

2.6. OPJ Enhances Neuronal Resilience and Survival by Activation of AMPKα-SIRT1-PGC1α and CREB-BDNF-TrkB Signaling Pathways

To investigate whether OPJ-mediated neuroprotection is associated with regulation of metabolic and neurotrophic signaling, the AMPKα–SIRT1–PGC1α and CREB–BDNF–TrkB pathways were analyzed in H_2O_2-exposed HT22 neuronal cells.

OPJ significantly counteracted H_2O_2-induced downregulation of AMPKα-SIRT1-PGC1α signaling, which is associated with energy biogenesis and antioxidant responses. OPJ increased phosphorylated AMPKα (p-AMPKα) by 2.5-fold, 5.0-fold, and 7.6-fold at 12.5, 25, and 50 μg/mL, respectively, compared to the H_2O_2-treated control (Figure 5A,B). This was accompanied by increased expression of other proteins within the same signaling axis, including SIRT1, PGC1α, and TFAM, as shown in the corresponding Western blot and quantitative analyses (Figure 5A–E). SIRT1 abundance was also elevated with OPJ treatment, increasing by 10.8% and 28.9% at 25 and 50 μg/mL, respectively (Figure 5C). OPJ enhanced PGC1α and its downstream target, TFAM, by 2.0-fold to 2.3-fold at 12.5 to 50 μg/mL and 1.8-fold to 2.1-fold at 25 to 50 μg/mL, respectively (Figure 5D,E). These results indicate that OPJ activates the AMPKα-SIRT1-PGC1α pathway, contributing to neuronal protection against oxidative stress. Additionally, OPJ increased activation of the CREB–BDNF–TrkB signaling pathway. As shown in Figure 5F, OPJ markedly increased CREB phosphorylation under H_2O_2-induced oxidative stress conditions in HT22 neuronal cells. This change was accompanied by increased expression of the neurotrophic factor BDNF (Figure 5G,H) and its receptor TrkB (Figure 5I), which are commonly associated with neuronal survival and synaptic function. Consistent with these observations, OPJ increased phospho-CREB by 38.6% and 40.8% at 25 and 50 μg/mL, respectively, while upregulating BDNF protein levels by 63.6% to 75.8% at 25 to 50 μg/mL (Figure 5G,H). TrkB protein levels were also enhanced by 1.6-fold to 2.1-fold at 12.5 to 50 μg/mL (Figure 5I).

 https://doi.org/10.3390/molecules31020336

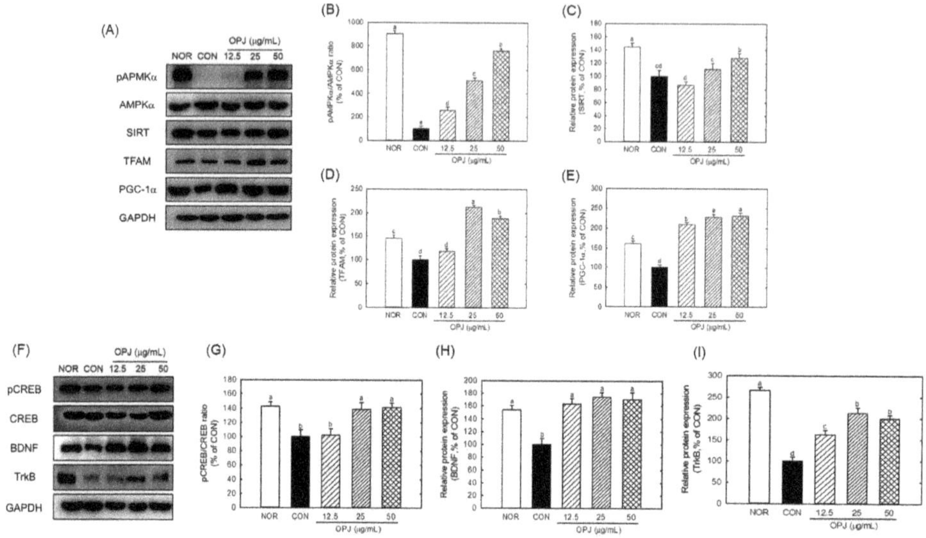

Figure 5. Effect of OPJ on AMPKa-SIRT1-PGC1a and CREB-BDNF-TrkB signaling pathways in hydrogen peroxide-exposed HT22 cells. The components of signaling pathways were determined by Western blot (**A**,**F**). p-AMPKα/total AMPKα (**B**), SIRT1 (**C**), TFAM (**D**), PGC1α (**E**), p-CREB/total CREB (**G**), BDNF (**H**), and TrkB (**I**) protein levels were quantified by Image J (Version 1.53t). All values represent mean ± SEM. Means with different letter showed a significant difference at $p < 0.05$. NOR, normal untreated group; CON, H_2O_2 (800 μM) treated group.

Collectively, these findings demonstrate that OPJ is associated with protection of neuronal cells from oxidative stress-induced damage, accompanied by modulation of both AMPKα-SIRT1-PGC1α and CREB-BDNF-TrkB signaling pathways. Figure 5 integrates the concurrent modulation of metabolic (AMPKα–SIRT1–PGC1α) and neurotrophic (CREB–BDNF–TrkB) signaling pathways, providing associative mechanistic insight for the neuroprotective effects of OPJ under oxidative stress conditions.

The simultaneous modulation of metabolic stability (AMPK) and neurotrophic signaling (CREB) represents a notable observation, as OPJ is associated with changes in multiple signaling pathways relevant to oxidative stress and neuronal resilience. AMPK functions as a cellular energy sensor, enhancing mitochondrial function and antioxidant defenses through downstream effectors such as SIRT1 and PGC1α [33]. SIRT1 stabilizes Nrf2 by deacetylation, increasing its DNA-binding affinity, while PGC1α contributes to transcriptional regulation of antioxidant genes [34].

The concurrent upregulation of Nrf2, HO-1, and AMPKα–SIRT1–PGC1α components suggests that OPJ enhances antioxidant defenses via both direct oxidative stress signaling and upstream metabolics, linking energy homeostasis with neuroprotection. This restoration of the AMPK axis highlights a potential therapeutic advantage of OPJ: its ability to address energetic failure, which is a critical upstream deficit in neurodegeneration, distinguishing it from agents that only act as downstream antioxidant scavengers. Beyond its role in regulating oxidative stress, OPJ appears to enhance neuronal resilience by activating the CREB–BDNF–TrkB signaling pathway. CREB (cAMP response element-binding protein) is a transcription factor that promotes the expression of brain-derived neurotrophic factor (BDNF), which in turn binds to its receptor TrkB to support synaptic plasticity, neurogenesis, and neuronal survival [35]. In this study, OPJ treatment enhanced CREB phosphorylation and upregulated both BDNF and TrkB

https://doi.org/10.3390/molecules31020336

in H_2O_2-stressed HT22 cells, consistent with suppressed apoptosis observed in earlier results (Figure 1) [36]. Together, these findings indicate that OPJ exerts neuroprotective effects through the parallel modulation of antioxidative (Nrf2–HO-1), metabolic (AMPKα–SIRT1–PGC1α), and neurotrophic (CREB–BDNF–TrkB) pathways. In this study, modulation of oxidative stress is framed as the primary mechanistic axis, while changes in metabolic, neurotrophic, and inflammatory signaling pathways are interpreted as associated or supportive responses rather than independent primary mechanisms. Such multi-pathway modulation has been infrequently reported for a single natural product, suggesting that OPJ may influence several stress-responsive mechanisms in neuronal cells. Such modulation further supports the therapeutic potential of OPJ as a functional food-derived intervention against neurodegeneration.

2.7. *OPJ Suppresses Neuroinflammation in BV2 Microglia by Inhibiting NF-κB and MAPK Signaling Pathways*

To evaluate the anti-inflammatory effects of OPJ under neuroinflammatory conditions, BV2 microglial cells were stimulated with lipopolysaccharide (LPS) and inflammatory mediator production and NF-κB/MAPK signaling were subsequently analyzed. Normal untreated group without LPS was defined as NOR, while LPS-treated cells served as the inflammatory control group (CON). OPJ did not exhibit cytotoxicity in BV2 cells at concentrations ranging from 3.12 to 200 μg/mL in the absence of LPS, whereas a modest reduction in cell viability (approximately 15%) was observed only at 400 μg/mL. In the presence of LPS, OPJ was non-toxic to BV2 cells up to 50 μg/mL, whereas higher concentrations (100–400 μg/mL) resulted in a concentration-dependent reduction in cell viability of approximately 20–30% compared to the NOR group (Figure 6A). OPJ effectively decreased LPS-induced nitric oxide (NO) production, with a 54% reduction in 50 μg/mL (Figure 6B).

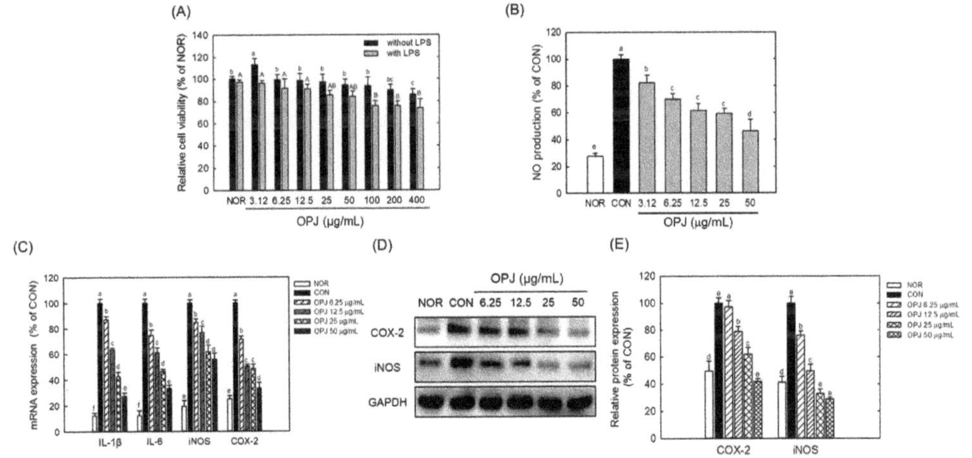

Figure 6. Effect of OPJ on cell viability and inflammatory responses in BV2 cells. BV2 cells were incubated with OPJ (3.125, 6.25, 12.5, 25, 50, 100, 200, 400 μg/mL) with or without lipopolysaccharide (LPS, 100 ng/mL) for 24 h, and then cell viability was measure by MTT assay (**A**). NO generation was determined using Griess reagent (**B**). BV2 cells were treated with OPJ (6.25, 12.5, 25, 50 μg/mL) 2 h pre-treatment before LPS stimulation and incubated for 24 h. mRNA levels of inducible NO (iNOS), COX-2, IL-6, and IL-1β was analyzed by real-time PCR (**C**). Protein abundance of iNOS and COX-2 was determined using Western blot and their levels were quantified using by Image J (Version 1.53t) (**D**,**E**). All values represent mean ± SEM. Means with different letter showed a significant difference at $p < 0.05$. NOR, normal untreated group; CON, only LPS (100 ng/mL) treated group.

 https://doi.org/10.3390/molecules31020336

OPJ inhibited the inflammatory response in LPS-stimulated BV2 cells. At concentrations of 12.5, 25, and 50 μg/mL, OPJ significantly downregulated the mRNA expression of key inflammatory mediators, including inducible nitric oxide synthase (iNOS) and cyclooxygenase-2 (COX-2), by approximately 20–60% in LPS-stimulated BV2 cells (Figure 6C). Similarly, mRNA levels of the pro-inflammatory cytokines IL-6 and IL-1β, which were elevated under LPS treatment, were reduced by 38.9%, 43.4%, and 66.8%, and 36.3%, 57.4%, and 72.7%, respectively, following OPJ treatment (Figure 6C). In addition to its effects on mRNA expression, OPJ also decreased the protein abundance of iNOS and COX-2. COX-2 protein levels were reduced by 21.1%, 38.0%, and 58.1%, while iNOS protein levels decreased by 50.4%, 67.3%, and 71.3% at 12.5, 25, and 50 μg/mL, respectively (Figure 6D,E). At the lowest concentration tested (6.25 μg/mL), OPJ exhibited detectable but weaker effects on inflammatory marker expression compared with the more pronounced responses observed at concentrations of 12.5 μg/mL and above. These findings indicate that OPJ effectively suppresses the LPS-induced inflammatory response by downregulating both the gene expression and protein abundance of key inflammatory markers.

OPJ further suppressed key inflammatory signaling pathways, including nuclear factor kappa B (NF-κB) and mitogen-activated protein kinases (MAPKs). The LPS-induced increase in total NF-κB protein levels was significantly reduced by OPJ treatment, with decreases of 76.9%, 79.2%, and 89.4% at concentrations of 12.5, 25, and 50 μg/mL, respectively (Figure 7A,B). Additionally, OPJ reduced the phosphorylated levels of three major components of the MAPK pathway: extracellular signal-regulated kinase (ERK), c-Jun N-terminal kinase (JNK), and p38 (Figure 7C). Specifically, phosphorylated ERK levels were reduced by 16.2%, 54.1%, and 65.5%; phosphorylated JNK levels decreased by 34.6%, 43.8%, and 54.7%; and phosphorylated p38 levels were lowered by 29.5%, 64.0%, and 66.4% at 12.5, 25, and 50 μg/mL, respectively, compared to the control group (Figure 7D).

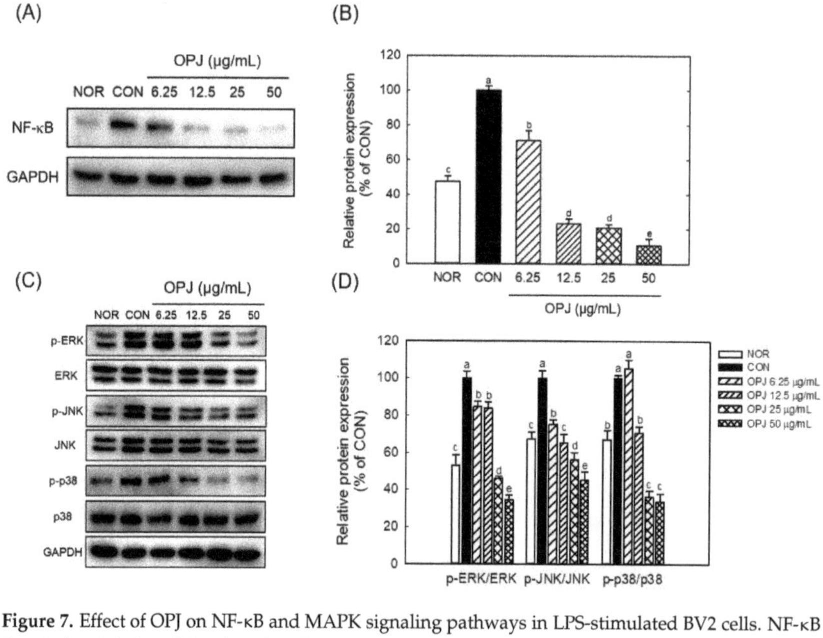

Figure 7. Effect of OPJ on NF-κB and MAPK signaling pathways in LPS-stimulated BV2 cells. NF-κB Protein levels (**A**), and the phosphorylation ratios of ERK, JNK, and p38 were analyzed using Western blot (**A**,**C**) and their levels were quantified using Image J (Version 1.53t) (**B**,**D**). All values represent mean ± SEM. Means with different letter showed a significant difference at $p < 0.05$. NOR, normal untreated group; CON, only LPS treated group (100 ng/mL).

https://doi.org/10.3390/molecules31020336

These findings demonstrate that OPJ is associated with suppression of inflammatory responses, accompanied by reduced NF-κB protein abundance and decreased phosphorylation of MAPK signaling components. The anti-inflammatory effects of OPJ are founded on its ability to directly suppress the central signaling pathways of inflammation (Figure 7). NF-κB and MAPK family (ERK, JNK, p38) play critical, interconnected roles in cellular stress responses and neuroinflammation. NF-κB is a master transcription factor regulating the expression of inflammatory mediators, which is often hyperactivated by oxidative stress and cytokines in neurodegenerative conditions, leading to chronic inflammation that directly harms neurons. Our findings demonstrated that OPJ effectively counteracted this inflammatory cascade. The reduction in inflammatory output (NO, iNOS, COX-2, and cytokines) is associated with decreased NF-κB protein levels and concurrent attenuation of MAPK pathway activation (Figure 7). Specifically, OPJ markedly reduced the phosphorylation of all three MAPK components (ERK, JNK, p38). The attenuation of p38 and JNK signaling is considered important, as these pathways are closely associated with pro-inflammatory and pro-apoptotic signaling processes involved in neurodegeneration, including JNK's role in neuronal apoptosis and p38's role in cytokine production [37–39]. Critically, the suppression of ERK (up to 65.5% reduction) is also highly relevant. Although ERK typically supports neuronal survival, its chronic over-activation is linked to pathological outcomes such as Tau hyperphosphorylation and neurofibrillary tangle formation [38]. Therefore, OPJ-mediated modulation of MAPK family members is consistent with attenuation of signaling processes involved in oxidative stress and neuroinflammation [37–39]. Furthermore, this mechanism complements OPJ's other neuroprotective effects: Nrf2, a key antioxidative factor activated by OPJ, is known to mitigate oxidative stress and reduce the upstream triggers of both NF-κB and MAPKs signalings [40]. Overall, the observed increase in Nrf2-related responses, together with modulation of NF-κB protein levels and MAPK phosphorylation, is consistent with reduced oxidative and inflammatory signaling, supporting the potential relevance of OPJ as a functional food-derived bioactive.

2.8. Comparative Biological Effects of Hydroxytyrosol and Tyrosol in Relation to OPJ-Mediated Neuroprotection

To compare the relative cellular activities of the major OPJ-derived phenolic compounds, hydroxytyrosol and tyrosol were assessed under equivalent experimental conditions for their effects on oxidative stress-induced neuronal viability in HT22 cells and on LPS-induced inflammatory responses in BV2 microglial cells. It should be noted that hydroxytyrosol and tyrosol were tested at equivalent concentrations to enable direct comparison of their intrinsic cellular activities, rather than to replicate their exact relative abundance within OPJ. Therefore, these results should be interpreted as comparative activity profiles rather than quantitative contributions within the extract. Hydroxytyrosol and tyrosol were non-toxic at concentrations of 100 μM and 400 μM, respectively, in HT22 cells (Figure 8A,B). Both compounds enhanced neuronal cell viability under H_2O_2-induced oxidative stress conditions, increasing cell viability by 76.1% and 24.2%, respectively, at a concentration of 100 μg/mL compared to the CON group (Figure 8C,D). This indicates that hydroxytyrosol exhibits greater protective activity than tyrosol under these conditions, while the magnitude and breadth of the OPJ-mediated neuroprotective effects observed in earlier sections suggest additional contributions from other OPJ constituents beyond individual phenolic compounds. In BV2 cell viability assays, both compounds showed non-toxic effects at concentrations up to ~200 μM (Figure 8E,F). Hydroxytyrosol and tyrosol decreased NO production by 40.0% and 17.6%, respectively, at a 100 μM concentration (Figure 8G,H), suggesting that hydroxytyrosol suppresses the LPS-induced inflamma-

 https://doi.org/10.3390/molecules31020336

tory response more effectively. While hydroxytyrosol is identified as the more bioactive compound in these comparative assays, OPJ contains tyrosol together with a substantial proportion of other unidentified polyphenols (21.46%), which may act cooperatively to shape the overall biological responses observed at the extract level. This suggests that the comprehensive protective effects of OPJ might be due to a 'cocktail effect' rather than the action of a single compound alone. Consistent with this interpretation, previous studies have reported that olive-derived phenolics, including hydroxytyrosol and oleuropein, modulate amyloid aggregation and neuronal survival, supporting the view that individual phenolic activities may partially explain, but do not fully account for, the broader biological effects of olive-based extracts [41,42]. Moreover, a recent study introduced a hybrid molecule combining hydroxytyrosol with donepezil, a well-established Alzheimer's drug, underscoring its therapeutic potential [43]. In addition to hydroxytyrosol and oleuropein, oleocanthal has also been reported as a potent olive-derived phenolic compound with strong biological effects at the neuronal level, including modulation of neuroinflammatory signaling and attenuation of amyloid-related neurotoxicity [44,45]. In the context of the present study, these observations support the interpretation that hydroxytyrosol contributes to, but does not solely define, the neuroprotective and anti-inflammatory effects of OPJ. Although phenolic compounds such as oleuropein are often considered key antioxidant and anti-inflammatory agents in olive-based products, our HPLC analysis did not detect oleuropein.

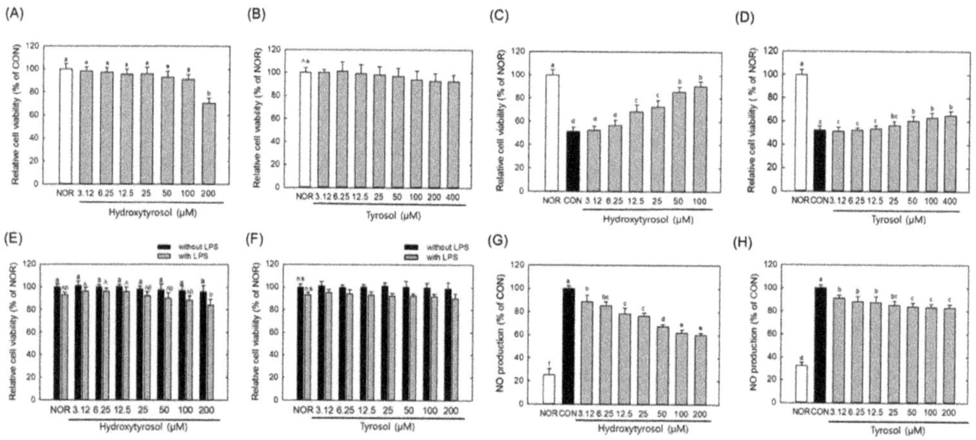

Figure 8. Comparative effects of hydroxytyrosol and tyrosol on cell viability and inflammatory response in HT22 and BV2 cells. HT22 cells were incubated with hydroxytyrosol or tyrosol (3.125 to 400 μM) (**A**,**B**), in the absence and presence of H_2O_2 (**C**,**D**) for 24 h, and then cell viability was measure by MTT assay. BV2 cells were incubated with hydroxytyrosol or tyrosol (3.125 to 200 μM) with or without lipopolysaccharide (LPS) for 24 h, and then cell viability was measure by MTT assay (**E**,**F**). BV2 cells were treated with hydroxytyrosol or tyrosol (6.25 to 200 μg/mL) 2 h before LPS stimulation and incubated for 24 h. NO generation was determined using Griess reagent (**G**,**H**). All values represent mean ± SEM. Means with different letter showed a significant difference at $p < 0.05$. NOR, normal untreated group; CON, H_2O_2-treated group (HT22 cells) or LPS-treated group (BV2 cells). n.s indicates no statistically significant difference.

Because OPJ-derived compounds must access the central nervous system to exert direct neuroprotective effects, blood–brain barrier (BBB) permeability is a relevant consideration. Hydroxytyrosol, a small (154.16 Da), lipophilic molecule, is reported to readily penetrate the BBB, whereas tyrosol exhibits comparable permeability but lower bioactiv-

 https://doi.org/10.3390/molecules31020336

ity, consistent with its weaker cellular effects observed in this study [46,47]. While other unidentified phenolic compounds in OPJ may contribute to neuroprotection, our findings indicate that hydroxytyrosol is the predominant phenolic compound in OPJ and a major contributor to its observed neuroprotective and anti-inflammatory effects. Although hydroxytyrosol was included in comparative cellular assays as a reference compound, the majority of mechanistic experiments in this study were conducted using the OPJ extract. Therefore, the observed neuroprotective and anti-inflammatory effects should be interpreted as extract-level phenomena rather than as effects attributable to a single compound. The potential contributions of tyrosol and other unidentified polyphenols, as well as synergistic interactions among OPJ constituents, should also be considered.

Limitations

The present study has several limitations that should be acknowledged. First, all findings were derived from in vitro models of oxidative stress-induced neuronal damage (HT22 cells) and LPS-stimulated microglial activation (BV2 cells), which do not fully recapitulate the complexity of neurodegenerative diseases in vivo. Second, although hydroxytyrosol has been reported to cross the blood–brain barrier, its oral bioavailability is limited, as it is rapidly metabolized into glucuronide and sulfate conjugates before reaching the systemic circulation and the brain. The neuroprotective efficacy and BBB permeability of these OPJ-derived metabolites may differ substantially from those of the parent compound tested in this in vitro study. Finally, the concentrations used in vitro were selected to investigate relative cellular responses and mechanistic associations, and therefore do not directly reflect achievable brain concentrations in vivo. Consequently, the present findings should be interpreted as providing mechanistic insight into oxidative stress– and inflammation-associated neurodegenerative processes, rather than as evidence of disease prevention or therapeutic efficacy, highlighting the need for future in vivo and pharmacokinetic studies.

3. Materials and Methods

To investigate the neuroprotective and anti-inflammatory potential of olive pomace juice (OPJ), an in vitro experimental framework was established using two complementary cell models. Mouse hippocampal neuronal HT22 cells were employed to evaluate oxidative stress-induced neuronal injury and associated signaling pathways, whereas BV2 microglial cells were used to assess lipopolysaccharide (LPS)-induced neuroinflammatory responses. OPJ was selected as the primary test material to reflect extract-level biological effects, while hydroxytyrosol and tyrosol were included as reference compounds for comparative analysis of major OPJ-derived phenolics. Oxidative stress and inflammatory conditions were induced using hydrogen peroxide (H_2O_2) and LPS, respectively, based on preliminary optimization experiments. The overall experimental sequence comprised OPJ compositional analysis, determination of non-cytotoxic concentration ranges, induction of oxidative stress or inflammation, and subsequent evaluation of cellular viability, signaling pathways, and functional biomarkers.

A schematic overview of the experimental workflow is provided in Figure 1. Based on this design, the Results and Discussion are organized to first describe OPJ composition, followed by its neuroprotective effects in HT22 neuronal cells and anti-inflammatory effects in BV2 microglial cells.

3.1. Preparation of OPJ

Olive pomace juice (OPJ) used in this study was obtained from JYGlobalfoods Co., Ltd. (Seoul, Republic of Korea) and was prepared through a three-step process. First,

 https://doi.org/10.3390/molecules31020336

ripe olive fruits (*Olea europaea* L.) were washed with water to remove impurities before being milled to facilitate oil extraction. The resulting mixture then underwent horizontal centrifugation, which separated it into two primary components: olive oil, the main product for consumption, and alpeorujo, a byproduct containing residual oil, water, and solids. In the second step, the alpeorujo was further processed through centrifugation, yielding three key components: solids, pomace oil, and olive vegetation water (also known as olive pomace juice). This olive vegetation water was then subjected to filtration and concentration to refine its composition. In the final step, the filtered and concentrated olive vegetation water underwent further processing through an exchange column process using osmosis water for enhanced purification. The purified liquid was then subjected to evaporation for heat sterilization. Following this, a non-GMO dextrin (as a carrier) and a small amount of calcium salts (as an anticaking agent) were added before the mixture entered a vacuum drying stage. The dried material was then milled into a fine powder (OPJ). OPJ was standardized to contain more than 10.0% hydroxytyrosol and 2.0% tyrosol (w/w) based on HPLC analysis (see Section 3.9) across three independent batches.

3.2. MTT Assay

HT22 or BV2 cells (1.0×10^5 cells/mL, a gift from Korea University) were cultured in Dulbecco's Modified Eagle's Medium (DMEM) containing 10% fetal bovine serum and 1% penicillin-streptomycin. When the cell density reached 70%, the cells were subcultured to maintain the culture. The cells were seeded in a 96-well plate and incubated for 24 h. Afterward, OPJ with indicated concentrations was introduced to the cells, and the culture was maintained for 24 h. After 24 h, the medium was replaced with a serum-free medium containing 3-(4,5-dimethylthiazol-2-yl)-2,5-diphenyltetrazolium bromide (MTT) reagent (0.5 mg/mL, Sigma-Aldrich, St. Louis, MO, USA), and the cells were incubated at 37 °C for 1 h. The medium was then removed, and dimethyl sulfoxide (DMSO) was added to each well to dissolve the MTT formazan. The absorbance was measured at 550 nm using a microplate reader (Spectra Max M3, Molecular Devices, San Jose, CA, USA). Untreated cells without OPJ, H_2O_2, or LPS exposure were used as the normal control (NOR), whereas cells treated with H_2O_2 alone (HT22 cells) or LPS alone (BV2 cells) served as the respective experimental control (CON). Culture medium without cells was used as the blank for background subtraction.

3.3. Cell Culture

The HT22 or BV2 cells (1.0×10^5 cells/mL) were initially seeded in a 6-well plate and cultured for 24 h in Dulbecco's Modified Eagle's Medium (DMEM, Welgene, Gyeongsan-si, Republic of Korea) containing 3.7 g/L sodium bicarbonate, 10% fetal bovine serum (FBS, Welgene, Gyeongsan-si, Republic of Korea), and 1% penicillin-streptomycin (PS, Welgene, Gyeongsan-si, Republic of Korea). The cells were then incubated at 37 °C with 5% CO_2 for an additional 24 h. To investigate the influence of OPJ on H_2O_2-induced cell death, cells were treated with OPJ for 2 h. Subsequently, H_2O_2 (800 μM, St. Louis, MO, USA) was administered to induce oxidative stress. To evaluate the effect of OPJ on the inflammatory response, cells were pretreated with OPJ for 2 h before stimulation with LPS (100 ng/mL) and incubated for 24 h. Cells cultured without H_2O_2 or LPS treatment were used as the normal control (NOR), while cells exposed to H_2O_2 (800 μM) or LPS (100 ng/mL) alone were used as the oxidative stress or inflammatory control (CON), respectively.

https://doi.org/10.3390/molecules31020336

3.4. Determination of NO Production

Cells were seeded in 6-well plates at a concentration between 1×10^5 and 2×10^5 cells/well and cultured for 24 h. Afterward, OPJ is treated for 2 h, followed by additional 24 h of incubation with LPS. Then, cell culture medium (100 μL) and Griess reagent (100 μL) were added to a 96-well plate and allowed to react at 25 °C for 5 min. The amount of nitric oxide (NO) is then determined by measuring the absorbance at 550 nm using a microplate reader (INNO M, LTek, Seongnam, Republic of Korea). Cells treated with culture medium alone were used as the normal control (NOR), and cells treated with LPS alone were used as the inflammatory control (CON). Culture medium without cells was used as the blank for NO measurement.

3.5. Determination of Intracellular Amyloid Fibrils

Thioflavin (Sigma-Aldrich, St. Louis, MO, USA), a cell-permeable benzothiazole dye, was prepared in PBS buffer at a concentration of 10 μM. After removing the culture medium, the dye was applied to the cells and incubated at 37 °C for 2 h in the dark. Fluorescence was then measured using a SpectraMax M3 fluorometer (Molecular Devices, Sunnyvale, VA, USA) with an excitation wavelength of 437 nm and an emission wavelength of 485 nm. Untreated cells were used as the normal control (NOR), whereas cells treated with H_2O_2 alone served as the oxidative stress control (CON).

3.6. Western Blot

After treating the cells with the sample and incubating for 24 h, the medium was removed. Lysis buffer containing protease inhibitors (aprotinin, leupeptin, phenylmethylsulfonyl fluoride [PMSF], Thermo Fisher Scientific Inc., Waltham, MA, USA) and phosphatase inhibitors was added to lyse the cells. The lysed cells were collected and centrifuged at 10,000× *g* for 10 min at 4 °C. The supernatant was collected to obtain the protein extract. Using the Bicinchoninic Acid (BCA) assay, an amount of the extract corresponding to 50 μg of protein was separated by electrophoresis on a gel containing 10% acrylamide (GeneAll Biotechnology Co., Ltd., Seoul, Republic of Korea). The proteins on the gel were transferred onto a polyvinylidene fluoride (PVDF) membrane. The membrane was blocked with 5% skim milk for 30 min, followed by incubation with a primary antibody at a 1:1000 dilution at 4 °C for 24 h. The membrane was washed twice with Tris-buffered saline with 0.1% Tween 20 (TBST) for 10 min each. It was then incubated with a secondary antibody at 25 °C for 2 h. After additional washes with TBST, the target proteins on the membrane were visualized using enhanced chemiluminescence (ECL) reagents. Protein bands were detected using the Alliance Q9 system (UVITECH, Cambridge, UK) and analyzed with Alliance Q9 software. The relative intensity of protein bands was quantified using ImageJ software. Primary antibodies used in this study are follows: BAX, BCL-2, procaspase-3, cleaved caspase-3, p-ERK, ERK, p-JNK, JNK, p-p38, p38, COX-2, iNOS, NF-κB, phospho-AMPKα, total AMPKα, SIRT1, PGC1α, phospho-CREB, total CREB, BDNF, and TrkB were purchased from Cell Signaling Technology, Inc (Denvers, MA, USA). Nrf2, heme oxygenase-1 (HO-1), and glyceraldehyde-3-phosphate dehydrogenase (GAPDH) were obtained from Santa Cruz Biotechnology, Inc. (Dallas, TX, USA). The secondary antibody (anti-rabbit IgG, horseradish peroxidase conjugated) was obtained from Cell Signaling Technology, Inc (Danvers, MA, USA). Protein expression levels were compared relative to untreated cells (NOR) or cells treated with H_2O_2 or LPS alone (CON), depending on the experimental condition.

https://doi.org/10.3390/molecules31020336

3.7. RNA Extraction and Real-Time PCR

Total RNA was extracted from collected cells using TRIzol reagent (Thermo Fisher Scientific) following the manufacturer's protocol. After centrifugation at 12,000 rpm for 8 min at 4 °C, the supernatant was mixed with isopropanol, incubated for 1 h, and centrifuged again to obtain an RNA pellet. The pellet was washed with 70% ethanol, air-dried at 25 °C, and dissolved in 30–50 μL of diethyl pyrocarbonate (DEPC)-treated water. cDNA was synthesized from 1 μg of RNA using the Maxime RT PreMix kit (iNtRON Biotechnology, Seongnam, Republic of Korea). qPCR was performed using SYBR Green PCR master mix (Applied Biosystems, Waltham, MA, USA) and the AriaMX real-time PCR system (Agilent Technology, Santa Clara, CA, USA). The primers used are listed in Table 1. Gene expression levels were normalized to GAPDH and compared relative to untreated control cells (NOR) or cells treated with H_2O_2 or LPS alone (CON).

Table 1. Primers used in this study.

Name	Forward (5′ to 3′)	Reverse (5′ to 3′)
iNOS	CAGCCAAGCACTCCAATGTA	CCTGAAGGAGCTTTGTCCAC
COX2	TCTCCCCTCTCTACGCATTC	GGCAGAACGACTCGGTTATC
IL-1β	ACTCATTGTGGCTGTGGAGA	TTGTTCATCTCGGAGCCTGT
IL-6	CTGCTGCACTTTGGAGTGAT	GCCTCTTTGCTGCTTTCACAC
Gpx	AGGGTAGAGGC CGGATAAG	AGAAGGCATACACGGTGGAC
Catalase	ACATGGTCTGGGACTTCTGG	CAAGTTTTTGATGCCCTGGT
SOD1	GAGACCTGGGCAATGTGACT	GTTTACTGCGCAATCCCAAT
GAPDH	CTGCGACTTCAACAGCAACT	GAGTTGGGATAGGGCCTCTC

3.8. Determination of β-Secretase Production and Acetylcholineesterase (AChE) Activity

The activity of β-secretase and acetylcholinesterase (AChE) in HT22 cells was evaluated using the β-secretase activity kit (R&D Systems, Wiesbaden, Germany) and the acetylcholinesterase activity kit (BM-ACH-100, Biomax, Guri, Republic of Korea), respectively, in accordance with the manufacturers' instructions. Untreated HT22 cells were used as the normal control (NOR), and cells treated with H_2O_2 alone served as the oxidative stress control (CON).

3.9. HPLC Analysis of Hydroxytyrosol and Tyrosol

Hydroxytyrosol and tyrosol samples were analyzed using an Alliance 1260 HPLC system (Waters, Milford, MA, USA) equipped with an autosampler, quaternary pump, and UV detector. Samples were prepared by dissolving 5 mg of each compound per mL of dimethyl sulfoxide (DMSO) and filtering through a 0.45 μm nylon membrane. The filtered supernatant was injected into a LIChrospher 100-C18 column (250 × 4.0 mm, 5 μm particle size) at 30 °C. Detection was performed at a wavelength of 280 nm. A gradient elution was carried out with a mixture of acetic acid (2.5%) and acetonitrile in the following ratios: 95:5 (at 0 min), 75:25 (20 min), 50:50 (40 min), 20:80 (50 min), and returning to 95:5 (60 min). The flow rate was maintained at 1.0 mL/min with an injection volume of 20 μL. HPLC analyses were performed on three independent batches of OPJ to confirm consistent content of key bioactive compounds. Quantification of hydroxytyrosol and tyrosol in OPJ was carried out by comparing peak areas to those of pure standards. Calibration curves were constructed using serial dilutions of each standard compound at appropriate concentrations. Each analyte showed excellent linearity within its respective range with coefficients of determination (R^2) exceeding 0.99. As a result, the final OPJ contains 10.92% hydroxytyrosol and 2.18% tyrosol (*w*/*w*). Hydroxytyrosol (4999S) and tyrosol (4949S)

 https://doi.org/10.3390/molecules31020336

standards were obtained from EXTRAXYNTHESE (Genay Cedex, France). In addition to HPLC analysis, the total polyphenol content of OPJ was determined using a colorimetric assay based on the Folin–Ciocalteu method. Total polyphenol content was expressed as gallic acid equivalents (GAE) based on a calibration curve constructed using gallic acid as the standard, and the results were calculated as percentage (w/w) relative to the dry weight of OPJ.

3.10. Statistical Analysis

Each experiment was performed at least three times, and the data were analyzed using one-way analysis of variance (ANOVA) with Statistical Package for the Social Sciences (SPSS; version 27.0; IBM-SPSS Inc., Chicago, IL, USA). Statistical significance among mean values was determined at the $p < 0.05$ level followed by Duncan's multiple range test. Results are expressed as the mean $\pm$ standard error of the mean (SEM). Changes less than 100% relative to control are expressed as percent change, whereas changes equal to or greater than 100% are expressed as fold change relative to control.

4. Conclusions

Our findings highlight the potential of olive pomace juice (OPJ) as a functional food-derived material with neuroprotective and anti-inflammatory properties under conditions of oxidative stress and neuroinflammation. OPJ treatment was associated with improved neuronal survival, attenuation of apoptosis, enhancement of antioxidant defense, and suppression of inflammatory signaling in HT22 neuronal cells and BV2 microglia. Hydroxytyrosol, identified as a principal bioactive component of OPJ, contributes substantially to the observed cellular protective effects; however, the majority of mechanistic analyses were conducted using the OPJ extract. Accordingly, the observed biological activities should be interpreted as extract-level phenomena, potentially involving synergistic interactions among multiple polyphenolic constituents. Importantly, the present findings are derived exclusively from in vitro models of oxidative stress-induced neuronal damage and LPS-stimulated microglial activation. Therefore, these results should be interpreted as providing mechanistic insight into processes associated with neurodegeneration rather than as evidence for disease prevention or therapeutic efficacy. Further in vivo and pharmacokinetic studies are required to clarify bioavailability, metabolic transformation, and pathway hierarchy underlying OPJ-mediated effects.

Beyond its biological relevance, OPJ—derived from olive pomace, an agro-industrial by-product—represents a promising candidate for upcycling into value-added ingredients for the food and nutraceutical industries, supporting circular economy principles [48–50] while offering potential health-related benefits.

Supplementary Materials: The following supporting information can be downloaded at: https://www.mdpi.com/article/10.3390/molecules31020336/s1. Figure S1. HPLC chromatogram of OPJ and standards. Standards (hydroxytyrosol and tyrosol) (A,B) and OPJ (C) were analyzed using HPLC. The HPLC chromatogram represents the relative distribution of major and minor phenolic constituents and was not used to calculate the total poly-phenol content.

Author Contributions: Y.-L.Y.: Investigation, Validation, Resources, Methodology, Writing—Review and Editing, Formal analysis. H.-J.B.: Investigation, Validation, Resources, Visualization, Methodology. N.K.: Investigation, Methodology, Conceptualization. M.S.L., J.-I.L. and I.P.: Investigation, Methodology. H.-S.C.: Conceptualization, Methodology, Supervision, Writing—Original draft preparation, Project administration, Writing—Review and Editing. All authors have read and agreed to the published version of the manuscript.

 https://doi.org/10.3390/molecules31020336

Funding: This work was supported by the Technology development Program (RS-2024-00427643) funded by the Ministry of SMEs and Startups (MSS, Republic of Korea). The APC was funded by the same grant.

Data Availability Statement: The original contributions presented in the study are included in the article.

Conflicts of Interest: Namgil Kang, Min Soo Lee, and Jeong-In Lee were employed by JY Globalfoods Co., Ltd. Author Ilbum Park was employed by Kwangdong Pharm Co., Ltd. The companies had no role in the study design, data collection and analysis, decision to publish, or preparation of the manuscript. The remaining authors declare that the research was conducted in the absence of any commercial or financial relationships that could be construed as a potential conflict of interest.

Abbreviations

AChE, Acetylcholineesterase; AMPK, AMP-activated protein kinase; BACE-1, Beta-site amyloid precursor protein cleaving enzyme-1 (beta-secretase); BAX, BCL associated X; BBB, Blood brain barrier; BCL2, B-cell lymphoma; BDNF, brain-derived neurotrophic factor; COX-2, Cyclooxigenase-2; CREB, Cyclic AMP Response Element Binding Protein; ELISA, Enzyme-Linked Immunosorbent Assay; ERK, Extracellular Signal-Regulated Kinase; Gpx, Glutathione peroxidase; HO-1, Hemoxygenase-1; iNOS, inducible nitricoxide synthase; JNK, c-Jun N-terminal Kinase; Keap1, Kelch-like ECH-associated protein 1; IL-1β/6, Interleukin-1beta; NF-κB, Nuclear Factor kappa-light-chain-enhancer of activated B cells; Nrf2, Nuclear factor erythroid 2-related factor 2; MAPKs, Mitogen-Activated Protein Kinases; OPJ, Olive pomace extract; PGC1a, Peroxisome proliferator-activated receptor gamma coactivator 1-alpha; ROS, Reactive oxygen species; SIRT1, NAD-dependent deacetylase sirtuin-1; SOD1, superoxide dismutase 1; TrkB, tropomyosin receptor kinase B.

References

1. Tsoupras, A.; Panagopoulou, E.; Kyzas, G.Z. Olive pomace bioactives for functional foods and cosmetics. *AIMS Agric. Food* **2024**, *9*, 743–766. [CrossRef]
2. Madureira, J.; Margaça, F.M.A.; Santos-Buelga, C.; Ferreira, I.C.F.R.; Barros, S.C.V.L. Applications of bioactive compounds extracted from olive industry wastes: A review. *Compr. Rev. Food Sci. Food Saf.* **2022**, *21*, 453–476. [CrossRef] [PubMed]
3. Cabrera, L.; Xavier, L.; Zecchi, B. Extraction of phenolic compounds with antioxidant activity from olive pomace using natural deep eutectic solvents: Modelling and optimization by response surface methodology. *Discov. Food* **2024**, *4*, 29. [CrossRef]
4. Valls-Pedret, C.; Sala-Vila, A.; Serra-Mir, M.; Corella, D.; de la Torre, R.; Martínez-González, M.Á.; Martínez-Lapiscina, E.H.; Fitó, M.; Pérez-Heras, A.; Salas-Salvadó, J.; et al. Mediterranean diet and age-related cognitive decline: A randomized clinical trial. *JAMA Intern. Med.* **2015**, *175*, 1094–1103. [CrossRef] [PubMed]
5. Bucciantini, M.; Leri, M.; Nardiello, P.; Casamenti, F.; Stefani, M. Olive polyphenols: Antioxidant and anti-inflammatory properties. *Antioxidants* **2021**, *10*, 1044. [CrossRef]
6. Al-Addous, M.; Alnaief, M.; Class, C.; Nsair, A.; Kuchta, K.; Alkasrawi, M. Technical possibilities of biogas production from olive and date waste in Jordan. *BioRes* **2017**, *12*, 9383–9395. [CrossRef]
7. Chait, S.; Chater, E.A.; Laaroussi, N.; Asbik, M. Theoretical simulation of olive pomace pyrolysis for biomass energy: A sustainable approach to waste reduction and energy efficiency. *E3S Web Conf.* **2023**, *469*, 00084. [CrossRef]
8. Bhatia, L.; Jha, H.; Sarkar, T.; Sarangi, P.K. Food waste utilization for reducing carbon footprints towards sustainable and cleaner environment: A review. *Int. J. Environ. Res. Public Health* **2023**, *20*, 2318. [CrossRef]
9. Nutrizio, M.; Dukić, J.; Sabljak, I.; Samardžija, A.; Fučkar, V.B.; Djekić, I.; Jambrak, A.R. Upcycling of food by-products and waste: Nonthermal green extractions and life cycle assessment approach. *Sustainability* **2024**, *16*, 9143. [CrossRef]
10. Gugel, I.; Marchetti, F.; Costa, S.; Baldini, E.; Vertuani, S.; Manfredini, S. 2G-lactic acid from olive oil supply chain waste: Olive leaves upcycling via *Lactobacillus casei* fermentation. *Appl. Microbiol. Biotechnol.* **2024**, *108*, 379. [CrossRef]
11. GBD 2021 Nervous System Disorders Collaborators. Global, regional, and national burden of disorders affecting the nervous system, 1990–2021: A systematic analysis for the Global Burden of Disease Study 2021. *Lancet Neurol.* **2024**, *23*, 344–381.

 https://doi.org/10.3390/molecules31020336

12. Lamptey, R.N.L.; Chaulagain, B.; Trivedi, R.; Gothwal, A.; Layek, B.; Singh, J. A review of the common neurodegenerative disorders: Current therapeutic approaches and the potential role of nanotherapeutics. *Int. J. Mol. Sci.* **2022**, *23*, 1851. [CrossRef] [PubMed]
13. Angeloni, C.; Malaguti, M.; Prata, C.; Freschi, M.; Barbalace, M.C.; Hrelia, S. Mechanisms underlying neurodegenerative disorders and potential neuroprotective activity of agrifood by-products. *Antioxidants* **2022**, *12*, 94. [CrossRef] [PubMed]
14. Zhong, F.; Liang, S.; Zhong, Z. Emerging role of mitochondrial DNA as a major driver of inflammation and disease progression. *Trends Immunol.* **2019**, *40*, 1120–1133. [CrossRef] [PubMed]
15. Praticò, D. Evidence of oxidative stress in Alzheimer's disease brain and antioxidant therapy: Lights and shadows. *Ann. N. Y. Acad. Sci.* **2008**, *1147*, 70–78. [CrossRef]
16. Berrechid, A.G. Neurological disorders of aging. *J. Neurol. Sci.* **2023**, *455*, 121046. [CrossRef]
17. Gonçalves, M.; Vale, N.; Silva, P. Neuroprotective Effects of Olive Oil: A Comprehensive Review of Antioxidant Properties. *Antioxidants* **2024**, *13*, 762. [CrossRef]
18. Micheli, L.; Bertini, L.; Bonato, A.; Villanova, N.; Caruso, C.; Caruso, M.; Bernini, R.; Tirone, F. Role of Hydroxytyrosol and Oleuropein in the Prevention of Aging and Related Disorders: Focus on Neurodegeneration, Skeletal Muscle Dysfunction and Gut Microbiota. *Nutrients* **2023**, *15*, 1767. [CrossRef]
19. Wei, L.; Li, Z.; Shi, M.; Song, W.; Teng, Z.; Zhang, C. Neuroprotective properties of extra virgin olive oil polyphenols in Alzheimer's disease: A multi-target mechanistic review. *Front. Nutr.* **2025**, *12*, 1736633. [CrossRef]
20. Herzig, S.; Shaw, R.J. AMPK: Guardian of metabolism and mitochondrial homeostasis. *Nat. Rev. Mol. Cell Biol.* **2018**, *19*, 121–135. [CrossRef]
21. Angelova, P.R.; Abramov, A.Y. Role of mitochondrial ROS in the brain: From physiology to neurodegeneration. *FEBS Lett.* **2018**, *592*, 692–702. [CrossRef] [PubMed]
22. Chen, Z.; Zhong, C. Oxidative stress in Alzheimer's disease. *Neurosci. Bull.* **2014**, *30*, 271–281. [CrossRef] [PubMed]
23. Maksimović, T.; Gădău, C.; Antal, G.; Čoban, M.; Eșanu, O.; Atyim, E.; Mioc, A.; Șoica, C. Polyphenol-based therapeutic strategies for mitochondrial dysfunction in aging. *Biomolecules* **2025**, *15*, 1116. [CrossRef] [PubMed]
24. He, F.; Ru, X.; Wen, T. NRF2, a transcription factor for stress response and beyond. *Int. J. Mol. Sci.* **2020**, *21*, 4777. [CrossRef]
25. Saha, S.; Buttari, B.; Profumo, E.; Tucci, P.; Saso, L. A perspective on Nrf2 signaling pathway for neuroinflammation: A potential therapeutic target in Alzheimer's and Parkinson's diseases. *Front. Cell. Neurosci.* **2022**, *15*, 787258. [CrossRef]
26. Xiong, W.; MacColl Garfinkel, A.E.; Li, Y.; Benowitz, L.I.; Cepko, C.L. NRF2 promotes neuronal survival in neurodegeneration and acute nerve damage. *J. Clin. Investig.* **2015**, *125*, 1433–1445. [CrossRef]
27. Klomparens, E.A.; Ding, Y. The neuroprotective mechanisms and effects of sulforaphane. *Brain Circ.* **2019**, *5*, 74–83. [CrossRef]
28. ALHaithloul, H.A.S.; Alotaibi, M.F.; Bin-Jumah, M.; Elgebaly, H.; Mahmoud, A.M. *Olea europaea* leaf extract up-regulates Nrf2/ARE/HO-1 signaling and attenuates cyclophosphamide-induced oxidative stress, inflammation and apoptosis in rat kidney. *Biomed. Pharmacother.* **2019**, *111*, 676–685. [CrossRef]
29. Cioffi, F.; Adam, R.H.I.; Broersen, K. Molecular mechanisms and genetics of oxidative stress in Alzheimer's disease. *J. Alzheimers Dis.* **2019**, *72*, 981–1017. [CrossRef]
30. Alavi Naini, S.M.; Soussi-Yanicostas, N. Tau hyperphosphorylation and oxidative stress, a critical vicious circle in neurodegenerative tauopathies? *Oxid. Med. Cell. Longev.* **2015**, *2015*, 151979. [CrossRef]
31. Terracina, S.; Petrella, C.; Francati, S.; Lucarelli, M.; Barbato, C.; Minni, A.; Ralli, M.; Greco, A.; Tarani, L.; Fiore, M.; et al. Antioxidant intervention to improve cognition in the aging brain: The example of hydroxytyrosol and resveratrol. *Int. J. Mol. Sci.* **2022**, *23*, 15674. [CrossRef]
32. Yu, H.; Yamashita, T.; Hu, X.; Bian, Z.; Hu, X.; Feng, T.; Tadokoro, K.; Morihara, R.; Abe, K. Protective and anti-oxidative effects of curcumin and resveratrol on Aβ-oligomer-induced damage in the SH-SY5Y cell line. *J. Neurol. Sci.* **2022**, *441*, 120356. [CrossRef] [PubMed]
33. Gao, Y.; Chu, S.; Li, J.; Zhang, Z.; Xu, D. Antioxidant and anti-inflammatory effects of curcumin in a Parkinson's disease rat model through the activation of SIRT1–PGC-1α signaling pathway. *Neurol. Res.* **2020**, *42*, 3–13.
34. Gureev, A.P.; Shaforostova, E.A.; Popov, V.N. Regulation of mitochondrial biogenesis as a way for active longevity: Interaction between the NRF2 and PGC-1α signaling pathways. *Front. Genet.* **2019**, *10*, 435. [CrossRef] [PubMed]
35. Esvald, E.E.; Tuvikene, J.; Sirp, A.; Patil, S.; Bramham, C.R.; Timmusk, T. CREB family transcription factors are major mediators of BDNF transcriptional autoregulation in cortical neurons. *J. Neurosci.* **2020**, *40*, 1405–1426. [CrossRef]
36. Li, Y.Z.; Wu, Z.Y.; Zhu, B.Q.; Wang, Y.X.; Kan, Y.Q.; Zeng, H.C. The BDNF-TrkB-CREB signalling pathway is involved in bisphenol S-induced neurotoxicity in male mice by regulating methylation. *Toxics* **2022**, *10*, 413. [CrossRef]
37. Bendotti, C.; Borsello, T. Targeting stress activated protein kinases, JNK and p38, as new therapeutic approach for neurodegenerative diseases. *Cent. Nerv. Syst. Agents Med. Chem.* **2006**, *6*, 109–117. [CrossRef]

 https://doi.org/10.3390/molecules31020336

38. Zhang, W.; Sun, H.; Zhao, W.; Li, J.; Meng, H. Suppression of JNK pathway protects neurons from oxidative injury via attenuating parthanatos in glutamate-treated HT22 neurons. *Sci. Rep.* **2024**, *14*, 25793. [CrossRef]
39. Colucci-D'Amato, L.; Perrone-Capano, C.; di Porzio, U. Chronic activation of ERK and neurodegenerative diseases. *Bioessays* **2003**, *25*, 1085–1095. [CrossRef]
40. Li, Y.; Han, N.; Hou, P.; Zhao, F.-Q.; Liu, H. Roles of MAPK and Nrf2 signaling pathways in quercetin alleviating redox imbalance induced by hydrogen peroxide in mammary epithelial cells. *Anim. Nutr.* **2024**, *1*, e1. [CrossRef]
41. Chen, C.; Ai, Q.; Wei, Y. Potential role of hydroxytyrosol in neuroprotection. *J. Funct. Foods* **2021**, *82*, 104506. [CrossRef]
42. Leri, M.; Natalello, A.; Bruzzone, E.; Stefani, M.; Bucciantini, M. Oleuropein aglycone and hydroxytyrosol interfere differently with toxic Aβ(1–42) aggregation. *Food Chem. Toxicol.* **2019**, *129*, 1–12. [CrossRef]
43. Costanzo, P.; Oliverio, M.; Maiuolo, J.; Bonacci, S.; De Luca, G.; Masullo, M.; Arcone, R.; Procopio, A. Novel hydroxytyrosol-donepezil hybrids as potential antioxidant and neuroprotective agents. *Front. Chem.* **2021**, *9*, 741444. [CrossRef] [PubMed]
44. Barbalace, M.C.; Freschi, M.; Rinaldi, I.; Zallocco, L.; Malaguti, M.; Manera, C.; Ortore, G.; Zuccarini, M.; Ronci, M.; Cuffaro, D.; et al. Unraveling the Protective Role of Oleocanthal and Its Oxidation Product, Oleocanthalic Acid, against Neuroinflammation. *Antioxidants* **2024**, *13*, 1074. [CrossRef] [PubMed]
45. Qosa, H.; Batarseh, Y.S.; Mohyeldin, M.M.; El Sayed, K.A.; Keller, J.N.; Kaddoumi, A. Oleocanthal enhances amyloid-β clearance from the brains of TgSwDI mice and in vitro across a human blood-brain barrier model. *ACS Chem. Neurosci.* **2015**, *6*, 1849–1859. [CrossRef] [PubMed]
46. Velásquez-Jiménez, D.; Corella-Salazar, D.A.; Zuñiga-Martínez, B.S.; Domínguez-Avila, J.A.; Montiel-Herrera, M.; Salazar-López, N.J.; Rodrigo-Garcia, J.; Villegas-Ochoa, M.A.; González-Aguilar, G.A. Phenolic compounds that cross the blood-brain barrier exert positive health effects as central nervous system antioxidants. *Food Funct.* **2021**, *12*, 10356–10369. [CrossRef]
47. Fan, L.; Peng, Y.; Li, X. Brain regional pharmacokinetics of hydroxytyrosol and its molecular mechanism against depression assessed by multi-omics approaches. *Phytomedicine* **2023**, *112*, 154712. [CrossRef]
48. De Oliveira, C.O.; Roll, A.A.P.; Medeiros Gonçalves, F.M.; Lopes, D.C.N.; Xavier, E.G. Olive pomace for the feeding of commercial poultry: Effects on performance, meat and eggs quality, haematological parameters, microbiota and immunity. *World Poult. Sci. J.* **2021**, *77*, 363–376. [CrossRef]
49. María del Mar, C.; Inmaculada, R.; Manuel, M.; Eulogio, C. Olive-derived biomass as a renewable source of value-added products. *Process Biochem.* **2020**, *97*, 43–56.
50. Rodrigues, R.; Alves, R.C.; Oliveira, M.B.P.P. Exploring olive pomace for skincare applications: A review. *Cosmetics* **2023**, *10*, 35. [CrossRef]

 https://doi.org/10.3390/molecules31020336

molecules

MDPI

Article

Standardization of Germinated Oat Extracts and Their Neuroprotective Effects Against $A\beta_{1-42}$ Induced Cytotoxicity in SH-SY5Y Cells

Yu-Young Lee [1], In-Su Na [2], Jeong-Eun Kim [2], Jae-Gwang Song [3], Chae-Eun Han [3], Hyung-Wook Kim [3] and Soon-Mi Shim [2,*]

[1] Department of Food Science, National Institute of Crop Science, Rural Development Administration, Suwon 16429, Republic of Korea; leeyy260@korea.kr
[2] Department of Food Science and Biotechnology, Sejong University, Seoul 05006, Republic of Korea; 23110112@sju.ac.kr (I.-S.N.); 24110208@sju.ac.kr (J.-E.K.)
[3] Department of Biological Science and Technology, Sejong University, Seoul 05006, Republic of Korea; songjg@sejong.ac.kr (J.-G.S.); ghwnsdlvv@sejong.ac.kr (C.-E.H.); kimhyung@sejong.ac.kr (H.-W.K.)
* Correspondence: soonmishim@sejong.ac.kr; Tel.: +82-2-3408-3229; Fax: +82-2-3408-4319

Abstract: The present study aimed to standardize germinated oat extracts (GOEs) by profiling avenanthramides (AVNs) and phenolic acids and evaluate their neuroprotective effects against $A\beta_{1-42}$-induced cytotoxicity in human neuroblastoma (SH-SY5Y) cells. GOEs were standardized to contain 1652.56 ± 3.37 μg/g dry weight (dw) of total AVNs, including 468.52 ± 17.69 μg/g AVN A, 390.33 ± 10.26 μg/g AVN B, and 641.22 ± 13.89 μg/g AVN C, along with 490.03 ± 7.83 μg/g dw of ferulic acid, using a validated analytical method. Treatment with AVN C and GOEs significantly inhibited $A\beta_{1-42}$-induced cytotoxicity ($p < 0.05$). Furthermore, both AVNs and GOEs markedly reduced $A\beta_{1-42}$-induced reactive oxygen species (ROS) generation in SH-SY5Y cells, showing significant scavenging activity at concentrations of 25 μg/mL (AVNs) and 50 μg/mL (GOEs) ($p < 0.05$). RT-PCR analysis revealed that AVNs and GOEs effectively downregulated the expression of inflammation- and apoptosis-related genes triggered by $A\beta_{1-42}$ exposure. These findings suggest that GOEs rich in AVNs may serve as a potential functional ingredient for enhancing memory function through the inhibition of neuroinflammation and oxidative stress.

Keywords: germinated oat extracts; avenanthramides; $A\beta_{1-42}$; human neuroblastoma cell; memory improvement

1. Introduction

Alzheimer's disease is a progressive neurodegenerative disorder characterized by memory loss and cognitive decline, primarily associated with the formation of senile plaques, neurofibrillary tangles, and the accumulation of β-amyloid (Aβ) fibrils (Uwishema, Mahmoud [1]. The percentage of people with Alzheimer's disease increases with age, and the annual number of new cases of Alzheimer's and other forms of dementia is expected to double by 2050 [2]. Several hypotheses have been proposed to explain its pathogenesis, including the cholinergic hypothesis, which attributes cognitive decline to the loss of cholinergic neurons and reduced acetylcholine levels; the tau hypothesis, which involves the aggregation of hyperphosphorylated tau proteins into neurofibrillary tangles; and the neuroinflammatory hypothesis, which emphasizes chronic neuroinflammation driven by activated microglia and astrocytes [3–5]. It is well-known that the primary trigger for the growth of Alzheimer's disease is the deviant metabolism of the amyloid precursor glycoprotein and the subsequent accumulation of soluble oligomers $A\beta_{1\text{-}40}$ and $A\beta_{1\text{-}42}$ [6,7]. It was also confirmed that $A\beta_{1\text{-}42}$ correlated with the formation of a neuroinflammatory mediator, such as reactive oxygen species (ROS) and the tumor necrosis factor (TNF)-α, which can cause neuronal cell apoptosis [8]. Many studies suggest, for this reason, that the prevention of oxidation and activation of anti-inflammation could be critical factors in order to improve Alzheimer's disease [9,10]. For instance, flavonoids, such as quercetin, which are well-known antioxidants, provide neuroprotective effects by inhibiting the apoptotic cell death of the SH-SY5Y cell against H_2O_2-induced oxidative stress [11,12]. The study also found that oral intake of quercetin before intracerebroventricular (i.c.v.) injection of aggregated $A\beta_{25\text{-}35}$ improved memory function in mice via a behavioral test.

Oats (*Avena sativa* L.) are the most cultivated grains after corn, beans, rice, wheat, and barley, and they are a sufficient source of nutrients, which include soluble fibers, beta-glucan, vitamin B, unsaturated fatty acids, proteins, and minerals [13,14]. Oats also contain various types of polyphenols, such as *p*-coumaric acid, ferulic acid, and caffeic acid, which are considered to be antioxidants [15]. Avenanthramides (AVNs), which are a special group of phenolic alkaloids, are in particular commonly found in oats [16]. AVNs themselves are known as antipathogens (phytoalexin), which are created in response to exposure to pathogens, due to their strong antioxidant and anti-inflammatory activity [17,18]. There are twenty different forms of AVNs that are identified in oat extracts, and three types are major forms, which consist of a 5-hydroxyanthranilic acid and a cinnamic acid, such as *p*-coumaric acid (AVN A), ferulic acid (AVN B), and caffeic acid (AVN C) [18,19]. There is a considerable evidence to suggest that oats protect neuronal cells against inflammation as well as improve memory loss due to dementia [20–22]. It was found that oat extract enriched AVNs increased the reduced form of glutathione in plasma by antioxidant activity, which resulted in the improvement of CNS differentiation and proliferation [18,23]. The inhibitory effects of AVN-C on ROS generation were observed in PC12 cells exposed to H_2O_2 and 6-hydroxydopamine hydrobromide, where AVN-C attenuated ROS production in a dose-dependent manner. These cells, derived from adrenal pheochromocytoma, showed a significant reduction in oxidative stress upon AVN-C treatment [24]. A previous study demonstrated the inhibitory effects on $A\beta$ deposition by treating H2-22 cells and the brains of Tg-5XFAD Alzheimer's disease mice with AVN A and oat seed extract [25]. Furthermore, it was also reported that the content and bioavailability of AVNs increased with germinating cereals [26,27]. In this context, "germinating cereals" refer to grains that have undergone controlled soaking and incubation to induce sprouting, whereas "non-germinating cereals" indicate grains that have not been subjected to such treatment [28]. The oral intake of oat extracts in particular promoted the peak plasma concentration of AVNs after 2 h, which indicated the highest bioavailability [18,29]. However, an accurate mechanism in regard to protecting the $A\beta$-impaired synaptic function by AVN-enriched germinated oat extracts (GOEs), which shows promise as a functional ingredient, is not fully understood. Thus, the current study firstly standardized the GOEs as a functional ingredient by profiling the AVNs and phenolic acids using the validated analytical method with high-performance liquid chromatography (HPLC) with an ultra-violet (UV) detector.

The human neuroblastoma SH-SY5Y cell line is a commonly used model in regard to screening neuroprotective therapeutic agents [17,30]. We investigated the AVN and GOEs in order to determine their efficacy in alleviating apoptosis and the ROS production against $A\beta$-induced SH-SY5Y cell damage. In addition, we measured the expression of the nuclear factor kappa-light-chain-enhancer of activated B cells (NF-κB), the anti-apoptotic enzyme Bcl-2, and the pro-apoptotic enzyme Bax in SH-SY5Y treated with $A\beta_{1-42}$ in order to examine the effects of the GOEs treatment on inflammation and apoptosis.

2. Materials and Methods

2.1. Chemicals and Reagents

The analytical standards of avenanthramide (AVN) A, B, and C were purchased from Sigma-Aldrich (≥98%, St. Louis, MO, USA). The analytical grades of vanillic acid, caffeic acid, ferulic acid, and *p*-coumaric acid were obtained from Sigma-Aldrich (≥98%, MO, USA). The formic acid (HCOOH) was purchased from Fisher Scientific (Morristown, NJ, USA), and the acetonitrile (ACN), methanol, and water for HPLC grade were obtained from J.T. Baker (Phillipsburg, NJ, USA). The distilled water was obtained from Joylife (Busan, Republic of Korea), and the ethanol was purchased from Daehan Ethanol Life (Hwaseong, Republic of Korea). All the chemicals were liquid chromatography (LC) grade. The minimum essential medium (MEM) and the fetal bovine serum (FBS) were purchased from HyClone™ (Marlborough, MA, USA). The phosphate-buffered saline (PBS) was obtained from Biowest (Nuaille, France), and the penicillin-streptomycin was obtained from Corning Inc. (Corning, NY, USA). The 3-[4,5-dimethylthiazol-2yl]-2,5-diphenyl-tetrazolium bromide and the 2′,7′-Dichlorodihydrofluorescein diacetate (DCFH-DA) were purchased from Sigma-Aldrich (MO, USA). The human β-amyloid ($A\beta_{1-42}$) peptide was obtained from Abcam (Cambridge, UK). TRIzol reagent was obtained from Invtrogen (Waltham, MA, USA), the isopropanol was acquired from Duksan (Seoul, Republic of Korea), and the nuclease-free water (AM9937) was purchased from Bio Basic Inc. (Markham, ON, Canada).

2.2. Sample Preparation for Germinated Oat Extracts (GOEs)

The *Daeyang* germinated oats were harvested from Gyeongju-gun, South Korea, which is Region 1, during the spring of 2021, and they were kindly provided by Doobo Food Company (Seoul, Republic of Korea). The avenanthramides from *Daeyang* germinated oats were extracted, which is described in a previous study, with slight modifications [31]. The *Daeyang* germinated oats were ground using a food grinder (Hanil, Bucheon, Republic of Korea), and an aliquot amount of the ground sample (100 g) was mixed with 30% ethanol to 10 times the ratio of the sample weight, which was then extracted at 40 °C for 6 h. The extracts were filtered using 8 µm of Whatman No. 2 filter paper (Whatman International Ltd., Maidstone, UK). They were then concentrated using a vacuum rotary evaporator (Sunileyela, Seongnam, Republic of Korea) until the ethanol solvent was fully removed by evaporation. The concentrated samples were sterilized using autoclaves at 90 °C for 30 min, which was followed by spray drying at 140 °C of the inlet and 45 °C of the outlet temperature using a two-fluid nozzle spray-dryer (SD1000, Eyela, Tokyo, Japan). The dried powder was then stored at 4 °C for further use.

2.3. Qualitative and Quantitative Analysis of Avenanthramides (AVNs) in Daeyang Germinated Oat Extracts (GOEs) by HPLC-UV

Identification of the avenanthramides in the GOEs was performed using High-Performance Liquid Chromatography-Ultraviolet (HPLC-UV) (Ultimate 3000, Thermofisher, Waltham, MA, USA), which was slightly modified from a previous method [31]. The systems were utilized with a reverse phase column of YMC-PACK ODS-A (150 × 4 mm, 5.0 µm, YMC, Kyoto, Japan), which was maintained at 35 °C. An aliquot amount of the dried sample (10 mg) was dissolved in 1 mL of 50% MeOH, which was diluted using 100% acetonitrile (ACN) to a 10:1 (*v*/*v*%) ratio for the HPLC-UV analysis that was followed by filtering with 0.45 µm polytetrafluoroethylene (PTFE) syringe filters. The mobile phase consisted of solvent A, which was water that contained 5% ACN with 0.1% formic acid, and solvent B, which was ACN with 0.1% formic acid, and they were eluted at a flow rate of 1 mL/min. The mobile phase gradient began with 100% solvent A, which followed the following order: 0–32.5 min with 100–87% of solvent A, 32.5–69 min with 87–70% of solvent A, 69–75 min with 70–90% of solvent A, and 75–80 min with 90–100% of solvent A, and it was then equilibrated back with 100% of solvent A within 5 min. All the analyses were conducted with a 10 μL injection and were monitored at a wavelength of 330 nm. AVN A, B, and C were quantified by comparisons of their peaks with the external standard curves of those analytical standards.

All analyses were conducted with a 10 μL injection volume. Each sample was analyzed in triplicate, and results are expressed as mean ± standard deviation (SD) from three independent experiments.

Analytical Method Validation

The HPLC-UV analytical method of AVN A, B, and C was validated for linearity, accuracy, precision, and sensitivity, which included the limit of detection (LOD) and the limit of quantification (LOQ). The linearity was identified from three replicate analyses at five concentrations, which included 1, 5, 10, 25, and 50 μg/mL, of AVN A, B, and C. The correlation coefficient (r^2) was calculated using the regression equation in Microsoft Excel 365. The accuracy was determined by the degree of agreement between a measured concentration and a nominal concentration. In total, 10 to 100 μg/mL of AVN A, B, and C were analyzed for five replicates, which were calculated using the formula that is provided below.

$$\%\text{Error} = \frac{\text{Measured concentration} - \text{Nominal concentration}}{\text{Nominal concentration}} \times 100 \tag{1}$$

$$\text{Accuracy (\%, recovery rate)} = (100 - \text{Error}) \tag{2}$$

The precision was expressed by repeatability, which was the approximation of the results that were obtained by the same operator, measuring system, condition, and laboratory over a brief period. The repeatability was conducted by using 10 to 100 μg/mL of AVN A, B, and C, which was replicated 5 times. The repeatability of the analytical method was calculated by the coefficient of variation (%CV), which is the same as the percent of relative standard deviation (%RSD).

$$\%\text{CV} = \frac{\text{S.D. of measured concentration}}{\text{Mean of measured concentration}} \times 100 \tag{3}$$

S.D. is the standard deviation.

The sensitivity was evaluated by the limit of detection and quantification (LOD and LOQ) by the calibration curve.

$$\text{LOD (Limit of dectection)} = 3.3 \times \frac{S_y}{S} \tag{4}$$

$$\text{LOQ (Limit of quantification)} = 10 \times \frac{S_y}{S} \tag{5}$$

S_y is the standard deviation of the y-intercepts of the regression curves, and S is the slope of the calibration curve.

2.4. Identification of Phenolic Acids in Daeyang Germinated Oat Extracts (GOEs)

An optimized HPLC analysis system was modified from a previous method for the identification of phenolic acids in the GOEs [15]. The samples were analyzed in an Ultimate

3000 HPLC system (Thermofisher, Waltham, MA, USA) on a reversed-phase column (YMC-PACK ODS-A, 150 × 4.6 mm, 5.0 μm, YMC, Kyoto, Japan). The solvent system consisted of 1% acetic acid in water (A) and 1% acetic acid in 100% methanol. The column oven was maintained at 30 °C, and the flow rate was 0.9 mL/min. The injection volume was 10 μL, and the linear gradient conditions were as follows: 0–1.5 min with 100–100% of solvent A, 1.5–7 min with 100–80% of solvent A, 7–13 min with 80–80% of solvent A, 12–44.5 min with 80–50% of solvent A, 44.5–47 min with 50–100% of solvent A, and 47–50 min with 100–100% of solvent A. The initial condition of the gradient was adjusted for 3 min by equilibration.

2.5. Human Neuroblastoma Cell (SH-SY5Y) Culture

The human neuroblastoma cell line, SH-SY5Y, was purchased from the Korean Cell Line Bank (KCLB, Seoul, Republic of Korea). The cells were seeded into 100 pi culture plates at a density of approximately 1–2 × 10^6 cells per plate, allowing them to reach 70–80% confluency within 2–3 days. Cells were cultivated in minimum essential medium (MEM) containing 10% fetal bovine serum (FBS) and 1% penicillin streptomycin (P/S) in a humidified atmosphere of 5% CO_2 in 95% air at 37 °C. The medium was refreshed every other day, and it was sub-cultured when it reached 70 to 80% confluency. All the samples were treated at a confluence of up to 90%.

2.6. Preparation and Aggregation of $A\beta_{1\text{-}42}$ Peptide

The aggregation of $A\beta_{1\text{-}42}$ peptide was performed according to a previous method [7]. The lyophilized human $A\beta_{1\text{-}42}$ peptide was dissolved to 200 μM with a 1% ammonia solution. The peptide was diluted with PBS to a concentration of 1 mg/mL for a longer preservation of $A\beta_{1\text{-}42}$. The stock solution was kept at −80 °C for further use. The aggregated $A\beta_{1\text{-}42}$ was diluted with a cell culture medium for treatments.

2.7. Cell Viability Using an MTT Assay

The MTT assay was performed using the specifications from a previous study with slight modifications [32]. Cell viability was assessed using the MTT assay to evaluate the protective effects of AVNs (A, B, and C) and GOEs against $A\beta_{1\text{-}42}$-induced cytotoxicity [33]. The 1 × 10^5 cells were seeded in each well of the 96-well cell culture plate with a cell culture medium with 10% fetal bovine serum (FBS) and 1% penicillin-streptomycin, and they were incubated for 24 h at 5% CO_2 and 37 °C. After that, a 1 to 25 μg/mL AVNs standard and 1 to 50 μg/mL GOEs were diluted in the culture medium (MEM), and they were treated with the control group, which included only MEM for 24 h. $A\beta_{1\text{-}42}$ (20 μM) was treated in order to identify the protective effects of the AVNs and the GOEs against the β-amyloid-induced apoptosis. The treatment was removed after a 24 h incubation, and a 10-fold diluted MTT solution (5 mg/mL) was added for 2 h. The MTT solution was removed after confirming the formation of the formazan, and it was diluted with 200 μL of DMSO. The cell viability was obtained by using a microplate (Varioskan Flash, Thermo Scientific, San Jose, CA, USA) at 570 nm, which was calculated using the equation that is provided below.

$$\text{Cell viability (\%)} = \frac{\text{Absorbance of treatment}}{\text{Absorbance of control}} \times 100 \tag{6}$$

2.8. Measurement of Reactive Oxygen Species (ROS) Induced by $A\beta_{1\text{-}42}$

The intracellular ROS levels were measured using a method from a previous study with slight modifications [33]. The 2′7′-DCFH-DA assay was conducted in order to identify the ROS levels that were induced by $A\beta_{1\text{-}42}$, and it was reduced by the AVNs, the phenolic acids standard, and the GOEs. The SH-SY5Y cells were seeded in 96-well black cell culture plates at a density of 1×10^5 cells per well for 24 h. Each well was washed using DPBS, and the cells were then treated using 1 to 25 μg/mL AVNs, the phenolic acids standard, and 1 to 50 μg/mL of the GOEs for 24 h in incubation. $A\beta_{1\text{-}42}$ (20 μM) was treated for 24 h in order to identify the protective effects of the AVNs and phenolic acids standard and the GOEs against β-amyloid-induced oxidative stress. A 200 μL 2′7′-DCFH-DA solution was added to each well after 24 h of incubation, and it was incubated for 30 min at 5% CO_2 and 37 °C. After that, each well was washed 2 times with PBS. The intracellular ROS levels were measured using a microplate at 488 nm for excitation and 525 nm for emission, which was calculated using the equation that is provided below.

$$\text{ROS}\ (\%) = \frac{\text{Average of treatment}}{\text{Average of control}} \times 100 \tag{7}$$

The DCFH-DA assay reflects the total intracellular oxidative stress rather than distinguishing individual ROS species, as ROS are short-lived and indirectly measured through the formation of the stable fluorescent compound DCF [34].

2.9. Sample Preparation for Reverse Transcription Polymerase Chain Reaction

Any pre-existing aggregates were removed by dissolving the $A\beta_{1\text{-}42}$ into 5 mM of hexafluoro-2-propanol (HFIP) in order to prepare the oligomeric $A\beta_{1\text{-}42}$, which was aliquoted into microcentrifuge tubes. It was then allowed to evaporate in a fume hood for 16 h and stored at −20 °C. $A\beta_{1\text{-}42}$ was diluted to 100 μM in a serum-free and phenol-free cell culture medium for the oligomeric samples, and it was incubated at 4 °C for 24 h. The 3.25×10^5 cells were seeded in 6-well cell culture plates and incubated with the cell culture medium for 24 h at 5% CO_2 and 37 °C. The medium was then removed, and the cells were pre-treated by 1 and 50 μg/mL GOEs and 25 μg/mL AVNs for 24 h. The oligomeric $A\beta_{1\text{-}42}$ was treated with 1 μM for 24 h in incubation in order to identify the protective effects of the AVNs and GOEs against oligomeric $A\beta_{1\text{-}42}$-induced neuroinflammation and the disruption of the neurotransmitter.

2.10. RNA Isolation and Reverse Transcription Polymerase Chain Reaction (RT-PCR) Analysis

The total RNA in the SH-SY5Y cells was extracted using TRIzol (Invitrogen, Carlsbad, CA, USA). The RNA concentration was analyzed using a NanoDrop 2000 spectrophotometer (Thermo Fisher Scientific, Waltham, MA, USA). A total of 500 ng of the SH-SY5Y RNA was reverse-transcribed with High-Capacity RNA to the cDNA Kit (Applied Biosystems, Foster City, CA, USA) by following the manufacturer's instructions. The RT-PCR was performed using a PowerUp™ SYBR™ Green Master Mix (Applied Biosystems, CA, USA) and a QuantStudio™ 1 Real-Time PCR instrument (Applied Biosystems, CA, USA). The relative expression levels were calculated by using the $2^{-\Delta\Delta CT}$ method. GAPDH was used as the endogenous control. The primer sequences are shown in Table 1.

Table 1. Primer lists for RT-PCR.

Gene	Forward Sequence	Reverse Sequence
NF-κB (p50)	GCAGCACTACTTCTTGACCACC	TCTGCTCCTGAGCATTGACGTC
NF-κB (p65)	TGAACCGAAACTCTGGCAGCTG	CATCAGCTTGCGAAAAGGAGCC
BCL-2	ATCGCCCTGTGGATGACTGAGT	GCCAGGAGAAATCAAACAGAGGC
BAX	TCAGGATGCGTCCACCAAGAAG	TGTGTCCACGGCGGCAATCATC
GAPDH	GTCTCCTCTGACTTCAACAGCG	ACCACCCTGTTGCTGTAGCCAA

2.11. Statistical Analysis

All the data was expressed as mean ± SD in triplicate. A one-way analysis of variance (ANOVA) was applied in order to measure the significant differences among the samples at $p < 0.05$ using GraphPad Prism 6.01 software (GraphPad, La Jolla, CA, USA). A p-value below 0.05 was considered statistically significant, and the post hoc Tukey test was conducted at the 95% confidence level. Data were analyzed using one-way or two-way analysis of variance (ANOVA), as appropriate, followed by Tukey's honestly significant difference (HSD) post hoc test for pairwise comparisons. Statistical significance was set at $p < 0.05$.

3. Results and Discussion

3.1. Bioanalytical Methods Validation of AVN A, B, and C by HPLC-UV

The method validation was conducted in order to set the optimal analysis using 1 to 100 μg/mL of the AVNs standard solution. AVN A, B, and C were well separated, which were eluted at 47.16 min, 49.95 min, and 40.29 min, respectively. This indicated that the analytical condition provided specificity, which is shown in Figure 1C. The correlation coefficient (r^2) for each standard solution with the range of 1 to 50 μg/mL was over 0.999. This means the regression equation performed well, as shown in Table 2. Both accuracy and precision of the analytical method were evaluated using 10, 50, and 100 μg/mL standard solutions via repeated measurements for five replicates during a single day, as shown in Table 2. The accuracy was between 98.82~100.47, 100.17~102.70, and 99.70~101.59 for AVN A, B, and C, respectively. This indicates that it is less than a 10% error, which is an accepted reference value. The precision ranged from 1.40 to 4.85% CV, which was estimated by the ratio of the mean and standard deviation for all standard solutions. Table 2 shows that the LOD and LOQ, which were determined in the standard solution, ranged between 0.610 and 0.733 μg/mL for the LOD and 1.848 and 2.223 μg/mL for the LOQ. These results confirmed that the developed analytical method for AVN A, B, and C provided acceptable specificity, linearity, accuracy, precision, and sensitivity.

Table 2. Analytical methods validation of AVN A, B, and C.

Parameters	AVN A	AVN B	AVN C
Linearity			
Concentration range (μg/mL)	1–50	1–50	1–50
Regression Equation	y = 55608x − 22077	y = 45200x − 14845	y = 35929x − 19394
Correlation coefficient, r^2	0.9997	0.9998	1
Accuracy and Precision			
Accuracy (%)	98.82–100.47	100.17–102.70	99.70–101.59
Precision (CV%)	1.40–4.51	2.65–4.85	0.91–2.30
Sensitivity			
Limit of detection (LOD, μg/mL)	0.733	0.616	0.610
Limit of quantification (LOQ, μg/mL)	2.223	1.866	1.848

3.2. Standardization of GOEs by Quantifying AVN and Phenolic Acids

It is well documented that avenanthramides function as phytoalexins in oats and are induced by fungal infection, particularly *Puccinia coronata*, as well as by chemical elicitors [35]. The AVNs and phenolic acids from the GOEs, which were followed by extraction, concentration, sterilization, and spray drying, were identified by using the validated analytical method that is described in Table 2. The contents of AVN A, B, and C from the GOEs were 468.52 ± 17.69, 390.33 ± 10.26, and 641.22 ± 13.89 μg/g dw after the spray drying, which are shown in Figure 1C ($p < 0.05$). In addition, the targeted phenolic acids, which included the ferulic acid and *p*-coumaric acid, were identified and quantified from the oat extracts. The *p*-coumaric acid was not detected from the GOEs, but it had 490.03 ± 7.83 µg/g dw of ferulic acid, which is displayed in Figure 1D. The yield (%) of the oat extracts, which was determined by the ratio of the weight of dried powder to the initial weight of germinated oats, was 6.81% at the GOEs.

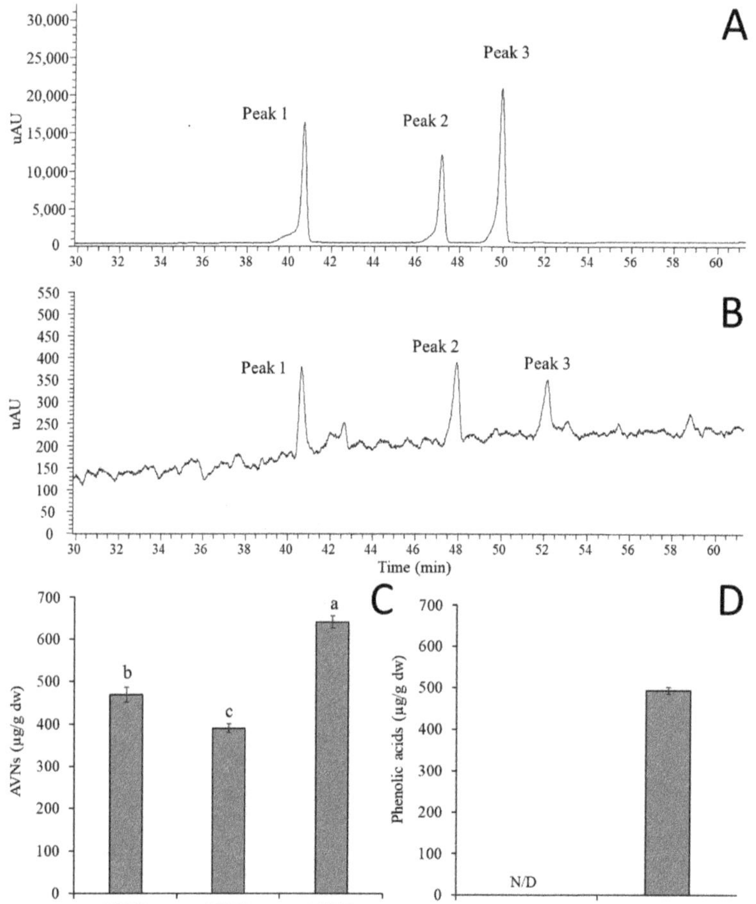

Figure 1. UV chromatograms of AVNs in the standard solution (**A**) and oat extracts (**B**), along with the quantification of AVNs (**C**) and phenolic acids (**D**) from the oat extracts, where peaks 1, 2, and 3 correspond to AVN C, A, and B, respectively. Data are presented as mean ± standard deviation (SD) of [n = 3] independent experiments. Statistical significance among treatment groups was determined using one-way ANOVA (or two-way ANOVA where applicable), followed by Tukey's honestly significant difference (HSD) post hoc test. Different letters (a, b, c) above the bars indicate statistically significant differences among groups ($p < 0.05$). N/D means not detected, which is below the limit of detection.

AVN A, B, and C are widely known as the major components in oats [31]. A previous study found that the AVNs accumulation in oat groat was mostly influenced by the growing location (68 to 77%), which was compared with six Canadian oat cultivars in different locations, and it was the highest when it was grown in crown rust-infected locations [36]. The study also suggested that high temperatures and precipitation tend to exhibit the improved growth of fungal pathogens, which are known to elicit AVNs biosynthesis. The total amount of AVNs in the germinated oat extracts was found to be significantly increased. It was also found that the contents of the AVNs remarkably increased over a 48 h germination process, which is similar to our findings, and it increased AVN C in non-germinated oats from 84.3 ± 3.1 µg/g to 216.0 ± 1.9 µg/g dw by germinating after extracting it with 80% ethanol [37]. A previous study reported that ferulic acid was significantly reduced by the germination process, which is in contrast to the AVNs [38].

AVN C restored damaged memory by relieving memory impairment and reinforcing synapses [39]. The findings from the current study, where germination increased the AVNs and certain types of phenolics in oats, imply that germinated oat extracts (GOEs) could be potent in regards to improving brain function. Thus, we conducted a further study with standardized germinated oat extract (GOE)-enriched AVNs in order to investigate the effects of oat extracts in regards to improving memory function.

3.3. Effects of AVNs and GOEs on $A\beta_{1\text{-}42}$-Induced Cytotoxicity of SH-SY5Y Cells

$A\beta_{1\text{-}42}$ ranged from 100 nM to 1 µM in a preliminary study, and it induced a significant decrease in cell viability, which showed 95.15 ± 3.87 to 64.96 ± 1.68% of the control in a dose-dependent manner. An $A\beta_{1\text{-}42}$ concentration of 20 µM showed the most toxicity, which was based on this data, and it was chosen for further study. Figure 2 shows the inhibitory effects of various concentrations of AVNs, which were from 1 to 25 µg/mL, and GOEs, which were from 1 to 50 µg/mL, against the $A\beta_{1\text{-}42}$-induced cytotoxicity of SH-SY5Y cells. The cell viability was significantly declined to 80.18 ± 3.30% at 1 µg/mL AVN A, 80.16 ± 2.18% at 12.5 µg/mL AVN A, and 92.15 ± 6.72% at 25 µg/mL AVN A ($p < 0.05$), which was 86.07 ± 5.23%, 86.64 ± 5.56%, and 89.66 ± 11.52% at 1, 12.5 and 25 µg/mL AVN B, respectively ($p < 0.05$). The cell viability was 98.90 ± 0.32%, 98.95 ± 9.40%, and 100.04 ± 11.83% at 1, 12.5 and 25 µg/mL, respectively, in regards to AVN C ($p < 0.05$). Furthermore, the cell viability increased 95.45 ± 2.46% at 1 µg/mL AVNs, 101.45 ± 10.55% at 12.5 µg/mL AVNs, and 106.66 ± 3.22% at 25 µg/mL total AVNs ($p < 0.05$), as displayed in Figure 2A. The overall co-treatment of the GOEs increased the cell viability against the cytotoxicity with values of 94.39 ± 3.67%, 97.81 ± 2.75%, 118.30 ± 7.66%, and 88.07 ± 2.01% at 1, 10, 25, and 50 µg/mL GOEs, respectively ($p < 0.05$), as shown in Figure 2B. The inhibitory effects of the AVNs against the $A\beta_{1\text{-}42}$-induced cytotoxicity were shown in a dose-dependent manner at AVN A, and it was significantly effective ($p < 0.05$). The co-treatment of AVN C, the total AVNs, and the GOEs, which included 1, 10, and 25 µg/mL, with $A\beta_{1\text{-}42}$ was significantly more protective than AVN A and AVB B ($p < 0.05$).

A previous study observed that the AVNs provided a protective effect on the PC12 cell that induced H_2O_2 damage. In particular, 10 to 40 µM AVN C displayed a strong protective effect [24]. A similar finding showed the cytotoxicity of $A\beta_{1\text{-}42}$ on the SH-SY5Y cells highly decreased at AVN C even at a low dose. The results from the current study imply that the low doses of AVNs and GOEs could mitigate the apoptosis of SH-SY5Y cells induced by $A\beta_{1\text{-}42}$.

3.4. Effects of SH-SY5Y Cell by AVNs and GOEs from $A\beta_{1\text{-}42}$-Induced Oxidative Stress

The intracellular ROS generation was quantified using DCFH-DA in order to investigate whether the AVNs and GOEs are accountable for the neuroprotection against the $A\beta_{1\text{-}42}$-induced generation of reactive oxygen species (ROS). The cellular ROS levels were significantly elevated to approximately 167.50 ± 12.39% compared to the control by the addition of 20 µM $A\beta_{1\text{-}42}$ into the SH-SY5Y cells, which is displayed in Figure 3A. The generated ROS by $A\beta_{1\text{-}42}$ was overall effectively eliminated, which is shown in Figure 3A, when the AVNs that ranged from 1 to 25 µg/mL were co-treated with $A\beta_{1\text{-}42}$. The intracellular ROS level was diminished to 114.46 ±

13.44% at 25 μg/mL AVN A, 82.93 ± 12.30% at 25 μg/mL AVN B, 37.91 ± 7.59% at 25 μg/mL AVN C, and 23.21 ± 3.64% at 25 μg/mL total AVNs, which is illustrated in Figure 3A. It also showed a dose-dependent manner. The ROS production tends to be decreased ($p < 0.05$) in regard to the co-treatment of the GOEs (1 to 50 μg/mL) with $A\beta_{1\text{-}42}$ for 24 h in the SH-SY5Y cells. It was particularly reduced to 155.85 ± 18.68% at 1 μg/mL, 131.39 ± 16.42% at 10 μg/mL, 108.36 ± 15.81% at 25 μg/mL, and 91.78 ± 15.89% at 50 μg/mL GOEs, which are depicted in Figure 3B. A remarkable elimination in the ROS induced by $A\beta_{1\text{-}42}$ occurred with a co-treatment of 25 μg/mL AVNs and at 50 μg/mL GOEs. However, 1 and 10 μg/mL GOEs did not show any significant differences from only the $A\beta_{1\text{-}42}$ treatment ($p < 0.05$).

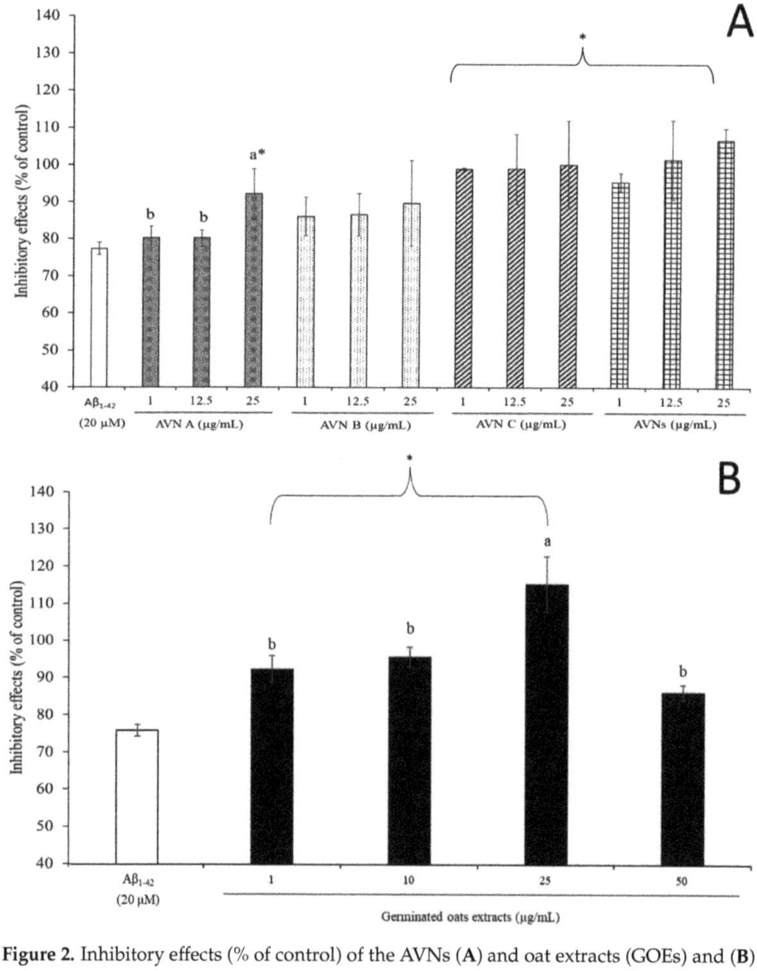

Figure 2. Inhibitory effects (% of control) of the AVNs (**A**) and oat extracts (GOEs) and (**B**) against $A\beta_{1\text{-}42}$-induced cytotoxicity in the SH-SY5Y cell. The different letters (a, b) indicate significant differences among concentrations ($p < 0.05$). An asterisk (*) indicates significant differences between the $A\beta_{1\text{-}42}$ and each treatment group ($p < 0.05$). Significant difference (HSD) post hoc test.

A previous study revealed that the generation of ROS or free radical-induced oxidative stress can cause damage to cells and tissues, which is parallel with a loss in the functional efficiency of various cellular processes that results in aging and age-related memory loss [40]. According to the hypothesis about Aβ fibril, $A\beta_{1\text{-}42}$ fibrils, which were entrapped by metal ions, such as copper ions, during the aggregation procedure, are considered toxic agents due to their contribution to the ROS generation that involves the oxidation of lipids, proteins, and nucleic acids [41]. ROS, which was in the mitochondria, leads to attacking the membrane phospholipids, and it contributed to the loss of the mitochondrial membrane by producing the intermembrane protein, such as cytochrome c triggers with activation of caspase-3 [42,43]. It was recently found that AVN C especially influenced the decrease in cleaved-caspase-3 activity in mouse hippocampus, which was reacted with the activated-caspase-3 that causes apoptotic cell death [37]. Similar results were found in a previous study, which evaluated the protective effects of AVN A and oat seeding extracts on the ROS production to lipopolysaccharide (LPS)-stimulated BV2 microglial cells, in which the levels of the ROS were significantly diminished [25]. The current study showed that the $A\beta_{1\text{-}42}$-induced ROS generation was partly diminished by the AVNs and GOEs.

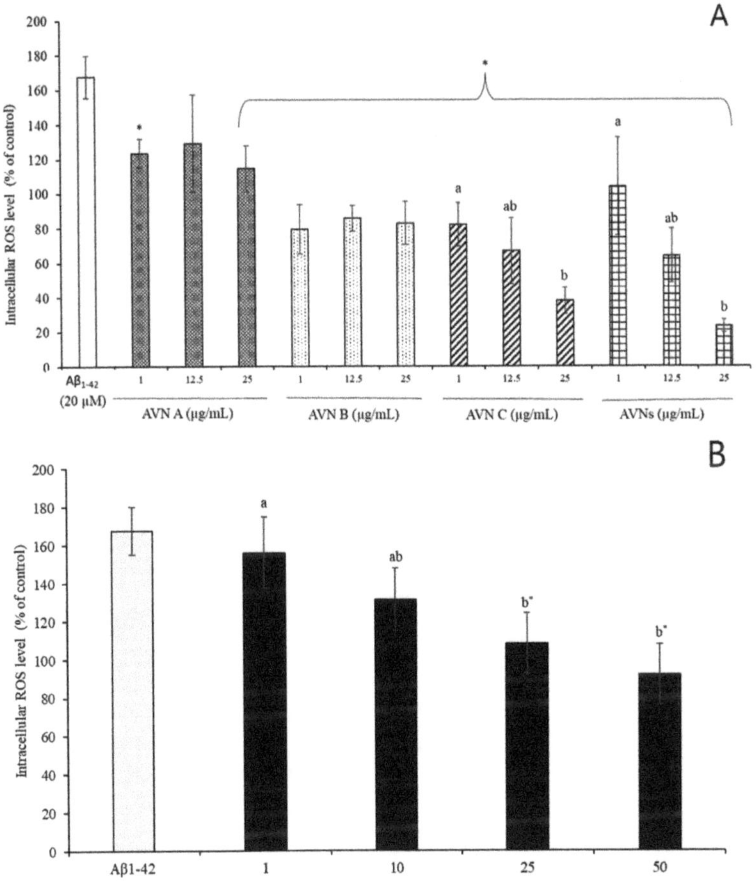

Figure 3. Inhibitory effect of AVNs (**A**) and oat extracts (GOEs) (**B**) on intracellular ROS production induced by $A\beta_{1\text{-}42}$. The different letters (a, b) indicate significant differences among concentrations ($p < 0.05$). An asterisk (*) indicates significant differences between the $A\beta_{1\text{-}42}$ and each treatment group ($p < 0.05$). Data are presented as mean ± standard deviation (SD) of [n = 3] independent experiments. Statistical significance among treatment groups was determined using one-way ANOVA (or two-way ANOVA where applicable), followed by Tukey's honestly significant difference (HSD) post hoc test.

3.5. *Effects of AVNs and GOEs Against $A\beta_{1\text{-}42}$-Induced Inflammation and Apoptosis of SH-SY5Y Cell*

The current study evaluated the mRNA expression levels of NF-kB p50 and p65 in order to investigate biomarkers in regard to preventing the effect of AVNs and GOEs against $A\beta_{1\text{-}42}$-induced neuroinflammation. The mRNA expression levels of NF-kB p50 were significantly increased to 2.38 ± 0.49 at $A\beta_{1\text{-}42}$, as shown in Figure 4A. On the other hand, the AVN and GOEs treatment groups showed p50 mRNA levels of 1.17 ± 0.18 at OAT1, 1.29 ± 0.45 at OAT50, 0.97 ± 0.05 at AVN A, 1.08 ± 0.1 at AVN B, 0.89 ± 0.16 at AVN C, and 1.06 ± 0.4 at AVNs, which significantly recovered the expression levels ($p < 0.05$). The $A\beta_{1\text{-}42}$ induced a higher p65 expression level of 1.49 ± 0.15, but OAT 1 and AVN B treatment groups tended to slightly decrease p65 mRNA levels of 1.21 ± 0.07 and 1.25 ± 0.09, respectively, which are illustrated in Figure 4B. The AVN and GOEs treatment groups showed p65 mRNA levels of 1.14 ± 0.09 at OAT50, 1.15 ± 0.04 at AVN A, 1.12 ± 0.07 at AVN C, and 1.05 ± 0.2 at AVNs, which revealed that the expression levels of these groups were significantly recovered ($p < 0.05$). The current study also investigated whether the expression of Bcl-2 and Bax was affected by AVN and GOEs. $A\beta_{1\text{-}42}$ significantly increased the expression levels of Bax by 2.87 ± 0.73, as shown in Figure 4C. On the other hand, a significant decrease in the Bax levels of the AVN and GOEs treatment groups were observed, which showed 1.46 ± 0.62 at OAT1, 1.46 ± 0.72 at OAT50, 0.96 ± 0.15 at AVN A, 0.88 ± 0.16 at AVN B, 0.86 ± 0.11 at AVN C, and 1.18 ± 0.26 at AVNs compared to $A\beta_{1\text{-}42}$, which are shown in Figure 4C. The Bcl-2 levels tended to decrease in all groups compared to CTL, 0.9 ± 0.06 at $A\beta_{1\text{-}42}$, 0.78 ± 0.13 at OAT1, 0.65 ± 0.17 at OAT50, 0.67 ± 0.08 at AVN A, 0.67 ± 0.04 at AVN B, 0.59 ± 0.02 at AVN C, and 0.64 ± 0.13 at AVNs, which are displayed in Figure 4D. The ratio between Bax and Bcl-2 proteins determines the mitochondrial apoptosis pathway. The Bax/Bcl-2 ratio also significantly increased in $A\beta_{1\text{-}42}$ compared to CTL, as shown in Figure 4E. The GOEs and AVNs tended to decrease compared to $A\beta_{1\text{-}42}$ at 1.83 ± 0.54 at OAT1, 2.15 ± 0.52 at OAT50, and 1.94 ± 0.86 at AVNs. AVN A, AVN B, and AVN C significantly decreased compared to $A\beta_{1\text{-}42}$ at 1.46 ± 0.34 at AVN A, 1.31 ± 0.25 at AVN B, and 1.46 ± 0.21 at AVN C.

Inflammation is important for the host to recover from damage when the inflammation response is broken, but it can cause tissue damage, which results in the development of inflammatory diseases [44]. Aβ accumulation causes neurotoxicity, which triggers the release of inflammatory mediators [45]. Activation of the NF-κB pathway stimulates the release of inflammatory cytokines, such as TNF-α, IL-6, and IL-1β by inflammatory mediators [46]. NF-κB is a crucial nuclear transcription factor in all cells in order to regulate the gene expression that encodes various immune and inflammatory response mediators [47]. Apoptosis is a natural process in the development, but abnormal apoptosis contributes to the progression of neurodegenerative conditions [48,49]. Defects of the mitochondria, which are involved in Aβ-induced neuronal apoptosis, increased the production of ROS that causes oxidative stress, which is one of the major causes of AD pathology [50,51]. Bax coordinates with Bcl-2 in order to activate the mitochondrial apoptotic pathway [52], so the current study measured its biomarkers. These results indicate that AVNs and GOEs prevent against Aβ-induced inflammation in SH-SY5Y. The previous studies reported that Bcl-2 family proteins inhibited ROS-induced apoptosis, whereas the overexpression of pro-apoptotic Bax increased the ROS production [53,54]. The results demonstrate that AVN and GOEs exhibit the neuroprotective effect by attenuating Aβ-induced mitochondrial-dependent apoptosis in SH-SY5Y.

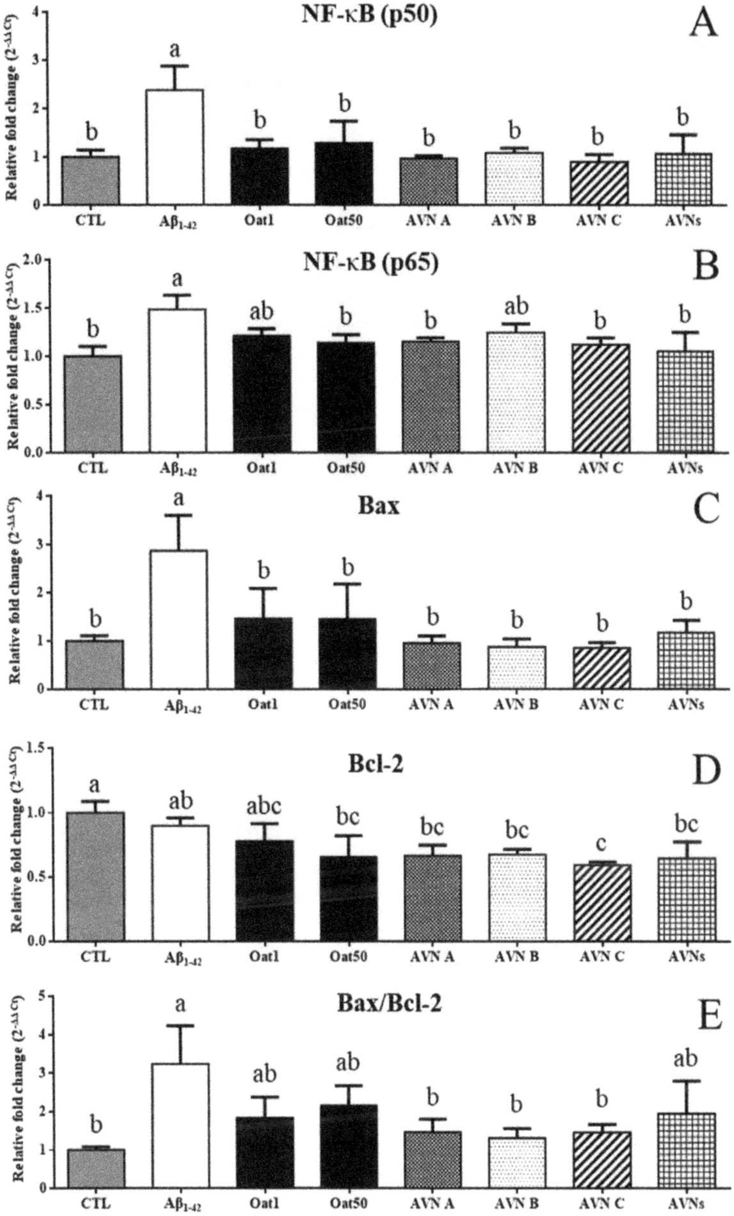

Figure 4. Expression levels of NF-κB, Bax, and Bcl-2 mRNA in SH-SY5Y cells with the sample treatment for 24 h, and $A\beta_{1\text{-}42}$ (1 μM) was inoculated for a further 24 h. (**A–E**) The relative expression levels of NF-κB (p50), NF-kB (p65), Bax, Bcl-2, and Bax/Bcl-2 ratio. The data was presented as mean ± SD. The different letters (a, b, c) indicate significant differences among each treatment group ($p < 0.05$). Data are presented as mean ± standard deviation (SD) of [n = 3] independent experiments. Statistical significance among treatment groups was determined using one-way ANOVA (or two-way ANOVA where applicable), followed by Tukey's honestly significant difference (HSD) post hoc test.

4. Conclusions

In conclusion, the findings in the current study suggest that germinated oat extracts (GOEs) enriched in AVNs could inhibit A$\beta_{1\text{-}42}$-induced cytotoxicity and oxidative stress in SH-SY5Y cells. The protective effects of GOEs against cytotoxicity were observed. The ROS production that was induced from A$\beta_{1\text{-}42}$ was highly observed from 25 μg/mL AVNs and at 50 μg/mL GOEs, which implies that the GOEs that contain AVNs and phenolic acids, such as p-coumaric acid and ferulic acid, could be effective in regard to protecting the ROS generation in SH-SY5Y against A$\beta_{1\text{-}42}$. Our results revealed that AVNs and GOEs mitigated the A$\beta_{1\text{-}42}$-induced cell damage by rescuing inflammation and apoptotic pathways, which thereby increased the cell viability. However, additional biomarkers that were related to the apoptosis, including cytochrome C, caspase 9, and caspase 3, and inflammation-related cytokines, such as TNF- α IL-6, and IL-1β, need to be further investigated via an in vivo model in order to use GOEs for health functional ingredients for improving memory function.

Author Contributions: Conceptualization, Y.-Y.L. and S.-M.S.; methodology, Y.-Y.L., I.-S.N., J.-E.K. and S.-M.S.; software, Y.-Y.L., I.-S.N., J.-E.K. and S.-M.S.; validation, Y.-Y.L., I.-S.N., J.-E.K. and S.-M.S.; formal analysis, Y.-Y.L., I.-S.N., J.-E.K. and S.-M.S.; investigation, J.-G.S. and C.-E.H.; resources, Y.-Y.L., H.-W.K. and S.-M.S.; data curation, Y.-Y.L. and I.-S.N.; writing—original draft preparation, I.-S.N., J.-E.K.; writing—review and editing, S.-M.S.; visualization, H.-W.K.; supervision, H.-W.K. and S.-M.S.; project administration, Y.-Y.L., J.-G.S. and C.-E.H. All authors have read and agreed to the published version of the manuscript.

Funding: This study was supported by an intramural research grant from Rural Development Administration (PJ016728, RS-2022RD010027).

Institutional Review Board Statement: Not applicable.

Informed Consent Statement: Not applicable.

Data Availability Statement: Data is contained within the article.

Conflicts of Interest: The authors declare no conflicts of interest.

References

1. Uwishema, O.; Mahmoud, A.; Sun, J.; Correia, I.F.S.; Bejjani, N.; Alwan, M.; Nicholas, A.; Oluyemisi, A.; Dost, B. Is Alzheimer's disease an infectious neurological disease? A review of the literature. *Brain Behav.* **2022**, *12*, e2728. [CrossRef] [PubMed]
2. Alzheimer's Association. 2019 Alzheimer's disease facts and figures. *Alzheimer's Dement.* **2019**, *15*, 321–387. [CrossRef]
3. Lin, L.; Zheng, L.J.; Zhang, L.J. Neuroinflammation, Gut Microbiome, and Alzheimer's Disease. *Mol. Neurobiol.* **2018**, *55*, 8243–8250. [CrossRef]
4. Peng, Y.; Tao, H.X.; Wang, S.P.; Xiao, J.B.; Wang, Y.T.; Su, H.X. Dietary intervention with edible medicinal plants and derived products for prevention of Alzheimer's disease: A compendium of time-tested strategy. *J. Funct. Foods* **2021**, *81*, 104463. [CrossRef]
5. Garbuz, D.; Zatsepina, O.; Evgen'ev, M. Beta amyloid, tau protein, and neuroinflammation: An attempt to integrate different hypotheses of Alzheimer's disease pathogenesis. *Mol. Biol.* **2021**, *55*, 670–682. [CrossRef]
6. Cosentino, S.A.; Stern, Y.; Sokolov, E.; Scarmeas, N.; Manly, J.J.; Tang, M.X.; Schupf, N.; Mayeux, R.P. Plasma ss-amyloid and cognitive decline. *Arch. Neurol.* **2010**, *67*, 1485–1490. [CrossRef]
7. Nedaei, H.; Rezaei-Ghaleh, N.; Giller, K.; Becker, S.; Karami, L.; Moosavi-Movahedi, A.A.; Griesinger, C.; Saboury, A.A. The calcium-free form of atorvastatin inhibits amyloid-beta(1–42) aggregation in vitro. *J. Biol. Chem.* **2022**, *298*, 101662. [CrossRef]
8. Lee, Y.J.; Han, S.B.; Nam, S.Y.; Oh, K.W.; Hong, J.T. Inflammation and Alzheimer's disease. *Arch. Pharm. Res.* **2010**, *33*, 1539–1556. [CrossRef]
9. Chen, S.-Y.; Gao, Y.; Sun, J.-Y.; Meng, X.-L.; Yang, D.; Fan, L.-H.; Xiang, L.; Wang, P. Traditional Chinese medicine: Role in reducing β-amyloid, apoptosis, autophagy, neuroinflammation, oxidative stress, and mitochondrial dysfunction of Alzheimer's disease. *Front. Pharmacol.* **2020**, *11*, 497. [CrossRef]

10. Juszczyk, G.; Mikulska, J.; Kasperek, K.; Pietrzak, D.; Mrozek, W.; Herbet, M. Chronic stress and oxidative stress as common factors of the pathogenesis of depression and Alzheimer's disease: The role of antioxidants in prevention and treatment. *Antioxidants* **2021**, *10*, 1439. [CrossRef]
11. Suematsu, N.; Hosoda, M.; Fujimori, K. Protective effects of quercetin against hydrogen peroxide-induced apoptosis in human neuronal SH-SY5Y cells. *Neurosci. Lett.* **2011**, *504*, 223–227. [CrossRef]
12. Kim, J.H.; Lee, J.; Lee, S.; Cho, E.J. Quercetin and quercetin-3-β-d-glucoside improve cognitive and memory function in Alzheimer's disease mouse. *Appl. Biol. Chem.* **2016**, *59*, 721–728. [CrossRef]
13. Chen, C.Y.; Milbury, P.E.; Collins, F.W.; Blumberg, J.B. Avenanthramides are bioavailable and have antioxidant activity in humans after acute consumption of an enriched mixture from oats. *J. Nutr.* **2007**, *137*, 1375–1382. [CrossRef] [PubMed]
14. Biel, W.; Bobko, K.; Maciorowski, R. Chemical composition and nutritive value of husked and naked oats grain. *J. Cereal. Sci.* **2009**, *49*, 413–418. [CrossRef]
15. Hitayezu, R.; Baakdah, M.M.; Kinnin, J.; Henderson, K.; Tsopmo, A. Antioxidant activity, avenanthramide and phenolic acid contents of oat milling fractions. *J. Cereal. Sci.* **2015**, *63*, 35–40. [CrossRef]
16. Boz, H. Phenolic Amides (Avenanthramides) in Oats—A review. *Czech. J. Food Sci.* **2015**, *33*, 399–404. [CrossRef]
17. Matsukawa, T.; Isobe, T.; Ishihara, A.; Iwamura, H. Occurrence of avenanthramides and hydroxycinnamoyl-CoA:hydroxyanthranilate N-hydroxycinnamoyltransferase activity in oat seeds. *Z. Naturforsch. C J. Biosci.* **2000**, *55*, 30–36. [CrossRef]
18. Meydani, M. Potential health benefits of avenanthramides of oats. *Nutr. Rev.* **2009**, *67*, 731–735. [CrossRef]
19. Pridal, A.A.; Bottger, W.; Ross, A.B. Analysis of avenanthramides in oat products and estimation of avenanthramide intake in humans. *Food Chem.* **2018**, *253*, 93–100. [CrossRef]
20. Burton-Freeman, B.M.; Sandhu, A.K.; Edirisinghe, I. Mangos and their bioactive components: Adding variety to the fruit plate for health. *Food Funct.* **2017**, *8*, 3010–3032. [CrossRef]
21. Xu, F.; Mu, J.; Teng, Y.; Zhang, X.; Sundaram, K.; Sriwastva, M.K.; Kumar, A.; Lei, C.; Zhang, L.; Liu, Q.M. Restoring Oat Nanoparticles Mediated Brain Memory Function of Mice Fed Alcohol by Sorting Inflammatory Dectin-1 Complex Into Microglial Exosomes. *Small* **2022**, *18*, 2105385. [CrossRef]
22. Kennedy, D.O. Polyphenols and the human brain: Plant "secondary metabolite" ecologic roles and endogenous signaling functions drive benefits. *Adv. Nutr.* **2014**, *5*, 515–533. [CrossRef] [PubMed]
23. Iskusnykh, I.Y.; Zakharova, A.A.; Pathak, D. Glutathione in Brain Disorders and Aging. *Molecules* **2022**, *27*, 324. [CrossRef] [PubMed]
24. Hou, Y.; Peng, S.; Song, Z.; Bai, F.; Li, X.; Fang, J. Oat polyphenol avenanthramide-2c confers protection from oxidative stress by regulating the Nrf2-ARE signaling pathway in PC12 cells. *Arch. Biochem. Biophys.* **2021**, *706*, 108857. [CrossRef] [PubMed]
25. Lee, W.S.; Lee, H.J.; Yang, J.Y.; Shin, H.L.; Choi, S.W.; Kim, J.K.; Seo, W.D.; Kim, E.H. The Potential Neuroprotective Effects of Extracts from Oat Seedlings against Alzheimer's Disease. *Nutrients* **2022**, *14*, 4103. [CrossRef]
26. Skoglund, M.; Peterson, D.M.; Andersson, R.; Nilsson, J.; Dimberg, L.H. Avenanthramide content and related enzyme activities in oats as affected by steeping and germination. *J. Cereal. Sci.* **2008**, *48*, 294–303. [CrossRef]
27. Kaukovirta-Norja, A.; Wilhelmson, A.; Poutanen, K. Germination: A means to improve the functionality of oat. *Agric. Food Sci.* **2004**, *13*, 100–112. [CrossRef]
28. Tian, B.; Xie, B.; Shi, J.; Wu, J.; Cai, Y.; Xu, T.; Xue, S.; Deng, Q. Physicochemical changes of oat seeds during germination. *Food Chem.* **2010**, *119*, 1195–1200. [CrossRef]
29. Chen, C.-Y.; Milbury, P.E.; Kwak, H.-K.; Collins, F.W.; Samuel, P.; Blumberg, J.B. Avenanthramides and phenolic acids from oats are bioavailable and act synergistically with vitamin C to enhance hamster and human LDL resistance to oxidation. *J. Nutr.* **2004**, *134*, 1459–1466. [CrossRef]
30. Lee, A.Y.; Choi, J.M.; Lee, M.H.; Lee, J.; Lee, S.; Cho, E.J. Protective effects of perilla oil and alpha linolenic acid on SH-SY5Y neuronal cell death induced by hydrogen peroxide. *Nutr. Res. Pract.* **2018**, *12*, 93–100. [CrossRef]
31. Woolman, M.; Liu, K. Simplified Analysis and Expanded Profiles of Avenanthramides in Oat Grains. *Foods* **2022**, *11*, 560. [CrossRef]
32. Sereia, A.L.; de Oliveira, M.T.; Baranoski, A.; Marques, L.L.M.; Ribeiro, F.M.; Isolani, R.G.; de Medeiros, D.C.; Chierrito, D.; Lazarin-Bidoia, D.; Zielinski, A.A.F.; et al. In vitro evaluation of the protective effects of plant extracts against amyloid-beta peptide-induced toxicity in human neuroblastoma SH-SY5Y cells. *PLoS ONE* **2019**, *14*, e0212089. [CrossRef]
33. Wang, Y.; Zhang, Y.; Zhang, X.; Yang, T.; Liu, C.; Wang, P. Alcohol Dehydrogenase 1B Suppresses beta-Amyloid-Induced Neuron Apoptosis. *Front. Aging Neurosci.* **2019**, *11*, 135. [CrossRef]

34. Aranda, A.; Sequedo, L.; Tolosa, L.; Quintas, G.; Burello, E.; Castell, J.; Gombau, L. Dichloro-dihydro-fluorescein diacetate (DCFH-DA) assay: A quantitative method for oxidative stress assessment of nanoparticle-treated cells. *Toxicol. Vitr.* **2013**, *27*, 954–963. [CrossRef]
35. Miyagawa, H.; Ishihara, A.; Nishimoto, T.; Ueno, T.; Mayama, S. Induction of avenanthramides in oat leaves inoculated with crown rust fungus, *Puccinia coronata* f. sp. *avenae*. *Biosci. Biotechnol. Biochem.* **1995**, *59*, 2305–2306. [CrossRef]
36. Nkhata Malunga, L.; Ames, N.; Mitchell Fetch, J.; Netticadan, T.; Joseph Thandapilly, S. Genotypic and environmental variations in phenolic acid and avenanthramide content of Canadian oat (*Avena sativa*). *Food Chem.* **2022**, *388*, 132904. [CrossRef] [PubMed]
37. Lee, Y.Y.; Wang, M.; Son, Y.; Yang, E.J.; Kang, M.S.; Kim, H.J.; Kim, H.S.; Jo, J. Oat Extract Avenanthramide-C Reverses Hippocampal Long-Term Potentiation Decline in Tg2576 Mice. *Molecules* **2021**, *26*, 6105. [CrossRef] [PubMed]
38. Donkor, O.N.; Stojanovska, L.; Ginn, P.; Ashton, J.; Vasiljevic, T. Germinated grains--sources of bioactive compounds. *Food Chem.* **2012**, *135*, 950–959. [CrossRef]
39. Ramasamy, V.S.; Samidurai, M.; Park, H.J.; Wang, M.; Park, R.Y.; Yu, S.Y.; Kang, H.K.; Hong, S.; Choi, W.S.; Lee, Y.Y.; et al. Avenanthramide-C Restores Impaired Plasticity and Cognition in Alzheimer's Disease Model Mice. *Mol. Neurobiol.* **2020**, *57*, 315–330. [CrossRef]
40. Muller, F.L.; Lustgarten, M.S.; Jang, Y.; Richardson, A.; Van Remmen, H. Trends in oxidative aging theories. *Free Radic. Biol. Med.* **2007**, *43*, 477–503. [CrossRef]
41. Cheignon, C.; Tomas, M.; Bonnefont-Rousselot, D.; Faller, P.; Hureau, C.; Collin, F. Oxidative stress and the amyloid beta peptide in Alzheimer's disease. *Redox Biol.* **2018**, *14*, 450–464. [CrossRef] [PubMed]
42. Pelicano, H.; Feng, L.; Zhou, Y.; Carew, J.S.; Hileman, E.O.; Plunkett, W.; Keating, M.J.; Huang, P. Inhibition of mitochondrial respiration: A novel strategy to enhance drug-induced apoptosis in human leukemia cells by a reactive oxygen species-mediated mechanism. *J. Biol. Chem.* **2003**, *278*, 37832–37839. [CrossRef] [PubMed]
43. Mei, Z.; Yan, P.; Situ, B.; Mou, Y.; Liu, P. Cryptotanshinione inhibits beta-amyloid aggregation and protects damage from beta-amyloid in SH-SY5Y cells. *Neurochem. Res.* **2012**, *37*, 622–628. [CrossRef]
44. Liu, T.; Zhang, L.; Joo, D.; Sun, S.-C. NF-κB signaling in inflammation. *Signal Transduct. Target. Ther.* **2017**, *2*, 17023. [CrossRef]
45. Agostinho, P.; Cunha, R.A.; Oliveira, C. Neuroinflammation, oxidative stress and the pathogenesis of Alzheimer's disease. *Curr. Pharm. Des.* **2010**, *16*, 2766–2778. [CrossRef]
46. Baez-Jurado, E.; Guio-Vega, G.; Hidalgo-Lanussa, O.; González, J.; Echeverria, V.; Ashraf, G.M.; Sahebkar, A.; Barreto, G.E. Mitochondrial Neuroglobin Is Necessary for Protection Induced by Conditioned Medium from Human Adipose-Derived Mesenchymal Stem Cells in Astrocytic Cells Subjected to Scratch and Metabolic Injury. *Mol. Neurobiol.* **2019**, *56*, 5167–5187. [CrossRef]
47. Barnes, P.J.; Karin, M. Nuclear factor-kappaB: A pivotal transcription factor in chronic inflammatory diseases. *N. Engl. J. Med.* **1997**, *336*, 1066–1071. [CrossRef]
48. Xu, T.; Niu, C.; Zhang, X.; Dong, M. β-Ecdysterone protects SH-SY5Y cells against β-amyloid-induced apoptosis via c-Jun N-terminal kinase- and Akt-associated complementary pathways. *Lab. Investig.* **2018**, *98*, 489–499. [CrossRef]
49. Radi, E.; Formichi, P.; Battisti, C.; Federico, A. Apoptosis and oxidative stress in neurodegenerative diseases. *J. Alzheimers Dis.* **2014**, *42*, S125–S152. [CrossRef]
50. Mancuso, M.; Coppedè, F.; Murri, L.; Siciliano, G. Mitochondrial Cascade Hypothesis Alzheimer's disease: Myth or reality? *Antioxid. Redox Signal.* **2007**, *9*, 1631–1646. [CrossRef]
51. Bosetti, F.; Brizzi, F.; Barogi, S.; Mancuso, M.; Siciliano, G.; Tendi, E.A.; Murri, L.; Rapoport, S.I.; Solaini, G. Cytochrome c oxidase and mitochondrial F1F0-ATPase (ATP synthase) activities in platelets and brain from patients with Alzheimer's disease. *Neurobiol. Aging* **2002**, *23*, 371–376. [CrossRef]
52. Martinou, J.C.; Youle, R.J. Mitochondria in apoptosis: Bcl-2 family members and mitochondrial dynamics. *Dev. Cell* **2011**, *21*, 92–101. [CrossRef]
53. Wang, Q.; Zhang, L.; Yuan, X.; Ou, Y.; Zhu, X.; Cheng, Z.; Zhang, P.; Wu, X.; Meng, Y.; Zhang, L. The Relationship between the Bcl-2/Bax Proteins and the Mitochondria-Mediated Apoptosis Pathway in the Differentiation of Adipose-Derived Stromal Cells into Neurons. *PLoS ONE* **2016**, *11*, e0163327. [CrossRef]
54. Zhang, L.; Yuan, X.; Wang, S.; Ou, Y.; Zheng, X.; Wang, Q. The relationship between mitochondrial fusion/fission and apoptosis in the process of adipose-derived stromal cells differentiation into astrocytes. *Neurosci. Lett.* **2014**, *575*, 19–24. [CrossRef]

molecules

MDPI

Article

Characteristics of Novel Fermented Cloudy Fruit Juices Produced Using *Lactiplantibacillus plantarum* and Lactic Acid-Producing *Lachancea* spp. Yeasts

Paweł Satora *, Magdalena Skotniczny and Martyna Maziarek

Department of Fermentation Technology and Microbiology, Faculty of Food Technology, University of Agriculture, Balicka Street 122, 30-149 Kraków, Poland; magdalena.skotniczny@urk.edu.pl (M.S.)
* Correspondence: pawel.satora@urk.edu.pl

Abstract: Fermented fruit juices are considered functional beverages because they contain bioactive compounds derived from plant materials and produced by the microorganisms involved in fermentation. The composition of these beverages can vary depending on the strain used. This study aimed to determine the effect of different microorganisms conducting lactic acid fermentation on the chemical composition and bioactive component content of naturally cloudy fermented pear and plum juices. The process used *Lactiplantibacillus plantarum* K7 bacteria, which were isolated during sauerkraut fermentation, as well as *Lachancea thermotolerans* PYCC6375 and *Lachancea fermentati* PYCC5883 yeast cultures, which have poor ethanol fermentation capabilities. The pH, acidity, sugars (HPLC), free amino nitrogen, selected organic acids (HPLC), color (CIELAB), polyphenols (HPLC), volatiles (GC-MS), aroma-active volatiles (GC-MS-O), and sensory characteristics were analyzed. The fermented juices obtained were rich in organic acids (of plant and microbial origin), polyphenols, and had a reduced sugar content (with polyols replacing glucose and fructose), as well as a low alcohol content (<0.2%). At the same time, all three microorganisms significantly enhanced the fruity aroma of the juices. *Lachancea* yeasts proved to be a viable alternative to lactic acid bacteria for producing fermented juices and were significantly better suited to fermenting plum juices. The highest polyphenol content and highest consumer preference rating were obtained with plum juices fermented with *L. fermentati* yeast.

Keywords: functional beverages; pear; plum; polyphenols; volatiles

1. Introduction

Pears and plums are popular stone fruits. In 2022, world production amounted to 26,314,506.67 tonnes for pears and 12,564,237.8 tonnes for plums and sloes. Their harvests occupied 1,329,245 hectares and 2,629,126 hectares, respectively. China is the world's largest producer of both fruits, accounting for over half of global output in 2022 (73% and 55%, respectively). In Poland, however, their harvest and crop area were mainly of regional importance, amounting to 80,600 tonnes and 5500 hectares, and 133,200 tonnes and 16,500 hectares, respectively [1]. Nevertheless, after apples, they are among the most commonly consumed native fruits in Poland. In November 2024, 46% of surveyed Poles declared consuming pears and 31% declared consuming plums [2].

Pears and plums are highly valued fruits due to their high content of bioactive compounds that contribute to their health-promoting properties. However, the composition and concentration of these compounds can vary depending on the species and variety of fruit, its ripeness, and pre- and postharvest conditions [3,4]. Therefore, the region in which the fruit is produced may also play an important role. The dominant bioactive group in these fruits is phenolic compounds, particularly phenolic acids and flavonoids [5,6]. In traditional Eastern medicine, pears (*Pyrus* spp.) have been used to relieve respiratory symptoms, treat fever and inflammation, and alleviate alcohol hangovers [7]. These properties have been confirmed by scientific research, which has also demonstrated their anti-diabetic, anti-obesity, anti-hyperlipidemic, anti-mutagenic, anti-carcinogenic and cardio-protective activity [7]. In turn, plums (*Prunus domestica*) are used in traditional medicine in various regions of the world mainly to support proper digestive function. This has also been confirmed in various scientific studies, which have also demonstrated antioxidant, anti-carcinogenic, anti-hyperlipidemic and anxiolytic activity [8,9]. However, it is worth mentioning that neither fruit may be suitable for people with carbohydrate malabsorption.

The lactic acid fermentation of fruit juices enhances their bioactive properties by releasing beneficial metabolites and increasing the bioavailability of certain nutrients. This also leads to a reduction in sugar content, an extension to the shelf life, and changes to sensory characteristics [10,11]. The effects may vary depending on the strain used. Lactic acid bacteria, such as *Lactiplantibacillus plantarum*, are traditionally involved in this process [12–17]. Using unconventional microorganisms instead of these bacteria could lead to new opportunities for developing functional beverages, one of the fastest-growing segments of the food market.

In the research, the non-*Saccharomyces* yeasts belonging to the genus *Lachancea*, which are capable of producing lactic acid were used. So far, they have been used in winemaking for the bioacidification of wine [18–21] and in brewing for the production of sour and low-alcohol beers [22–27]. The *L. thermotolerans* PYCC6375 and *L. fermentati* PYCC5883 strains were selected for the study as they produce high levels of lactic acid and low levels of ethanol during wort fermentation [27]. The aim of the study was to investigate the impact of these strains on the chemical and bioactive properties of naturally cloudy pear and plum juices, compared to lactic acid bacteria. The *L. plantarum* K7 strain used in the experiment was isolated from the sauerkraut fermentation process [28], and previous unpublished experiments have shown that it has excellent properties for fermenting plant material. To date, no such studies have been published in the literature. The idea was to create fermented juices with increased functional potential and minimal sensory changes; therefore, the fermentation was brief, and salt was not added. Such products could gain recognition among increasingly health-conscious consumers.

2. Results and Discussion

2.1. Fermentation Kinetics and Basic Physicochemical Parameters

The kinetics of fruit juice fermentation varied depending on the strain and matrix used (Figure 1). In the case of pear juices, the most efficient fermentation was performed by *L. thermotolerans* PYCC6375 (LT), while in the case of plum juices, it was performed by *L. plantarum* K7 (LP). *L. fermentati* PYCC5883 (LF) showed poor fermentation abilities in both types of juice. The same *Lachancea* strains produced similar results during wort fermentation [27].

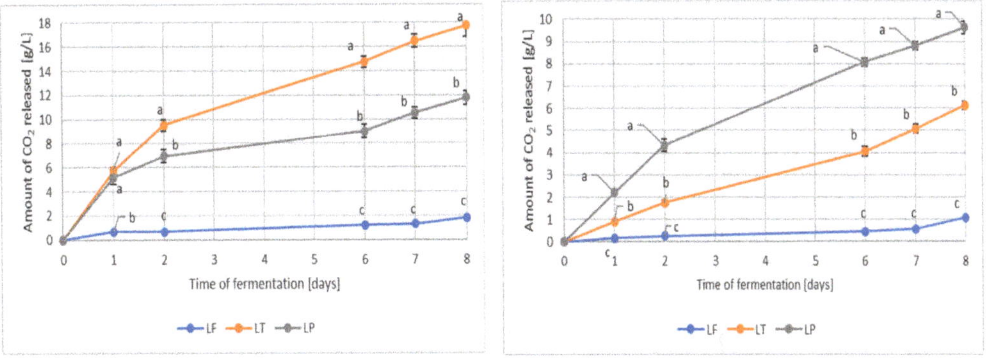

(**a**) (**b**)

Figure 1. The kinetics of fruit juice fermentation ((**a**) pear juice; (**b**) plum juice) depend on the microorganism used: LP—*Lactiplantibacillus plantarum* K7, LT—*Lachancea thermotolerans* PYCC6375, LF—*Lachancea fermentati* PYCC5883. Values with different Roman letters (a–c) indicate statistically significant differences at $p < 0.05$ according to the Tukey test.

The growth of microorganisms in juices was possible due to the sufficient amount of sugars and free amino nitrogen (FAN; Table 1). Fructose dominated in pear juice and glucose in plum juice, accounting for approximately 75% and 64% of total sugars, respectively. These results are consistent with previous studies, which state that in addition to glucose and fructose, pears and plums also contain sucrose and sorbitol [29,30]. Glucose was found to be the preferred carbon source for microorganisms in both fruit juices, as evidenced by its higher consumption than fructose. All strains were also able to completely break down sucrose. This may explain why plum juice fermented by the LP strain contained significantly more fructose than unfermented juice. Furthermore, the tested strains were able to produce polyols as a by-product of fermentation. Glycerol appears to be produced most efficiently from glucose. In the case of sorbitol and mannitol, which are isomers, the analytical method used does not allow them to be distinguished. Nevertheless, their concentration in fermented pear juice increased significantly. This suggests that mannitol was formed as a by-product of fructose fermentation [31]. The polyols in fermented juices can compensate for the loss of sweetness caused by the reduction in sugar content. In terms of free amino nitrogen (FAN) compounds, plum juice contained larger amounts than pear juice, reflecting their respective fruit content. The amount of FAN in grapes has been shown to depend on fruit yield and growing conditions, and its concentration may decrease during the storage or processing of the fruit [32]. During fermentation, the tested strains utilized nitrogen compounds, but these differences were not statistically significant. Fermented pear juices had lower FAN values than fermented plum juices, making them more microbiologically stable, since nitrogen is essential for microbial growth.

Table 1. The content of sugar, polyols, and free amino nitrogen (FAN) in unfermented and fermented pear and plum juices depending on the microorganism used.

Fruit Juice	Sample	Glucose	Fructose	Sucrose	Glycerol	Sorbitol + Mannitol	FAN
		[g/L]					
Pear	J	29.7 ^{d} ± 0.5	98.4 ^{e} ± 1.6	3.7 ^{b} ± 1.2	0.00 ^{a} ± 0.00	31.4 ^{b} ± 2.0	65.4 ± 1.8
	LP	4.5 ^{b} ± 0.4	45.6 ^{cd} ± 5.4	0.0 ^{a} ± 0.0	4.26 ^{b} ± 0.68	39.8 ^{c} ± 3.3	24.6 ± 8.9
	LT	1.2 ^{a} ± 0.3	37.5 ^{bcd} ± 5.0	0.0 ^{a} ± 0.0	4.82 ^{bc} ± 1.70	43.4 ^{c} ± 1.1	15.1 ± 2.2
	LF	4.0 ^{b} ± 0.2	48.3 ^{d} ± 4.7	0.0 ^{a} ± 0.0	5.99 ^{bc} ± 0.74	39.1 ^{c} ± 0.4	36.9 ± 10.4

Table 1. *Cont.*

Fruit Juice	Sample	Glucose	Fructose	Sucrose	Glycerol	Sorbitol + Mannitol	FAN
		[g/L]					
Plum	J	69.9 ^{e} ± 0.1	27.6 ^{ab} ± 0.8	11.7 ^{c} ± 0.8	0.00 ^{a} ± 0.00	30.7 ^{ab} ± 0.3	165 ± 2.9
	LP	9.9 ^{c} ± 0.8	36.5 ^{bc} ± 3.1	0.0 ^{a} ± 0.0	9.05 ^{d} ± 1.12	26.9 ^{ab} ± 0.4	129 ± 20.3
	LT	4.4 ^{b} ± 0.7	18.1 ^{a} ± 4.8	0.0 ^{a} ± 0.0	7.62 ^{cd} ± 1.56	25.1 ^{a} ± 1.6	134 ± 25.0
	LF	4.8 ^{b} ± 1.5	19.6 ^{a} ± 2.1	0.0 ^{a} ± 0.0	6.21 ^{bcd} ± 0.78	27.8 ^{ab} ± 3.9	120 ± 21.0
Sig.	Fruit juice	***	***	***	***	***	***
	Fermentation	***	***	***	***	ns	***
	Fruit juice × fermentation	***	***	***	**	***	ns

Legend: J—unfermented juice, LP—*Lactiplantibacillus plantarum* K7, LT—*Lachancea thermotolerans* PYCC6375, LF—*Lachancea fermentati* PYCC5883; Sig.—significance; ns—not significant. ** and *** indicate significance at levels of 0.01–0.005 and <0.005, respectively, using the least significant difference test. Values with different superscript Roman letters (a–e) in the same column indicate statistically significant interactions between fruit juice and fermentation (Tukey HSD, $p < 0.05$).

As is well known, during lactic acid fermentation, microorganisms convert sugars into organic acids, which are the primary by-products. These acids lower the pH and increase the total acidity of the medium. This process is typically performed by lactic acid bacteria. Due to the presence of the lactate dehydrogenase enzyme, *Lachancea* yeast strains also possess this property [22]. All of the tested microorganisms were able to produce organic acids which caused changes in the pH and total acidity of the juices (Table 1). The initial pH of pear juice was 4.5 and of plum juice was 3.3, with average total acidity of 6.17 g/L and 15.25 g/L, respectively. In both juices, all strains significantly reduced the pH to between 3.21 and 3.92 in pear juice and to between 3.07 and 3.11 in plum juice. Concurrently, the total acidity of the fermented juices increased to 6.89–12.9 g/L in pear juice and 18.17–24.12 g/L in plum juice. In both cases, the LP strain proved to be the most efficient. Lowering the pH of juices extends their shelf life and stabilizes their color by preventing browning and inhibiting the activity of polyphenol oxidase, as well as stabilizing plum anthocyanins [33,34].

During fermentation, microorganism produced succinic, lactic and acetic acids, while citric acid was metabolized, but only in pear juice. The level of malic acid did not change after the fermentation of any of the juices. Ethanol was found in all fermented juices, but its content was below 0.2%, which allows the beverages to be classified as non-alcoholic (Table 2). Surprisingly, the LP strain was the most efficient in this respect. During pear juice fermentation, its quantity was higher than that of lactic acid. Ethanol can be produced by *L. plantarum* spp. As a by-product of heterolactic acid fermentation or citrate metabolism, but only in small quantities [35]. Certain strains of *L. plantarum* have also been shown to possess a bifunctional alcohol/aldehyde dehydrogenase gene (AdhE), which converts acetaldehyde to ethanol during mannitol and sorbitol utilization [36,37]. The higher amount of ethanol produced in pear juice may be due to fructose conversion into mannitol via the mannitol dehydrogenase enzyme, which is present in heterofermentative bacteria [38]. Perhaps similar mechanisms occurred in the *Lachancea* strains. Ethanol fermentation in both juices can be confirmed by the increased content of glycerol and, in the case of pear juice, succinic acid. The greater production of ethanol than lactic acid in the fermented pear juice samples suggests that lactic acid fermentation was partially inhibited, probably due to the high fructose content.

Table 2. The pH, total acidity, organic acids, and ethanol content of unfermented and fermented pear and plum juices depending on the microorganism used.

Fruit Juice	Sample	pH	Total Acidity	Citric Acid	Malic Acid	Succinic Acid	Lactic Acid	Acetic Acid	Ethanol
			[g/L]						
Pear	J	4.50 ^{a} ± 0.02	6.17 ± 0.29	1.91 ^{a} ± 0.21	3.64 ± 0.19	0.23 ^{a} ± 0.16	0.00 ^{a} ± 0.00	0.00 ^{a} ± 0.00	0.00 ^{a} ± 0.00
	LP	3.21 ^{c} ± 0.02	12.9 ± 2.36	1.54 ^{b} ± 0.25	3.49 ± 0.37	1.18 ^{b} ± 0.75	0.29 ^{ab} ± 0.02	1.50 ^{c} ± 0.21	1.32 ^{e} ± 0.29
	LT	3.92 ^{b} ± 0.06	6.84 ± 2.00	0.80 ^{b} ± 0.87	2.79 ± 0.27	0.88 ^{b} ± 0.96	0.96 ^{b} ± 0.33	1.32 ^{bc} ± 0.06	1.14 ^{de} ± 0.20
	LF	3.81 ^{c} ± 0.06	7.08 ± 1.70	1.30 ^{b} ± 0.66	3.01 ± 0.47	1.82 ^{c} ± 0.30	0.45 ^{ab} ± 0.12	1.02 ^{b} ± 0.29	0.31 ^{ab} ± 0.06
Plum	J	3.30 ^{c} ± 0.03	15.25 ± 0.43	3.13 ^{c} ± 0.11	2.35 ± 0.12	0.02 ^{a} ± 0.01	0.00 ^{a} ± 0.00	0.00 ^{a} ± 0.00	0.00 ^{a} ± 0.00
	LP	3.07 ^{d} ± 0.01	24.12 ± 0.62	3.14 ^{c} ± 0.21	2.32 ±0.66	0.31 ^{a} ± 0.03	2.71 ^{c} ± 0.37	0.07 ^{a} ±0.04	0.86 ^{cd} ± 0.21
	LT	3.11 ^{d} ± 0.00	18.17 ± 0.29	3.27 ^{c} ± 0.09	2.30 ± 0.68	0.27 ^{a} ± 0.05	2.79 ^{c} ± 0.57	0.14 ^{a} ± 0.04	0.36 ^{ab} ± 0.09
	LF	3.10 ^{d} ± 0.02	20.52 ± 4.38	3.21 ^{c} ± 0.32	2.36 ± 0.20	0.29 ^{a} ± 0.02	2.25 ^{c} ± 0.39	0.13 ^{a} ± 0.07	0.45 ^{bc} ± 0.04
Sig.	Fruit juice	***	***	***	***	***	***	***	***
	Fermentation	***	***	**	ns	***	***	***	***
	Fruit juice × fermentation	***	ns	***	ns	***	***	***	***

Legend: J—unfermented juice, LP—*Lactiplantibacillus plantarum* K7, LT—*Lachancea thermotolerans* PYCC6375, LF—*Lachancea fermentati* PYCC5883; Sig.—significance; ns—not significant. ** and *** indicate significance at levels of 0.01–0.005 and <0.005, respectively, using the least significant difference test. Values with different superscript Roman letters (a–e) in the same column indicate statistically significant interactions between fruit juice and fermentation (Tukey HSD, $p < 0.05$).

As mentioned previously, the type and concentration of organic acids directly influence the pH, but they could also affect the color of the juices. Furthermore, there is a close relationship between color parameters and pigments [39]. The yellow color of pears is mainly due to their carotenoid content, while the purple color of plums is due to their anthocyanin content [40,41]. The results of the CIELab analysis are presented in Table 3. The fermented pear juices were slightly darker (parameter L*) than the control due to an increase in red pigment (parameter a*). This was most likely due to changes in the phenolic composition (phenolic oxidation) [42]. Conversely, the fermented plum juices were lighter than the control due to an increase in yellow pigment (parameter b*). This may be due to the breakdown of anthocyanins forming yellowish compounds [43]. In both juices, the LP strain caused the greatest changes in color (ΔE).

Table 3. CIELab color parameters (L*, a*, b*, C*, h*) and color differences (ΔE) in unfermented and fermented pear and plum juices depending on the microorganism used.

Fruit Juice	Sample	L*	a*	b*	C*	h*	ΔE
Pear	J	70.9 ^{d} ± 1.9	11.6 ± 0.8	55.5 ^{bc} ± 1.7	56.7 ^{a} ± 1.8	78.2 ^{d} ± 0.5	
	LP	63.8 ^{cd} ± 4.0	24.0 ± 2.1	64.6 ^{d} ± 1.5	69.0 ^{cd} ± 0.7	69.6 ^{c} ± 2.1	15.7 ± 2.4
	LT	68.9 ^{d} ± 5.0	17.4 ± 2.5	58.8 ^{cd} ± 6.7	61.4 ^{ab} ± 6.8	73.5 ^{cd} ± 2.1	9.4 ± 2.1
	LF	64.6 ^{cd} ± 2.5	19.7 ± 2.2	62.5 ^{d} ± 1.4	65.5 ^{bc} ± 1.8	72.6 ^{c} ± 1.6	10.8 ± 3.1
Plum	J	55.8 ^{b} ± 2.6	35.0 ± 3.5	62.4 ^{d} ± 1.5	71.7 ^{de} ± 0.7	60.7 ^{b} ± 3.0	
	LP	39.8 ^{a} ± 1.5	46.5 ± 2.2	47.4 ^{a} ± 1.6	66.4 ^{bcd} ± 2.0	45.5 ^{a} ± 1.7	22.3 ± 2.2
	LT	43.6 ^{a} ± 2.9	46.2 ± 2.6	52.0 ^{ab} ± 3.4	69.6 ^{cd} ± 1.9	48.4 ^{a} ± 3.2	17.0 ± 4.9
	LF	57.4 ^{bc} ± 4.5	41.7 ± 6.0	64.5 ^{d} ± 3.1	77.0 ^{e} ± 1.1	57.1 ^{b} ± 5.0	9.4 ± 2.8
Sig.	Fruit juice	***	***	***	***	***	
	Fermentation	***	***	***	***	***	
	Fruit juice × fermentation	***	ns	***	***	***	

Legend: J—unfermented juice, LP—*Lactiplantibacillus plantarum* K7, LT—*Lachancea thermotolerans* PYCC6375, LF—*Lachancea fermentati* PYCC5883; Sig.—significance; ns—not significant. *** indicates significance at a level of <0.005 using the least significant difference test. Values with different Roman numerals (a–e) in the same column indicate statistically significant interactions between fruit juice and fermentation (Tukey HSD, $p < 0.05$).

2.2. Volatile Compounds

In addition to color, volatile compounds significantly impact the sensory evaluation of the product. Table 4 lists the volatile compounds detected in unfermented and fermented pear and plum juices. Only aroma-active compounds detected by olfactometric analysis and/or exceeding the threshold in any sample are included. These compounds are classified into four groups based on their chemical structure: alcohols, carbonyl compounds, esters, and terpenes.

Table 4. The volatile compounds in unfermented and fermented pear and plum juices depending on the microorganism used.

Compound [μg/L]	*m/z*	LRI	THR	Pear Juice				Plum Juice				Sig.		
				J	LP	LT	LF	J	LP	LT	LF	Fruit Juice	Fermentation	Fruit Juice x Fermentation
Alcohols														
3-Methyl-1-butanol	42, 55, 70	724	71	2 ^{a}	111 ^{ab}	358 ^{cd}	256 ^{bc}	2 ^{a}	341 ^{cd}	410 ^{cd}	422 ^{d}	***	***	*
(Z)-3-Heksen-1-ol	41, 55, 67	837	13.0	1.5 ^{a}	0.1 ^{a}	0.2 ^{a}	0.1 ^{a}	30.1 ^{b}	3.3 ^{a}	3.3 ^{a}	3.2 ^{a}	***	***	***
Heksan-1-ol	41, 43, 56	852	100	297 ^{c}	45 ^{a}	50 ^{a}	41 ^{a}	185 ^{b}	29 ^{a}	28 ^{a}	27 ^{a}	ns	***	***
1-Heptanol	41, 56, 70	953	5.4	0.0 ^{a}	4.4 ^{bc}	7.8 ^{c}	7.7 ^{c}	0.0 ^{a}	3.5 ^{ab}	2.3 ^{ab}	0.8 ^{ab}	***	***	***
1-Octen-3-ol	43, 57, 72	961	0.01 [1] 0.5 [2,3]	5.0 ^{bc}	7.1 ^{cde}	9.4 ^{e}	8.4 ^{de}	0.2 ^{a}	5.9 ^{bcd}	3.9 ^{bc}	3.0 ^{ab}	***	***	***
Carbonyl compounds														
2-Methylbutanal	42, 55, 86	640	1.0	0.0 ^{a}	0.0 ^{a}	0.0 ^{a}	0.0 ^{a}	2.3 ^{b}	0.0 ^{a}	0.0 ^{a}	0.0 ^{a}	***	***	***
3-Hexenal	41, 55, 69	773	0.25	0.0 ^{a}	0.0 ^{a}	0.0 ^{a}	0.0 ^{a}	9.4 ^{b}	0.0 ^{a}	0.0 ^{a}	0.0 ^{a}	***	***	***
Hexanal	41, 44, 56	776	10.0	287 ^{c}	0.0 ^{a}	1.0 ^{a}	1.0 ^{a}	160 ^{b}	0.0 ^{a}	0.0 ^{a}	0.0 ^{a}	***	***	***
2-Hexenal	41, 55, 69	825	17.0	44.5 ^{c}	0.0 ^{a}	0.0 ^{a}	0.0 ^{a}	25.1 ^{b}	0.0 ^{a}	0.0 ^{a}	0.0 ^{a}	***	***	***
1-Octen-3-one	55, 70, 97	954	0.01 [1] 0.5 [2,3]	2.2 ^{b}	2.6 ^{b}	3.6 ^{c}	2.1 ^{b}	0.0 ^{a}	0.2 ^{a}	0.3 ^{a}	0.5 ^{a}	***	***	***
Octanal	44, 56, 81	978	0.7	0.0 ^{a}	1.1 ^{d}	0.9 ^{c}	0.2 ^{b}	0.3 ^{b}	0.0 ^{a}	0.0 ^{a}	0.0 ^{a}	***	***	***
Phenylacetaldehyde	65, 91, 120	1005	2.0	0.3 ^{a}	0.1 ^{a}	0.1 ^{a}	0.1 ^{a}	12.0 ^{b}	0.1 ^{a}	0.1 ^{a}	0.2 ^{a}	***	***	***
(E)-2-Octenal	41, 55, 70	1036	0.2	6.7 ^{c}	0.9 ^{b}	1.7 ^{b}	0.9 ^{b}	0.0 ^{a}	0.2 ^{a}	0.2 ^{a}	0.3 ^{a}	***	***	***
Nonanal	41, 55, 70	1084	1.0	3.8 ^{b}	0.4 ^{a}	0.5 ^{a}	0.3 ^{a}	3.4 ^{b}	0.3 ^{a}	0.2 ^{a}	0.2 ^{a}	ns	***	ns
(E)-2-Nonenal	41, 55, 70	1134	0.1	3.0 ^{b}	0.1 ^{a}	0.2 ^{a}	0.1 ^{a}	0.0 ^{a}	0.0 ^{a}	0.0 ^{a}	0.0 ^{a}	***	***	***
Decanal	41, 55, 70	1181	0.1	1.0 ^{b}	0.2 ^{a}	0.2 ^{a}	0.1 ^{a}	1.5 ^{c}	0.0 ^{a}	0.1 ^{a}	0.0 ^{a}	***	***	ns
(Z,Z)-2,4-Decadienal	41, 67, 81	1264	0.2	4.5 ^{b}	0.0 ^{a}	0.0 ^{a}	0.0 ^{a}	0.0 ^{a}	0.0 ^{a}	0.0 ^{a}	0.0 ^{a}	***	***	***
(E,E)-2,4-Decadienal	41, 67, 81	1284	0.4	3.2 ^{b}	0.0 ^{a}	0.0 ^{a}	0.0 ^{a}	0.0 ^{a}	0.0 ^{a}	0.0 ^{a}	0.0 ^{a}	***	***	***
(E)-2-Undecenal	41, 57, 70	1332	0.8	0.0 ^{a}	0.4 ^{b}	1.0 ^{d}	0.7 ^{c}	0.0 ^{a}	0.0 ^{a}	0.0 ^{a}	0.0 ^{a}	***	***	***
Esters														
Ethyl butyrate	43, 71, 88	785	1.0	0.0 ^{a}	0.6 ^{ab}	0.6 ^{ab}	0.8 ^{bc}	0.0 ^{a}	1.4 ^{c}	0.8 ^{bc}	0.7 ^{abc}	ns	***	ns
Ethyl L-lactate	43, 45, 75	795	50.0	0.0 ^{a}	63.3 ^{c}	65.0 ^{c}	59.2 ^{c}	0.0 ^{a}	56.3 ^{bc}	52.6 ^{bc}	32.2 ^{b}	**	***	ns
Butyl acetate	43, 56, 73	798	10.0	2.6 ^{ab}	16.5 ^{d}	3.7 ^{b}	12.4 ^{c}	1.4 ^{ab}	0.0 ^{a}	0.0 ^{a}	0.0 ^{a}	***	***	***
Ethyl 2-methylbutyrate	57, 85, 102	836	0.1	0.5 ^{a}	0.1 ^{a}	0.3 ^{a}	0.1 ^{a}	1.4 ^{b}	0.3 ^{a}	0.4 ^{a}	0.2 ^{a}	**	***	***
Ethyl isovalerate	41, 57, 102	838	0.1	0.0 ^{a}	0.0 ^{a}	0.1 ^{b}	0.0 ^{a}	0.0 ^{a}	0.4 ^{c}	0.4 ^{c}	0.4 ^{c}	***	***	***
Isoamyl acetate	55, 43, 70	860	12.0	0.0 ^{a}	27.2 ^{c}	12.4 ^{b}	15.3 ^{b}	0.0 ^{a}	18.5 ^{bc}	16.3 ^{b}	10.0 ^{b}	ns	***	*
2-Methylbutyl acetate	43, 55, 70	862	5.0	0.0 ^{a}	1.4 ^{abc}	2.5 ^{abc}	1.2 ^{ab}	0.0 ^{a}	5.2 ^{c}	4.7 ^{bc}	3.2 ^{abc}	***	***	ns
Ethyl hexanoate	43, 88, 99	979	1.0	0.3 ^{a}	3.5 ^{b}	5.4 ^{b}	3.6 ^{b}	0.3 ^{a}	5.5 ^{b}	5.1 ^{b}	3.3 ^{ab}	ns	***	ns
Hexyl acetate	43, 56, 61	992	2.0	5.7 ^{e}	2.2 ^{d}	1.1 ^{bc}	1.3 ^{c}	0.5 ^{ab}	0.0 ^{a}	0.0 ^{a}	0.0 ^{a}	ns	***	***
Methyl salicylate	92, 120, 152	1170	35	31.3 ^{b}	0.1 ^{a}	0.1 ^{a}	0.1 ^{a}	83.2 ^{c}	2.7 ^{a}	2.4 ^{a}	5.0 ^{a}	***	***	***
Methyl (E,Z)-2,4-decadienoate	67, 81, 111	1367	0.1	0.8 ^{c}	0.1 ^{ab}	0.2 ^{b}	0.1 ^{ab}	0.0 ^{a}	0.0 ^{a}	0.0 ^{a}	0.0 ^{a}	***	***	***
Ethyl decanoate	70, 88, 101	1370	5.0	3.1 ^{ab}	4.3 ^{bc}	3.6 ^{abc}	3.2 ^{ab}	0.9 ^{a}	6.1 ^{cd}	7.6 ^{d}	3.6 ^{abc}	***	***	*
Ethyl (E,Z)-2,4-decadienoate	67, 81, 125	1434	0.1	0.6 ^{b}	0.1 ^{a}	0.0 ^{a}	0.0 ^{a}	0.0 ^{a}	0.0 ^{a}	0.0 ^{a}	0.0 ^{a}	**	***	***

Table 4. *Cont.*

Compound [μg/L]	m/z	LRI	THR	Pear Juice J	Pear Juice LP	Pear Juice LT	Pear Juice LF	Plum Juice J	Plum Juice LP	Plum Juice LT	Plum Juice LF	Sig. Fruit Juice	Sig. Fermentation	Sig. Fruit Juice x Fermentation
Terpenes and others														
Acetal	45, 47, 73	721	4.9	0.0 ^{a}	1.8 ^{a}	6.5 ^{ab}	9.4 ^{b}	0.0 ^{a}	20.6 ^{c}	9.3 ^{b}	4.6 ^{ab}	***	***	***
Linalool	43, 71, 93	1086	1.0	4.4 ^{c}	0.7 ^{ab}	1.2 ^{ab}	0.4 ^{a}	4.1 ^{c}	1.4 ^{b}	1.3 ^{ab}	1.1 ^{ab}	**	***	ns
(Z)-Rose oxide	55, 69, 139	1095	0.1	77.4 ^{c}	0.1 ^{a}	0.1 ^{a}	0.1 ^{a}	45.8 ^{b}	0.2 ^{a}	0.1 ^{a}	0.1 ^{a}	**	***	***
(E)-Rose oxide	55, 69, 139	1113	0.1	34.6 ^{c}	0.1 ^{a}	0.1 ^{a}	0.1 ^{a}	20.4 ^{b}	0.1 ^{a}	0.1 ^{a}	0.1 ^{a}	**	***	***
(E)-β-Damascenone	69, 105, 121	1365	0.002 ^{1} 0.1 ^{2,3}	0.0 ^{a}	1.5 ^{bc}	0.4 ^{a}	0.6 ^{ab}	2.6 ^{cd}	2.7 ^{d}	2.6 ^{d}	2.3 ^{cd}	***	*	***
α-Ionone	43, 93, 121	1411	0.4	0.5 ^{b}	0.0 ^{a}	0.0 ^{a}	0.0 ^{a}	0.3 ^{ab}	0.0 ^{a}	0.0 ^{a}	0.0 ^{a}	ns	***	ns
β-Ionone	43, 73, 177	1475	0.007	1.9 ^{b}	0.1 ^{a}	0.1 ^{a}	0.1 ^{a}	1.4 ^{b}	0.1 ^{a}	0.1 ^{a}	0.1 ^{a}	ns	***	ns
Volatile organic acids [μg/L]														
Butyric acid	41, 60, 73	780	10.0	0.0 ^{a}	7.3 ^{b}	15.7 ^{c}	32.7 ^{d}	0.0 ^{a}	0.0 ^{a}	0.0 ^{a}	3.4 ^{ab}	***	***	***
Hexanoic acid	41, 60, 73	962	36.0	6.5 ^{c}	1.7 ^{a}	3.2 ^{a}	6.8 ^{c}	2.1 ^{a}	5.5 ^{bc}	4.5 ^{bc}	6.6 ^{c}	ns	*	***
Octanoic acid	43, 60, 73	1162	910	17.5 ^{b}	4.4 ^{a}	5.9 ^{a}	7.4 ^{a}	26.9 ^{c}	4.5 ^{a}	6.7 ^{a}	8.8 ^{a}	*	***	*
Decanoic acid	41, 60, 73	1339	130	6.8 ^{b}	1.6 ^{ab}	1.4 ^{a}	30.8 ^{d}	13.1 ^{c}	2.0 ^{ab}	2.6 ^{ab}	2.2 ^{ab}	***	***	***
Dodecanoic acid	43, 60, 73	1530	7200	1.9 ^{bc}	0.0 ^{a}	5.2 ^{d}	4.4 ^{d}	2.5 ^{c}	1.8 ^{bc}	1.3 ^{b}	1.2 ^{ab}	***	***	***

Legend: LRI—linear retention index, THR—threshold, [1]—value for unfermented juices, [2]—value for fermented pear juice, [3]—value for fermented plum juice, J—unfermented juice, LP—*Lactiplantibacillus plantarum* K7, LT—*Lachancea thermotolerans* PYCC6375, LF—*Lachancea fermentati* PYCC5883, Sig.—significance; ns—not significant. *, **, and *** indicate significance at levels of 0.05–0.01, 0.01–0.005, and <0.005, respectively, as determined by the least significant difference test. Values with different superscript Roman letters (a–e) in the same row indicate statistically significant interactions between fruit juice and fermentation (Tukey HSD, $p < 0.05$). Green color indicates a concentration above the threshold.

The aroma of plums is shaped by a complex mixture of volatile compounds, with esters, aldehydes, alcohols, ketones, terpenes and lactones being the most significant contributors [44]. In unfermented plum must, 17 aroma-active components were identified (Table 4). Olfactometric analysis (Table 5) showed that the most pronounced aromas were terpenes, such as rose oxides, (E)-β-damascenone, and β-ionone; six-carbon carbonyl compounds, such as 3-hexenal and hexanal; and ethyl 2-methylbutyrate. These introduced floral (rose), plum, green, and fruity notes. Other compounds typical of plum aroma were also detected in the samples, but at a lower intensity. Key aroma-active components identified across various plum cultivars include esters such as acetic acid hexyl ester, hexyl acetate and butanoic acid butyl ester; aldehydes such as hexanal, (E)-2-hexenal, and 3-hexenal; alcohols such as 1-hexanol and (E)-2-hexen-1-ol; and terpenes such as linalool and limonene [44,45]. Other compounds that contribute significantly to the characteristic 'plum' aroma include benzaldehyde, methyl cinnamate, γ-decalactone, and β-damascenone, with the latter being particularly potent in some plum wines [46]. The specific aroma profile varies by plum variety and ripening stage, with certain esters and aldehydes (notably C6 compounds) being closely linked to the perception of fresh, fruity and green notes [44]. Sensory and chemical analyses demonstrate that the diversity and concentration of these volatiles can distinguish between different cultivars and maturity stages, rendering them valuable markers for quality and variety identification [47].

Table 5. The aroma-active components of unfermented and fermented pear and plum juices depend on the microorganism used.

Compound	LRI	Pear Juice				Plum Juice				Aroma
		J	LP	LT	LF	J	LP	LT	LF	
2-Methylbutanal	640	0	0	0	0	1.5	0	0	0	malty fermented
Acetal	721	0	0	0.5	1	0	2	1	0.5	ether green
3-Methyl-1-butanol	724	0	1	2	2	0	2	2	2	fusel alcoholic
3-Hexenal	773	0	0	0	0	2.5	0	0	0	leafy green
Hexanal	776	2.5	0	0	0	2.5	0	0	0	green aldehydic
Butyric acid	780	0	0	0.5	1.5	0	0	0	0.5	buttery cheesy
Ethyl butyrate	785	0	0	0	0	0	0.5	0	0	fruity pineapple
Ethyl L-lactate	795	0	0.5	0.5	0.5	0	0.5	0.5	0	fruity buttery
Butyl acetate	798	0	1	0	0.5	0	0	0	0	fruity banana
2-Hexenal	825	1.5	0	0	0	0.5	0	0	0	almond green
Ethyl 2-methylbutyrate	836	2	0.5	1.5	0.5	2.5	1.5	2	1	apple fruity
(Z)-3-Heksen-1-ol	837	0	0	0	0	1.5	0	0	0	green grassy
Ethyl isovalerate	838	0	0	0.5	0	0	2	2	2	fruity apple
Heksan-1-ol	852	1.5	0	0	0	1	0	0	0	pungent green
Isoamyl acetate	860	0	1.5	0.5	0.5	0	1	0.5	0	fruity banana
2-Methylbutyl acetate	862	0	0	0	0	0	0.5	0.5	0	sweet banana
1-Heptanol	953	0	0	0.5	0.5	0	0	0	0	herbal green
1-Octen-3-one	954	3	2	2.5	2	0	0	0	0.5	herbal mushroom
1-Octen-3-ol	960	3	2.5	2.5	2.5	2.5	2.5	2.5	2	mushroom earthy
Octanal	978	0	1	0.5	0	0	0	0	0	aldehydic citrus
Ethyl hexanoate	979	0	2	2	2	0	2	2	2	fruity apple
Hexyl acetate	992	1.5	0.5	0	0	0	0	0	0	fruity green apple
Phenylacetaldehyde	1005	0	0	0	0	2	0	0	0	green floral
(E)-2-Octenal	1035	2.5	2	2.5	2	0	0.5	0.5	0.5	cucumber green
Nonanal	1084	2	0	0	0	2	0	0	0	aldehydic rose
Linalool	1086	2	0	0.5	0	2	0.5	0.5	0.5	citrus floral
(Z)-Rose oxide	1095	3	0.5	0.5	0.5	3	1	0.5	0.5	red rose
(E)-Rose oxide	1113	3	0.5	0.5	0.5	3	0.5	0.5	0.5	herbal
(E)-2-Nonenal	1134	2.5	0.5	1	0.5	0	0	0	0	green cucumber
Methyl salicylate	1170	0	0	0	0	1.5	0	0	0	wintergreen mint
Decanal	1181	2.5	1	1	0.5	2.5	0	0.5	0	aldehydic orange peel
(Z,Z)-2,4-Decadienal	1264	2.5	0	0	0	0	0	0	0	seaweed
(E,E)-2,4-Decadienal	1284	2.5	0	0	0	0	0	0	0	oily cucumber
(E)-2-Undecenal	1332	0	0	0.5	0	0	0	0	0	fruity citrus
(E)-β-Damascenone	1365	0	2.5	2	2	3	2.5	2.5	2.5	sweet plum
Methyl (E,Z)-2,4-decadienoate	1367	2.5	0.5	1	0.5	0	0	0	0	fruity pear
Ethyl decanoate	1370	0	0	0	0	0	0.5	1	0	fruity apple brandy
α-Ionone	1411	0.5	0	0	0	0	0	0	0	woody floral
Ethyl (E,Z)-2,4-decadienoate	1434	2	0.5	0	0	0	0	0	0	green pear
β-Ionone	1475	3	2.5	2.5	2.5	3	2.5	2.5	2.5	floral woody

Legend: LRI—linear retention index, J—unfermented juice, LP—*Lactiplantibacillus plantarum* K7, LT—*Lachancea thermotolerans* PYCC6375, LF—*Lachancea fermentati* PYCC5883. The lowest values in the row are colored dark red, while the highest values are colored dark green.

Regardless of the microorganism used, fermentation significantly altered the composition and content of the aroma components. This is particularly noticeable in the case of C6 components, which were not detected in fermented juices. During fermentation, the concentration of C6 compounds—mainly C6 aldehydes (such as hexanal and (E)-2-hexenal) and C6 alcohols (such as 1-hexanol)—typically decreases. This reduction is primarily due to yeast enzymes converting C6 aldehydes into their corresponding alcohols and further metabolic transformations reducing their overall concentration. This transformation is important because C6 compounds contribute green, grassy aromas, and their reduction during fermentation leads to a softer, fruitier aroma profile in the final product. The extent of C6 compound reduction varies depending on fermentation conditions, yeast strain, and the presence of fruit solids or skins [48,49]. A similar phenomenon occurred with most other carbonyl compounds. The content of terpenes derived from the fruit also decreased. However, β-damascenone and β-ionone remained significant components of the aroma of the fermented juice. Terpenes, which are important aroma compounds that contribute floral and fruity notes, often decrease in concentration during fermentation, especially in products such as wine and fruit beverages. This reduction is mainly due to the conversion of glycosidically bound terpenes into free forms, which are then further metabolized, volatilized, or lost through other chemical reactions during fermentation [50]. The extent of terpene loss can vary depending on the fermentation conditions, the species of microorganisms present, and the presence of other compounds in the matrix, such as phenolic acids, which can inhibit terpene hydrolysis and volatilization [51]. Finally, the aroma of fermented plum juices was influenced by the alcohols and esters formed by microorganisms during fermentation. The most important of these were ethyl hexanoate and 3-methyl-1-butanol, which were produced more by yeasts than bacteria. Ethyl hexanoate is a key ester responsible for the fruity aromas found in fermented beverages such as beer, wine, and, in particular, the strongly aromatic Chinese liquor Baijiu. Its formation during fermentation primarily involves the action of yeast, particularly through the alcohol acyltransferase pathway, where enzymes such as EEB1 and Eht1p catalyse the synthesis of ethyl hexanoate from hexanoyl-CoA and ethanol [52]. Lactic acid bacteria (LAB) can synthesize ethyl hexanoate during milk fermentation, mainly through esterification reactions between free fatty acids and ethanol. There is significant variability among strains, with higher activity generally observed for esterification than for alcoholysis [53]. There is little evidence regarding the production of ethyl hexanoate by LAB during the fermentation of fruit juices. Fermenting melon and cashew apple juices with *Lacticaseibacillus casei* resulted in decreased ethyl hexanoate levels, suggesting that LAB may degrade or transform this ester rather than synthesize it in fruit juice matrices [54]. Our research has shown that the LP strain produces similar amounts of ethyl hexanoate to the *Lachancea* sp. Yeast. Due to this compound's low aroma detection threshold, it significantly influences the aroma of fermented fruit juice. Furthermore, 3-Methyl-1-butanol, also known as isoamyl alcohol, is a key higher alcohol formed during the fermentation of fruit juices that plays a significant role in the aroma of fermented beverages. It is primarily formed through the catabolism of the amino acid leucine via the Ehrlich pathway, whereby yeast and certain bacteria convert leucine into 3-methyl-1-butanol via transamination, decarboxylation and reduction steps [55]. Studies on apple and cherry juice fermentations have identified 3-methyl-1-butanol as a dominant volatile compound in the final product. Its concentration is influenced by the type of fruit, the yeast strain used, and fermentation conditions, such as temperature, pH, and sugar content [56].

The key aroma components of pears are primarily volatile compounds belonging to the ester, aldehyde, and alcohol classes, which together create the fruit's characteristic fruity and floral scent [57,58]. Esters such as hexyl acetate, butyl acetate, ethyl butanoate and, in particular, ethyl (E,Z)-2,4-decadienoate—often referred to as the 'pear ester'—are major contributors to the typical pear-like aroma. Ethyl (E,Z)-2,4-decadienoate is especially important in Bartlett pears and pear-based products [59]. Aldehydes such as hexanal and (E)-2-hexenal provide green and fresh notes, while alcohols like 1-hexanol and 2-phenylethanol impart grassy and floral nuances [57,58]. The specific composition and concentration of these volatiles varies widely among pear cultivars and is influenced by factors such as species, ripening stage, and storage conditions [57,58]. In addition to esters, aldehydes and alcohols, smaller amounts of acids, ketones, terpenes and lactones also contribute to the overall aromatic complexity [60]. The balance and abundance of these compounds determine the sensory quality and consumer appeal of different pear varieties [57,58]. The pear musts we examined contained 20 aroma-active components (Tables 4 and 5). The most important compounds for the pear aroma were C8 compounds such as 1-octen-3-one, 1-octen-3-ol and (E)-2-octenal, C9 and C10 aldehydes—(E)-2-nonenal, decanal, and 2,4-decadienals, as well as terpenes—rose oxides, and β-ionone. As previously mentioned, we detected two esters of 2,4-decadienoic acid, which are characteristic of the pear aroma; the methyl ester was found to have a stronger effect on the aroma than the ethyl ester.

As with plum juice, the aroma of pear juice changed after fermentation, with the strongest aroma found in juices fermented by LP and LT. However, similarly to fresh juice, the presence of C8 components such as 1-octen-3-one, 1-octen-3-ol and (E)-2-octenal was important for the aroma. The components 1-Octen-3-ol and 1-octen-3-one are important volatile compounds known for their characteristic mushroom-like aroma, and their formation is primarily linked to the oxidative breakdown of linoleic acid, with its levels influenced by both the initial composition of the juice and the fermentation matrix [61]. Although there have been few direct studies on these components in fruit juice fermentation, related research on fermented juices (such as kiwifruit and litchi) has shown that fermentation by lactic acid bacteria or yeast can increase the concentrations of various ketones and alcohols, including C8 compounds like 1-octen-3-ol, which is closely related to the formation of 1-octen-3-one [62,63]. The formation of 1-octen-3-one is generally associated with the enzymatic oxidation of linoleic acid and other unsaturated fatty acids during fermentation. Its levels can be affected by the specific strains used, as well as by the presence of glycosidic precursors [61]. (E)-2-Octenal is a volatile aldehyde that is commonly formed during fermentation processes. It contributes to the aroma and flavor profiles of various fermented foods. During juice fermentation, (E)-2-octenal and other aldehydes generally decrease in concentration. In watermelon juice fermented with various lactic acid bacteria, the levels of aldehydes including (E)-2-octenal decreased after fermentation, while alcohols, ketones, acids, and other volatiles increased. This reduction in aldehydes was associated with improved sensory qualities, such as a sweeter and more natural flavor, and a decrease in off-flavors and bitterness. Similarly, lactic acid bacteria fermentation of kiwifruit juice led to a significant decrease in aldehyde concentrations, including (E)-2-octenal, while increasing the levels of esters, ketones and alcohols, which contributed to a more desirable aroma profile [64]. The decrease in (E)-2-octanal during juice fermentation is primarily due to microbial metabolism whereby aldehydes are reduced or transformed into other compounds, resulting in an improvement in the flavor and aroma of the fermented juice [62]. As previously discussed in the context of plum juice, the content of most terpene compounds decreased after fermentation. The exception was β-damascenone, which was not present in the fresh juice but became one of the most important aromatic components after fermentation. B-Damascenone is a key aroma compound found in pear-derived products. In

Bartlett pear brandies, (E)-β-damascenone has been identified as one of the main contributors to the characteristic fruity, pear-like aroma, with high odor activity values indicating its strong sensory impact [65]. The concentration of β-damascenone can change during processing; for instance, fermentation and distillation can influence its levels, enhancing or modifying the overall aroma profile [59]. During grape wine fermentation, damascenone concentrations rise from low or undetectable levels to several parts per billion as fermentation progresses, with further increases observed during barrel ageing. This formation is linked to the transformation of specific ketone precursors by yeast activity, with different yeast strains influencing the amount produced [66]. Various microorganisms, including specific yeast strains, can increase the production of damascenone during fermentation by converting carotenoid-derived precursors into this aromatic compound. The efficiency of damascenone formation depends on the microbial strain and fermentation conditions used, such as pH, sugar content, and fermentation time [67]. As with plum juices, the metabolites produced during fermentation that had the greatest impact on aroma in fermented pear juices were ethyl hexanoate and 3-methyl-1-butanol. Once again, LT produced the greatest quantities of these components, while LF produced the least.

Due to the presence of a buttery aroma in selected fermented juices, the content of volatile organic acids in the samples was also analyzed (Table 4). It was found that this aroma was related to butyric acid, which was produced in quantities greater than the threshold for aroma in pear juice fermented with both *Lachancea* yeast strains. Current research shows that butyric acid production during fermentation is primarily associated with certain bacteria, especially *Clostridium* species, rather than yeasts. These bacteria can utilize C4 compounds (characteristic of pear juices), such as acetate and butyrate precursors, to efficiently generate butyric acid. The metabolic pathway involves converting acetyl-CoA to butyryl-CoA, which is then transformed into butyric acid [68]. These bacteria produce much higher amounts of butyric acid (up to 40 g/L), which is higher than that found in our studies (Table 4). Due to the lack of reports regarding the production of butyric acid by yeasts, this topic is interesting and should be investigated. This is especially important since butyric acid plays several important roles in human health. It serves as a primary energy source for colon cells, supports the maintenance of the intestinal barrier, and has anti-inflammatory effects that can help to protect against conditions such as colitis and other inflammatory bowel diseases [69].

2.3. *Polyphenols*

Both plums and pears are rich sources of phenolic antioxidants, which contribute significantly to their health benefits. The total phenolic content of plums varies widely among cultivars, ranging from 125 to 372.6 mg/100 g, and is strongly correlated with antioxidant capacity as measured by FRAP, TEAC and ORAC assays [70]. The plum skin contains much higher levels of phenolic compounds and flavonoids than the pulp, making it a particularly potent antioxidant source [71]. The major antioxidant compounds found in plums include neochlorogenic acid, chlorogenic acid, anthocyanins (such as cyanidin derivatives), and flavonols such as quercetin [70]. While pears also contain valuable phenolic antioxidants, they generally have lower antioxidant activity than plums. They are rich in hydroxycinnamates, such as caffeic acid and p-coumaric acid, but their overall antioxidant capacity is lower than that of anthocyanin-rich fruits like plums [72].

As shown in Table 6, plum juice contains significantly higher amounts of polyphenols than pear juice. Neochlorogenic acid and its isomer chlorogenic acid were dominant in plums, while the highest amounts of chlorogenic, shikimic, and caffeic acids were found in pear must. Neochlorogenic acid is a major phenolic compound in plums, found in both the skin and pulp but typically in higher concentrations in the skin [73]. It is one of the predominant phenolic compounds in both fresh and dried plums, often exceeding the levels of related compounds such as chlorogenic acid [74]. The content of neochlorogenic acid can vary depending on plum variety, the stage of fruit development, and even the rootstock used for cultivation [75]. Its concentration generally decreases as the fruit matures, with higher levels observed in less ripe plums. Chlorogenic acid is a major phenolic compound found in both plums and pears, though its concentration is typically much higher in plums. In plums, chlorogenic acid and its isomer neochlorogenic acid are both abundant, contributing significantly to the total phenolic content. Some studies recommend using chlorogenic acid as a standard for measuring phenolic compounds in chlorogenic acid-rich plum cultivars [76]. In pears, chlorogenic acid also plays a role in enzymatic browning and fruit quality. The content of chlorogenic acid in pears decreases as the fruit matures, and its synthesis is regulated by specific genes, with higher levels sometimes found in dwarf pear varieties [77]. The presence of chlorogenic acid in both fruits is associated with antioxidant activity and potential health benefits, such as reducing oxidative stress and providing anxiolytic effects [78].

Table 6. The polyphenol and related compound contents of unfermented and fermented pear and plum juices depending on the microorganism used.

Compound [mg/L]	Pear Juice				Plum Juice				Sig.		
	J	LP	LF	LT	J	LP	LF	LT	Fruit Juice	Fermentation	Fruit Juice × Fermentation
Phenolic acids											
Neochlorogenic acid	0.00 ^{a}	0.00 ^{a}	0.00 ^{a}	0.00 ^{a}	129 ^{b}	371 ^{c}	946 ^{d}	458 ^{c}	***	***	***
Chlorogenic acid	3.09 ^{a}	1.90 ^{a}	5.02 ^{a}	1.50 ^{a}	13.37 ^{a}	36.09 ^{b}	91.25 ^{c}	45.10 ^{b}	***	***	***
Caffeic acid	2.22 ^{bc}	0.88 ^{a}	1.14 ^{ab}	0.72 ^{a}	0.06 ^{a}	0.37 ^{a}	2.76 ^{c}	0.82 ^{a}	ns	***	***
Vanillic acid	0.25 ^{a}	1.46 ^{c}	2.93 ^{d}	1.59 ^{c}	0.34 ^{a}	0.55 ^{ab}	4.23 ^{e}	1.08 ^{bc}	ns	***	***
Salicylic acid	1.72 ^{ab}	0.99 ^{a}	0.98 ^{a}	1.39 ^{ab}	2.43 ^{b}	1.63 ^{ab}	4.74 ^{c}	4.24 ^{c}	***	***	***
Gentisic acid	0.05 ^{ab}	0.01 ^{a}	0.05 ^{ab}	0.02 ^{ab}	0.07 ^{bc}	0.06 ^{bc}	0.22 ^{d}	0.11 ^{c}	***	***	***
Gallic acid	0.96 ^{ab}	0.20 ^{ab}	0.25 ^{ab}	0.00 ^{a}	8.94 ^{d}	2.45 ^{bc}	12.18 ^{e}	3.96 ^{c}	***	***	***
Ferulic acid	0.02 ^{a}	0.00 ^{a}	0.00 ^{a}	0.00 ^{a}	0.00 ^{a}	0.39 ^{b}	0.34 ^{b}	1.39 ^{c}	***	***	***
p-Coumaric acid	0.49 ^{b}	0.00 ^{a}	0.00 ^{a}	0.00 ^{a}	0.17 ^{a}	0.24 ^{ab}	0.97 ^{c}	1.07 ^{c}	***	***	***
Ellagic acid	0.00 ^{a}	0.00 ^{a}	0.00 ^{a}	0.00 ^{a}	1.26 ^{e}	0.31 ^{b}	1.00 ^{d}	0.67 ^{c}	***	***	***
Flavonoids											
Catechin	0.31 ^{a}	0.46 ^{a}	0.61 ^{a}	0.41 ^{a}	6.87 ^{a}	22.52 ^{b}	69.16 ^{c}	27.34 ^{b}	***	***	***
Quercetin	1.40 ^{bc}	1.31 ^{abc}	1.46 ^{bc}	0.97 ^{ab}	1.70 ^{c}	1.64 ^{c}	1.46 ^{bc}	0.73 ^{a}	ns	**	***
Anthocyanins											
Cyanidin-3-O-rutinoside	0.71 ^{c}	0.00 ^{a}	0.00 ^{a}	0.00 ^{a}	0.65 ^{bc}	0.13 ^{a}	0.75 ^{c}	0.45 ^{b}	***	***	***
Cyanidin-3-O-glucoside	1.08 ^{ab}	0.97 ^{ab}	1.56 ^{c}	1.00 ^{ab}	0.74 ^{a}	1.12 ^{ab}	1.26 ^{bc}	1.22 ^{bc}	ns	***	***
Others											
Shikimic acid	361 ^{d}	141 ^{b}	201 ^{c}	247 ^{c}	48 ^{a}	7 ^{a}	10 ^{a}	9 ^{a}	***	***	***

Legend: J—unfermented juice, LP—*Lactiplantibacillus plantarum* K7, LT—*Lachancea thermotolerans* PYCC6375, LF—*Lachancea fermentati* PYCC5883, Sig.—significance; ns—not significant. ** and *** indicate significance at levels of 0.01–0.005 and <0.005, respectively, as determined by the least significant difference test. Values with different Roman letters (a–e) in the same row indicate statistically significant interactions between fruit juice and fermentation (Tukey HSD, $p < 0.05$).

Shikimic acid is a naturally occurring organic acid found in pears and plums. While shikimic acid itself is not classified as a polyphenol, it is a key metabolite in the shikimate pathway, which plays a crucial role in the biosynthesis of numerous plant polyphenols and phenolic compounds [79]. In pears, shikimic acid is consistently identified as one of the main organic acids, although its concentration is generally lower than that of malic and citric acids. Studies of various pear cultivars demonstrate that shikimic acid content can fluctuate depending on the species, cultivar, and developmental stage and is positively correlated with other acids, such as quinic acid [80]. In plums, shikimic acid is also present as part of the organic acid profile. Its content typically increases during early fruit development and then decreases as the fruit ripens [81]. Shikimic acid has several documented influences on human health, including antioxidant, anti-inflammatory, antibacterial, and analgesic effects. It has been shown to suppress lipid accumulation and reduce the expression of genes involved in fat synthesis, suggesting a potential role in preventing or treating fatty liver disease and hyperlipidemia [82].

Juice fermentation increased the amount of analyzed polyphenolic and related compounds, depending on the type of microorganism used (Table 6). The best results were obtained with the yeast *L. fermentati*, which was particularly evident in plum juices fermented with this microorganism. The juices used for fermentation were naturally cloudy, especially the plum juices, and contained a large number of solid particles. It is assumed that substances with antioxidant activity were released from these particles during fermentation with the participation of microbiological enzymes. Fermentation often leads to an increase in polyphenol concentration in various foods and beverages. This increase is primarily due to the release of bound polyphenols and the transformation of polyphenolic compounds by microbial enzymes, which makes them more bioavailable and enhances their antioxidant activity [83]. However, the effect can vary depending on the food matrix and fermentation conditions. For example, some polyphenols may decrease while others increase during apple cider fermentation [84]. Taking into account the dominant polyphenols in the tested juices (Table 6), research shows that chlorogenic acid concentration can either increase or decrease after fermentation, depending on the substrate, fermentation conditions, and microorganisms used. Studies on blueberry and blackberry juices fermented with various lactic acid bacteria and probiotics report a significant reduction in chlorogenic acid levels following fermentation, as the compound is metabolized by microorganisms or transformed into other phenolic compounds [85]. For instance, chlorogenic acid often disappeared or decreased in cider apple juice and mixed fruit juices during fermentation, while other acids such as quinic acid increased, suggesting microbial metabolism of chlorogenic acid [86]. In solid-state fermentation of soybeans with *L. casei*, the yield of chlorogenic acid was significantly higher in the fermented samples than in the unfermented ones. This indicates that fermentation can enhance the extraction and availability of chlorogenic acid by breaking down cell wall components and releasing bound phenolics [87].

2.4. Sensory Analysis

The research concluded with a sensory analysis, during which the aroma and taste of both unfermented and fermented musts were assessed (Figures 2 and 3). The aroma and taste of various samples of pear juice showed significantly greater variability than those of plum juice. At the same time, unfermented juices had a lower aroma intensity and were rated lower than fermented juices. Using all three microorganisms significantly enhanced the fruity aroma of the juices. This was due not only to the creation of new components, such as esters [53], but also to the conversion of compounds in the fruit from bound (non-volatile) to free (volatile) forms [50]. Juices fermented with the LP strain were rated highest in terms of aroma (Figure 2). The only drawback of pear juices fermented with *Lachancea* yeast was their buttery flavor, which resulted from higher levels of butyric acid, as mentioned. This lowered the overall rating of these juices. However, butyric acid has also been reported to have positive effects on the human body [69]. No sensory evidence of butyric acid was found in plum juices; therefore, all fermented juices were rated equally highly.

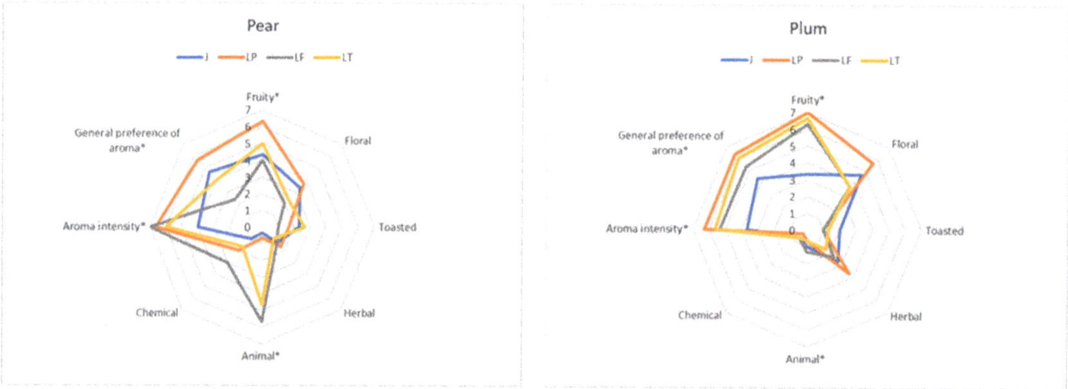

(a) (b)

Figure 2. Spider web diagrams showing the aroma qualities and impressions of unfermented (J) pear (**a**) and plum (**b**) juices and of juices produced using different fermentation microorganisms: LP—*Lactiplantibacillus plantarum* K7, LT—*Lachancea thermotolerans* PYCC6375, LF—*Lachancea fermentati* PYCC5883). Parameters that are statistically significantly different according to a two-way ANOVA ($p < 0.05$) are marked with an asterisk (*).

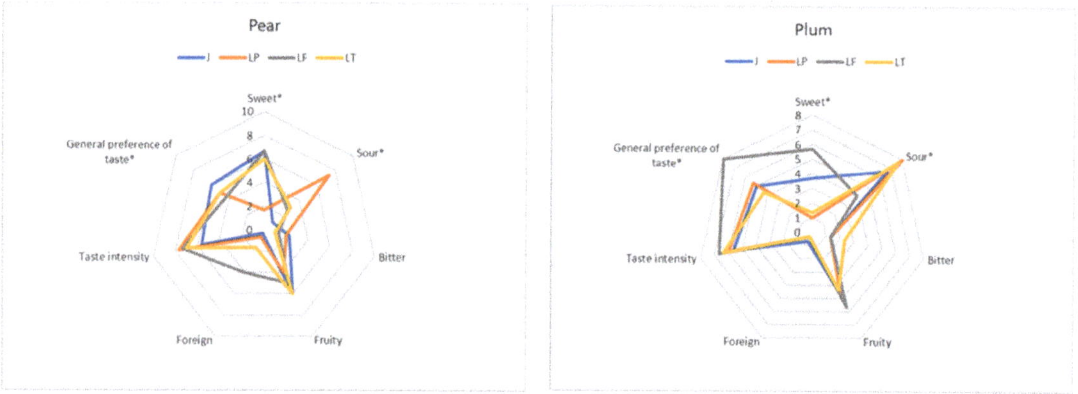

(a) (b)

Figure 3. Spider web diagrams showing the taste qualities and impressions of unfermented (J) pear (**a**) and plum (**b**) juices and of juices produced using different fermentation microorganisms: LP—*Lactiplantibacillus plantarum* K7, LT—*Lachancea thermotolerans* PYCC6375, LF—*Lachancea fermentati* PYCC5883). Parameters that are statistically significantly different according to a two-way ANOVA ($p < 0.05$) are marked with an asterisk (*).

In terms of taste, juice scores were most strongly influenced by sourness (related to the production of lactic acid) and sweetness (related to the use of sugars during fermentation). Research consistently shows that sourness can suppress the perception of sweetness in juices and other beverages. Studies using both human panels and biohybrid taste systems demonstrate that as the concentration of sour compounds (such as citric and tartaric acids) increases, the perceived sweetness of sugars decreases, and vice versa—sweetness can also suppress perceived sourness [88]. Therefore, the juices fermented with LAB had the strongest sour taste and were the least sweet. Yeast positively contributed to increasing the intensity of the juices' taste. In the case of pear juices, the juice fermented with the LF strain was rated the lowest (mainly due to a foreign aftertaste, probably related to the presence of butyric acid), whereas plum juices received the highest scores by far. This juice was characterized by a balanced feeling of sweetness and sourness, and a strong, fruity plum flavor.

3. Materials and Methods

3.1. Microbial Strains and Media Used

The *Lactiplantibacillus plantarum* K7 strain used in the research was isolated from sauerkraut and is deposited in the collection of pure cultures of the Department of Fermentation Technology and Microbiology, University of Agriculture in Krakow, Poland. The yeasts *Lachancea thermotolerans* PYCC6375 and *Lachancea fermentati* PYCC5883 were obtained from the Portuguese Yeast Culture Collection (PYCC), Caparica, Portugal. Microorganisms were cultivated on MRS Broth (LAB; Biomaxima, Lublin, Poland) or Sabouraud Dextrose Broth (yeasts; Biomaxima, Lublin, Poland).

3.2. Preparation and Fermentation of Juices

The Conference variety of pears (*Pyrus communis* L.) and the Węgierka zwykła variety of plums (*Prunus domestica* L.) were purchased from a local market in Kraków, Poland. The fruit was washed in boiling water, dried, cut into quarters, and juiced using a slow juicer (ZC420E38; Tefal, Rumilly, France) to obtain naturally cloudy juice. Three hundred milliliters of juice were poured into sterile fermentation flasks and pasteurized in a water bath (AJL electronic, Krakow, Poland) at 90 °C for ten minutes.

The juices were aseptically inoculated with a final concentration of 10^8 CFU/mL of lactic acid bacteria or selected yeast strains. The number of bacteria was determined using a densitometer (Biosan, Jozefow, Poland) based on a standard curve developed for the LAB strain. The number of yeast cells was determined through direct counting using a Thom chamber. The flasks were then closed with rubber stoppers and fitted with fermentation tubes filled with glycerol. For each variety, assays were performed in triplicate. Fermentation was conducted for eight days at 22 °C. Weight losses were measured daily to monitor the process. Once fermentation was complete, the samples were centrifuged at 179× g (MPW-350R; CONBEST, Krakow, Poland) and then pasteurized at 85 °C for 20 min.

3.3. pH and Acidity Measurement

The pH of the juices was measured using a CP-505 pH meter (Elmetron, Zabrze, Poland). The total acidity of the juices was determined by potentiometric titration of 25 mL of the sample with 0.1 N NaOH until a pH of 7 was reached.

3.4. Organic Acids' Analysis

The organic acids were analyzed using a UV–Vis detector at a wavelength of 210 nm on a NEXERA XR apparatus (Shimadzu, Kyoto, Japan), with the REZEX ROA-Organic Acid H+ (8%) Phenomenex (USA) column. The juices were centrifuged, and the resulting supernatant was filtered using a 0.45 µm syringe filter. The samples were diluted fivefold with distilled water. Concentrations of the individual components were determined using previously prepared standard curves for citric, malic, succinic, lactic, and acetic acids (all Sigma-Aldrich, St. Louis, MO, USA).

3.5. Sugar and Ethanol Analysis

Sugar compounds and ethanol were analyzed using a Shimadzu NEXERA XR apparatus (Kyoto, Japan) fitted with an RF-20A refractometric detector and a Shodex Asahipak NH2P-50 4.6 × 250 mm column (Showa Denko Europe, Munich, Germany), which was thermostatted at 30 °C. The mobile phase was a 70% aqueous acetonitrile solution, and the isocratic elution programme lasted 16 min at a flow rate of 0.8 mL/min. Concentrations of the individual components were determined based on standard curves previously prepared for glucose, fructose, sucrose, glycerol, sorbitol and ethanol (all Sigma-Aldrich, St. Louis, MO, USA).

3.6. Free Amino Nitrogen Analysis

Free amino nitrogen (FAN) was quantified using the ninhydrin method. Ninhydrin reacts with NH_3 upon exposure to heat to form a colored complex. Therefore, the ninhydrin reagent was added to water (blank), the test samples, and the reference sample (glycine standard). The samples were then boiled for 10 min, and the absorbance was measured at 575 nm using a DU-650 UV–Vis spectrophotometer (Beckman, Brea, CA, USA).

3.7. Color Analysis

The color of the juices was measured according to the CIELab method using a CM-3500d colorimeter (Konica Minolta, Inc., Tokyo, Japan). A D65 light source and an observation angle of 10° were used. The following parameters were measured: L˙ (brightness) (L˙ = 0 black, L* = 100 white), a* (proportion of green) (a* < 0) or red (a* > 0), b* (proportion of blue) (b* < 0) or yellow (b* > 0), C* (color saturation) and h* (color angle). Unfermented juices were compared with fermented juices. In order to determine the color difference between the samples, the parameter ΔE was calculated from the following formula:

$$\Delta E = \sqrt{\left(〖(\Delta L)〗^2 + 〖(\Delta a)〗^2 + 〖(\Delta b)〗^2 \right)}.$$

3.8. Volatile Compounds' Analysis (HS-SPME-GC-MS)

To analyse the volatile compounds, 0.05 g of juice, 2 mL of distilled water, and 0.1 mL of an internal standard solution (containing 0.57 mg/L of 4-methyl-2-pentanol, 0.2 mg/L of anethol and 1.48 mg/L of ethyl nonanoate, all from Sigma-Aldrich, St. Louis, MO, USA) were placed in a 10 mL vial containing 1 g of NaCl. A conditioned (250 °C for 1 h) SPME device (Supelco Inc., Bellefonte, PA, USA) coated with 100 µm PDMS fibers was used for sampling. This was placed in the headspace, with stirring at 300 rpm for 40 min at 40 °C. Next, the SPME device was desorbed in the injector port of the chromatograph system for 3 min. These analyses were performed using a Shimadzu GC-2010 Plus gas chromatograph coupled with a GCMS-QP2020 system. Separation of the analyzed volatiles was performed using a Rxi®-1 ms capillary column (Crossbond 100% dimethyl polysiloxane; 30 m × 0.53 mm × 0.5 µm). The column was heated using the following programme: 35 °C for 4 min, increasing by 5 °C/min to 110 °C; then increasing by 20 °C/min to 230 °C; then maintaining a constant temperature for 4 min. Helium was used as the carrier gas at a constant flow rate of 1.0 mL/min. The analyte was transferred in splitless mode. Mass spectra were recorded in SEM mode. Compound identification was performed using mass spectral libraries and linear retention indices derived from the C6 to C30 n-alkane series. Volatiles (Sigma-Aldrich) were identified qualitatively and quantitatively by comparing the retention times and peak areas of the sample and standard chromatograms. The results obtained were analyzed using the NIST database.

3.9. *Odor-Active Volatile Components (HS-SPME-GC-O)*

The odor-active volatile compounds in the juices were identified by olfactometry. A 2 mL sample of beer containing 1 g of NaCl was placed in a 10 mL vial and exposed to a divinylbenzene/carboxen/polydimethylsiloxane (DVB/CAR/PDMS) SPME fibre (50/30 μm, Supelco/Sigma-Aldrich, Bellafonte, PA, USA) for 40 min at 40 °C. The SPME device was then introduced into the injector port of the Hewlett Packard 5890 Series II chromatograph system and left there for 3 min. The components were then separated using an Rxi-1 ms capillary column (Crossbond 100% dimethyl polysiloxane, 30 m × 0.53 mm × 0.5 μm). The detector temperature was set to 250 °C, and the column was heated using the following programme: 35 °C for four minutes at an increment of 5 °C/min up to 110 °C, followed by an increment of 20 °C/min up to 230 °C, and then maintaining a constant temperature for four minutes. The carrier gas was helium at a constant flow of 1.0 mL/min. For analysis, an olfactory detection port (ODP-3, Gerstel, Linthicum, MD, USA) was used. A trained GC-O analyst was asked to describe the perceived odors and their intensity. Odor intensity was assessed using a 4-point scale (not detected, weak, moderate, and strong).

3.10. *Determination of Volatile Organic Acids by Derivatization with Benzyl Bromide*

One milliliter of the sample was transferred into the reaction vial, along with one milliliter of distilled water, one milliliter of buffer and 15 microliters of benzyl bromide. The sample was stirred at 50 °C for a defined period of time (210 min). After derivatization, the vial was cooled in an ice water bath, and the solution was saturated with 1 g of NaCl. It was then thermostatted at 30 °C and the analyte extracted by headspace SPME using a PA fibre (85 μm, Supelco Inc., Bellefonte, PA, USA) for 60 min. Next, the SPME device was desorbed in the injector port of the chromatograph system for 3 min. A Shimadzu GC-2010 Plus gas chromatograph coupled with a GCMS-QP2020 system was used for the analysis. An Rxi®-1 ms capillary column (Crossbond 100% dimethyl polysiloxane, 30 m × 0.53 mm × 0.5 μm) was used to separate the analyzed volatiles. The initial temperature was 40 °C for 5 min. It was then raised to 250 °C at a rate of 4 °C per minute. The transfer line temperature was 250 °C, and the mass spectrometer was operated in electron impact mode (70 eV). Helium was used as the carrier gas at a constant flow rate of 1.0 mL/min. Mass spectra were recorded in EI mode at an ionization voltage of 70 eV, with a transfer line and ion source temperature of 250 °C. Standard curves were prepared using volatile acids (butyric, hexanoic, octanoic, decanoic, dodecanoic; Sigma-Aldrich) by derivatization under the conditions described above.

3.11. *Polyphenol Analysis*

Polyphenol analysis was performed using a Shimadzu Nexera XR system (Shimadzu, Kyoto, Japan), which consists of a CBM-20A system controller, an LC-20AD VP pump, an SIL-10AC VP autoinjector, a DGU-20A degasser, an SPD-M20A VP UV–VIS detector and LabSolutions ver. 5.93, as well as a Shim-pack VP-ODS 250 × 4.6 (5 μm) column. The injection volume was 10 μL. Solutions were filtered through a 0.45 μm nylon membrane prior to HPLC injection. A reverse-phase HPLC assay was carried out using a gradient system with a flow rate of 0.7 mL/min and a column temperature of 25 °C. The mobile phase consisted of phosphoric acid at a concentration of 0.1% (solution A) and acetonitrile (solution B). The elution programme was as follows: a linear increase in the concentration of solution B from 14% to 19% over 8 min; isocratic elution with 19% solution B over 6 min; a linear increase in the concentration of solution B from 19% to 31% over 20 min; a linear increase in the concentration of solution B from 31% to 90% over 5 min; a linear decrease in the concentration of solution B to 14% over 1 min; and isocratic elution with 14% solution B over 9 min. Shikimic acid was detected at $\lambda = 210$ nm; vanillic acid at $\lambda = 250$ nm; salicylic acid, gallic acid, ellagic acid, and catechin at $\lambda = 280$ nm; neochlorogenic acid, chlorogenic acid, caffeic acid, gentisic acid, p-coumaric acid, and ferulic acid at $\lambda = 325$ nm; quercetin at $\lambda = 360$ nm; and cyanidin-3-O-glucoside and cyanidin-3-O-rutinoside at $\lambda = 520$ nm.

Standard curves were prepared using the appropriate standards for quantitative determinations: shikimic acid, vanillic acid, salicylic acid, ellagic acid, (+)-catechin, neochlorogenic acid, chlorogenic acid, caffeic acid, gentisic acid, ferulic acid, quercetin, cyanidin-3-O-glucoside and cyanidin-3-O-rutinoside (all from Sigma-Aldrich), gallic acid, p-coumaric acid, and ferulic acid (all from Extrasynthese, Genay, France).

3.12. *Sensory Analysis (QDA)*

A sensory assessment of juice samples was performed using quantitative descriptive analysis (QDA). The evaluation was conducted by ten trained testers (five men and five women) from the Department of Fermentation Technology and Microbiology, aged between 30 and 50 years. Fifteen sensory qualities, selected in an initial panel discussion, were evaluated on a ten-point scale; these related to aroma (fruity, floral, toasted, herbal, animal, chemical, intensity, general preference) and taste (sweet, sour, bitter, fruity, foreign, intensity, general preference).

3.13. *Statistical Analysis*

All analyses were performed in triplicate, and their results are presented in the paper as arithmetic means with standard deviation. Statistical analysis of pH, acidity, content of individual components, and juice color was performed using a two-way analysis of variance (ANOVA) using Tukey's HSD post hoc test for $p < 0.05$ (Statistica, v. 13.3; StatSoft Inc., Tulsa, OK, USA).

4. Conclusions

The food market is looking for new natural products with the highest possible health-promoting value and sensory properties that consumers will find acceptable. Such products may be fruit juices fermented with various microorganisms (without added salt) that contain microbiological metabolites and living microbial cells, as well as components derived from plant material such as polyphenols, organic acids, and volatile compounds. In our research, we used cloudy pear and plum juices, which are widely consumed in various parts of the world in their natural state; lactic acid bacteria derived from plant fermentation; and lactic acid-producing yeasts belonging to the genus *Lachancea*. The resulting juices contained high levels of health-promoting compounds (polyphenols were more bioavailable), lower levels of sugars (which were metabolized during fermentation), and potentially toxic substances (such as ethanol). They also had modified sensory characteristics which were often rated higher than unfermented juices. *Lachancea* yeast proved to be a viable alternative to lactic acid bacteria, producing similar quantities and qualities of components, and being resistant to low pH. They performed well in juices that naturally contained high levels of organic acids. However, further research into using these microorganisms to ferment fruit musts with significant acidity (such as blackcurrant and cherry)—in order to soften their taste, stabilize them, and introduce their own characteristic components—is required. The practical nature of our research is also significant, as it provides a basis for product recipes that can be implemented in production and brought to market, thereby increasing the range of beverages that have a positive impact on our health.

Author Contributions: Conceptualization, P.S.; methodology, P.S.; validation, P.S.; formal analysis, P.S. and M.M.; investigation, P.S., M.M. and M.S.; resources, P.S. and M.S.; data curation, P.S.; writing—original draft preparation, P.S. and M.S.; writing—review and editing, P.S. and M.S.; visualization, P.S. and M.S.; supervision, P.S.; funding acquisition, P.S. All authors have read and agreed to the published version of the manuscript.

Funding: This research was financed by The Ministry of Science and Higher Education of Poland, as a part of the Science Subsidy no. 070013-D020.

Institutional Review Board Statement: Not applicable.

Data Availability Statement: The original contributions presented in this study are included in the article. Further inquiries can be directed to the corresponding authors.

Conflicts of Interest: The authors declare no conflicts of interest.

Abbreviations

The following abbreviations are used in this manuscript:

J	Raw juice
LP	Juice fermented by *Lactilantibacillus plantarum* K7
LT	Juice fermented by *Lachancea thermotolerans* PYCC6375
LF	Juice fermented by *Lachancea fermentati* PYCC5883

References

1. FAOSTAT. Available online: https://www.fao.org/faostat/en/#data/QCL (accessed on 24 September 2025).
2. Core Team. Available online: https://coreteam.pl/369449-ulubione-owoce-listopada-kantar-polska (accessed on 25 September 2025).
3. Francini, A.; Pintado, M.; Manganaris, G.A.; Ferrante, A. Bioactive compounds biosynthesis and metabolism in fruit and vegetables. *Front. Plant Sci.* **2020**, *11*, 129. [CrossRef] [PubMed]
4. Skrovankova, S.; Sumczynski, D.; Mlcek, J.; Jurikova, T.; Sochor, J. Bioactive compounds and antioxidant activity in different types of berries. *Int. J. Mol. Sci.* **2015**, *16*, 24673–24706. [CrossRef] [PubMed]
5. Ji, Y.; Zhang, N.; Wang, J.; Cheng, Y.; Guan, J. Health Benefits and Bioactive Compounds of Pear and Its Products: A Comprehensive Review. *Food Sci. Hum. Wellness* **2025**. [CrossRef]
6. Ayub, H.; Nadeem, M.; Mohsin, M.; Ambreen, S.; Khan, F.A.; Oranab, S.; Rahim, M.A.; Khalid, M.Z.; Zongo, E.; Zarlasht, M.; et al. A comprehensive review on the availability of bioactive compounds, phytochemicals, and antioxidant potential of plum (*Prunus domestica*). *Int. J. Food Prop.* **2023**, *26*, 2388–2406. [CrossRef]
7. Hong, S.Y.; Lansky, E.; Kang, S.S.; Yang, M. A review of pears (*Pyrus* spp.), ancient functional food for modern times. *BMC Complement. Med. Ther.* **2021**, *21*, 219. [CrossRef]
8. Mahboubi, M. *Prunus domestica* as effective and acceptable treatment for stool softening and relief of constipation symptoms. *Songklanakarin J. Sci. Technol.* **2021**, *43*, 1183–1189. [CrossRef]
9. Mishra, S.; Vyas, S. Therapeutic and pharmacological potential of Prunus domestica: A Comprehensive Review. *Int. J. Pharm. Sci. Res.* **2021**, *12*, 3034–3041. [CrossRef]
10. Saud, S.; Xiaojuan, T.; Fahad, S. The consequences of fermentation metabolism on the qualitative qualities and biological activity of fermented fruit and vegetable juices. *Food Chem. X* **2024**, *21*, 101209. [CrossRef]
11. Hu, Q.-Y.; Tang, X.-X.; Li, Z.; Wei, L.-F.; Wu, X.-P.; Zhao, H. Effects of lactic acid bacteria fermentation on antioxidant activity and sensory quality of *Rosa sterilis* S D Shi. *Food Med. Homol.* **2025**, *2*, 9420026. [CrossRef]
12. Li, Z.; Teng, J.; Lyu, Y.; Hu, X.; Zhao, Y.; Wang, M. Enhanced antioxidant activity for apple juice fermented with *Lactobacillus plantarum* ATCC14917. *Molecules* **2018**, *24*, 51. [CrossRef]

13. Rizzi, F.; Juan, B.; Espadaler-Mazo, J.; Capellas, M.; Huedo, P. *Lactiplantibacillus plantarum* KABP051: Stability in Fruit Juices and Production of Bioactive Compounds During Their Fermentation. *Foods* **2024**, *13*, 3851. [CrossRef]
14. Zeng, H.; Shuai, Y.; Zeng, X.; Xin, B.; Huang, M.; Li, B.; Qiao, J.; Wang, Y.; Qui, X.; Wang, C. Evaluation of health-related composition and bioactivity of five fruit juices following *Lactobacillus plantarum* fermentation and simulated digestion. *Int. J. Food Sci. Technol.* **2021**, *56*, 648–660. [CrossRef]
15. Yang, J.; Sun, Y.; Gao, T.; Wu, Y.; Sun, H.; Zhu, Q.; Liu, C.; Zhou, C.; Han, Y.; Tao, Y. Fermentation and storage characteristics of "Fuji" apple juice using *Lactobacillus acidophilus, Lactobacillus casei* and *Lactobacillus plantarum*: Microbial growth, metabolism of bioactives and in vitro bioactivities. *Front. Nutr.* **2022**, *9*, 833906. [CrossRef] [PubMed]
16. Zhang, Y.; Liu, W.; Wei, Z.; Yin, B.; Man, C.; Jiang, Y. Enhancement of functional characteristics of blueberry juice fermented by *Lactobacillus plantarum*. *LWT* **2021**, *139*, 110590. [CrossRef]
17. Cheng, Z.; Wang, S.-H.; Li, L.-Y.; Zhou, Y.C.; Zhang, C.Y.; Liu, H.; Pan, T.M.; Liu, G.-M.; Liu, Q.-M. Fermented *Porphyra haitanensis* polysaccharides inhibit the degranulation of mast cell and passive cutaneous anaphylaxis. *Food Med. Homol.* **2025**. [CrossRef]
18. Fairbairn, S.; Engelbrecht, L.; Setati, M.E.; Du Toit, M.; Bauer, F.F.; Divol, B.; Rossouw, D. Combinatorial analysis of population dynamics, metabolite levels and malolactic fermentation in *Saccharomyces cerevisiae/Lachancea thermotolerans* mixed fermentations. *Food Microbiol.* **2021**, *96*, 103712. [CrossRef]
19. Fernández-Vázquez, D.; Sunyer-Figueres, M.; Vázquez, J.; Puxeu, M.; Nart, E.; De Lamo, S.; Andorrà, I. Selection and Use of Wild *Lachancea thermotolerans* Strains from Rioja AOC with Bioacidificant Capacity as Strategy to Mitigate Climate Change Effects in Wine Industry. *Beverages* **2025**, *11*, 70. [CrossRef]
20. Izquierdo-Cañas, P.M.; Del Fresno, J.M.; Malfeito-Ferreira, M.; Mena-Morales, A.; García-Romero, E.; Heras, J.M.; Loira, I.; González, C.; Morata, A. Wine bioacidification: Fermenting Airén grape juices with *Lachancea thermotolerans* and *Metschnikovia pulcherrima* followed by sequential *Saccharomyces cerevisiae* inoculation. *Int. J. Food Microbiol.* **2025**, *427*, 110977. [CrossRef]
21. Sizzano, F.; Bianconi, V.; Blackford, M.; Bieri, S.; Vuichard, F.; Monnard, C.; Amiet, L.; Spring, J.-L.; Dorsaz, E.; Pfenninger-Bridy, N.; et al. Use of *Lachancea thermotolerans* for the Bioacidification of White Grape Musts: Assays from the Bench to the Cellar Scale. *Fermentation* **2024**, *10*, 458. [CrossRef]
22. Bellut, K.; Michel, M.; Hutzler, M.; Zarnkow, M.; Jacob, F.; De Schutter, D.P.; Daenen, L.; Lynch, K.M.; Zannini, E.; Arendt, E.K. Investigation into the potential of *Lachancea fermentati* strain KBI 12.1 for low alcohol beer brewing. *J. Am. Soc. Brew. Chem.* **2019**, *77*, 157–169. [CrossRef]
23. Bellut, K.; Krogerus, K.; Arendt, E.K. *Lachancea fermentati* strains isolated from kombucha: Fundamental insights, and practical application in low alcohol beer brewing. *Front. Microbiol.* **2020**, *11*, 764. [CrossRef]
24. Domizio, P.; House, J.F.; Joseph, C.L.; Bisson, L.F.; Bamforth, C.W. *Lachancea thermotolerans* as an alternative yeast for the production of beer. *J. Inst. Brew.* **2016**, *122*, 599–604. [CrossRef]
25. Fu, X.; Guo, L.; Li, Y.; Chen, X.; Song, Y.; Li, S. Transcriptional Analysis of Mixed-Culture Fermentation of *Lachancea thermotolerans* and *Saccharomyces cerevisiae* for Natural Fruity Sour Beer. *Fermentation* **2024**, *10*, 180. [CrossRef]
26. Postigo, V.; Esteban, S.; Arroyo, T. *Lachancea thermotolerans*, an innovative alternative for sour beer production. *Beverages* **2023**, *9*, 20. [CrossRef]
27. Zdaniewicz, M.; Satora, P.; Kania, P.; Florkiewicz, A. The Impact of Selected *Lachancea* Yeast Strains on the Production Process, Chemical Composition and Aroma Profiles of Beers. *Molecules* **2024**, *29*, 5674. [CrossRef] [PubMed]
28. Satora, P.; Skotniczny, M.; Strnad, S.; Piechowicz, W. Chemical composition and sensory quality of sauerkraut produced from different cabbage varieties. *LWT* **2021**, *136*, 110325. [CrossRef]
29. Willems, J.L.; Low, N.H. Authenticity analysis of pear juice employing chromatographic fingerprinting. *J. Agric. Food Chem.* **2014**, *62*, 11737–11747. [CrossRef]
30. Cupic, T.; Josipovic, M.; Viljevac, M.; Sudar, R.; Dugalic, K. Sorbitol and sugar composition in plum fruits influenced by climatic conditions. *J. Agr. Sci. Technol.* **2014**, *16*, 1145–1155.
31. Martínez-Miranda, J.G.; Chairez, I.; Durán-Páramo, E. Mannitol production by heterofermentative lactic acid bacteria: A review. *Appl. Biochem. Biotechnol.* **2022**, *194*, 2762–2795. [CrossRef]
32. Ma, S.; Neilson, A.P.; Lahne, J.; Peck, G.M.; O'Keefe, S.F.; Stewart, A.C. Free amino acid composition of apple juices with potential for cider making as determined by UPLC-PDA. *J. Inst. Brew.* **2018**, *124*, 467–476. [CrossRef]
33. Moon, K.M.; Kwon, E.B.; Lee, B.; Kim, C.Y. Recent trends in controlling the enzymatic browning of fruit and vegetable products. *Molecules* **2020**, *25*, 2754. [CrossRef] [PubMed]
34. Xue, H.; Zhao, J.; Wang, Y.; Shi, Z.; Xie, K.; Liao, X.; Tan, J. Factors affecting the stability of anthocyanins and strategies for improving their stability: A review. *Food Chem. X* **2024**, *24*, 101883. [CrossRef] [PubMed]
35. Gänzle, M.G. Lactic metabolism revisited: Metabolism of lactic acid bacteria in food fermentations and food spoilage. *Curr. Opin. Food Sci.* **2015**, *2*, 106–117. [CrossRef]

36. Yang, X.; Teng, K.; Su, R.; Li, L.; Zhang, T.; Fan, K.; Zhang, J.; Zhong, J. AcrR and Rex Control Mannitol and Sorbitol Utilization through Their Cross-Regulation of Aldehyde-Alcohol Dehydrogenase (AdhE) in *Lactobacillus plantarum*. *Appl. Environ. Microbiol.* **2019**, *85*, e02035-18. [CrossRef]
37. Ijoma, G.N.; Adegbenro, G.; Rashama, C.; Matambo, T.S. Peculiar response in the co-culture fermentation of *Leuconostoc mesenteroides* and *Lactobacillus plantarum* for the production of ABE solvents. *Fermentation* **2021**, *7*, 212. [CrossRef]
38. von Weymarn, N.; Hujanen, M.; Leisola, M. Production of D-mannitol by heterofermentative lactic acid bacteria. *Process Biochem.* **2002**, *37*, 1207–1213. [CrossRef]
39. Chen, W.; Xie, C.; He, Q.; Sun, J.; Bai, W. Improvement in color expression and antioxidant activity of strawberry juice fer-mented with lactic acid bacteria: A phenolic-based research. *Food Chem. X* **2023**, *17*, 100535. [CrossRef]
40. Charoenchongsuk, N.; Ikeda, K.; Itai, A.; Oikawa, A.; Murayama, H. Comparison of the expression of chlorophyll-degrada-tion-related genes during ripening between stay-green and yellow-pear cultivars. *Sci. Hortic.* **2015**, *181*, 89–94. [CrossRef]
41. Usenik, V.; Štampar, F.; Veberič, R. Anthocyanins and fruit colour in plums (*Prunus domestica* L.) during ripening. *Food Chem.* **2009**, *114*, 529–534. [CrossRef]
42. Wang, L.; Zhang, H.; Lei, H. Phenolics profile, antioxidant activity and flavor volatiles of pear juice: Influence of lactic acid fermentation using three *Lactobacillus* strains in monoculture and binary mixture. *Foods* **2021**, *11*, 11. [CrossRef]
43. Srinivasan, L.V.; Rana, S.S. Anthocyanins: A promising source of natural colorants and nutraceuticals. *Discov. Appl. Sci.* **2025**, *7*, 694. [CrossRef]
44. Sun, H.; Lu, X.; Wang, Y.; Li, J.; Liu, S. Study on Evaluation of Fruit Aroma of Plum Variety Resources Based on Headspace Solid-Phase Microextraction Combined with Gas Chromatography–Mass Spectrometry. *Foods* **2024**, *13*, 3515. [CrossRef]
45. Ismail, H.; Williams, A.; Tucknott, O. The flavour of plums (*Prunus domestica* L.). An examination of the aroma components of plum juice from the cultivar victoria. *J. Sci. Food and Agri.* **1981**, *32*, 613–619. [CrossRef]
46. Niimi, J.; Guixer, B.; Splivallo, R. Odour active compounds determined in the headspace of yellow and black plum wines (*Prunus domestica* L.). *LWT* **2020**, *130*, 109702. [CrossRef]
47. Taiti, C.; Pandolfi, C.; Caparrotta, S.; Dei, M.; Giordani, E.; Mancuso, S.; Nencetti, V. Fruit aroma and sensorial characteristics of traditional and innovative Japanese plum (*Prunus salicina* Lindl.) cultivars grown in Italy. *Eur. Food Res. Technol.* **2019**, *245*, 2655–2668. [CrossRef]
48. Yao, X.; Zhang, H.X.; Xia, N.; Duan, C.; Yang, W.; Pan, Q. Leaching and evolution of anthocyanins and aroma compounds during Cabernet Sauvignon wine fermentation with whole-process skin-seed contact. *Food Chem.* **2023**, *436*, 137727. [CrossRef] [PubMed]
49. Herraiz, T.; Herraiz, M.; Reglero, G.; Martín-Alvarez, P.; Cabezudo, M. Changes in the composition of alcohols and aldehydes of C6 chain length during the alcoholic fermentation of grape must. *J. Agric. Food Chem.* **1990**, *38*, 969–972. [CrossRef]
50. Zoecklein, B.; Marcy, J.; Williams, J.; Jasinski, Y. Effect of Native Yeasts and Selected Strains of *Saccharomyces cerevisiae* on Glycosyl Glucose, Potential Volatile Terpenes, and Selected Aglycones of White Riesling (*Vitis vinifera* L.) Wines. *J. Food Compos. Anal.* **1997**, *10*, 55–65. [CrossRef]
51. Wang, X.; Li, Y.; Song, H.; Tao, Y.; Russo, N. Phenolic matrix effect on aroma formation of terpenes during simulated wine fermentation—Part I: Phenolic acids. *Food Chem.* **2021**, *341*, 128288. [CrossRef]
52. Yin, H.; Liu, L.; Yang, M.; Ding, X.; Jia, S.; Dong, J.; Zhong, C. Enhancing Medium-Chain Fatty Acid Ethyl Ester Production During Beer Fermentation Through EEB1 and ETR1 Overexpression in *Saccharomyces pastorianus*. *J. Agric. Food Chem.* **2019**, *67*, 5607–5613. [CrossRef]
53. Mukdsi, M.; Medina, R.; Alvarez, M.; González, S. Ester synthesis by lactic acid bacteria isolated from goat's and ewe's milk and cheeses. *Food Chem.* **2009**, *117*, 241–247. [CrossRef]
54. De Godoy Alves Filho, E.; Rodrigues, T.; Fernandes, F.; Pereira, A.; Narain, N.; De Brito, E.; Rodrigues, S. Chemometric evaluation of the volatile profile of probiotic melon and probiotic cashew juice. *Food Res. Int.* **2017**, *99*, 461–468. [CrossRef] [PubMed]
55. Filannino, P.; Cardinali, G.; Rizzello, C.; Buchin, S.; De Angelis, M.; Gobbetti, M.; Di Cagno, R. Metabolic Responses of *Lactobacillus plantarum* Strains during Fermentation and Storage of Vegetable and Fruit Juices. *Appl. Environ. Microbiol.* **2014**, *80*, 2206–2215. [CrossRef] [PubMed]
56. Braga, C.; Zielinski, A.; Da Silva, K.; De Souza, F.; De Arruda Moura Pietrowski, G.; Couto, M.; Granato, D.; Wosiacki, G.; Nogueira, A. Classification of juices and fermented beverages made from unripe, ripe and senescent apples based on the aromatic profile using chemometrics. *Food Chem.* **2013**, *141*, 967–974. [CrossRef]
57. Wang, X.; Chen, Y.; Zhang, J.; Wang, Z.; Qi, K.; Li, H.; Tian, R.; Wu, X.; Qiao, X.; Zhang, S.; et al. Comparative analysis of volatile aromatic compounds from a wide range of pear (*Pyrus* L.) germplasm resources based on HS-SPME with GC-MS. *Food Chem.* **2023**, *418*, 135963. [CrossRef]

58. Soledad, M.; Moya, P.; Leahu, A.; Ghinea, C.; Zhang, W.; Yan, M.; Zheng, X.; Chen, Z.; Li, H.; Mao, J.; et al. Exploring the Aroma Fingerprint of Various Chinese Pear Cultivars through Qualitative and Quantitative Analysis of Volatile Compounds Using HS-SPME and GC×GC-TOFMS. *Molecules* **2023**, *28*, 4794. [CrossRef]
59. Zierer, B.; Schieberle, P.; Granvogl, M. Aroma-Active Compounds in Bartlett Pears and Their Changes during the Manufacturing Process of Bartlett Pear Brandy. *J. Agric. Food Chem.* **2016**, *64*, 9515–9522. [CrossRef]
60. Wang, C.; Zhang, W.; Li, H.; Mao, J.; Guo, C.; Ding, R.; Wang, Y.; Fang, L.; Chen, Z.; Yang, G. Analysis of Volatile Compounds in Pears by HS-SPME-GC×GC-TOFMS. *Molecules* **2019**, *24*, 1795. [CrossRef]
61. Delcros, L.; Collas, S.; Hervé, M.; Blondin, B.; Roland, A. Evolution of Markers Involved in the Fresh Mushroom Off-Flavor in Wine During Alcoholic Fermentation. *J. Agric. Food Chem.* **2023**, *71*, 14687–14696. [CrossRef]
62. Lan, T.; Lv, X.; Zhao, Q.; Lei, Y.; Gao, C.; Yuan, Q.; Sun, X.; Liu, X. Optimization of strains for fermentation of kiwifruit juice and effects of mono- and mixed culture fermentation on its sensory and aroma profiles. *Food Chem. X* **2023**, *17*, 100595. [CrossRef]
63. Wang, D.; Deng, Y.; Chen, X.; Wang, K.; Zhao, L.; Wang, Z.; Liu, X.; Hu, Z. Elucidating the effects of *Lactobacillus plantarum* fermentation on the aroma profiles of pasteurized litchi juice using multi-scale molecular sensory science. *CRFS* **2023**, *6*, 100481. [CrossRef]
64. Mandha, J.; Shumoy, H.; Devaere, J.; Schouteten, J.; Gellynck, X.; De Winne, A.; Matemu, A.; Raes, K. Effect of lactic acid fermentation of watermelon juice on its sensory acceptability and volatile compounds. *Food Chem.* **2021**, *358*, 129809. [CrossRef] [PubMed]
65. Willner, B.; Granvogl, M.; Schieberle, P. Characterization of the key aroma compounds in Bartlett pear brandies by means of the sensomics concept. *J. Agric. Food Chem.* **2013**, *61*, 9583–9593. [CrossRef] [PubMed]
66. Lloyd, N.; Capone, D.; Ugliano, M.; Taylor, D.; Skouroumounis, G.; Sefton, M.; Elsey, G. Formation of Damascenone under both commercial and model fermentation conditions. *J. Agric. Food Chem.* **2011**, *59*, 1338–1343. [CrossRef] [PubMed]
67. Tang, J.; Lin, B.; Jiang, W.; Li, Q.; Zhu, L.; Zhang, G.; Chen, Q.; Yang, Q.; Yang, S.; Chen, S. Screening of β-damascenone-producing strains in light-flavor Baijiu and its production optimization via response surface methodology. *Front. Microbiol.* **2022**, *13*, 1067671. [CrossRef]
68. Chang, Y.; Reddy, M. Butyric Acid and Caproic Acid Production Using Single and Mixed Bacterial Cultures. *Eng. Proc.* **2023**, *56*, 73. [CrossRef]
69. Park, Y.; Kim, T.; Ham, J.; Choi, J.; Lee, H.; Yeon, Y.; Choi, S.; Kim, N.; Kim, Y.; Seok, Y. Physiological activity of *E. coli* engineered to produce butyric acid. *Microb. Biotechnol.* **2021**, *15*, 832–843. [CrossRef]
70. Kim, D.; Chun, O.; Kim, Y.; Moon, H.; Lee, C. Quantification of polyphenolics and their antioxidant capacity in fresh plums. *J. Agric. Food Chem.* **2003**, *51*, 6509–6515. [CrossRef]
71. Cosmulescu, S.; Trandafir, I.; Nour, V.; Botu, M. Total Phenolic, Flavonoid Distribution and Antioxidant Capacity in Skin, Pulp and Fruit Extracts of Plum Cultivars. *J. Food Biochem.* **2015**, *39*, 64–69. [CrossRef]
72. Manzoor, M.; Mahmood, A.; Ajaz, M.; Rasool, W.; Shabbir, M. Juicy Gems of Nutrition: Exploring the Nutrient Profile and Antioxidant Activity of Rosaceae Fruits. *DIET FACTOR (J. Nutr. Food Sci.)* **2023**, *4*, 08–19. [CrossRef]
73. Cabrera-Bañegil, M.; Rodas, N.; Losada, M.; Cipollone, F.; Espino, M.; Peña, A.; Durán-Merás, I. Evolution of polyphenols content in plum fruits (*Prunus salicina*) with harvesting time by second-order excitation-emission fluorescence multivariate calibration. *Microchem J.* **2020**, *158*, 105299. [CrossRef]
74. DiNardo, A.; Subramanian, J.; Singh, A. Investigation of Antioxidant Content and Capacity in Yellow European Plums. *Int. J. Fruit Sci.* **2018**, *18*, 116–199. [CrossRef]
75. Zhang, H.; Pu, J.; Tang, Y.; Wang, M.; Tian, K.; Wang, Y.; Luo, X.; Deng, Q. Changes in Phenolic Compounds and Antioxidant Activity during Development of 'Qiangcuili' and 'Cuihongli' Fruit. *Foods* **2022**, *11*, 3198. [CrossRef] [PubMed]
76. Chun, O.; Kim, D. Consideration on equivalent chemicals in total phenolic assay of chlorogenic acid-rich plums. *Food Res. Int.* **2004**, *37*, 337–342. [CrossRef]
77. He, J.; Cheng, Y.; Guan, J.; Ge, W.; Zhao, Z. Changes of chlorogenic acid content and its synthesis-associated genes expression in Xuehua pear fruit during development. *J. Integr. Agric.* **2017**, *16*, 471–477. [CrossRef]
78. Bouayed, J.; Rammal, H.; Dicko, A.; Younos, C.; Soulimani, R. Chlorogenic acid, a polyphenol from *Prunus domestica* (Mirabelle), with coupled anxiolytic and antioxidant effects. *J. Neurol. Sci.* **2007**, *262*, 77–84. [CrossRef]
79. Zagoskina, N.; Zubova, M.; Nechaeva, T.; Kazantseva, V.; Goncharuk, E.; Katanskaya, V.; Baranova, E.; Aksenova, M. Polyphenols in Plants: Structure, Biosynthesis, Abiotic Stress Regulation, and Practical Applications (Review). *Int. J. Mol. Sci.* **2023**, *24*, 13874. [CrossRef] [PubMed]
80. Wu, J.; Fan, J.; Li, Q.; Jia, L.; Xu, L.; Wu, X.; Wang, Z.; Li, H.; Qi, K.; Qiao, X.; et al. Variation of Organic Acids in Mature Fruits of 193 Pear (*Pyrus* spp.) Cultivars. *J. Food Compos. Anal.* **2022**, *109*, 104483. [CrossRef]

81. Bae, H.; Yun, S.; Yoon, I.; Nam, E.; Kwon, J.; Jun, J. Assessment of organic acid and sugar composition in apricot, plumcot, plum, and peach during fruit development. *J. Appl. Bot. Food Qual.* **2014**, *87*, 24–29. [CrossRef]
82. Kim, M.; Sim, D.; Lee, H.; Lee, H.; Kim, S. Hypolipogenic Effect of Shikimic Acid Via Inhibition of MID1IP1 and Phosphorylation of AMPK/ACC. *Int. J. Mol. Sci.* **2019**, *20*, 582. [CrossRef]
83. Liu, S.; He, Y.; He, W.; Song, X.; Peng, Y.; Hu, X.; Bian, S.; Li, Y.; Nie, S.; Yin, J.; et al. Exploring the Biogenic Transformation Mechanism of Polyphenols by *Lactobacillus plantarum* NCU137 Fermentation and Its Enhancement of Antioxidant Properties in Wolfberry Juice. *J. Agric. Food Chem.* **2024**, *72*, 12752–12761. [CrossRef] [PubMed]
84. Ye, M.; Yue, T.; Yuan, Y. Evolution of polyphenols and organic acids during the fermentation of apple cider. *J. Sci. Food Agric.* **2014**, *94*, 2951–2957. [CrossRef] [PubMed]
85. Wu, Y.; Li, S.; Tao, Y.; Li, D.; Han, Y.; Show, P.; Wen, G.; Zhou, J. Fermentation of blueberry and blackberry juices using *Lactobacillus plantarum, Streptococcus thermophilus* and *Bifidobacterium bifidum*: Growth of probiotics, metabolism of phenolics, antioxidant capacity in vitro and sensory evaluation. *Food Chem.* **2021**, *348*, 129083. [CrossRef]
86. Mantzourani, I.; Nikolaou, A.; Kourkoutas, Y.; Alexopoulos, A.; Dasenaki, M.; Mastrotheodoraki, A.; Proestos, C.; Thomaidis, N.; Plessas, S. Chemical Profile Characterization of Fruit and Vegetable Juices after Fermentation with Probiotic Strains. *Foods* **2024**, *13*, 1136. [CrossRef] [PubMed]
87. Akpabli-Tsigbe, N.; Ma, Y.; Ekumah, J.; Osabutey, J.; Hu, J.; Xu, M.; Johnson, N.; Quaisie, J. Two-step optimization of solid-state fermentation conditions of heilong48 soybean variety for maximum chlorogenic acid extraction yield with improved antioxidant activity. *Ind. Crops Prod.* **2021**, *168*, 113565. [CrossRef]
88. Junge, J.Y.; Bertelsen, A.S.; Mielby, L.A.; Zeng, Y.; Sun, Y.-X.; Byrne, D.V.; Kidmose, U. Taste Interactions between Sweetness of Sucrose and Sourness of Citric and Tartaric Acid among Chinese and Danish Consumers. *Foods* **2020**, *9*, 1425. [CrossRef]

molecules

MDPI

Review

Classical Food Fermentations as Modern Biotechnological Platforms: Alcoholic, Acetic, Butyric, Lactic and Propionic Pathways and Applications

Anna Rymuszka [1,*] and Wiktoria Gorczynska [2]

[1] Department of Physiology and Toxicology, Faculty of Medicine, The John Paul II Catholic University of Lublin, Konstantynów Str. 1i, 20-708 Lublin, Poland
[2] Independent Researcher, 20-950 Lublin, Poland; gorczynska.wikt@gmail.com
* Correspondence: anrym@kul.pl

Abstract

Fermentation remains central to food manufacturing and to the bio-based production of organic acids, solvents, and functional metabolites. This review integrates the biochemical pathways, key microorganisms, and application space of five major industrial fermentations—alcoholic, acetic, butyric, lactic, and propionic. We summarize the principal metabolic routes (EMP/ED glycolysis; oxidative ethanol metabolism; butyrate-forming pathways; and the Wood–Werkman, acrylate, and 1,2-propanediol routes to propionate) and relate them to the dominant microbial groups involved, including yeasts, acetic acid bacteria, lactic acid bacteria, clostridia, and propionibacteria. We highlight how the resulting metabolite spectra—ethanol, acetic acid, butyrate, lactate, propionate, and associated secondary metabolites—underpin product quality and safety in fermented foods and beverages and enable the industrial synthesis of platform chemicals, polymers, and biofuels. Finally, we discuss current challenges and opportunities for sustainable fermentation, including waste stream valorization, process intensification, and the integration of systems biology and metabolic engineering within circular economy frameworks.

Keywords: alcoholic fermentation; acetic acid bacteria; butyric acid; lactic acid bacteria; propionic acid; fermented foods; bioprocessing; bio-based chemicals; functional foods

1. Introduction

Fermentation is one of the oldest biotechnological processes, in which microorganisms, through their metabolic activity, transform plant- or animal-based raw materials into products with modified physicochemical and sensory properties [1–4]. From a biochemical standpoint, fermentation is a metabolic process in which energy is generated from organic compounds in the absence of an exogenous electron acceptor [5]. During fermentation, microorganisms catabolize carbohydrates into alcohols, carbon dioxide and/or organic acids, while simultaneously utilizing nutrients present in raw materials, such as carbon, nitrogen, vitamins and trace elements, for growth and reproduction [1]. As a result of microbial enzymatic activity, complex organic molecules undergo chemical and physical transformations and are converted into simpler, bioactive, functional and nutritionally beneficial compounds [6]. The metabolic reactions occurring during fermentation not only alter food composition but also enhance its nutritional, sensory and functional characteristics [7]. This biochemical capacity has been exploited empirically since antiquity, long before the microbial basis of fermentation was understood.

The origins of fermentation can be traced back to early human societies, with archaeological evidence indicating its use as early as the Late Epipaleolithic period. Residues of fermented cereal-based beverages discovered at Raqefet Cave in present-day Israel provide the earliest known proof of deliberate fermentation, highlighting its role in early food processing practices. Over subsequent millennia, fermentation became a fundamental method of food production and preservation across ancient civilizations, including Mesopotamia, Egypt, and China, where it was used for the manufacture of bread, beer, wine, and other fermented products. Initially, fermentation relied on spontaneous processes driven by indigenous microbiota associated with raw materials and the environment. Through empirical observation, humans gradually learned to manipulate fermentation conditions to improve product quality and stability. The biological basis of fermentation remained unknown until the 19th century, when Louis Pasteur demonstrated its microbial nature. This discovery enabled intentional strain selection and process optimization, ultimately laying the foundation for modern industrial biotechnology [8–10]. Today, fermentation is increasingly integrated with systems biology and metabolic engineering to improve productivity, robustness, and sustainability of bioprocesses. In practice, these tools enable the identification of metabolic bottlenecks and stress-response limitations and support rational strain and process design to improve yields, robustness, and substrate utilization.

Accordingly, modern fermentation practice is often distinguished by the degree of microbial and process control applied during production. Two primary approaches to food fermentation are recognized. The first is spontaneous fermentation, driven by autochthonous microorganisms naturally present in raw materials or the processing environment [3,11]. Due to the undefined composition and dynamics of the microbial communities involved, spontaneous fermentation requires strict monitoring and stabilization [6]. The second approach involves the use of starter cultures, which dominate contemporary industrial practice. Starter cultures consist of selected and well-characterized strains of bacteria, yeasts or molds that initiate and guide fermentation in a controlled, reproducible manner, suppressing undesirable microorganisms and enhancing food safety [12]. These cultures accelerate fermentation, convert available carbohydrates into alcohols and organic acids that act as natural preservatives, and impart desirable sensory attributes. Their application standardizes product quality, extends shelf life and reduces technological variability. In addition, many strains exhibit antagonistic activity toward pathogens, supporting microbial safety. For these reasons, ongoing research focuses on isolating and characterizing strains with optimal technological and probiotic potential for use as functional starter cultures [6,12].

While spontaneous fermentations are increasingly valued for their contribution to product diversity, regional identity, and complex sensory profiles, their inherent microbial variability limits their applicability in large-scale industrial production. In contrast, defined starter cultures enable reproducible fermentation kinetics, improved safety through rapid acidification or ethanol production, and consistent product quality. Consequently, modern food fermentation often combines traditional practices with controlled starter cultures or selected mixed consortia to balance authenticity with industrial reliability, particularly in processes requiring scalability and compliance with safety standards [9].

Fermentation technologies therefore continue to play a key role in the food sector, improving preservation, ensuring microbiological safety, enhancing sensory quality, and enabling the transformation of perishable raw materials into stable, shelf-stable, and nutritionally valuable products [4]. Because these pathways are metabolically connected and often occur sequentially or in mixed-culture systems, a comparative overview is

https://doi.org/10.3390/molecules31020333

essential to capture fermentation as an integrated biotechnological platform rather than isolated processes. Depending on the substrates and microorganisms involved, five major types of fermentation can be distinguished: alcoholic, acetic, butyric, lactic and propionic.

The aim of this review is to provide a comprehensive and integrative overview of the biochemical pathways, microbial ecology, and industrial applications of these five major fermentation types. The article synthesizes current knowledge on the metabolic mechanisms underlying each process, the microorganisms involved, their technological roles and their relevance in the food, pharmaceutical and biotechnological industries. Additionally, the review highlights recent advances, emerging trends and future prospects in fermentation-based bioprocessing, with particular emphasis on sustainability, functional food development and the valorization of biological raw materials. Key barriers to sustainable implementation include heterogeneous and seasonally variable waste streams, inhibitory compounds in low-cost substrates, and the need for robust strains and scalable downstream processing. In addition to summarizing classical pathways and applications, this review addresses cross-cutting industrial challenges such as process scale-up, yield optimization, and economic feasibility, and discusses emerging trends in sustainable and data-driven fermentation technologies. In the following sections, each fermentation type is discussed within a consistent framework, progressing from core biochemical reactions to the principal microorganisms involved and, finally, to representative industrial applications. By jointly analyzing five archetypal fermentation types rather than a single product, this review highlights fermentation as a modular and interconnected biotechnological platform, in which distinct pathways collectively support food production, platform chemicals, biofuels, and functional biomaterials.

2. Alcoholic Fermentation

2.1. Biochemical Basis

Alcoholic (ethanolic) fermentation is an anaerobic biochemical process in which sugars such as glucose, fructose, and sucrose are converted into ethyl alcohol (ethanol) and carbon dioxide [1,3,8,10,13,14]. The process begins with glycolysis—a series of enzyme-catalyzed reactions that degrade one molecule of glucose into two molecules of pyruvate, accompanied by the formation of two molecules of ATP and two molecules of NADH. In the next step, pyruvate undergoes decarboxylation catalyzed by pyruvate decarboxylase, producing acetaldehyde and releasing carbon dioxide. Subsequently, acetaldehyde is reduced to ethanol in a reaction catalyzed by alcohol dehydrogenase, during which NADH is oxidized back to NAD^+. The regeneration of NAD^+ enables the continuation of glycolysis and energy production under anaerobic conditions [7,8,13,15–17]. The overall reaction can be expressed as follows:

$$C_6H_{12}O_6 \rightarrow 2\ C_2H_5OH + 2\ CO_2$$

While the overall stoichiometry is conserved, ethanol yield, productivity, and stress tolerance depend strongly on the microorganism and the glycolytic route used, as outlined below.

2.2. Microorganisms and Metabolic Pathways

Alcoholic fermentation is carried out primarily by yeasts and, to a lesser extent, by certain bacterial species [18–20]. The most commonly employed microorganism is *Saccharomyces cerevisiae*, which ferments glucose via the classical glycolytic pathway known as the Embden–Meyerhof–Parnas (EMP) pathway. In this route, each molecule of glucose is converted into two molecules of pyruvate and subsequently—according to the mechanism described above—into ethanol and carbon dioxide [15,20–22].

 https://doi.org/10.3390/molecules31020333

Certain bacteria, such as *Zymomonas mobilis*, utilize an alternative sugar catabolic pathway—the Entner–Doudoroff (ED) pathway. Similarly to the EMP pathway, it leads to the formation of pyruvate but is characterized by a lower energetic yield, producing only one ATP molecule per molecule of glucose, and by reduced biomass formation, thereby channeling a higher proportion of sugar toward ethanol production [15,18,19,21].

As illustrated in Figure 1, the ED pathway differs from the classical EMP route in both reaction sequence and enzymatic machinery. In the EMP pathway, glucose is metabolized to pyruvate with the net production of two ATP and two NADH molecules. A key enzyme, fructose-1,6-bisphosphate aldolase, cleaves fructose-1,6-bisphosphate into glyceraldehyde-3-phosphate and dihydroxyacetone phosphate. In contrast, in the ED pathway, glucose-6-phosphate is converted into 6-phosphogluconate and subsequently into 2-keto-3-deoxy-6-phosphogluconate (KDPG), which is cleaved by KDPG aldolase directly into pyruvate and glyceraldehyde-3-phosphate—yielding lower energy output [21,23,24]. Despite differences in the early steps of glucose metabolism, the downstream reactions—conversion of glyceraldehyde-3-phosphate to pyruvate—proceed identically in both pathways, involving the same enzymes and resulting in the same end products. In each case, pyruvate is further converted to ethanol and carbon dioxide [15,22].

Although the EMP pathway yields more ATP per mole of glucose than the ED pathway, this higher energy conservation in *S. cerevisiae* is generally associated with increased biomass formation and stress tolerance rather than maximal ethanol yield [16,23,25]. In contrast, the lower ATP yield of the ED pathway in *Z. mobilis* can limit biomass synthesis and direct a higher fraction of carbon toward ethanol production, resulting in higher specific ethanol productivity despite lower energetic efficiency [18,23,24].

The choice of microorganism and metabolic route depends on physiology, ethanol yield, fermentation rate, substrate spectrum and tolerance to inhibitory compounds. Yeasts such as *S. cerevisiae* are valued for their high ethanol tolerance, robustness under industrial conditions, and capacity to withstand osmotic, thermal, and inhibitory stresses. These features, together with its genetic stability, extensive industrial know-how, and availability of advanced molecular tools, explain why *S. cerevisiae* remains the dominant microorganism in industrial alcoholic fermentation [25,26]. In contrast, *Z. mobilis* exhibits high ethanol productivity, rapid fermentation rates, and low biomass formation due to the lower ATP yield of the ED pathway, making strain selection strongly dependent on the specific requirements of a given bioprocess [19]. These metabolic features directly translate into technological performance, which explains why specific strains are preferred for beverage, baking, and biofuel applications.

In recent years, alcoholic fermentation has become a major target of metabolic engineering and synthetic biology aimed at improving ethanol yield, productivity, and stress tolerance under industrial conditions [25,27,28]. In *S. cerevisiae*, engineering strategies have focused on redirecting carbon flux toward ethanol by minimizing by-product formation (e.g., glycerol and organic acids), enhancing cofactor balance, and increasing tolerance to ethanol, temperature, and osmotic stress [16,25,29]. Genome-scale metabolic models and CRISPR/Cas-based genome editing have enabled systematic modification of glycolytic regulation, redox metabolism, and membrane composition, resulting in strains with improved robustness and fermentation performance [25,30–32].

In parallel, *Z. mobilis* has been extensively engineered to expand its substrate spectrum beyond glucose to pentoses such as xylose and arabinose, facilitating the efficient conversion of lignocellulosic hydrolysates into ethanol [24,33]. Synthetic biology approaches, including pathway modularization, dynamic gene regulation, and adaptive laboratory evolution, are increasingly integrated with process engineering to optimize large-scale bioethanol production and reduce overall process costs [18,34,35]. These advances highlight the transition of alcoholic

https://doi.org/10.3390/molecules31020333

fermentation from a classical bioprocess to a data-driven and design-oriented platform for sustainable biofuel and biochemical production [25,27].

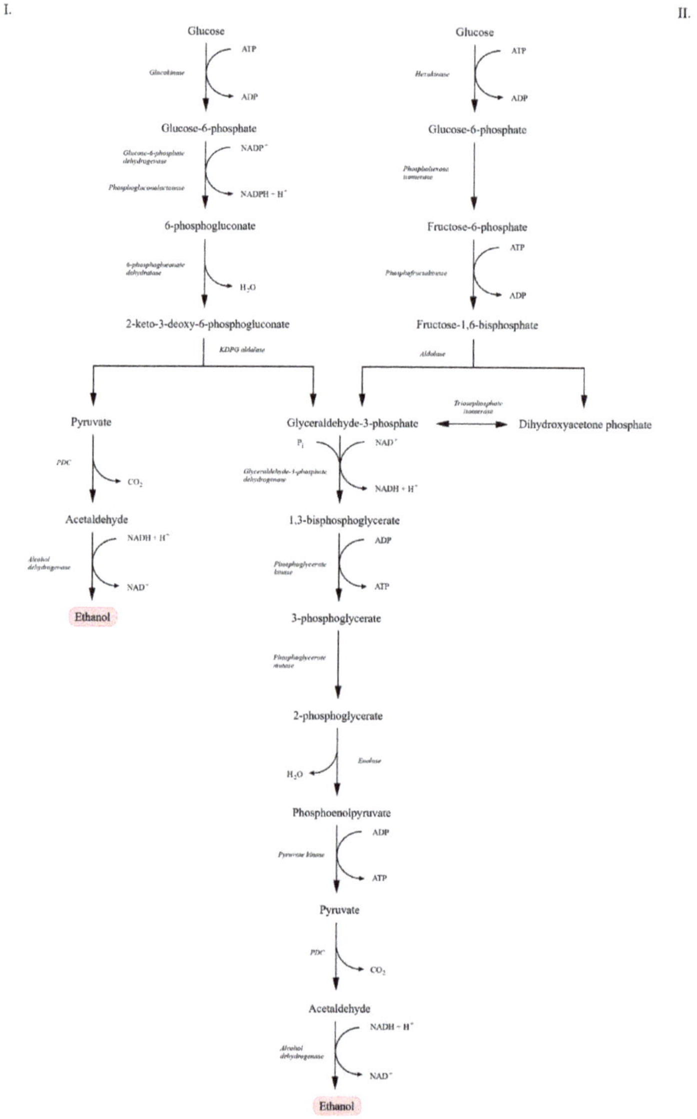

Figure 1. Comparison of the Entner–Doudoroff (ED) and Embden–Meyerhof–Parnas (EMP) pathways leading to alcoholic fermentation. In the ED pathway (**I**), glucose is converted through glucose-6-phosphate, 6-phosphogluconate, and 2-keto-3-deoxy-6-phosphogluconate (KDPG), yielding pyruvate and glyceraldehyde-3-phosphate. In the EMP pathway (**II**), glucose is metabolized through glucose-6-phosphate, fructose-1,6-bisphosphate, and glyceraldehyde-3-phosphate. In both pathways, pyruvate is decarboxylated to acetaldehyde by pyruvate decarboxylase (PDC), followed by reduction to ethanol by alcohol dehydrogenase (ADH). Scheme adapted from [21].

https://doi.org/10.3390/molecules31020333

2.3. Industrial Applications

Alcoholic fermentation is a biotechnological process with a wide range of industrial applications, encompassing the food, biotechnological, chemical, and energy sectors (Table 1). Its ability to convert carbohydrates into ethanol and other bioactive metabolites makes it a cornerstone of both traditional and modern bioprocessing technologies [9,26].

This type of fermentation represents one of the oldest and most fundamental biotechnological processes applied in the food industry [7,36–38]. The classical examples include the production of wine, beer, cider, mead, and distilled beverages such as whisky, rum, and vodka. In these processes, yeasts—primarily *S. cerevisiae*—convert sugars present in raw materials such as grape must, barley mash, apple juice, or honey solutions into ethanol and carbon dioxide. For distilled beverages, fermentation serves as a preliminary stage before distillation, which increases ethanol concentration. In addition to ethanol, numerous aromatic compounds, such as esters, aldehydes, and higher alcohols are generated, contributing to the sensory complexity and quality of the final beverages. For example, ethyl esters (e.g., ethyl acetate, ethyl hexanoate) impart fruity and floral notes, higher alcohols contribute complexity and mouthfeel, and aldehydes influence freshness and balance, collectively shaping the characteristic sensory profiles of fermented beverages [8,39–42].

Alcoholic fermentation also plays an important role in the production of low-alcohol fermented beverages, including water kefir, kombucha, kvass, and ginger beer. In the initial stage of these processes, yeasts carry out alcoholic fermentation, resulting in small amounts of ethanol, which is subsequently oxidized to acetic acid by acetic acid bacteria. The final product therefore contains only trace amounts of alcohol, and its sensory and nutritional value arises from fermentation metabolites and microorganisms with potential probiotic properties [9,37].

Beyond beverage production, alcoholic fermentation has significant applications in biotechnology for the biosynthesis of natural flavors, enzymes, and aromatic components. During fermentation, yeasts generate ethyl esters, aldehydes, and ketones that impart characteristic aroma profiles and are widely utilized in the dairy, bakery, and confectionery industries [16,36,43,44]. In bakery processes, *S. cerevisiae* metabolizes sugars in flour into ethanol and carbon dioxide, leading to dough leavening and crumb formation. Although ethanol evaporates during baking, intermediate metabolites formed during fermentation contribute to the desirable flavor and aroma of baked goods [45,46].

In the chemical and pharmaceutical sectors, ethanol obtained via yeast fermentation is a valuable raw material and universal solvent. It serves as a substrate for the synthesis of various organic compounds such as ethylene, ethyl acetate, diethyl ether, acetic acid, glycerol, and organic acids (e.g., lactic and succinic), which are employed in the production of pharmaceuticals, cosmetics, plastics, solvents, and preservatives [7,29,47,48]. In pharmacy, ethanol functions as a carrier for active substances, a preservative, a disinfectant, and a technological agent in extraction and stabilization processes of bioactive compounds [36,43,49].

In modern bioenergy technologies, alcoholic fermentation is used for the production of bioethanol—a renewable and low-emission fuel that serves as a sustainable alternative to fossil fuels [14,19,25,48,50]. Fermentation-derived ethanol also plays an important role in green chemistry, acting as a feedstock for the synthesis of bioplastics, bioesters, and other bio-based fuel components, thus supporting the development of a circular and sustainable economy [9,19].

Alcoholic fermentation further holds cultural and historical importance. In many regions worldwide, it forms part of culinary heritage and serves as the basis for traditional beverages such as Japanese sake, Eastern European kvass, South American chicha, and Asian plant-based drinks such as pulque, tuba, and toddy. These processes, often carried out by wild yeasts or naturally occurring microbial consortia, give products unique sensory and cultural characteristics [36,51–54].

https://doi.org/10.3390/molecules31020333

Table 1. Representative industrial applications of alcoholic fermentation in the food, chemical and energy sectors.

Application Area	Example Products	Main Microorganisms	Fermentation Products/Effects	Technological Significance	Typical Product Concentration	Representative References
Alcoholic beverages	Wine, beer, cider, mead, spirits (whisky, rum, vodka)	*Saccharomyces cerevisiae*, *S. bayanus*, *Schizosaccharomyces pombe*	Ethanol, CO_2, esters, higher alcohols, aldehydes	Alcohol production; development of aroma and flavor; natural preservation	Beer: 3–8% (*v/v*); Wine: 10–15% (*v/v*); Spirits (after distillation): >40% (*v/v*)	[8,26,36,46,55–57]
Low-alcohol fermented drinks	Kombucha, kvass, water kefir, ginger beer	*Saccharomyces*, *Zygosaccharomyces*, *Acetobacter*, *Lactobacillus*	Ethanol → acetic acid, CO_2, organic acids	Light carbonation; improved flavor and probiotic potential	<1% (*v/v*) ethanol	[9,12,58–66]
Baking and confectionery	Bread, rolls, yeast doughs	*Saccharomyces cerevisiae*	CO_2, volatile ethanol, esters	Dough leavening; improved texture, flavor, and aroma	Ethanol transient; CO_2 for leavening (ethanol evaporates during baking)	[14,36,43,46,67–69]
Flavor and aroma production	Natural flavors, fruit and dairy aroma compounds	*Saccharomyces*, *Kluyveromyces*, *Torulaspora* spp.	Esters, aldehydes, ketones, higher alcohols	Development of natural flavoring ingredients for food and beverages	Trace levels ($mg{\cdot}L^{-1}$ range)	[8,36,40,43,57,70–73]
Pharmaceutical and chemical industry	Solvents, acetic acid, ethyl acetate, diethyl ether, glycerol	*S. cerevisiae*, *Candida*, *Kluyveromyces* spp.	Ethanol, organic acids, secondary metabolites	Production of solvents, preservatives, and pharmaceutical intermediates	Ethanol typically >95% (*v/v*) after purification	[7,14,25,36,47,74]
Bioenergy sector	Bioethanol fuel, bioplastics, bioesters	*S. cerevisiae*, *Zymomonas mobilis*	Ethanol, CO_2	Renewable energy generation; sustainable fuel and biocomponent production	80–120 $g{\cdot}L^{-1}$ ethanol	[14,25,33,36,47,74,75]
Cultural and traditional fermentation	Sake, chicha, kvass, pulque, toddy, tuba	Wild yeasts, mixed consortia (*Saccharomyces*, *Lactobacillus*)	Ethanol, CO_2, esters, organic acids	Preservation of traditional food heritage; unique sensory profiles	Typically 2–8% (*v/v*) ethanol	[26,36,39,46,53,54,76–79]

Typical ethanol concentrations range from approximately 3–8% (*v/v*) in beer, 10–15% (*v/v*) in wine, <1% (*v/v*) in low-alcohol fermented beverages, and >80–100 $g{\cdot}L^{-1}$ in industrial bioethanol fermentations, reflecting differences in microbial physiology, pathway utilization (EMP versus ED), and process objectives [9,18,24–26,41,64].

3. Acetic Fermentation

Following alcoholic fermentation, acetic fermentation represents a distinct yet functionally connected process in which ethanol is further oxidized by specialized bacteria under aerobic conditions, linking primary sugar catabolism with organic acid production.

3.1. Biochemical Pathway and Key Microorganisms

The production of acetic acid represents a complex, two-step metabolic pathway involving different groups of microorganisms and specialized enzymatic systems. The first stage comprises alcoholic fermentation, occurring under anaerobic conditions, during which simple sugars, such as glucose are converted into ethanol and carbon dioxide by the action of yeasts, mainly of the genus *Saccharomyces*. The subsequent stage, proceeding in the presence of oxygen, involves acetic acid bacteria (AAB) that oxidize ethanol to acetic acid and water [80], according to the reaction:

$$C_2H_5OH + O_2 \rightarrow CH_3COOH + H_2O$$

Both stages require microorganisms with high adaptive capacity, enabling their activity under variable and often unfavorable environmental conditions. During alcoholic

https://doi.org/10.3390/molecules31020333

fermentation, the use of yeast strains with high alcohol tolerance is essential to ensure efficient ethanol production. In the acetic acid fermentation phase, the selection of bacterial strains capable of withstanding both high ethanol concentrations and the accumulation of acetic acid is crucial, as this process demands dual tolerance to both compounds [81].

The most important acetic acid–producing bacteria used in biotechnological vinegar production belong to the genera *Acetobacter*, *Gluconacetobacter*, *Gluconobacter*, and *Komagataeibacter*. These strains exhibit high efficiency in ethanol oxidation to acetic acid and substantial tolerance to product accumulation in the fermentation environment [81,82].

The mechanism of acetic acid fermentation is based on a specialized membrane-bound enzymatic system located in the cytoplasmic membrane of acetic acid bacteria. Two key dehydrogenases are involved in this process: the pyrroloquinoline quinone (PQQ)-dependent alcohol dehydrogenase (ADH) and the acetaldehyde dehydrogenase (ALDH). The first enzyme catalyzes the oxidation of ethanol to acetaldehyde, while the second converts acetaldehyde into acetic acid. Both dehydrogenases are membrane-bound and oriented toward the periplasmic space, allowing direct oxidation of substrates located outside the cytoplasm and the release of products into the extracellular medium [83–86].

Electrons released during ethanol and acetaldehyde oxidation are transferred from PQQ to ubiquinone (UQ), which acts as a mobile electron carrier. Reduced ubiquinone (ubiquinol, UQH_2) donates electrons to ubiquinol oxidase (UOX), which transfers them to molecular oxygen, the terminal electron acceptor. This reaction yields water, while proton translocation across the membrane generates a proton motive force that drives ATP synthesis via ATP synthase [83–86].

In addition to the membrane-bound system, AAB possess a cytoplasmic dehydrogenase complex using $NAD^+/NADP^+$ as cofactors. NAD-dependent alcohol dehydrogenase (NAD-ADH) and NADP-dependent acetaldehyde dehydrogenase (NADP-ALDH) oxidize ethanol that diffuses into the cytoplasm to acetic acid; the latter can be converted into acetyl-CoA, which enters the tricarboxylic acid (TCA) cycle and is fully oxidized to CO_2 and water (Figure 2) [23,83,87]. Thus, the coupling of periplasmic ethanol oxidation to proton motive force generation enables efficient ATP synthesis under high ethanol and acidic conditions, while the combined ethanol and acid tolerance of yeasts and acetic acid bacteria directly underpins their robustness and efficiency in industrial acetic fermentation systems [83,87–89].

Under conditions relevant for acetic acid production, the membrane-bound PQQ-ADH and ALDH dominate, whereas cytoplasmic NAD-ADH and NADP-ALDH are largely inhibited. High ethanol concentrations favor the membrane pathway. As ethanol is depleted, cytoplasmic dehydrogenase activity increases and the contribution of the membrane system declines. This dynamic regulation allows AAB to optimize energy conservation and adapt to shifting environmental conditions [83,85,87,90]. Together, these metabolic adaptations explain why acetic acid bacteria are uniquely suited for industrial vinegar production and other acetic acid-based bioprocesses.

Moreover, with traditional strain selection, acetic acid bacteria have increasingly become targets of metabolic engineering and systems-level optimization aimed at improving acetic acid productivity, acid tolerance, and process robustness [83,87,91,92]. Recent studies have focused on enhancing ethanol and acetic acid tolerance through modification of membrane lipid composition, overexpression of stress response regulators, and optimization of respiratory chain components involved in periplasmic ethanol oxidation [83,85,90]. Advances in genome sequencing and comparative genomics of *Acetobacter*, *Komagataeibacter*, and *Gluconacetobacter* species have enabled the identification of key genes associated with acid resistance, oxidative stress defense, and efficient energy conservation [93].

 https://doi.org/10.3390/molecules31020333

Emerging synthetic biology approaches, including targeted gene deletions, controlled overexpression of PQQ-dependent dehydrogenases, and adaptive laboratory evolution, have further improved strain performance under high-acidity and high-ethanol conditions relevant to industrial vinegar production [94]. Coupled with bioprocess intensification strategies and real-time monitoring of dissolved oxygen and redox status, these developments highlight the transition of acetic fermentation from a largely empirical process to a rationally engineered biotechnological platform [84,95].

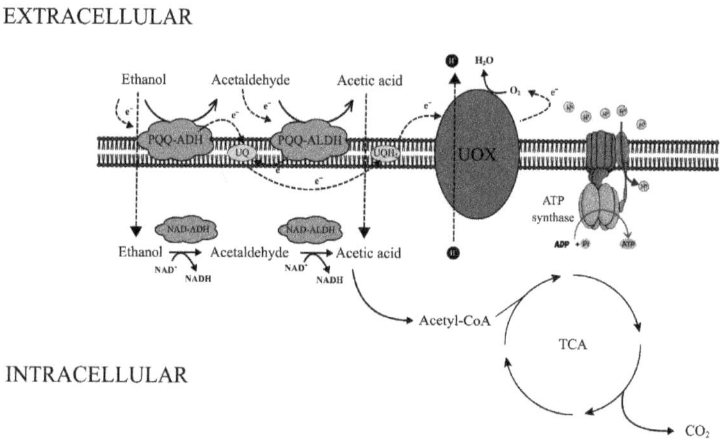

Figure 2. Schematic representation of the membrane-bound and cytosolic respiratory chains involved in ethanol oxidation in acetic acid bacteria. In this system, periplasmic PQQ-dependent dehydrogenases catalyze the oxidation of ethanol and acetaldehyde, transferring electrons through ubiquinone to the terminal ubiquinol oxidase (UOX), which drives proton translocation and ATP formation. Intracellularly, acetate is further converted to acetyl-CoA and enters the tricarboxylic acid (TCA) cycle. Scheme adapted from [83].

3.2. *Industrial Applications*

Building on the biochemical mechanisms described above, acetic fermentation has been widely exploited in both food and non-food industries due to its ability to generate acetic acid and related metabolites with preservative, sensory, and functional properties. While vinegar production remains the most recognized application, acetic fermentation should be viewed as a platform bioprocess supporting multiple industrial sectors.

Acetic fermentation plays a crucial role across the food industry, extending beyond conventional vinegar production. The most widespread application is the manufacture of edible vinegars, which are aqueous acetic acid solutions typically containing 4–15% (*w*/*v*) acetic acid. Depending on the raw material and technology used, spirit vinegar (from ethanol), wine vinegar, apple cider vinegar, and rice vinegar can be distinguished, each with characteristic sensory and nutritional properties [61,82]. Acetic acid content determines preservative, antimicrobial and organoleptic attributes. Vinegar functions as both a condiment and a natural preservative due to its ability to lower pH and inhibit undesirable microflora. Commercial vinegar production commonly involves a two-stage fermentation: alcoholic fermentation by yeasts (e.g., *S. cerevisiae*), followed by acetic oxidation by AAB [36,54]. A summary of the major industrial applications of acetic fermentation and acetic acid across different sectors is provided in Table 2.

Acetic fermentation also underpins the production of low-alcohol fermented beverages such as kombucha, water kefir, and kvass. In these systems, AAB form symbiotic

https://doi.org/10.3390/molecules31020333

consortia with yeasts (SCOBY), producing not only acetic and gluconic acids but also bacterial cellulose (e.g., *Komagataeibacter xylinus*). These metabolites contribute to acidity, carbonation, antioxidant capacity, and microbiota-modulating effects, enhancing the functional value of these beverages [9,54].

In the food industry, acetic fermentation is further exploited for the biosynthesis of natural flavors and esters. Acetic acid, acetaldehyde, and ethyl esters derived from fermentation are widely used in seasoning, dairy, and bakery applications as natural aromatic components. Acidification through acetic acid production also sup-ports biopreservation by inhibiting spoilage and pathogenic microorganisms, extending the shelf life of pickles, sauces, dressings, and marinades [43,61,81,82,96]. In this context, acetic fermentation provides a natural alternative to synthetic preservatives, aligning with consumer demand for clean-label and minimally processed foods.

At the industrial scale, acetic acid represents a high-volume platform chemical with global production exceeding 15 million tons annually, primarily for the synthesis of vinyl acetate monomer (VAM), purified terephthalic acid (PTA), and acetic anhydride [86,97]. While bulk acetic acid is still predominantly produced via petrochemical methanol carbonylation, food-grade and specialty acetic acid is exclusively obtained through microbial fermentation, and bio-based acetic acid is increasingly explored as a renewable alternative for polymer, textile, and pharmaceutical applications [86,98]. In the chemical industry, acetic acid serves as a precursor for the synthesis of polyethylene terephthalate (PET), cellulose acetate, and polyvinyl acetate, which are used in beverage packaging, photographic materials, coatings, and adhesives [99–101].

In the textile industry, acetic acid is used in both fiber production and fabric processing. Owing to its chemical properties, it plays a key role in dyeing processes, acting as a mordant that enhances color fastness and resistance to fading. It also functions as a pH regulator, adjusting the dye bath environment to meet the specific requirements of dyes and fiber types. Additionally, it is used in fabric cleaning, degreasing, and textile printing processes, where it facilitates dye binding and enhances pattern intensity and durability [24,34,87,90,102].

In cosmetics, acetic acid and its derivatives (esters and acetate salts) are incorporated into shampoos, conditioners, perfumes, and skincare formulations as pH adjusters, antimicrobial agents, and fragrance components [1,72,90,97]. In the pharmaceutical industry, acetic acid is used as a solvent, antiseptic, and synthetic intermediate in esterification and acetylation reactions, enabling the production of compounds with therapeutic properties. Due to its antimicrobial activity, it is also applied in wound cleansing and treatment of superficial infections [1,90].

Table 2. Industrial applications of acetic fermentation and acetic acid in the food, chemical, textile, cosmetic and pharmaceutical sectors.

Application Area	Example Products/ Applications	Main Microorganisms	Fermentation Products/Effects	Technological Significance/Outcomes	Representative References
Vinegar production	Spirit vinegar, wine vinegar, apple cider vinegar, rice vinegar	*Acetobacter aceti*, *Komagataeibacter xylinus*, *Gluconacetobacter europaeus*	Acetic acid, water	Conversion of ethanol to acetic acid; pH reduction; flavor and aroma development; natural food preservation	[36,61,81,84,85,103,104]
Low-alcohol fermented beverages	Kombucha, water kefir, kvass	Symbiotic consortia of *Acetobacter*, *Komagataeibacter*, *Saccharomyces*, *Zygosaccharomyces*	Acetic acid, gluconic acid, bacterial cellulose	Functional beverages with antioxidant, detoxifying, and probiotic properties; SCOBY formation	[9,12,58–63,65,66]

 https://doi.org/10.3390/molecules31020333

Table 2. *Cont.*

Application Area	Example Products/ Applications	Main Microorganisms	Fermentation Products/Effects	Technological Significance/Outcomes	Representative References
Flavor and ester biosynthesis	Natural flavor concentrates, fruit and dairy flavoring compounds	*Acetobacter*, *Gluconobacter*, *Saccharomyces* spp.	Acetic acid, acetaldehyde, ethyl acetate, ethyl lactate	Generation of natural aromatic compounds for the dairy, seasoning, and bakery industries	[8,36,40,43,70,72,105]
Food preservation and bioprotection	Pickles, sauces, salad dressings, marinades	*Acetobacter* spp., *Lactobacillus* spp.	Acetic acid (acidification)	Growth inhibition of spoilage and pathogenic microorganisms; natural biopreservation; shelf-life extension	[1,4,6,36,65,106–110]
Chemical industry	Synthesis of polyethylene terephthalate (PET), cellulose acetate, polyvinyl acetate	Industrial oxidation systems using *Acetobacter* spp. or catalytic pathways	Acetic acid	Precursor for plastics, adhesives, and synthetic fibers; solvent in esterification and acetylation processes	[14,47,75,111,112]
Textile industry	Fiber dyeing and finishing, textile printing, degreasing	Industrial-grade acetic acid (chemical product)	Acetic acid	pH regulation in dye baths; mordant improving color fixation and fastness; cleaning and degreasing agent	[7,61,113]
Cosmetic industry	Hair care products, perfumes, skincare formulations	(industrial acetic acid and its esters)	Acetic acid, acetate esters, salts	Ingredient in shampoos and conditioners; pH adjuster; perfume component; antimicrobial additive	[47,61,95,104,110]
Pharmaceutical industry	Antiseptics, drug synthesis intermediates, solvents	(industrial acetic acid)	Acetic acid, acetyl derivatives	Reactant in acetylation and esterification; disinfectant and antimicrobial agent in medicinal preparations	[7,47,74,95,103,112]
Household and cleaning products	Window cleaners, dishwashing liquids, descalers	—	Diluted acetic acid	Removal of limescale, grease, and mineral deposits; eco-friendly cleaning and descaling agent	[61,75,95,103]

In household and institutional applications, diluted acetic acid is widely used as an eco-friendly cleaning agent. It is a common component of window cleaners, dishwashing liquids, and descaling formulations, where it effectively removes grease, mineral deposits, limescale, and rust [61,90,98].

Taken together, these examples demonstrate that acetic fermentation is not limited to vinegar production but constitutes a versatile biotechnological platform linking food processing with chemical manufacturing, materials science, pharmaceuticals, and sustainable consumer products.

 https://doi.org/10.3390/molecules31020333

4. Butyric Fermentation

In contrast to acetic fermentation, which relies on aerobic ethanol oxidation, butyric fermentation proceeds strictly under anaerobic conditions and redirects carbohydrate-derived carbon toward reduced short-chain fatty acids.

4.1. Biochemical Pathway and Key Microorganisms

Butyric fermentation is an anaerobic pathway of carbohydrate degradation in which the main end product is butyric acid (butanoic acid, $CH_3CH_2CH_2COOH$), a short-chain volatile fatty acid. This process is carried out primarily by bacteria of the genera *Clostridium, Butyrivibrio* and *Butyribacterium*. Butyrate production has also been described in *Coprococcus, Eubacterium, Fusobacterium, Megasphaera, Roseburia* and *Sarcina*. Among these, *Clostridium* species are of particular importance due to their high fermentative potential and capacity to synthesize butyric acid from diverse substrates, including glucose, xylose, lactose and glycerol [114–116].

Metabolically, butyric fermentation begins with glycolysis via the EMP pathway, yielding pyruvate. Under strict anaerobic conditions, pyruvate is oxidatively decarboxylated to acetyl-CoA by pyruvate:ferredoxin oxidoreductase (PFOR), which functionally replaces the pyruvate dehydrogenase complex. This conversion is accompanied by the release of CO_2 and the generation of reduced ferredoxin, with excess reducing equivalents dissipated as molecular hydrogen (H_2). Acetyl-CoA serves as a central branching point. A fraction is converted to acetic acid via phosphotransacetylase (PTA) and acetate kinase (AK), generating ATP through substrate-level phosphorylation. The remaining acetyl-CoA condenses to acetoacetyl-CoA and is subsequently reduced to butyryl-CoA via β-hydroxybutyryl-CoA and crotonyl-CoA. Key enzymes include thiolase, β-hydroxybutyryl-CoA dehydrogenase, crotonase and butyryl-CoA dehydrogenase (linked to an electron-transferring flavoprotein, ETF) [114–117].

In the terminal step, butyryl-CoA is converted to butyric acid either via phosphotransbutyrylase (PTB) and butyrate kinase (BUK), through a butyryl phosphate intermediate, or via a CoA-transferase (CTF) that transfers CoA to acetate, producing butyrate and regenerating acetyl-CoA (Figure 3). Only some *Clostridium* strains possess both routes, and their relative contributions to total butyrate formation remain incompletely resolved [115,116,118]. Up to three moles of ATP per mole of glucose can be generated in butyric fermentation, which is relatively high for an anaerobic process. Most *Clostridium* species are heterofermentative, producing butyrate, acetate and, in smaller quantities, ethanol or isopropanol [117,118]. These metabolic characteristics explain both the technological potential and the process-related challenges associated with butyric fermentation in industrial systems.

Recent advances in metabolic engineering and synthetic biology have substantially expanded the biotechnological potential of butyric fermentation, particularly in solventogenic and butyrate-producing *Clostridium* species. Engineering efforts have focused on redirecting carbon flux toward butyrate and butanol by modulating key nodes such as the acetyl-CoA branch point (acetate vs. butyrate formation), strengthening reducing equivalent availability (ferredoxin/NADH balance), and reducing competing pathways that lead to lactate, acetone, or excessive acetate formation [115,116,119,120]. At the same time, CRISPR/Cas-based genome editing, together with systems-level analyses (transcriptomics, proteomics, and genome-scale metabolic models), has enabled targeted modification of acidogenesis/solventogenesis regulation, improved tolerance to butyric acid and butanol, and enhanced utilization of diverse substrates, including lignocellulosic sugars and glycerol [121,122]. Adaptive laboratory evolution and dynamic pathway control are

https://doi.org/10.3390/molecules31020333

increasingly combined with in situ product removal strategies to mitigate end-product inhibition and improve process productivity, supporting the transition of butyric fermentation from a classical anaerobic pathway to a rationally optimized platform for the sustainable production of short-chain fatty acids and bio-based solvents [123–125].

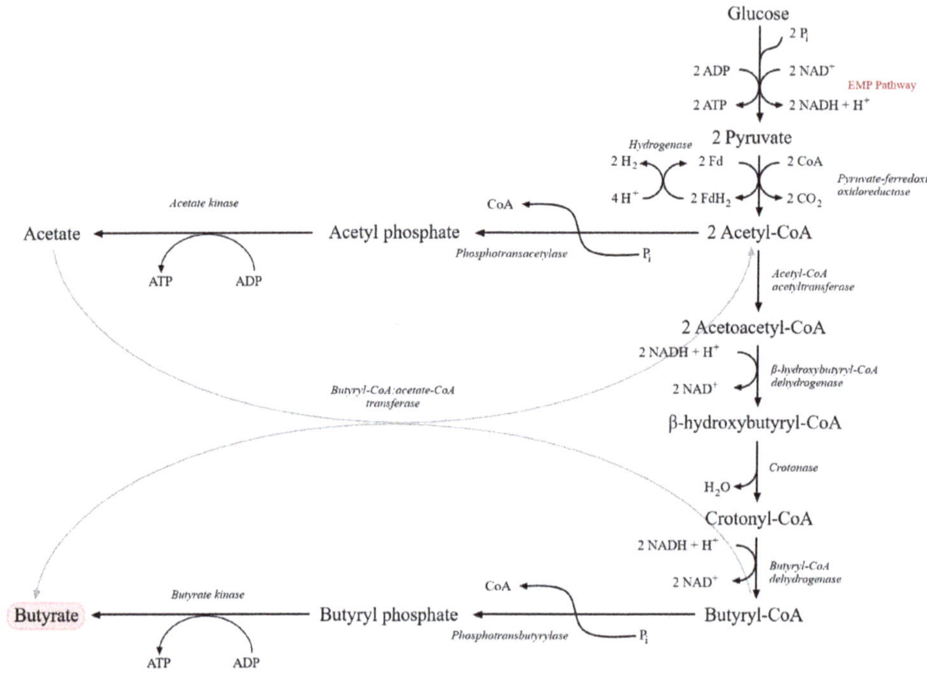

Figure 3. Metabolic pathway of butyric acid fermentation originating from glucose catabolism via the Embden–Meyerhof–Parnas (EMP) pathway. Glucose is converted to pyruvate, which is subsequently transformed into acetyl-CoA and further metabolized through acetoacetyl-CoA, β-hydroxybutyryl-CoA, crotonyl-CoA, and butyryl-CoA intermediates. Final conversion into butyrate occurs via butyryl phosphate. Parallel reactions involving acetate formation and ATP generation through substrate-level phosphorylation are also shown. Key enzymatic steps and electron transfer reactions involved in NADH oxidation and hydrogen production are indicated (adapted from [120]).

4.2. Industrial Applications

In industrial practice, butyric fermentation has been exploited across the food, chemical, energy, and pharmaceutical sectors due to its capacity to generate butyrate, solvents, and other value-added metabolites. Although pure butyric acid has a pungent odor, its esters exhibit pleasant fruity aromas and are widely used as flavoring agents in food, confectionery and perfumery. In dairy technology, uncontrolled butyric fermentation by *Clostridium tyrobutyricum* can cause late blowing in cheese, leading to quality defects and economic losses. Conversely, controlled butyrate formation contributes to characteristic flavor profiles in some traditional cheeses [126–128]. Representative industrial applications of butyric fermentation and butyric acid are summarized in Table 3.

In the chemical and materials industries, butyric acid and its derivatives are important intermediates in the synthesis of plastics and textile fibers, improving resistance to heat, light and mechanical stress. Cellulose acetate butyrate (CAB), a thermoplastic material

https://doi.org/10.3390/molecules31020333

derived from butyric acid, combines flexibility, UV stability and solubility in organic solvents, making it suitable for durable coatings, films and specialty polymers [47,114,116].

Butyric fermentation also underlies the acetone—butanol—ethanol (ABE) process, in which *Clostridium acetobutylicum* and *C. beijerinckii* produce solvents such as butanol, acetone and ethanol. Butanol is regarded as a promising advanced biofuel due to its high energy density and favorable combustion properties, emphasizing the importance of butyric fermentation in renewable energy systems [7,34,116].

Butyrate and its salts (e.g., sodium and calcium butyrate) are widely used in pharmaceuticals, animal nutrition, cosmetics and the synthesis of bioplastics and bioesters. In animal feed, butyrate-based additives serve as alternatives to antibiotic growth promoters, supporting gut health, nutrient absorption and immune function [75,114,115,129].

In environmental biotechnology, butyrate-producing bacteria participate in the degradation of biomass and organic waste, contributing to biogas generation and the production of organic acids that can be upgraded to biofuels or biochemical intermediates [75,123,130]. *Clostridium butyricum* is also considered a probiotic capable of producing butyrate in the colon, where it supports epithelial cell metabolism, exerts anti-inflammatory and anticarcinogenic effects, and promotes a beneficial microbiota composition [11,129,131].

In medicine, butyric acid has therapeutic potential in gastrointestinal diseases, colorectal cancer and hemoglobinopathies [132–134]. It induces differentiation and reduces proliferation of cancer cells, promotes regulatory T cell (Treg) generation and modulates immune responses [133,135,136]. By enhancing mitochondrial function, oxidative phosphorylation and fatty acid β-oxidation, butyrate also shows neuroprotective properties [137,138]. Several butyrate derivatives with antithyroid, anesthetic and vasoconstrictive activities have been developed, and new acyloxyalkyl butyrate prodrugs are under clinical investigation [139,140].

Table 3. Representative applications of butyric fermentation and butyric acid in food, energy, chemical and pharmaceutical industries.

Application Area	Example Products/ Applications	Main Microorganisms	Fermentation Products/Effects	Technological Significance/Outcomes	Representative References
Food and flavor industry	Butter flavoring, cheese production, aroma compounds, fruit esters	*Clostridium butyricum*, *Clostridium tyrobutyricum*, *Butyrivibrio fibrisolvens*	Butyric acid, ethyl butyrate, butyl butyrate	Natural flavor generation (buttery, fruity notes); improvement of aroma profiles; undesirable in cheese spoilage ("late blowing")	[43,114,115,117, 128,141–145]
Chemical and materials industry	Cellulose acetate butyrate (CAB), plasticizers, textile fibers	*Clostridium acetobutylicum*, *Clostridium beijerinckii*	Butyric acid, butanol, acetone, esters	Synthesis of thermoplastics, coatings, and resins; improved material flexibility, UV stability, and solvent resistance	[25,33,47,74,75,111, 114,143]
Biofuel and solvent production (ABE process)	Butanol, acetone, ethanol	*Clostridium acetobutylicum*, *C. beijerinckii*, *C. pasteurianum*	Butanol, acetone, ethanol	Production of renewable solvents and biofuels; butanol as a high-energy, low-volatility gasoline substitute	[14,33,47,75,111, 112,114,143]
Animal nutrition and feed additives	Livestock feed, poultry supplements	*Clostridium butyricum*, *Butyrivibrio fibrisolvens*	Sodium butyrate, calcium butyrate	Replacement for antibiotic growth promoters; enhancement of gut health, nutrient absorption, and immunity	[45,100,112,115, 128,134,146,147]

 https://doi.org/10.3390/molecules31020333

Table 3. *Cont.*

Application Area	Example Products/ Applications	Main Microorganisms	Fermentation Products/Effects	Technological Significance/Outcomes	Representative References
Pharmaceutical and medical applications	Therapeutics for gut disorders, cancer, hemoglobinopathies	*Clostridium butyricum*	Butyric acid, butyrate derivatives	Anti-inflammatory, anticarcinogenic, and neuroprotective effects; induction of cell differentiation; modulation of immune response	[43,100,115,128, 134,146,148–150]
Probiotic and microbiome modulation	Probiotic supplements, intestinal health products	*Clostridium butyricum*, *Butyrivibrio fibrisolvens*	Butyric acid (SCFA)	Regulation of gut microbiota; stimulation of epithelial regeneration; trophic effect on colonocytes	[45,100,110,112, 115,134,146,149, 151,152]
Environmental biotechnology	Biogas and biohydrogen production, waste valorization	*Clostridium butyricum*, *C. pasteurianum*	H_2, CO_2, volatile fatty acids	Conversion of organic waste into biogas and organic acids; sustainable bioenergy recovery and waste reduction	[33,50,75,109,111, 114,123,130,143, 153]
Cosmetic industry	Skin and hair care formulations	Industrially derived butyric esters	Butyric esters, butyrate salts	Use in fragrance formulations; moisturizing and conditioning properties; pH regulation	[95,100,110,115, 134,146]

5. Lactic Fermentation

Unlike butyric fermentation, which primarily directs carbon toward reduced short-chain fatty acids and solvents, lactic fermentation channels carbohydrate metabolism toward lactate formation and rapid acidification of the environment.

5.1. Metabolic Pathways and Key Microbial Groups

Lactic acid fermentation is carried out by both homo- and heterofermentative lactic acid bacteria (LAB) [4,54,154]. Homolactic LAB metabolize hexoses (primarily glucose) via the EMP pathway (Figure 4), yielding two molecules of lactate per molecule of glucose [7,36,155,156]. Glucose is converted to pyruvate, which is reduced to lactate by lactate dehydrogenase with concomitant re-oxidation of NADH to NAD^+, resulting in a net gain of two ATP per mole of glucose [157]. Strict homolactic fermentation of glucose does not generate CO_2 [158]. Representative homolactic LAB include *Lactiplantibacillus plantarum* (formerly *Lactobacillus plantarum*), *Pediococcus pentosaceus*, *Lactococcus lactis* and *Streptococcus thermophilus* [112,142,156,159].

Homolactic pathway:

$C_6H_{12}O_6$ (glucose) $\rightarrow$ 2 CH_3-CHOH-COOH (lactic acid)

In contrast, heterofermentative LAB can ferment both hexoses and pentoses. For hexoses, they employ the phosphoketolase branch of the pentose phosphate pathway (PPP; Figure 3), which requires ribulose-5-phosphate 3-epimerase and phosphoketolase. This route typically yields equimolar lactate, ethanol (or acetate) and CO_2 [106,157,158,160,161]. Obligate heterofermenters lack key EMP enzymes—most notably fructose-bisphosphate aldolase and triose phosphate isomerase—thus channeling carbon exclusively through the phosphoketolase pathway. Typical heterofermentative genera include *Leuconostoc*, *Oenococcus*, *Levilactobacillus* (e.g., *L. brevis*) and *Limosilactobacillus* (e.g., *L. fermentum*) [142,158,159,161].

Heterofermentative phosphoketolase pathway:

https://doi.org/10.3390/molecules31020333

$C_6H_{12}O_6$ (glucose) → CH_3-CHOH-COOH (lactic acid) + C_2H_5OH (ethanol) + CO_2 [106,157,161].

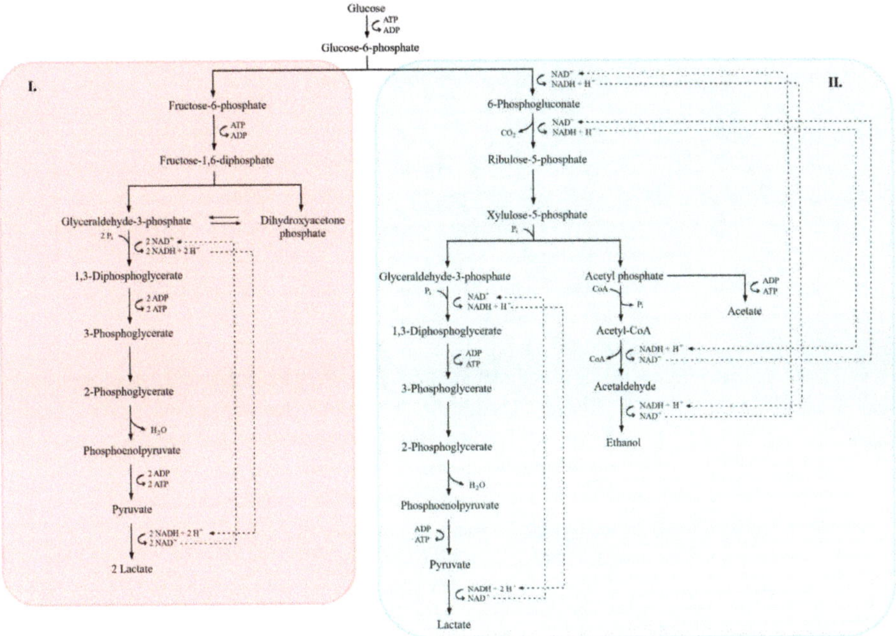

Figure 4. Comparative overview of homofermentative (**I**) and heterofermentative (**II**) lactic acid fermentation pathways. In the homofermentative route (**I**), glucose is metabolized via the Embden–Meyerhof–Parnas (EMP) pathway, yielding pyruvate and subsequently lactate as the main end product, along with ATP generation. In the heterofermentative route (**II**), glucose is processed through the 6-phosphogluconate/phosphoketolase pathway, leading to the formation of lactate, ethanol or acetate, and CO_2. Key intermediates, redox reactions, and ATP-yielding steps are indicated. Dashed lines denote $NADH/NAD^+$ electron transfer pathways, while the shaded areas highlight the distinct metabolic branches of each fermentation type (adapted from [7,36,156]).

(In the presence of external electron acceptors, ethanol can be replaced by acetate with additional ATP formation) [158,161]. The balance between homo- and heterofermentative metabolism, ATP yield, and by-product formation directly influences both the technological performance of lactic acid bacteria and the characteristics of fermented products [106,142,160,161].

In recent years, lactic acid bacteria have become prominent targets of metabolic engineering and synthetic biology aimed at improving lactic acid yield, optical purity, robustness, and industrial scalability [155,156,158]. Engineering strategies have focused on controlling the stereospecificity of lactate dehydrogenases to selectively produce optically pure L- or D-lactic acid, which is critical for high-performance polylactic acid (PLA) synthesis [162]. Additional approaches include redirection of carbon flux toward lactate by minimizing by-product formation (e.g., ethanol, acetate, or CO_2), enhancement of acid tolerance through membrane and stress-response engineering, and optimization of redox balance [112,157,159].

Advances in genome sequencing, CRISPR/Cas-based genome editing, and systems-level analyses have enabled precise modification of industrial LAB strains such as *Lactiplantibacillus plantarum*, *Lactococcus lactis*, and *Streptococcus thermophilus*. These tools

 https://doi.org/10.3390/molecules31020333

have facilitated the development of strains with improved substrate utilization, resistance to bacteriophages, and stable performance under low-pH and high-product conditions [27,158,163]. Together with adaptive laboratory evolution and process-integrated control strategies, these innovations support the transition of lactic fermentation from a traditional food process to a rationally engineered platform for sustainable production of organic acids, biopolymers, and functional food ingredients [123,130].

5.2. *Industrial and Health-Related Applications*

From an industrial and technological perspective, lactic fermentation has been widely applied in food processing, health-oriented products, and industrial biotechnology due to its efficiency, safety, and functional metabolite profile. LAB convert carbohydrates, mainly glucose and lactose, into lactic acid and secondary metabolites such as diacetyl, acetaldehyde, hydrogen peroxide and bacteriocins [7,36,157,158]. These compounds contribute to the sensory profile, stability and microbiological safety of fermented foods by lowering pH and inhibiting pathogenic and spoilage microorganisms [110,142]. The major applications of lactic fermentation and lactic acid across food, medical, and materials-related sectors are summarized in Table 4.

Lactic fermentation underlies the production of numerous fermented foods, including dairy products (yogurt, kefir, buttermilk, cheese), vegetables (sauerkraut, kimchi, pickled cucumbers), meats (dry-cured sausages, salami), cereal-based products (sourdough) and plant-based beverages (soy, oat and beetroot drinks) [9,79]. LAB such as *Lactobacillus delbrueckii* subsp. *bulgaricus*, *S. thermophilus*, *L. plantarum* and *Leuconostoc mesenteroides* are essential for improving quality, nutritional value and sensory characteristics [129,142,159].

Beyond traditional fermentation, LAB are central to the development of functional and probiotic foods. Strains such as *Lactobacillus rhamnosus* GG, *L. casei* and *Bifidobacterium bifidum* are associated with health benefits, including support of the intestinal microbiota, enhanced mineral absorption, synthesis of B-group vitamins and modulation of immune responses [110,129,146,149,154,164,165]. In the food industry, lactic acid acts as a natural preservative, acidulant, pH regulator, flavor enhancer, cryoprotectant and prebiotic component [106,166]. Owing to its antibacterial and antioxidant properties, it is also used as a disinfectant and stabilizer in fermented food production [71,102,110,155,158].

In the chemical industry, lactic acid is a renewable platform chemical for the production of biopolymers, particularly polylactic acid (PLA), which is an environmentally friendly alternative to petroleum-based plastics. PLA is used in biodegradable films, fibers and packaging materials, contributing to green chemistry and circular economy strategies [7,14,47,50]. Lactic acid is also used in the synthesis of lactate esters, acrylic acid, propylene oxide, acetaldehyde and propylene glycol, which are widely applied in cosmetic, pharmaceutical and chemical products [47,155].

In dermatology and cosmetology, lactic acid is valued for its moisturizing, exfoliating and brightening properties. It is incorporated into anti-acne, anti-ageing and hydrating formulations as well as intimate hygiene products. Acting as a humectant, it binds water in the stratum corneum and improves skin elasticity and appearance [49,155,166].

In medicine and pharmacy, lactic acid and its salts (lactates) are widely used due to their biocompatibility, low toxicity and involvement in natural metabolic pathways [7,155]. They form key components of infusion solutions such as Ringer's lactate, which help maintain acid–base balance and restore electrolytes in patients with dehydration or metabolic acidosis [7]. Calcium, magnesium and zinc lactates are used in dietary supplements and pharmaceutical formulations due to their high bioavailability and stabilizing properties [110,146,165].

A major pharmaceutical application of lactic acid is its role as a monomer for biodegradable polymers, including PLA and its copolymers such as poly(lactide-co-glycolide) (PLGA). These materials are used in biodegradable implants, surgical sutures

https://doi.org/10.3390/molecules31020333

and controlled drug delivery systems, enabling gradual and targeted release of active compounds, particularly in oncology, endocrinology and immunotherapy [7,14,36,47]. Lactic acid-derived materials are also used to fabricate biodegradable scaffolds for tissue engineering, supporting bone, muscle and skin regeneration [7,47,167]. Their biodegradability and renewable origin align with sustainable development principles and offer an eco-friendly alternative to conventional synthetic polymers [9,14,50,99,101].

Table 4. Lactic fermentation and lactic acid: major applications in food, cosmetics, pharmaceuticals and materials science.

Application Area	Function/Role of Lactic Fermentation or Lactic Acid	Key Microorganisms/ Compounds	Industrial or Health Relevance	Selected References
Food Fermentation and Preservation	Conversion of carbohydrates (mainly glucose and lactose) into lactic acid, diacetyl, acetaldehyde, hydrogen peroxide, and bacteriocins; acidification of the environment inhibits spoilage and pathogenic microorganisms.	*Lactobacillus delbrueckii* subsp. *bulgaricus*, *Streptococcus thermophilus*, *Lactiplantibacillus plantarum*, *Leuconostoc mesenteroides*	Production of fermented foods (yogurt, kefir, cheese, sauerkraut, kimchi, sourdough); enhanced safety, shelf-life, and sensory quality.	[5,13,27,36,71, 77,79,106,155, 156,159,168–174]
Functional and Probiotic Foods	Support of intestinal microbiota; synthesis of B-group vitamins; enhancement of mineral absorption; immune modulation.	*Lactobacillus rhamnosus* GG, *L. casei*, *Bifidobacterium bifidum*	Development of health-promoting foods with probiotic activity; improvement of gastrointestinal and immune health.	[6,11,96,110, 129,131,143, 146,147,149, 151,152,154, 165,175–185]
Food Industry (Technological Additive)	Acts as a natural preservative, acidulant, pH regulator, flavor enhancer, cryoprotectant, and prebiotic component; provides antimicrobial and antioxidant protection.	Lactic acid; bacteriocins (e.g., nisin); hydrogen peroxide	Improves food quality, safety, and texture; stabilizes emulsions; inhibits spoilage flora.	[4,17,22,28,36, 71,72,106,127, 155,157–159,161,186–188]
Biopolymer Production	Precursor for polylactic acid (PLA) synthesis, an eco-friendly biodegradable polymer replacing petrochemical plastics.	Lactic acid (from *Lactobacillus* fermentation)	PLA used in films, fibers, and packaging materials; supports circular economy and green chemistry.	[7,14,25,33,43, 47,74,75,124, 161]
Chemical Industry	Intermediate for synthesis of lactate esters, acrylic acid, propylene oxide, acetaldehyde, and propylene glycol.	Lactic acid and its derivatives	Production of solvents, adhesives, surfactants, and coatings for cosmetics, pharmaceuticals, and polymers.	[7,14,47,74,75, 155,161,166]
Cosmetics and Dermatology	Humectant, exfoliant, and brightening agent in skincare; promotes hydration, elasticity, and renewal of the stratum corneum.	Lactic acid and its salts	Used in anti-aging, anti-acne, and moisturizing formulations; improves skin tone and appearance.	[107,110,146, 148,155]
Medicine and Pharmacy	Ingredient in infusion solutions (e.g., Ringer's lactate); component of mineral supplements; stabilizer in drug formulations.	Calcium, magnesium, and zinc lactates	Rehydrates and maintains acid–base balance; enhances bioavailability of minerals.	[7,47,148,155]
Pharmaceutical Biotechnology	Building block for biodegradable polymers such as PLA and PLGA used in drug delivery systems, sutures, and implants.	Lactic acid monomers and copolymers	Enables controlled drug release, tissue compatibility, and gradual biodegradation in vivo.	[7,33,43,47, 112,161]
Tissue Engineering and Regenerative Medicine	Source material for biodegradable scaffolds supporting bone, muscle, and skin regeneration.	PLA and its copolymers; lactic acid-based composites	Sustainable alternative to petrochemical polymers; biocompatible matrices for tissue growth.	[7,9,14,33,43, 47,74,161]

https://doi.org/10.3390/molecules31020333

6. Propionic Fermentation

In contrast to lactic fermentation, propionic fermentation represents a metabolically linked process in which lactate and other reduced substrates are further converted into propionic acid, acetate, and CO_2 under anaerobic conditions.

6.1. Wood–Werkman (Dicarboxylic Acid) Pathway and Key Microorganisms

Propionic acid fermentation is an anaerobic metabolic process in which substrates such as glucose, glycerol or lactic acid are converted into propionic acid, accompanied by acetic acid and CO_2 formation [111,119]. This process is characteristic of several groups of anaerobic bacteria, among which *Propionibacterium* species have the highest industrial relevance. These microorganisms exhibit high fermentative efficiency and a broad substrate spectrum (Table 5) [189]. Other bacteria capable of propionic fermentation include *Clostridium propionicum* and selected species of *Veillonella, Selenomonas, Megasphaera, Fusobacterium* and *Bacteroides,* which typically show lower yields and a narrower substrate range [111,119].

Table 5. Propionate-producing microorganisms, their primary substrates and fermentation products.

Genus/Species	Primary Substrate(s)	Main Fermentation Products	References
Propionibacterium acidipropionici	Glucose, sucrose, lactose, lactate, glycerol	Propionate, acetate, succinate, CO_2	[111,144,189]
Propionibacterium freudenreichii	Lactate, glucose	Propionate, acetate, CO_2	[111,144,189,190]
Propionibacterium shermanii	Lactate, glucose	Propionate, acetate, CO_2	[111,144]
Acidipropionibacterium thoenii	Lactate, glucose	Propionate, acetate	[111,191]
Acidipropionibacterium jensenii	Lactate	Propionate, acetate	[111,191]
Acidipropionibacterium microaerophilum	Lactate	Propionate, acetate	[191]
Clostridium propionicum	Glycerol, lactate, alanine	Propionate, succinate, acetate, formate, n-propanol	[192,193]
Clostridium neopropionicum	Ethanol	Propionate, acetate	[121,193]
Clostridium homopropionicum	Glucose	Propionate	[121]
Bacteroides fragilis	Glucose	Acetate, propionate, lactate, succinate, formate	[194]
Bacteroides ruminicola	Glucose, cellobiose	Acetate, propionate	[194,195]
Prevotella ruminicola	Glucose, cellulose hydrolysates	Propionate, acetate, succinate	[195]
Veillonella parvula	Lactate	Propionate, acetate, CO_2, H_2	[196]
Veillonella alcalescens	Lactate	Propionate, acetate, CO_2	[196]
Selenomonas ruminantium	Lactate, glucose	Propionate, lactate, acetate	[195,197]
Selenomonas sputigena	Glucose	Propionate, acetate	[197]
Selenomonas lipolytica	Glycerol, fatty acids	Propionate, acetate	[198]
Megasphaera elsdenii	Lactate	Acetate, propionate, butyrate	[148,150,199]
Megasphaera micronuciformis	Lactate	Propionate, butyrate, acetate	[148]
Fusobacterium necrophorum	Lactate	Acetate, propionate, butyrate	[150]
Anaerovibrio lipolyticus	Lipids, glycerol	Propionate, acetate, succinate	[200]

Stoichiometrically, propionic acid fermentation can be expressed as follows:
For glucose,

$$C_6H_{12}O_6 \rightarrow 4/3\ CH_3CH_2COOH + 2/3\ CH_3COOH + 2/3\ CO_2 + 4/3\ H_2O + 4\ ATP$$

For lactic acid,

$$CH_3CHOHCOOH \rightarrow 2/3\ CH_3CH_2COOH + 1/3\ CH_3COOH + 1/3\ CO_2 + 2/3\ H_2O + ATP$$

https://doi.org/10.3390/molecules31020333

Propionic acid can be produced via three main metabolic routes: (1) the Wood–(dicarboxylic acid) cycle, (2) the acrylate pathway and (3) the 1,2-propanediol pathway [189]. In *Propionibacterium* spp., the predominant route is the Wood– pathway. It starts from pyruvate, generated via glycolysis, which can follow two alternative routes:

(i) Carboxylation to oxaloacetate: this initiates the Wood–Werkman cycle and leads to propionic acid. (ii) Decarboxylation to acetyl-CoA: this serves as the precursor for acetic acid.

In the first route, pyruvate is converted to oxaloacetate in a biotin-dependent carboxyl transfer reaction catalyzed by methylmalonyl-CoA carboxytransferase, which transfers a carboxyl group from methylmalonyl-CoA to pyruvate, yielding oxaloacetate and propionyl-CoA. Oxaloacetate is then reduced to malate, dehydrated to fumarate and reduced to succinate. Succinate is converted to succinyl-CoA by succinyl-CoA synthetase and rearranged via a vitamin B_{12}-dependent methylmalonyl-CoA mutase to methylmalonyl-CoA and back to propionyl-CoA. Finally, propionyl-CoA is converted to propionic acid with the release of CoA by a CoA-transferase [74,88,98,118,189,201].

Parallel to this, a portion of pyruvate may be converted into acetyl-CoA, followed by the formation of acetic acid. This occurs through pyruvate dehydrogenase complex–catalyzed decarboxylation of pyruvate to acetyl-CoA, which is subsequently converted into acetate via phosphotransacetylase (PTA) and acetate kinase (AK) reactions, yielding ATP as a byproduct. Consequently, acetic acid is produced as a secondary metabolite of propionic acid fermentation [88,98,118,189,201].

A schematic representation of the propionic acid fermentation pathway and the key enzymes involved is shown in Figure 5 [74,88,98,118,189]. While the Wood–Werkman pathway predominates in *Propionibacterium* species, alternative metabolic routes have evolved in other anaerobic bacteria to support propionate formation under different physiological and ecological conditions.

In recent years, propionibacteria have increasingly been explored as targets for metabolic engineering and systems-guided strain improvement to enhance propionic acid productivity, reduce by-product formation (especially acetate), and improve tolerance to product inhibition [111,119,189,201]. A major limitation of industrial propionate production is the strong inhibitory effect of propionic acid on cell growth and redox homeostasis; therefore, contemporary strategies focus on improving acid tolerance and export capacity, as well as optimizing intracellular redox balance and ATP conservation in the Wood–Werkman cycle [89,121,193]. In parallel, pathway-level interventions aimed at shifting carbon flux toward propionyl-CoA (and away from acetyl-CoA/acetate) have been investigated, including regulation of key nodes such as pyruvate utilization, succinate conversion steps, and CoA-transferase reactions [192,194,202].

Advances in genome sequencing, metabolic modeling, and multi-omics analyses have improved understanding of propionibacterial physiology and facilitated rational process optimization, particularly for *Propionibacterium freudenreichii* and *P. acidipropionici*, which are also major industrial producers of vitamin B_{12} [189,191,203]. Emerging synthetic biology approaches—including targeted genome editing, adaptive laboratory evolution, and engineered co-cultures that couple lactate-producing LAB with propionate producers—are increasingly used to improve stability, yields, and substrate flexibility, supporting the development of propionic fermentation as a sustainable platform for food preservation and bio-based chemicals [28,123,204].

https://doi.org/10.3390/molecules31020333

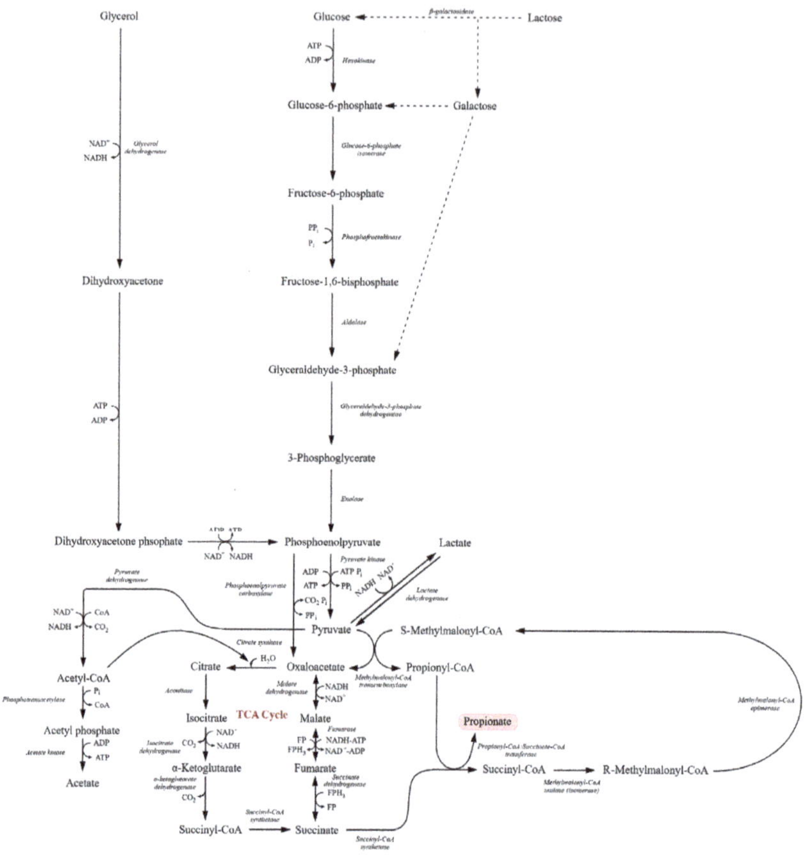

Figure 5. Integrated metabolic pathways involved in the conversion of carbohydrates and glycerol into propionate via the succinate–propionate pathway. Glucose, galactose, and glycerol are converted to dihydroxyacetone phosphate and subsequently processed through glycolysis to phosphoenolpyruvate and pyruvate. Pyruvate is then metabolized through the tricarboxylic acid (TCA) cycle to oxaloacetate, malate, fumarate, succinate, and succinyl-CoA. Further reactions lead to the formation of methylmalonyl-CoA intermediates and ultimately propionyl-CoA and propionate. Key enzymatic steps, redox reactions, ATP-generating conversions, and metabolic branch points linking glycolysis, glycerol utilization, and the TCA cycle are indicated (adapted from [88,189,201]).

6.2. Alternative Propionate Pathways (Acrylate and 1,2-Propanediol)

The acrylate pathway is an alternative propionic acid fermentation route used by some obligate anaerobes, notably *C. propionicum*. It begins with the conversion of pyruvate to lactate via L-lactate dehydrogenase. Propionyl-CoA transferase then transfers CoA from propionyl-CoA to lactate, producing lactoyl-CoA and releasing propionic acid, the main product of this pathway. Lactoyl-CoA is dehydrated by lactoyl-CoA dehydratase to acryloyl-CoA, a toxic intermediate. To minimize toxicity, acryloyl-CoA is rapidly reduced back to propionyl-CoA by acryloyl-CoA reductase. The pathway thus functions as a cyclic process with internal CoA recycling, in which propionyl-CoA transferase plays a central dual role (Figure 6) [88,189,193,201].

https://doi.org/10.3390/molecules31020333

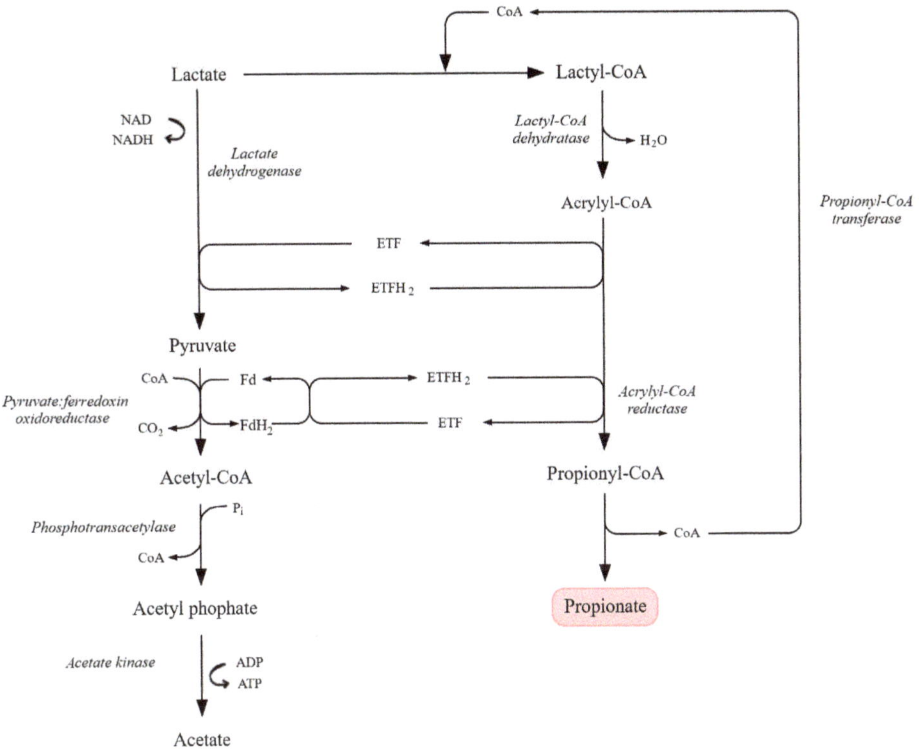

Figure 6. Lactate-based acrylate pathway leading to propionate formation. Lactate is oxidized to pyruvate via lactate dehydrogenase, with subsequent conversion to acetyl-CoA by pyruvate:ferredoxin oxidoreductase. In the acrylate branch, lactyl-CoA is dehydrated to acrylyl-CoA, which is further reduced to propionyl-CoA by acrylyl-CoA reductase in an electron-transferring flavoprotein (ETF/$ETFH_2$)-dependent system. Propionyl-CoA is then converted to propionate through CoA transferase activity. Parallel conversion of acetyl-CoA to acetate via acetyl phosphate also contributes to ATP generation. Key enzymatic reactions, redox cycles, and metabolite flow are indicated (adapted from [201]).

Because acryloyl-CoA is cytotoxic, bacteria must maintain its intracellular concentration at very low levels. This becomes particularly challenging at elevated pH, when redox balance is disturbed. Under such conditions, microorganisms redirect carbon flux toward acetic acid formation at the expense of propionic acid, rendering the acrylate pathway less efficient. The metabolic burden associated with rapid detoxification of acryloyl intermediates further reduces overall fermentation yield [89,189,201,202,205].

The 1,2-propanediol fermentation pathway constitutes another route to propionic acid. Several bacteria, including *Salmonella enterica, Roseburia inulinivorans* and representatives of *Lactobacillus*, *Ruminococcus obeum* and *Bacteroides thetaiotaomicron,* can synthesize 1,2-propanediol from various carbon sources such as deoxy sugars (e.g., fucose and rhamnose), dihydroxyacetone or lactic acid [122,206,207].

Three main microbial routes for 1,2-propanediol biosynthesis are described (Figure 7):

I. Deoxyhexose pathway—L-rhamnose or L-fucose is metabolized to 1,2-propanediol. For example,

L-fucose → L-fuculose → fuculose-1-phosphate → L-lactaldehyde → 1,2-propanediol, via L-fucose isomerase, L-fuculokinase, L-fuculose-1-phosphate aldolase and lactaldehyde reductase [122,207].

 https://doi.org/10.3390/molecules31020333

II. Methylglyoxal pathway—dihydroxyacetone phosphate (DHAP) is converted to methylglyoxal and subsequently reduced to 1,2-propanediol. Due to the cytotoxicity of methylglyoxal, this route is less favored [122,208].

III. Lactic acid pathway—lactic acid is converted to lactoyl-CoA, reduced to lactaldehyde and then to 1,2-propanediol. This pathway, often active under acidic conditions, is considered most attractive for industrial applications because it does not generate toxic intermediates and does not require expensive substrates [200,204].

After synthesis, 1,2-propanediol can be further converted to propionic acid. It is dehydrated to propionaldehyde by propanediol dehydratase, then oxidized to propionyl-CoA and reduced to propanol by propionaldehyde and propanol dehydrogenases, respectively. Propionyl-CoA is finally converted to propionyl phosphate and then to propionic acid by phosphotransacylase and propionate kinase [189,192,200].

The diversity of propionate-forming pathways, their energetic efficiency, and their dependence on substrate availability collectively determine the industrial and technological relevance of propionic fermentation.

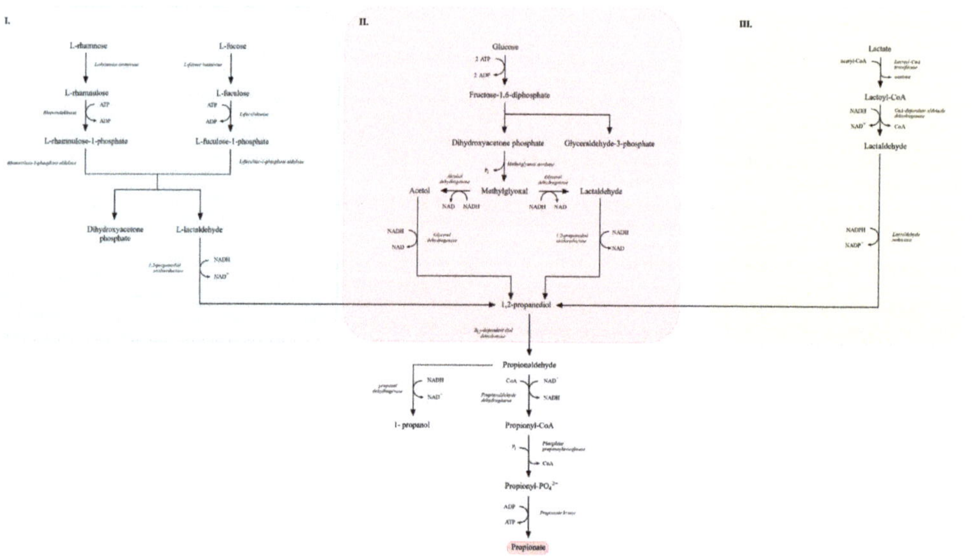

Figure 7. Metabolic pathways leading to the formation of 1,2-propanediol and propionate from sugars, sugar-derived deoxyhexoses, and lactate. Panel (**I**) shows the catabolism of L-rhamnose and L-fucose to dihydroxyacetone phosphate and L-lactaldehyde via L-rhamnulose-1-phosphate and L-fuculose-1-phosphate intermediates. Panel (**II**) presents the conversion of glucose-derived triose phosphates into acetol, methylglyoxal, and lactaldehyde, followed by their reduction to 1,2-propanediol. Panel (**III**) illustrates the lactate-dependent pathway in which lactoyl-CoA and lactaldehyde serve as precursors for 1,2-propanediol formation. The central pathway depicts the downstream conversion of 1,2-propanediol to propionaldehyde, propionyl-CoA, propionyl-phosphate, and finally propionate. Key enzymatic steps, redox reactions, and metabolic branch points between carbohydrate, deoxyhexose, and lactate catabolism are indicated (adapted from [189,201]).

6.3. Industrial Applications

Building on the metabolic frameworks described above, propionic fermentation has found widespread application in food processing, biotechnology, and chemical industries due to the preservative, functional, and platform-chemical properties of propionic acid [7,54,203].

https://doi.org/10.3390/molecules31020333

The best-known example is the ripening of Swiss-type cheeses (e.g., Emmental, Maasdam, Gruyère), in which *Propionibacterium freudenreichii* subsp. *shermanii* converts lactic acid into propionic and acetic acids and CO_2. Gas evolution forms the characteristic eyes in the cheese, while organic acids contribute to the nutty aroma and mild flavor. Propionic acid also inhibits molds and spoilage bacteria, extending shelf life [36,144]. An overview of the major industrial applications of propionic fermentation and propionic acid is provided in Table 6.

Table 6. Industrial applications of propionic fermentation and propionic acid across food, chemical, cosmetic, polymer and pharmaceutical sectors.

Application Area	Function/Role of Propionic Fermentation or Propionic Acid	Key Microorganisms/ Compounds	Industrial or Health Relevance	Selected References
Dairy Industry (Cheese Ripening)	Conversion of lactic acid to propionic and acetic acids and CO_2 during cheese maturation; contributes to flavor, aroma, and texture.	*Propionibacterium freudenreichii* subsp. *shermanii, P. acidipropionici*	Formation of characteristic eyes in Swiss-type cheeses (Emmental, Maasdam, Gruyère); development of nutty aroma; natural mold inhibition and extended shelf life.	[111,119,126, 128,141,144, 145,189,192]
Food Preservation	Natural preservative and antifungal agent; inhibits spoilage microorganisms and foodborne pathogens.	Propionic acid; sodium, calcium, potassium, and ammonium propionates	Preservation of bakery products, cheeses, and silage; suppression of *Aspergillus flavus*, *Bacillus* spp., *Salmonella* spp., *Listeria monocytogenes*; enhancement of product safety and shelf stability.	[98,103,106, 107,109–111, 119,165,189]
Biotechnological Production of Vitamin B_{12}	Microbial biosynthesis of cobalamin (vitamin B_{12}) through propionic fermentation.	*Propionibacterium freudenreichii, P. acidipropionici*	Used in pharmaceuticals and functional food fortification; essential for human metabolism.	[6,54,111,189, 192]
Chemical Industry	Intermediate for synthesis of organic compounds, herbicides, and propionate esters.	Propionic acid and esters	Production of protective coatings, industrial varnishes, and solvents; valued for volatility, durability, and film-forming ability.	[14,47,75,111, 112,119,189]
Construction and Cleaning Products	Antimicrobial additive in paints, adhesives, detergents, and impregnation agents.	Propionic acid and salts	Prevents microbial growth on surfaces, improving durability and hygiene in materials and environments.	[14,47,75,111, 119,189]
Plastics and Polymer Industry	Intermediate for synthesis of cellulose propionate and biodegradable polymers.	Cellulose propionate; propionate esters	Used for films, packaging, optical materials, and bioplastic components; combines functionality with environmental sustainability.	[14,43,47,74, 75,111,112, 119]
Cosmetic Industry	Natural preservative and fragrance base; stabilizer in cosmetic emulsions.	Propionic acid and salts	Improves texture and stability of formulations; antimicrobial effect reduces synthetic preservative need.	[47,110,111, 119,146,189]
Pharmaceutical and Veterinary Applications	Antibacterial and anti-inflammatory properties; therapeutic and auxiliary agent.	Sodium, calcium, and ammonium propionates	Used in treatment of skin infections, fungal diseases, and conjunctivitis; veterinary hygiene and hoof-care products.	[47,103,110, 111,119,189]
Green and Circular Biotechnology	Integration of propionic acid production into sustainable bioprocesses and bioeconomy frameworks.	*Propionibacterium* spp.; bio-based propionate systems	Key biocomponent linking food, chemical, and pharmaceutical sectors; aligned with green chemistry and circular economy principles.	[6,9,14,33,50, 75,111,119, 165,189,192]

https://doi.org/10.3390/molecules31020333

In the wider food industry, propionic acid and its salts—sodium, calcium, potassium and ammonium propionates—are used as natural preservatives due to strong antibacterial and antifungal properties. They inhibit species such as *Aspergillus flavus*, *Bacillus* spp., *Salmonella* spp. and yeasts, and in combination with lactic and acetic acids effectively suppress *Listeria monocytogenes* and other foodborne pathogens [74,98,189]. Propionates are therefore used in bakery products, confectionery, cheeses and silage to improve microbial stability and safety.

Propionic acid (CH_3CH_2COOH) is a three-carbon carboxylic acid, miscible with water and many organic solvents, and has been granted GRAS status by the U.S. Food and Drug Administration. Beyond food, it is used in chemical, cosmetic, polymer and pharmaceutical industries [111,119]. In biotechnology, propionic fermentation serves as the basis for industrial vitamin B_{12} (cobalamin) production by *P. freudenreichii* and *P. acidipropionici*. Microbial cobalamin is widely used in pharmaceuticals and functional foods [79,189].

In the chemical industry, propionic acid acts as an important intermediate in the synthesis of various organic compounds, including herbicides and pesticides used in crop protection. Due to its chemical reactivity, it is also used as a precursor for the synthesis of propionate esters, which serve as components in protective coatings, industrial varnishes, and organic solvents. These esters are valued for their volatility, film-forming properties, and durability, making them crucial in the paint and coatings industry [74,98].

In the construction materials and cleaning products sector, propionic acid is applied for its antimicrobial activity. It is added to paints, adhesives, detergents, and impregnating agents to prevent the growth of bacteria, molds, and fungi on surfaces, thereby enhancing material durability and improving hygienic safety [119].

In the plastics industry, propionic acid and its esters are key intermediates for cellulose-based fibers and biodegradable polymers. Cellulose propionate combines favorable mechanical, optical and moisture-resistance properties, making it suitable for films, packaging, photographic layers and decorative components, and it is considered a next-generation biopolymer [100,111,209].

In the cosmetic industry, propionate salts are used as fragrance bases and natural preservatives in skincare formulations. When combined with butyl rubber, they improve the texture, elasticity, and stability of cosmetic emulsions, while their antimicrobial activity reduces the need for synthetic preservatives [119].

In pharmaceutical and veterinary applications, propionic acid and its salts—especially sodium propionate—are used as auxiliary and therapeutic substances. Due to their antibacterial and anti-inflammatory properties, they are applied in the treatment of skin infections, fungal diseases, and mucosal inflammations. Propionates are also components of ophthalmic preparations used to treat conjunctivitis and are included in antiseptic and anti-inflammatory drugs. In veterinary medicine, they are used in hygiene preparations for livestock, protecting against infections of the skin and hooves [74,98,119]. The multifunctionality of propionic acid and its derivatives makes them a crucial link between biotechnology, chemical, pharmaceutical, and material industries. Owing to their potential integration into sustainable bioproduction processes, propionic acid is considered a key biocomponent of the future, aligning with the principles of green chemistry and the circular economy [9,111].

7. Comparative Roles of Key Microorganisms Across Fermentation Systems

Having reviewed the biochemical pathways, microbial ecology, and industrial applications of alcoholic, acetic, butyric, lactic, and propionic fermentations, it becomes evident that although certain microorganisms occur across multiple fermentation systems, their

 https://doi.org/10.3390/molecules31020333

metabolic roles and technological relevance differ substantially depending on ecological context and process conditions. *S. cerevisiae* exemplifies this functional plasticity, acting as the primary biocatalyst of alcoholic fermentation by converting sugars into ethanol and CO_2 via fermentative glycolysis, while simultaneously generating a wide spectrum of aroma-active secondary metabolites that define beverage quality [8,20,26,57]. In contrast, in lactic and mixed-culture fermentations, yeasts such as *S. cerevisiae*, *Kluyveromyces* spp., and *Torulaspora* spp. typically play auxiliary roles, contributing to redox balance, oxygen scavenging, vitamin release, and the formation of volatile compounds that support lactic acid bacteria growth and enhance sensory complexity rather than serving as dominant fermentative agents [46,54,142,173].

Lactic acid bacteria represent the principal acidifying microorganisms in lactic fermentation, where they drive rapid pH reduction, microbial stabilization, and product safety through lactic acid and bacteriocin production [37,106,157,159]. However, in alcoholic and acetic fermentations, LAB generally constitute secondary microbiota, influencing microbial succession, flavor development, and maturation processes, particularly in wine, beer, and traditional fermented beverages [41,57,64,87]. A further context-dependent functional shift is observed for acetic acid bacteria, which dominate ethanol oxidation during acetic fermentation but appear as minor or late-stage contributors in alcoholic or mixed fermentations, where they modulate acidity and aroma rather than serving as primary producers [83,87,95].

In propionic fermentation, *Propionibacterium* species assume a specialized role by converting lactate into propionic and acetic acids and CO_2 via the Wood–Werkman pathway, thereby linking lactic and propionic fermentations both metabolically and technologically, as exemplified by Swiss-type cheese ripening [126,189,193,195]. Collectively, these examples demonstrate that microbial functionality in fermentation cannot be interpreted solely at the species level but must be evaluated within a dynamic ecological framework shaped by substrate availability, interspecies interactions, and process design [27,54,173]. This integrative perspective provides a conceptual synthesis of microbial roles across fermentation systems and underpins the diverse applications discussed throughout this review.

8. Conclusions and Future Perspectives

Fermentation, one of humanity's oldest biotechnological tools, remains central to both traditional food production and cutting-edge industrial biotechnology. This review has summarized the biochemical foundations, microbial ecology and industrial relevance of the five major fermentation types: alcoholic, acetic, butyric, lactic and propionic fermentation. Despite their diversity, these processes share common features: they enable energy conservation under anaerobic or microaerophilic conditions, generate organic acids and alcohols as key metabolites, and transform raw materials into products with enhanced shelf life, safety, sensory quality and functional value.

The diversity of microbial communities plays a decisive role in shaping the sensory properties, quality, and consistency of fermented products. In spontaneous and mixed-culture fermentations, complex consortia of yeasts, lactic acid bacteria, acetic acid bacteria, and other microorganisms engage in cross-feeding, metabolic cooperation, and competitive interactions that generate rich and distinctive flavor profiles. At the same time, this microbial diversity introduces variability and challenges for process standardization and reproducibility. Understanding and managing community dynamics therefore represents a key frontier in balancing product authenticity with industrial consistency.

Despite their long history and technological maturity, fermentation-based processes face several challenges onat industrialon an industrial scale. Key limitations include

 https://doi.org/10.3390/molecules31020333

product inhibition by organic acids and solvents, trade-offs between yield and productivity, high downstream processing costs, and variability associated with complex feedstocks. Scale-up from laboratory to industrial bioreactors further requires careful control of mass transfer, redox balance, and microbial stability. Addressing these challenges relies on integrated strategies combining strain improvement, process optimization, in situ product recovery, and techno-economic assessment to ensure competitiveness with petrochemical alternatives. Many of these challenges are fermentation-specific and were highlighted in the individual sections above [28,33,121].

Recent advances in metabolic engineering and synthetic biology are reshaping classical fermentation processes. Engineered *S. cerevisiae* strains with optimized redox balance, inhibitor tolerance, and expanded substrate spectra are increasingly applied in alcoholic and mixed fermentations. Specifically, systems biology tools such as genome-scale metabolic models, flux balance analysis, and multi-omics integration have enabled the identification of metabolic bottlenecks, redox imbalances, and by-product formation pathways, which can then be rationally targeted by metabolic engineering. In lactic and propionic fermentations, genome-scale metabolic models and CRISPR-based tools enable targeted modulation of acid yields, cofactor regeneration, and by-product formation. Similarly, engineered *Clostridium* species are being developed to improve solvent selectivity and butyrate productivity while reducing acid toxicity. These approaches, supported by multi-omics integration and adaptive laboratory evolution, transform traditional fermentations into controllable microbial cell factories [24,25,156].

Alcoholic fermentation underpins the production of fermented beverages, baked goods and bioethanol, linking cultural traditions with modern energy and chemical industries. Acetic fermentation is essential for vinegar and low-alcohol beverages and provides acetic acid as a versatile platform chemical for polymers, textiles, cosmetics and pharmaceuticals. Butyric fermentation is both a challenge in some fermented foods and a powerful biotechnological route to short-chain fatty acids, solvents and advanced biofuels. Lactic fermentation remains the backbone of fermented foods and functional/probiotic products and supplies lactic acid for biodegradable polymers and medical biomaterials. Propionic fermentation closes the circle by supporting cheese ripening, vitamin B_{12} production and a wide range of chemical, cosmetic and pharmaceutical applications.

Looking forward, several trends are likely to shape the future of fermentation-based technologies:

- Strain engineering and synthetic biology will enable the design of microbial cell factories with enhanced productivity, substrate flexibility and stress tolerance, including engineered LAB, *Saccharomyces*, *Zymomonas*, *Clostridium* and *Propionibacterium* strains.
- Multi-omics and systems biology approaches will deepen understanding of microbial consortia in spontaneous and mixed-culture fermentations, allowing rational manipulation of community structure and function to optimize product profiles.
- Valorization of agro-industrial by-products will expand, using fermentation to convert low-value streams into high-value compounds such as organic acids, biopolymers, aroma compounds and biofuels, supporting circular bioeconomy strategies.
- From a sustainability perspective, fermentation technologies increasingly support circular and zero-waste bioeconomy concepts. Alcoholic and lactic fermentations are widely applied to valorize agro-industrial by-products such as whey, molasses, fruit pomace, and cereal residues into ethanol, lactic acid, and functional metabolites. Acetic acid bacteria enable the upgrading of ethanol-rich side streams into vinegar and gluconic acid, while propionic and butyric fermentations convert lactate- or glycerol-rich wastes into preservatives, biofuels, and platform chemicals. Integrated biorefineries combining

 https://doi.org/10.3390/molecules31020333

multiple fermentation steps exemplify how classical processes can be coupled to achieve near-complete carbon utilization, reduced waste generation, and lower environmental impact. Key challenges in transitioning toward sustainable bioprocessing include the heterogeneity and seasonal variability of waste streams, the presence of inhibitory compounds, logistical constraints in feedstock collection, and the need for robust strains capable of efficiently utilizing mixed and low-cost substrates.

- Development of functional and personalized foods will increasingly rely on tailored fermentations and starter cultures with targeted health effects, including modulation of the gut microbiota, immune system and metabolic homeostasis.

In parallel, technological innovation and digitalization are transforming fermentation monitoring and control. The integration of real-time sensing technologies for pH, redox potential, dissolved gases, and key metabolites, combined with soft sensors, digital twins, machine learning, and artificial intelligence-driven control systems, enables predictive modeling, early fault detection, and dynamic optimization of fermentation conditions. Such data-driven approaches are particularly valuable for large-scale and mixed-culture fermentations, where they support improved reproducibility, scalability, resource efficiency, and economic performance. Together with process intensification strategies such as membrane bioreactors, in situ product recovery, and high-pressure processing, these tools contribute to more sustainable, robust, and competitive fermentation-based bioprocesses [27,33,173].

Author Contributions: Conceptualization, A.R. and W.G.; methodology, A.R.; investigation, A.R. and W.G.; data curation, W.G.; writing—original draft preparation, A.R. and W.G.; writing—review and editing, A.R.; visualization, W.G.; supervision, A.R.; project administration, A.R.; funding acquisition, A.R. All authors have read and agreed to the published version of the manuscript.

Funding: This research was funded by the Ministry of Science and Higher Education, Poland, within the program "Nauka dla Społeczeństwa II", project number NdS-II/SN/0476/2023/01, with a total funding amount of 1,811,910 PLN.

Institutional Review Board Statement: Not applicable.

Informed Consent Statement: Not applicable.

Data Availability Statement: No new data were created or analyzed in this study. The data presented in this study are available within the article.

Conflicts of Interest: The authors declare no conflicts of interest.

References

1. Siddiqui, S.A.; Erol, Z.; Rugji, J.; Taşçı, F.; Kahraman, H.A.; Toppi, V.; Musa, L.; Di Giacinto, G.; Bahmid, N.A.; Mehdizadeh, M.; et al. An Overview of Fermentation in the Food Industry—Looking Back from a New Perspective. *Bioresour. Bioprocess.* **2023**, *10*, 85. [CrossRef]
2. Todorovic, S.; Akpinar, A.; Assunção, R.; Bär, C.; Bavaro, S.L.; Berkel Kasikci, M.; Domínguez-Soberanes, J.; Capozzi, V.; Cotter, P.D.; Doo, E.-H.; et al. Health Benefits and Risks of Fermented Foods—The PIMENTO Initiative. *Front. Nutr.* **2024**, *11*, 1458536. [CrossRef] [PubMed]
3. Voidarou, C.; Antoniadou, M.; Rozos, G.; Tzora, A.; Skoufos, I.; Varzakas, T.; Lagiou, A.; Bezirtzoglou, E. Fermentative Foods: Microbiology, Biochemistry, Potential Human Health Benefits and Public Health Issues. *Foods* **2020**, *10*, 69. [CrossRef]
4. Zapaśnik, A.; Sokołowska, B.; Bryła, M. Role of Lactic Acid Bacteria in Food Preservation and Safety. *Foods* **2022**, *11*, 1283. [CrossRef]
5. Ray, R.C.; Joshi, V.K. *Fermented Foods: Past, Present and Future*; Fermented Foods (CRC Press); CRC Press: Boca Raton, FL, USA, 2014; ISBN 9781466585769.
6. Mannaa, M.; Han, G.; Seo, Y.-S.; Park, I. Evolution of Food Fermentation Processes and the Use of Multi-Omics in Deciphering the Roles of the Microbiota. *Foods* **2021**, *10*, 2861. [CrossRef]

https://doi.org/10.3390/molecules31020333

7. El-Mansi, M.; Nielsen, J.; Mousdale, D.M.; Allman, T.; Carlson, R. (Eds.) *Fermentation Microbiology and Biotechnology*, 4th ed.; CRC Press: Boca Raton, FL, USA, 2019; ISBN 978-1-138-58102-9.
8. Maicas, S. The Role of Yeasts in Fermentation Processes. *Microorganisms* **2020**, *8*, 1142. [CrossRef] [PubMed]
9. Praveen, M.; Brogi, S. Microbial Fermentation in Food and Beverage Industries: Innovations, Challenges, and Opportunities. *Foods* **2025**, *14*, 114. [CrossRef]
10. Taveira, I.C.; Nogueira, K.M.V.; Oliveira, D.L.G.D.; Silva, R.D.N. Fermentation: Humanity's Oldest Biotechnological Tool. *Front. Young Minds* **2021**, *9*, 568656. [CrossRef]
11. Dimidi, E.; Cox, S.; Rossi, M.; Whelan, K. Fermented Foods: Definitions and Characteristics, Impact on the Gut Microbiota and Effects on Gastrointestinal Health and Disease. *Nutrients* **2019**, *11*, 1806. [CrossRef]
12. Schwan, R.F.; Bressani, A.P.P.; Martinez, S.J.; Batista, N.N.; Dias, D.R. The Essential Role of Spontaneous and Starter Yeasts in Cocoa and Coffee Fermentation. *FEMS Yeast Res.* **2023**, *23*, foad019. [CrossRef]
13. Amit, S.K.; Uddin, M.M.; Rahman, R.; Islam, S.M.R.; Khan, M.S. A Review on Mechanisms and Commercial Aspects of Food Preservation and Processing. *Agric. Food Secur.* **2017**, *6*, 51. [CrossRef]
14. Ruane, J.; Sonnino, A.; Agostini, A. Bioenergy and the Potential Contribution of Agricultural Biotechnologies in Developing Countries. *Biomass Bioenergy* **2010**, *34*, 1427–1439. [CrossRef]
15. Boumba, V.A.; Ziavrou, K.S.; Vougiouklakis, T. Biochemical Pathways Generating Post-Mortem Volatile Compounds Co-Detected during Forensic Ethanol Analyses. *Forensic Sci. Int.* **2008**, *174*, 133–151. [CrossRef]
16. Eknikom, S.; Nasuno, R.; Takagi, H. Molecular Mechanism of Ethanol Fermentation Inhibition via Protein Tyrosine Nitration of Pyruvate Decarboxylase by Reactive Nitrogen Species in Yeast. *Sci. Rep.* **2022**, *12*, 4664. [CrossRef]
17. Malakar, S.; Paul, S.K.; Jolvis Pou, K.R. Biotechnological Interventions in Beverage Production. In *Biotechnological Progress and Beverage Consumption*; Elsevier: Amsterdam, The Netherlands, 2020; pp. 1–37, ISBN 978-0-12-816678-9.
18. Aminian, A.; Motamedian, E. Investigating Ethanol Production Using the Zymomonas Mobilis Crude Extract. *Sci. Rep.* **2023**, *13*, 1165. [CrossRef]
19. Yan, S. The Biochemical Basis of Ethanol Fermentation and Its Industrial Applications. *Biol. Evid.* **2024**, *14*, 238–249. [CrossRef]
20. Zamora, F. Biochemistry of Alcoholic Fermentation. In *Wine Chemistry and Biochemistry*; Moreno-Arribas, M.V., Polo, M.C., Eds.; Springer: New York, NY, USA, 2009; pp. 3–26, ISBN 978-0-387-74116-1.
21. Guragain, Y.N.; Probst, K.V.; Vadlani, P.V. Fuel Alcohol Production. In *Encyclopedia of Food Grains*; Elsevier: Amsterdam, The Netherlands, 2016; pp. 235–244, ISBN 978-0-12-394786-4.
22. Kopp, D.; Sunna, A. Alternative Carbohydrate Pathways—Enzymes, Functions and Engineering. *Crit. Rev. Biotechnol.* **2020**, *40*, 895–912. [CrossRef] [PubMed]
23. Ward, B. Bacterial Energy Metabolism. In *Molecular Medical Microbiology*; Elsevier: Amsterdam, The Netherlands, 2015; pp. 201–233, ISBN 978-0-12-397169-2.
24. Xia, J.; Yang, Y.; Liu, C.-G.; Yang, S.; Bai, F.-W. Engineering Zymomonas Mobilis for Robust Cellulosic Ethanol Production. *Trends Biotechnol.* **2019**, *37*, 960–972. [CrossRef]
25. Nielsen, J.; Larsson, C.; Van Maris, A.; Pronk, J. Metabolic Engineering of Yeast for Production of Fuels and Chemicals. *Curr. Opin. Biotechnol.* **2013**, *24*, 398–404. [CrossRef] [PubMed]
26. Chambers, P.J.; Pretorius, I.S. Fermenting Knowledge: The History of Winemaking, Science and Yeast Research. *EMBO Rep.* **2010**, *11*, 914–920. [CrossRef] [PubMed]
27. De Filippis, F.; Parente, E.; Ercolini, D. Metagenomics Insights into Food Fermentations. *Microb. Biotechnol.* **2017**, *10*, 91–102. [CrossRef] [PubMed]
28. Lindemann, S.R.; Bernstein, H.C.; Song, H.-S.; Fredrickson, J.K.; Fields, M.W.; Shou, W.; Johnson, D.R.; Beliaev, A.S. Engineering Microbial Consortia for Controllable Outputs. *ISME J.* **2016**, *10*, 2077–2084. [CrossRef] [PubMed]
29. Maslanka, R.; Zadrag-Tecza, R. Reproductive Potential of Yeast Cells Depends on Overall Action of Interconnected Changes in Central Carbon Metabolism, Cellular Biosynthetic Capacity, and Proteostasis. *Int. J. Mol. Sci.* **2020**, *21*, 7313. [CrossRef] [PubMed]
30. Li, K.-J.; Qi, L.; Zhu, Y.-X.; He, M.; Xiang, Q.; Zheng, D.-Q. Spontaneous and Environment Induced Genomic Alterations in Yeast Model. *Cell Insight* **2025**, *4*, 100209. [CrossRef]
31. Wan, Z.; Hu, H.; Liu, K.; Qiao, Y.; Guo, F.; Wang, C.; Xin, F.; Zhang, W.; Jiang, M. Engineering Industrial Yeast for Improved Tolerance and Robustness. *Crit. Rev. Biotechnol.* **2024**, *44*, 1461–1477. [CrossRef]
32. Yook, S.; Alper, H.S. Recent Advances in Genetic Engineering and Chemical Production in Yeast Species. *FEMS Yeast Res.* **2025**, *25*, foaf009. [CrossRef]
33. Chandel, A.K.; Garlapati, V.K.; Singh, A.K.; Antunes, F.A.F.; Da Silva, S.S. The Path Forward for Lignocellulose Biorefineries: Bottlenecks, Solutions, and Perspective on Commercialization. *Bioresour. Technol.* **2018**, *264*, 370–381. [CrossRef]

 https://doi.org/10.3390/molecules31020333

34. Haldar, D.; Shabbirahmed, A.M.; Singhania, R.R.; Chen, C.-W.; Dong, C.-D.; Ponnusamy, V.K.; Patel, A.K. Understanding the Management of Household Food Waste and Its Engineering for Sustainable Valorization-A State-of-the-Art Review. *Bioresour. Technol.* **2022**, *358*, 127390. [CrossRef]
35. Hu, M.; Bao, W.; Peng, Q.; Hu, W.; Yang, X.; Xiang, Y.; Yan, X.; Li, M.; Xu, P.; He, Q.; et al. Metabolic Engineering of Zymomonas Mobilis for Co-Production of D-Lactic Acid and Ethanol Using Waste Feedstocks of Molasses and Corncob Residue Hydrolysate. *Front. Bioeng. Biotechnol.* **2023**, *11*, 1135484. [CrossRef]
36. Hui, Y.H.; Meunier-Goddik, L.; Josephsen, J.; Nip, W.-K.; Stanfield, P.S. (Eds.) *Handbook of Food and Beverage Fermentation Technology*; CRC Press: Boca Raton, FL, USA, 2004; ISBN 978-0-203-91355-0.
37. Tamang, J.P.; Watanabe, K.; Holzapfel, W.H. Review: Diversity of Microorganisms in Global Fermented Foods and Beverages. *Front. Microbiol.* **2016**, *7*, 377. [CrossRef]
38. Tamang, J.P.; Kailasapathy, K. *Fermented Foods and Beverages of the World*; Food Science and Technology; CRC Press: Boca Raton, FL, USA, 2010; ISBN 978-1-4200-9495-4.
39. Estreicher, S.K. Wine and France: A Brief History. *Eur. Rev.* **2023**, *31*, 91–179. [CrossRef]
40. Kaur, P.; Ghoshal, G.; Banerjee, U.C. Traditional Bio-Preservation in Beverages: Fermented Beverages. In *Preservatives and Preservation Approaches in Beverages*; Elsevier: Amsterdam, The Netherlands, 2019; pp. 69–113, ISBN 978-0-12-816685-7.
41. Mas, A.; Portillo, M.C. Strategies for Microbiological Control of the Alcoholic Fermentation in Wines by Exploiting the Microbial Terroir Complexity: A Mini-Review. *Int. J. Food Microbiol.* **2022**, *367*, 109592. [CrossRef] [PubMed]
42. Tzamourani, A.P.; Taliadouros, V.; Paraskevopoulos, I.; Dimopoulou, M. Developing a Novel Selection Method for Alcoholic Fermentation Starters by Exploring Wine Yeast Microbiota from Greece. *Front. Microbiol.* **2023**, *14*, 1301325. [CrossRef] [PubMed]
43. Martins, S.; Mussatto, S.I.; Martínez-Avila, G.; Montañez-Saenz, J.; Aguilar, C.N.; Teixeira, J.A. Bioactive Phenolic Compounds: Production and Extraction by Solid-State Fermentation. A Review. *Biotechnol. Adv.* **2011**, *29*, 365–373. [CrossRef]
44. Wittwer, A.; Howell, K. Rising Stars in the Bakery: Novel Yeasts for Modern Bread. *Microbiol. Aust.* **2022**, *43*, 75–78. [CrossRef]
45. Chen, A.; Pan, C.; Chen, J. Comparative Analysis of Bread Quality Using Yeast Strains from Alcoholic Beverage Production. *Microorganisms* **2024**, *12*, 2609. [CrossRef]
46. Lahue, C.; Madden, A.A.; Dunn, R.R.; Smukowski Heil, C. History and Domestication of Saccharomyces Cerevisiae in Bread Baking. *Front. Genet.* **2020**, *11*, 584718. [CrossRef]
47. Lee, J.W.; Kim, H.U.; Choi, S.; Yi, J.; Lee, S.Y. Microbial Production of Building Block Chemicals and Polymers. *Curr. Opin. Biotechnol.* **2011**, *22*, 758–767. [CrossRef] [PubMed]
48. Parapouli, M.; Vasileiadi, A.; Afendra, A.-S.; Hatziloukas, E. Saccharomyces Cerevisiae and Its Industrial Applications. *AIMS Microbiol.* **2020**, *6*, 1–31. [CrossRef]
49. Saha Turna, N.; Chung, R.; McIntyre, L. A Review of Biogenic Amines in Fermented Foods: Occurrence and Health Effects. *Heliyon* **2024**, *10*, e24501. [CrossRef]
50. Nadar, C.G.; Fletcher, A.; Moreira, B.R.D.A.; Hine, D.; Yadav, S. Waste to Protein: A Systematic Review of a Century of Advancement in Microbial Fermentation of Agro-industrial Byproducts. *Compr. Rev. Food Sci. Food Saf.* **2024**, *23*, e13375. [CrossRef]
51. Subandi, S.; Ratih, N.K.; Soka, S.; Suwanto, A. Effect of Tempeh Supplementation on the Profiles of Human Intestinal Immune System and Gut Microbiota. *Microbiol. Indones.* **2017**, *11*, 11–17. [CrossRef]
52. Ohya, Y.; Kashima, M. History, Lineage and Phenotypic Differentiation of Sake Yeast. *Biosci. Biotechnol. Biochem.* **2019**, *83*, 1442–1448. [CrossRef] [PubMed]
53. Romulo, A.; Surya, R. Tempe: A Traditional Fermented Food of Indonesia and Its Health Benefits. *Int. J. Gastron. Food Sci.* **2021**, *26*, 100413. [CrossRef]
54. Tamang, J.P.; Cotter, P.D.; Endo, A.; Han, N.S.; Kort, R.; Liu, S.Q.; Mayo, B.; Westerik, N.; Hutkins, R. Fermented Foods in a Global Age: East Meets West. *Compr. Rev. Food Sci. Food Saf.* **2020**, *19*, 184–217. [CrossRef] [PubMed]
55. Le Jeune, C.; Erny, C.; Demuyter, C.; Lollier, M. Evolution of the Population of Saccharomyces Cerevisiae from Grape to Wine in a Spontaneous Fermentation. *Food Microbiol.* **2006**, *23*, 709–716. [CrossRef]
56. Pretorius, I.S.; Bauer, F.F. Meeting the Consumer Challenge through Genetically Customized Wine-Yeast Strains. *Trends Biotechnol.* **2002**, *20*, 426–432. [CrossRef]
57. Swiegers, J.H.; Kievit, R.L.; Siebert, T.; Lattey, K.A.; Bramley, B.R.; Francis, I.L.; King, E.S.; Pretorius, I.S. The Influence of Yeast on the Aroma of Sauvignon Blanc Wine. *Food Microbiol.* **2009**, *26*, 204–211. [CrossRef]
58. Coton, M.; Pawtowski, A.; Taminiau, B.; Burgaud, G.; Deniel, F.; Coulloumme-Labarthe, L.; Fall, A.; Daube, G.; Coton, E. Unraveling Microbial Ecology of Industrial-Scale Kombucha Fermentations by Metabarcoding and Culture-Based Methods. *FEMS Microbiol. Ecol.* **2017**, *93*, fix048. [CrossRef]

59. Cousin, F.; Le Guellec, R.; Schlusselhuber, M.; Dalmasso, M.; Laplace, J.-M.; Cretenet, M. Microorganisms in Fermented Apple Beverages: Current Knowledge and Future Directions. *Microorganisms* **2017**, *5*, 39. [CrossRef]
60. Ivanišová, E.; Meňhartová, K.; Terentjeva, M.; Harangozo, Ľ.; Kántor, A.; Kačániová, M. The Evaluation of Chemical, Antioxidant, Antimicrobial and Sensory Properties of Kombucha Tea Beverage. *J. Food Sci. Technol.* **2020**, *57*, 1840–1846. [CrossRef]
61. Khalifa, S.A.M.; El-Shabasy, R.M.; Tahir, H.E.; Abo-Atya, D.M.; Saeed, A.; Abolibda, T.Z.; Guo, Z.; Zou, X.; Zhang, D.; Du, M.; et al. Vinegar—A Beneficial Food Additive: Production, Safety, Possibilities, and Applications from Ancient to Modern Times. *Food Funct.* **2024**, *15*, 10262–10282. [CrossRef]
62. Marsh, A.J.; O'Sullivan, O.; Hill, C.; Ross, R.P.; Cotter, P.D. Sequence-Based Analysis of the Bacterial and Fungal Compositions of Multiple Kombucha (*Tea Fungus*) Samples. *Food Microbiol.* **2014**, *38*, 171–178. [CrossRef]
63. Oliveira, Í.A.C.L.D.; Rolim, V.A.D.O.; Gaspar, R.P.L.; Rossini, D.Q.; De Souza, R.; Bogsan, C.S.B. The Technological Perspectives of Kombucha and Its Implications for Production. *Fermentation* **2022**, *8*, 185. [CrossRef]
64. Roselli, G.E.; Kerruish, D.W.M.; Crow, M.; Smart, K.A.; Powell, C.D. The Two Faces of Microorganisms in Traditional Brewing and the Implications for No- and Low-Alcohol Beers. *Front. Microbiol.* **2024**, *15*, 1346724. [CrossRef]
65. Tefon Öztürk, B.E.; Eroğlu, B.; Delik, E.; Çiçek, M.; Çiçek, E. Comprehensive Evaluation of Three Important Herbs for Kombucha Fermentation. *Food Technol. Biotechnol.* **2023**, *61*, 127–137. [CrossRef]
66. Yuniarto, A.; Anggadiredja, K.; Annisa Nur Aqidah, R. Antifungal Activity of Kombucha Tea against Human Pathogenic Fungi. *Asian J. Pharm. Clin. Res.* **2016**, *9*, 253. [CrossRef]
67. Brancoli, P.; Gmoser, R.; Taherzadeh, M.J.; Bolton, K. The Use of Life Cycle Assessment in the Support of the Development of Fungal Food Products from Surplus Bread. *Fermentation* **2021**, *7*, 173. [CrossRef]
68. Lau, S.W.; Chong, A.Q.; Chin, N.L.; Talib, R.A.; Basha, R.K. Sourdough Microbiome Comparison and Benefits. *Microorganisms* **2021**, *9*, 1355. [CrossRef] [PubMed]
69. Pashaei, M.; Zare, L.; Khalili Sadrabad, E.; Hosseini Sharif Abad, A.; Mollakhalili-Meybodi, N.; Abedi, A.-S. The Impacts of Salt Reduction Strategies on Technological Characteristics of Wheat Bread: A Review. *J. Food Sci. Technol.* **2022**, *59*, 4141–4151. [CrossRef]
70. Filannino, P.; Bai, Y.; Di Cagno, R.; Gobbetti, M.; Gänzle, M.G. Metabolism of Phenolic Compounds by Lactobacillus Spp. during Fermentation of Cherry Juice and Broccoli Puree. *Food Microbiol.* **2015**, *46*, 272–279. [CrossRef]
71. Nsogning Dongmo, S.; Procopio, S.; Sacher, B.; Becker, T. Flavor of Lactic Acid Fermented Malt Based Beverages: Current Status and Perspectives. *Trends Food Sci. Technol.* **2016**, *54*, 37–51. [CrossRef]
72. Singh, R.; Singh, A.; Sachan, S. Enzymes Used in the Food Industry: Friends or Foes? In *Enzymes in Food Biotechnology*; Elsevier: Amsterdam, The Netherlands, 2019; pp. 827–843, ISBN 978-0-12-813280-7.
73. Zhao, G.; Ding, L.-L.; Yao, Y.; Cao, Y.; Pan, Z.-H.; Kong, D.-H. Extracellular Proteome Analysis and Flavor Formation During Soy Sauce Fermentation. *Front. Microbiol.* **2018**, *9*, 1872. [CrossRef] [PubMed]
74. Wang, Y.; Ling, C.; Chen, Y.; Jiang, X.; Chen, G.-Q. Microbial Engineering for Easy Downstream Processing. *Biotechnol. Adv.* **2019**, *37*, 107365. [CrossRef] [PubMed]
75. Thi, N.B.D.; Lin, C.-Y.; Kumar, G. Waste-to-Wealth for Valorization of Food Waste to Hydrogen and Methane towards Creating a Sustainable Ideal Source of Bioenergy. *J. Clean. Prod.* **2016**, *122*, 29–41. [CrossRef]
76. Kusumoto, K.-I.; Yamagata, Y.; Tazawa, R.; Kitagawa, M.; Kato, T.; Isobe, K.; Kashiwagi, Y. Japanese Traditional Miso and Koji Making. *J. Fungi* **2021**, *7*, 579. [CrossRef]
77. Liu, L.; Wang, J.; Rosenberg, D.; Zhao, H.; Lengyel, G.; Nadel, D. Fermented Beverage and Food Storage in 13,000 y-Old Stone Mortars at Raqefet Cave, Israel: Investigating Natufian Ritual Feasting. *J. Archaeol. Sci. Rep.* **2018**, *21*, 783–793. [CrossRef]
78. Nout, M.J.R.; Kiers, J.L. Tempe Fermentation, Innovation and Functionality: Update into the Third Millenium. *J. Appl. Microbiol.* **2005**, *98*, 789–805. [CrossRef]
79. Tamang, J.P.; Shin, D.-H.; Jung, S.-J.; Chae, S.-W. Functional Properties of Microorganisms in Fermented Foods. *Front. Microbiol.* **2016**, *7*, 578. [CrossRef]
80. Deshmukh, G.; Manyar, H. Production Pathways of Acetic Acid and Its Versatile Applications in the Food Industry. In *Biotechnological Applications of Biomass*; Peixoto Basso, T., Olitta Basso, T., Carlos Basso, L., Eds.; IntechOpen: London, UK, 2021; ISBN 978-1-83881-180-8.
81. Ge, Y.; Wu, Y.; Aihaiti, A.; Wang, L.; Wang, Y.; Xing, J.; Zhu, M.; Hong, J. The Metabolic Pathways of Yeast and Acetic Acid Bacteria During Fruit Vinegar Fermentation and Their Influence on Flavor Development. *Microorganisms* **2025**, *13*, 477. [CrossRef]
82. Gomes, R.J.; Borges, M.D.F.; Rosa, M.D.F.; Castro-Gómez, R.J.H.; Spinosa, W.A. Acetic Acid Bacteria in the Food Industry: Systematics, Characteristics and Applications. *Food Technol. Biotechnol.* **2018**, *56*, 139–151. [CrossRef] [PubMed]
83. He, Y.; Xie, Z.; Zhang, H.; Liebl, W.; Toyama, H.; Chen, F. Oxidative Fermentation of Acetic Acid Bacteria and Its Products. *Front. Microbiol.* **2022**, *13*, 879246. [CrossRef]

 https://doi.org/10.3390/molecules31020333

84. Mota, J.; Vilela, A. Aged to Perfection: The Scientific Symphony behind Port Wine, Vinegar, and Acetic Acid Bacteria. *Fermentation* **2024**, *10*, 200. [CrossRef]
85. Qiu, X.; Zhang, Y.; Hong, H. Classification of Acetic Acid Bacteria and Their Acid Resistant Mechanism. *AMB Express* **2021**, *11*, 29. [CrossRef]
86. Vidra, A.; Németh, Á. Bio-Produced Acetic Acid: A Review. *Period. Polytech. Chem. Eng.* **2017**, *62*, 245–256. [CrossRef]
87. Lynch, K.M.; Zannini, E.; Wilkinson, S.; Daenen, L.; Arendt, E.K. Physiology of Acetic Acid Bacteria and Their Role in Vinegar and Fermented Beverages. *Compr. Rev. Food Sci. Food Saf.* **2019**, *18*, 587–625. [CrossRef] [PubMed]
88. Müller, V. Energy Conservation in Acetogenic Bacteria. *Appl. Environ. Microbiol.* **2003**, *69*, 6345–6353. [CrossRef]
89. Buckel, W. Energy Conservation in Fermentations of Anaerobic Bacteria. *Front. Microbiol.* **2021**, *12*, 703525. [CrossRef]
90. Shinjoh, M.; Toyama, H. Industrial Application of Acetic Acid Bacteria (Vitamin C and Others). In *Acetic Acid Bacteria*; Matsushita, K., Toyama, H., Tonouchi, N., Okamoto-Kainuma, A., Eds.; Springer: Tokyo, Japan, 2016; pp. 321–338, ISBN 978-4-431-55931-3.
91. Härer, L.; Hilgarth, M.; Ehrmann, M.A. Comparative Genomics of Acetic Acid Bacteria within the Genus Bombella in Light of Beehive Habitat Adaptation. *Microorganisms* **2022**, *10*, 1058. [CrossRef]
92. Nakano, S.; Ebisuya, H. Physiology of *Acetobacter* and *Komagataeibacter* spp.: Acetic Acid Resistance Mechanism in Acetic Acid Fermentation. In *Acetic Acid Bacteria*; Matsushita, K., Toyama, H., Tonouchi, N., Okamoto-Kainuma, A., Eds.; Springer: Tokyo, Japan, 2016; pp. 223–234, ISBN 978-4-431-55931-3.
93. Pelicaen, R.; Weckx, S.; Gonze, D.; De Vuyst, L. Application of Comparative Genomics of *Acetobacter* Species Facilitates Genome-Scale Metabolic Reconstruction of the *Acetobacter ghanensis* LMG 23848T and *Acetobacter* Senegalensis 108B Cocoa Strains. *Front. Microbiol.* **2022**, *13*, 1060160. [CrossRef]
94. Song, J.; Wang, J.; Wang, X.; Zhao, H.; Hu, T.; Feng, Z.; Lei, Z.; Li, W.; Zheng, Y.; Wang, M. Improving the Acetic Acid Fermentation of *Acetobacter pasteurianus* by Enhancing the Energy Metabolism. *Front. Bioeng. Biotechnol.* **2022**, *10*, 815614. [CrossRef]
95. Román-Camacho, J.J.; García-García, I.; Santos-Dueñas, I.M.; García-Martínez, T.; Mauricio, J.C. Latest Trends in Industrial Vinegar Production and the Role of Acetic Acid Bacteria: Classification, Metabolism, and Applications—A Comprehensive Review. *Foods* **2023**, *12*, 3705. [CrossRef]
96. Chakraborty, R.; Roy, S. Exploration of the Diversity and Associated Health Benefits of Traditional Pickles from the Himalayan and Adjacent Hilly Regions of Indian Subcontinent. *J. Food Sci. Technol.* **2018**, *55*, 1599–1613. [CrossRef] [PubMed]
97. Budiman, A.W.; Nam, J.S.; Park, J.H.; Mukti, R.I.; Chang, T.S.; Bae, J.W.; Choi, M.J. Review of Acetic Acid Synthesis from Various Feedstocks Through Different Catalytic Processes. *Catal. Surv. Asia* **2016**, *20*, 173–193. [CrossRef]
98. Xu, Z.; Shi, Z.; Jiang, L. Acetic and Propionic Acids. In *Comprehensive Biotechnology*; Elsevier: Amsterdam, The Netherlands, 2011; pp. 189–199, ISBN 978-0-08-088504-9.
99. Auras, R.; Harte, B.; Selke, S. An Overview of Polylactides as Packaging Materials. *Macromol. Biosci.* **2004**, *4*, 835–864. [CrossRef] [PubMed]
100. Jiang, Z.; Ngai, T. Recent Advances in Chemically Modified Cellulose and Its Derivatives for Food Packaging Applications: A Review. *Polymers* **2022**, *14*, 1533. [CrossRef]
101. Drumright, R.E.; Gruber, P.R.; Henton, D.E. Polylactic Acid Technology. *Adv. Mater.* **2000**, *12*, 1841–1846. [CrossRef]
102. Raveendran, S.; Parameswaran, B.; Ummalyma, S.B.; Abraham, A.; Mathew, A.K.; Madhavan, A.; Rebello, S.; Pandey, A. Applications of Microbial Enzymes in Food Industry. *Food Technol. Biotechnol.* **2018**, *56*, 16–30. [CrossRef]
103. Johnston, C.S.; Steplewska, I.; Long, C.A.; Harris, L.N.; Ryals, R.H. Examination of the Antiglycemic Properties of Vinegar in Healthy Adults. *Ann. Nutr. Metab.* **2010**, *56*, 74–79. [CrossRef]
104. Yanagihara, N.; Mayumi, M.; Yoshikawa, J.; Akuzawa, S.; Fujii, A.; Nagano, M.; Koizumi, Y.; Maehashi, K. Flavor Assessment of a Lactic Fermented Vinegar Described in Japanese Books from the Edo Period (1603–1867). *Heliyon* **2024**, *10*, e32344. [CrossRef]
105. Zhao, Z.; Zhang, Y.; Simpson, D.J.; Gänzle, M.G. A Phylogenetic Perspective on the Performance of Lactococcus Lactis as Starter Culture in Milk and Plant Milks. *Food Res. Int.* **2025**, *221*, 117562. [CrossRef]
106. Caplice, E. Food Fermentations: Role of Microorganisms in Food Production and Preservation. *Int. J. Food Microbiol.* **1999**, *50*, 131–149. [CrossRef] [PubMed]
107. Diez-Ozaeta, I.; Astiazaran, O.J. Fermented Foods: An Update on Evidence-Based Health Benefits and Future Perspectives. *Food Res. Int.* **2022**, *156*, 111133. [CrossRef]
108. Raak, C.; Ostermann, T.; Boehm, K.; Molsberger, F. Regular Consumption of Sauerkraut and Its Effect on Human Health: A Bibliometric Analysis. *Glob. Adv. Health Med.* **2014**, *3*, 12–18. [CrossRef]
109. Sabater, C.; Ruiz, L.; Delgado, S.; Ruas-Madiedo, P.; Margolles, A. Valorization of Vegetable Food Waste and By-Products Through Fermentation Processes. *Front. Microbiol.* **2020**, *11*, 581997. [CrossRef]
110. Şanlier, N.; Gökcen, B.B.; Sezgin, A.C. Health Benefits of Fermented Foods. *Crit. Rev. Food Sci. Nutr.* **2019**, *59*, 506–527. [CrossRef]

 https://doi.org/10.3390/molecules31020333

111. Gonzalez-Garcia, R.; McCubbin, T.; Navone, L.; Stowers, C.; Nielsen, L.; Marcellin, E. Microbial Propionic Acid Production. *Fermentation* **2017**, *3*, 21. [CrossRef]
112. Wang, Y.; Wu, J.; Lv, M.; Shao, Z.; Hungwe, M.; Wang, J.; Bai, X.; Xie, J.; Wang, Y.; Geng, W. Metabolism Characteristics of Lactic Acid Bacteria and the Expanding Applications in Food Industry. *Front. Bioeng. Biotechnol.* **2021**, *9*, 612285. [CrossRef]
113. Awad, H.M.; Diaz, R.; Malek, R.A.; Othman, N.Z.; Aziz, R.A.; El Enshasy, H.A. Efficient Production Process for Food Grade Acetic Acid by *Acetobacter Aceti* in Shake Flask and in Bioreactor Cultures. *J. Chem.* **2012**, *9*, 2275–2286. [CrossRef]
114. Huang, J.; Tang, W.; Zhu, S.; Du, M. Biosynthesis of Butyric Acid by *Clostridium Tyrobutyricum*. *Prep. Biochem. Biotechnol.* **2018**, *48*, 427–434. [CrossRef] [PubMed]
115. Jiang, L.; Fu, H.; Yang, H.K.; Xu, W.; Wang, J.; Yang, S.-T. Butyric Acid: Applications and Recent Advances in Its Bioproduction. *Biotechnol. Adv.* **2018**, *36*, 2101–2117. [CrossRef]
116. Zigová, J.; Šturdík, E. Advances in Biotechnological Production of Butyric Acid. *J. Ind. Microbiol. Biotechnol.* **2000**, *24*, 153–160. [CrossRef]
117. Wong, P.Y.Y.; Kitts, D.D. Chemistry of Buttermilk Solid Antioxidant Activity. *J. Dairy Sci.* **2003**, *86*, 1541–1547. [CrossRef] [PubMed]
118. Ciani, M.; Comitini, F.; Mannazzu, I. Fermentation. In *Encyclopedia of Ecology*; Elsevier: Amsterdam, The Netherlands, 2013; pp. 310–321, ISBN 978-0-444-64130-4.
119. Ahmadi, N.; Khosravi-Darani, K.; Mortazavian, A.M. An Overview of Biotechnological Production of Propionic Acid: From Upstream to Downstream Processes. *Electron. J. Biotechnol.* **2017**, *28*, 67–75. [CrossRef]
120. Bao, T.; Feng, J.; Jiang, W.; Fu, H.; Wang, J.; Yang, S.-T. Recent Advances in N-Butanol and Butyrate Production Using Engineered Clostridium Tyrobutyricum. *World J. Microbiol. Biotechnol.* **2020**, *36*, 138. [CrossRef]
121. Folch, P.L.; Bisschops, M.M.M.; Weusthuis, R.A. Metabolic Energy Conservation for Fermentative Product Formation. *Microb. Biotechnol.* **2021**, *14*, 829–858. [CrossRef] [PubMed]
122. Tao, Y.; Bu, C.; Zou, L.; Hu, Y.; Zheng, Z.-J.; Ouyang, J. A Comprehensive Review on Microbial Production of 1,2-Propanediol: Micro-Organisms, Metabolic Pathways, and Metabolic Engineering. *Biotechnol. Biofuels* **2021**, *14*, 216. [CrossRef]
123. Torres-León, C.; Chávez-González, M.L.; Hernández-Almanza, A.; Martínez-Medina, G.A.; Ramírez-Guzmán, N.; Londoño-Hernández, L.; Aguilar, C.N. Recent Advances on the Microbiological and Enzymatic Processing for Conversion of Food Wastes to Valuable Bioproducts. *Curr. Opin. Food Sci.* **2021**, *38*, 40–45. [CrossRef]
124. Lizardi-Jiménez, M.A.; Hernández-Martínez, R. Solid State Fermentation (SSF): Diversity of Applications to Valorize Waste and Biomass. *3 Biotech* **2017**, *7*, 44. [CrossRef] [PubMed]
125. Azimi, H.; Tezel, H.; Thibault, J. Optimization of the in Situ Recovery of Butanol from ABE Fermentation Broth via Membrane Pervaporation. *Chem. Eng. Res. Des.* **2019**, *150*, 49–64. [CrossRef]
126. Oštarić, F.; Antunac, N.; Cubric-Curik, V.; Curik, I.; Jurić, S.; Kazazić, S.; Kiš, M.; Vinceković, M.; Zdolec, N.; Špoljarić, J.; et al. Challenging Sustainable and Innovative Technologies in Cheese Production: A Review. *Processes* **2022**, *10*, 529. [CrossRef]
127. Puri, R.; Bot, F.; Singh, U.; O'Mahony, J.A. Influence of Transglutaminase Crosslinking on Casein Protein Fractionation during Low Temperature Microfiltration. *Foods* **2021**, *10*, 3146. [CrossRef]
128. Salque, M.; Bogucki, P.I.; Pyzel, J.; Sobkowiak-Tabaka, I.; Grygiel, R.; Szmyt, M.; Evershed, R.P. Earliest Evidence for Cheese Making in the Sixth Millennium Bc in Northern Europe. *Nature* **2013**, *493*, 522–525. [CrossRef] [PubMed]
129. Kok, C.R.; Hutkins, R. Yogurt and Other Fermented Foods as Sources of Health-Promoting Bacteria. *Nutr. Rev.* **2018**, *76*, 4–15. [CrossRef]
130. Tropea, A. Food Waste Valorization. *Fermentation* **2022**, *8*, 168. [CrossRef]
131. Kim, D.-H.; Jeong, D.; Kim, H.; Seo, K.-H. Modern Perspectives on the Health Benefits of Kefir in next Generation Sequencing Era: Improvement of the Host Gut Microbiota. *Crit. Rev. Food Sci. Nutr.* **2019**, *59*, 1782–1793. [CrossRef]
132. Canani, R.B. Potential Beneficial Effects of Butyrate in Intestinal and Extraintestinal Diseases. *World J. Gastroenterol.* **2011**, *17*, 1519. [CrossRef]
133. Hamer, H.M.; Jonkers, D.; Venema, K.; Vanhoutvin, S.; Troost, F.J.; Brummer, R.-J. Review Article: The Role of Butyrate on Colonic Function. *Aliment. Pharmacol. Ther.* **2008**, *27*, 104–119. [CrossRef]
134. Kuchta-Noctor, A.M.; Murray, B.A.; Stanton, C.; Devery, R.; Kelly, P.M. Anticancer Activity of Buttermilk Against SW480 Colon Cancer Cells Is Associated with Caspase-Independent Cell Death and Attenuation of Wnt, Akt, and ERK Signaling. *Nutr. Cancer* **2016**, *68*, 1234–1246. [CrossRef] [PubMed]
135. Arpaia, N.; Campbell, C.; Fan, X.; Dikiy, S.; Van Der Veeken, J.; deRoos, P.; Liu, H.; Cross, J.R.; Pfeffer, K.; Coffer, P.J.; et al. Metabolites Produced by Commensal Bacteria Promote Peripheral Regulatory T-Cell Generation. *Nature* **2013**, *504*, 451–455. [CrossRef] [PubMed]

 https://doi.org/10.3390/molecules31020333

136. Furusawa, Y.; Obata, Y.; Fukuda, S.; Endo, T.A.; Nakato, G.; Takahashi, D.; Nakanishi, Y.; Uetake, C.; Kato, K.; Kato, T.; et al. Commensal Microbe-Derived Butyrate Induces the Differentiation of Colonic Regulatory T Cells. *Nature* **2013**, *504*, 446–450. [CrossRef]
137. Bourassa, M.W.; Alim, I.; Bultman, S.J.; Ratan, R.R. Butyrate, Neuroepigenetics and the Gut Microbiome: Can a High Fiber Diet Improve Brain Health? *Neurosci. Lett.* **2016**, *625*, 56–63. [CrossRef] [PubMed]
138. Stilling, R.M.; Van De Wouw, M.; Clarke, G.; Stanton, C.; Dinan, T.G.; Cryan, J.F. The Neuropharmacology of Butyrate: The Bread and Butter of the Microbiota-Gut-Brain Axis? *Neurochem. Int.* **2016**, *99*, 110–132. [CrossRef] [PubMed]
139. Blank-Porat, D.; Gruss-Fischer, T.; Tarasenko, N.; Malik, Z.; Nudelman, A.; Rephaeli, A. The Anticancer Prodrugs of Butyric Acid AN-7 and AN-9, Possess Antiangiogenic Properties. *Cancer Lett.* **2007**, *256*, 39–48. [CrossRef]
140. Han, A.; Bennett, N.; Ahmed, B.; Whelan, J.; Donohoe, D.R. Butyrate Decreases Its Own Oxidation in Colorectal Cancer Cells through Inhibition of Histone Deacetylases. *Oncotarget* **2018**, *9*, 27280–27292. [CrossRef] [PubMed]
141. Fox, P.F.; Guinee, T.P.; Cogan, T.M.; McSweeney, P.L.H. *Fundamentals of Cheese Science*; Springer: Boston, MA, USA, 2017; ISBN 978-1-4899-7679-6.
142. Gobbetti, M. The Sourdough Microflora: Interactions of Lactic Acid Bacteria and Yeasts. *Trends Food Sci. Technol.* **1998**, *9*, 267–274. [CrossRef]
143. Huang, W.; Dong, A.; Pham, H.T.; Zhou, C.; Huo, Z.; Wätjen, A.P.; Prakash, S.; Bang-Berthelsen, C.H.; Turner, M.S. Evaluation of the Fermentation Potential of Lactic Acid Bacteria Isolated from Herbs, Fruits and Vegetables as Starter Cultures in Nut-Based Milk Alternatives. *Food Microbiol.* **2023**, *112*, 104243. [CrossRef]
144. Thierry, A.; Maillard, M.-B. Production of Cheese Flavour Compounds Derived from Amino Acid Catabolism by *Propionibacterium freudenreichii*. *Le Lait* **2002**, *82*, 17–32. [CrossRef]
145. Zheng, X.; Shi, X.; Wang, B. A Review on the General Cheese Processing Technology, Flavor Biochemical Pathways and the Influence of Yeasts in Cheese. *Front. Microbiol.* **2021**, *12*, 703284. [CrossRef]
146. Damián, M.R.; Cortes-Perez, N.G.; Quintana, E.T.; Ortiz-Moreno, A.; Garfias Noguez, C.; Cruceño-Casarrubias, C.E.; Sánchez Pardo, M.E.; Bermúdez-Humarán, L.G. Functional Foods, Nutraceuticals and Probiotics: A Focus on Human Health. *Microorganisms* **2022**, *10*, 1065. [CrossRef]
147. Turkmen, N.; Akal, C.; Özer, B. Probiotic Dairy-Based Beverages: A Review. *J. Funct. Foods* **2019**, *53*, 62–75. [CrossRef]
148. Chen, L.; Shen, Y.; Wang, C.; Ding, L.; Zhao, F.; Wang, M.; Fu, J.; Wang, H. Megasphaera Elsdenii Lactate Degradation Pattern Shifts in Rumen Acidosis Models. *Front. Microbiol.* **2019**, *10*, 162. [CrossRef]
149. Soemarie, Y.B.; Milanda, T.; Barliana, M.I. Fermented Foods as Probiotics: A Review. *J. Adv. Pharm. Technol. Res.* **2021**, *12*, 335–339. [CrossRef]
150. Thomas, T.A.; Deters, A.; Miller, A.C.; Loh, H.Y.; Engle, T.E.; Nagaraja, T.G. PSXII-29 Development of an in Vitro Rumen Fermentation Model to Assess *Fusobacterium necrophorum* Growth Inhibition. *J. Anim. Sci.* **2023**, *101*, 641–642. [CrossRef]
151. Leeuwendaal, N.K.; Stanton, C.; O'Toole, P.W.; Beresford, T.P. Fermented Foods, Health and the Gut Microbiome. *Nutrients* **2022**, *14*, 1527. [CrossRef] [PubMed]
152. Lisko, D.; Johnston, G.; Johnston, C. Effects of Dietary Yogurt on the Healthy Human Gastrointestinal (GI) Microbiome. *Microorganisms* **2017**, *5*, 6. [CrossRef]
153. Tropea, A.; Potortì, A.G.; Lo Turco, V.; Russo, E.; Vadalà, R.; Rando, R.; Di Bella, G. Aquafeed Production from Fermented Fish Waste and Lemon Peel. *Fermentation* **2021**, *7*, 272. [CrossRef]
154. Marco, M.L.; Heeney, D.; Binda, S.; Cifelli, C.J.; Cotter, P.D.; Foligné, B.; Gänzle, M.; Kort, R.; Pasin, G.; Pihlanto, A.; et al. Health Benefits of Fermented Foods: Microbiota and Beyond. *Curr. Opin. Biotechnol.* **2017**, *44*, 94–102. [CrossRef] [PubMed]
155. Abedi, E.; Hashemi, S.M.B. Lactic Acid Production—Producing Microorganisms and Substrates Sources-State of Art. *Heliyon* **2020**, *6*, e04974. [CrossRef]
156. Aguirre-Garcia, Y.L.; Nery-Flores, S.D.; Campos-Muzquiz, L.G.; Flores-Gallegos, A.C.; Palomo-Ligas, L.; Ascacio-Valdés, J.A.; Sepúlveda-Torres, L.; Rodríguez-Herrera, R. Lactic Acid Fermentation in the Food Industry and Bio-Preservation of Food. *Fermentation* **2024**, *10*, 168. [CrossRef]
157. Carr, F.J.; Chill, D.; Maida, N. The Lactic Acid Bacteria: A Literature Survey. *Crit. Rev. Microbiol.* **2002**, *28*, 281–370. [CrossRef] [PubMed]
158. Bintsis, T. Lactic Acid Bacteria as Starter Cultures: An Update in Their Metabolism and Genetics. *AIMS Microbiol.* **2018**, *4*, 665–684. [CrossRef] [PubMed]
159. Leroy, F.; De Vuyst, L. Lactic Acid Bacteria as Functional Starter Cultures for the Food Fermentation Industry. *Trends Food Sci. Technol.* **2004**, *15*, 67–78. [CrossRef]

 https://doi.org/10.3390/molecules31020333

160. Gänzle, M.G. Lactic Metabolism Revisited: Metabolism of Lactic Acid Bacteria in Food Fermentations and Food Spoilage. *Curr. Opin. Food Sci.* **2015**, *2*, 106–117. [CrossRef]
161. Liu, S. Practical Implications of Lactate and Pyruvate Metabolism by Lactic Acid Bacteria in Food and Beverage Fermentations. *Int. J. Food Microbiol.* **2003**, *83*, 115–131. [CrossRef]
162. Zheng, Z.; Sheng, B.; Ma, C.; Zhang, H.; Gao, C.; Su, F.; Xu, P. Relative Catalytic Efficiency of *ldhL*- and *ldhD*-Encoded Products Is Crucial for Optical Purity of Lactic Acid Produced by Lactobacillus Strains. *Appl. Environ. Microbiol.* **2012**, *78*, 3480–3483. [CrossRef]
163. Leenay, R.T.; Vento, J.M.; Shah, M.; Martino, M.E.; Leulier, F.; Beisel, C.L. Genome Editing with CRISPR-Cas9 in *Lactobacillus plantarum* Revealed That Editing Outcomes Can Vary Across Strains and Between Methods. *Biotechnol. J.* **2019**, *14*, 1700583. [CrossRef]
164. Leite, A.M.D.O.; Miguel, M.A.L.; Peixoto, R.S.; Rosado, A.S.; Silva, J.T.; Paschoalin, V.M.F. Microbiological, Technological and Therapeutic Properties of Kefir: A Natural Probiotic Beverage. *Braz. J. Microbiol.* **2013**, *44*, 341–349. [CrossRef]
165. Rastogi, Y.R.; Thakur, R.; Thakur, P.; Mittal, A.; Chakrabarti, S.; Siwal, S.S.; Thakur, V.K.; Saini, R.V.; Saini, A.K. Food Fermentation—Significance to Public Health and Sustainability Challenges of Modern Diet and Food Systems. *Int. J. Food Microbiol.* **2022**, *371*, 109666. [CrossRef]
166. Sadh, P.K.; Kumar, S.; Chawla, P.; Duhan, J.S. Fermentation: A Boon for Production of Bioactive Compounds by Processing of Food Industries Wastes (By-Products). *Molecules* **2018**, *23*, 2560. [CrossRef]
167. Gunatillake, P.; Adhikari, R. Biodegradable Synthetic Polymers for Tissue Engineering. *Eur. Cell. Mater.* **2003**, *5*, 1–16. [CrossRef]
168. Delgado, S.; Guadamuro, L.; Flórez, A.B.; Vázquez, L.; Mayo, B. Fermentation of Commercial Soy Beverages with Lactobacilli and Bifidobacteria Strains Featuring High β-Glucosidase Activity. *Innov. Food Sci. Emerg. Technol.* **2019**, *51*, 148–155. [CrossRef]
169. Fisberg, M.; Machado, R. History of Yogurt and Current Patterns of Consumption. *Nutr. Rev.* **2015**, *73*, 4–7. [CrossRef]
170. Pame, K.; Laskar, S.; Borah, S. Utilization of Processed Animal Byproducts as a Raw Material to Develop Shelf-Stable and Cost Effective Pet Food. *Int. J. Vet. Sci. Anim. Husb.* **2023**, *8*, 31–34. [CrossRef]
171. Read, J. Living Fermented Foods and Drinks. In *Oxford Research Encyclopedia of Food Studies*; Oxford University Press: Oxford, UK, 2024; ISBN 978-0-19-776253-0.
172. Robinson, R.K. (Ed.) *Modern Dairy Technology*; Springer: Boston, MA, USA, 1993; ISBN 978-1-4684-8174-7.
173. Smid, E.J.; Lacroix, C. Microbe–Microbe Interactions in Mixed Culture Food Fermentations. *Curr. Opin. Biotechnol.* **2013**, *24*, 148–154. [CrossRef]
174. Tsafrakidou, P.; Michaelidou, A.-M.; G. Biliaderis, C. Fermented Cereal-Based Products: Nutritional Aspects, Possible Impact on Gut Microbiota and Health Implications. *Foods* **2020**, *9*, 734. [CrossRef] [PubMed]
175. Behera, S.S.; El Sheikha, A.F.; Hammami, R.; Kumar, A. Traditionally Fermented Pickles: How the Microbial Diversity Associated with Their Nutritional and Health Benefits? *J. Funct. Foods* **2020**, *70*, 103971. [CrossRef]
176. Cárdenas, N.; Calzada, J.; Peirotén, Á.; Jiménez, E.; Escudero, R.; Rodríguez, J.M.; Medina, M.; Fernández, L. Development of a Potential Probiotic Fresh Cheese Using Two *Lactobacillus salivarius* Strains Isolated from Human Milk. *BioMed Res. Int.* **2014**, *2014*, 1–12. [CrossRef]
177. Conway, V.; Couture, P.; Gauthier, S.; Pouliot, Y.; Lamarche, B. Effect of Buttermilk Consumption on Blood Pressure in Moderately Hypercholesterolemic Men and Women. *Nutrition* **2014**, *30*, 116–119. [CrossRef] [PubMed]
178. Cuamatzin-García, L.; Rodríguez-Rugarcía, P.; El-Kassis, E.G.; Galicia, G.; Meza-Jiménez, M.D.L.; Baños-Lara, M.D.R.; Zaragoza-Maldonado, D.S.; Pérez-Armendáriz, B. Traditional Fermented Foods and Beverages from around the World and Their Health Benefits. *Microorganisms* **2022**, *10*, 1151. [CrossRef]
179. EL-Sayed, A.I.M.; El-Borai, A.M.; Akl, S.H.; EL-Aassar, S.A.; Abdel-Latif, M.S. Identification of Lactobacillus Strains from Human Mother Milk and Cottage Cheese Revealed Potential Probiotic Properties with Enzymatic Activity. *Sci. Rep.* **2022**, *12*, 22522. [CrossRef]
180. Kim, B.; Mun, E.-G.; Kim, D.; Kim, Y.; Park, Y.; Lee, H.-J.; Cha, Y.-S. A survey of research papers on the health benefits of kimchi and kimchi lactic acid bacteria. *J. Nutr. Health* **2018**, *51*, 1–13. [CrossRef]
181. Koehler, K.; Drenowatz, C. Integrated Role of Nutrition and Physical Activity for Lifelong Health. *Nutrients* **2019**, *11*, 1437. [CrossRef]
182. Santos, A.; San Mauro, M.; Sanchez, A.; Torres, J.M.; Marquina, D. The Antimicrobial Properties of Different Strains of Lactobacillus Spp. Isolated from Kefir. *Syst. Appl. Microbiol.* **2003**, *26*, 434–437. [CrossRef]
183. Sasaki, M.; Oba, C.; Nakamura, K.; Takeo, H.; Toya, H.; Furuichi, K. Milk-Based Culture of Penicillium Camemberti and Its Component Oleamide Affect Cognitive Function in Healthy Elderly Japanese Individuals: A Multi-Arm Randomized, Double-Blind, Placebo-Controlled Study. *Front. Nutr.* **2024**, *11*, 1357920. [CrossRef]

184. Shankar, V.; Mahboob, S.; Al-Ghanim, K.A.; Ahmed, Z.; Al-Mulhm, N.; Govindarajan, M. A Review on Microbial Degradation of Drinks and Infectious Diseases: A Perspective of Human Well-Being and Capabilities. *J. King Saud Univ. Sci.* **2021**, *33*, 101293. [CrossRef]
185. Tapsell, L.C. Fermented Dairy Food and CVD Risk. *Br. J. Nutr.* **2015**, *113*, S131–S135. [CrossRef] [PubMed]
186. Csatlos, N.-I.; Simon, E.; Teleky, B.-E.; Szabo, K.; Diaconeasa, Z.M.; Vodnar, D.-C.; Ciont, C.; Pop, O.-L. Development of a Fermented Beverage with Chlorella Vulgaris Powder on Soybean-Based Fermented Beverage. *Biomolecules* **2023**, *13*, 245. [CrossRef] [PubMed]
187. Dullius, A.; Rama, G.R.; Giroldi, M.; Goettert, M.I.; Lehn, D.N.; Volken De Souza, C.F. Bioactive Peptide Production in Fermented Foods. In *Current Developments in Biotechnology and Bioengineering*; Elsevier: Amsterdam, The Netherlands, 2022; pp. 47–72, ISBN 978-0-12-823506-5.
188. Kregiel, D. Health Safety of Soft Drinks: Contents, Containers, and Microorganisms. *BioMed Res. Int.* **2015**, *2015*, 1–15. [CrossRef]
189. Piwowarek, K.; Lipińska, E.; Hać-Szymańczuk, E.; Kieliszek, M.; Ścibisz, I. Propionibacterium Spp.—Source of Propionic Acid, Vitamin B12, and Other Metabolites Important for the Industry. *Appl. Microbiol. Biotechnol.* **2018**, *102*, 515–538. [CrossRef]
190. Dumas, E.; Feurtey, A.; Rodríguez De La Vega, R.C.; Le Prieur, S.; Snirc, A.; Coton, M.; Thierry, A.; Coton, E.; Le Piver, M.; Roueyre, D.; et al. Independent Domestication Events in the Blue-cheese Fungus *Penicillium roqueforti*. *Mol. Ecol.* **2020**, *29*, 2639–2660. [CrossRef]
191. Miyamoto, J.; Ando, Y.; Yamano, M.; Nishida, A.; Murakami, K.; Kimura, I. Acidipropionibacterium Acidipropionici, a Propionate-Producing Bacterium, Contributes to GPR41 Signaling and Metabolic Regulation in High-Fat Diet-Induced Obesity in Mice. *Front. Nutr.* **2025**, *12*, 1542196. [CrossRef] [PubMed]
192. Baur, T.; Dürre, P. New Insights into the Physiology of the Propionate Producers Anaerotignum Propionicum and Anaerotignum Neopropionicum (Formerly Clostridium Propionicum and Clostridium Neopropionicum). *Microorganisms* **2023**, *11*, 685. [CrossRef] [PubMed]
193. Seeliger, S.; Janssen, P.H.; Schink, B. Energetics and Kinetics of Lactate Fermentation to Acetate and Propionate via Methylmalonyl-CoA or Acrylyl-CoA. *FEMS Microbiol. Lett.* **2002**, *211*, 65–70. [CrossRef]
194. Reichardt, N.; Duncan, S.H.; Young, P.; Belenguer, A.; McWilliam Leitch, C.; Scott, K.P.; Flint, H.J.; Louis, P. Phylogenetic Distribution of Three Pathways for Propionate Production within the Human Gut Microbiota. *ISME J.* **2014**, *8*, 1323–1335. [CrossRef] [PubMed]
195. Russell, J.B.; Hespell, R.B. Microbial Rumen Fermentation. *J. Dairy Sci.* **1981**, *64*, 1153–1169. [CrossRef]
196. Zhang, S.-M.; Huang, S.-L. The Commensal Anaerobe Veillonella Dispar Reprograms Its Lactate Metabolism and Short-Chain Fatty Acid Production during the Stationary Phase. *Microbiol. Spectr.* **2023**, *11*, e03558-22. [CrossRef]
197. Nguyen, T.T.; Kim, Y.-K.; Nguyen, T.V.T.; Kwon, J.; Bang, Y.-J. Metabolic Profiling and Genetic Tool Development in the Mucosal Bacterium Selenomonas Sputigena. *Genes Genomics* **2025**, *47*, 997–1009. [CrossRef]
198. Dighe, A.S.; Shouche, Y.S.; Ranade, D.R. *Selenomonas lipolytica* Sp. Nov., an Obligately Anaerobic Bacterium Possessing Lipolytic Activity. *Int. J. Syst. Bacteriol.* **1998**, *48*, 783–791. [CrossRef]
199. Cabral, L.D.S.; Weimer, P.J. Megasphaera Elsdenii: Its Role in Ruminant Nutrition and Its Potential Industrial Application for Organic Acid Biosynthesis. *Microorganisms* **2024**, *12*, 219. [CrossRef]
200. Li, B.; Gao, W.; Pan, Y.; Yao, Y.; Liu, G. Progress in 1,3-Propanediol Biosynthesis. *Front. Bioeng. Biotechnol.* **2024**, *12*, 1507680. [CrossRef] [PubMed]
201. Vidra, A.; Németh, Á. Bio-Produced Propionic Acid: A Review. *Period. Polytech. Chem. Eng.* **2017**, *62*, 57–67. [CrossRef]
202. Selmer, T.; Willanzheimer, A.; Hetzel, M. Propionate CoA-transferase from *Clostridium propionicum*: Cloning of the Gene and Identification of Glutamate 324 at the Active Site. *Eur. J. Biochem.* **2002**, *269*, 372–380. [CrossRef]
203. Tripathi, A.; Pandey, V.K.; Panesar, P.S.; Taufeeq, A.; Mishra, H.; Rustagi, S.; Malik, S.; Kovács, B.; Suthar, T.; Shaikh, A.M. Fermentative Production of Vitamin B_{12} by *Propionibacterium shermanii* and *Pseudomonas denitrificans* and Its Promising Health Benefits: A Review. *Food Sci. Nutr.* **2024**, *12*, 8675–8691. [CrossRef]
204. Oude Elferink, S.J.W.H.; Krooneman, J.; Gottschal, J.C.; Spoelstra, S.F.; Faber, F.; Driehuis, F. Anaerobic Conversion of Lactic Acid to Acetic Acid and 1,2-Propanediol by *Lactobacillus buchneri*. *Appl. Environ. Microbiol.* **2001**, *67*, 125–132. [CrossRef]
205. Schink, B.; Pfennig, N. *Propionigenium modestum* Gen. Nov. Sp. Nov. a New Strictly Anaerobic, Nonsporing Bacterium Growing on Succinate. *Arch. Microbiol.* **1982**, *133*, 209–216. [CrossRef]
206. Jakobson, C.M.; Slininger, M.F.; Tullman-Ercek, D.; Mangan, N.M. A Systems-Level Model Reveals That 1,2-Propanediol Utilization Microcompartments Enhance Pathway Flux Through Intermediate Sequestration. *PLoS Comput. Biol.* **2017**, *13*, e1005525. [CrossRef]
207. Scott, K.P.; Martin, J.C.; Campbell, G.; Mayer, C.-D.; Flint, H.J. Whole-Genome Transcription Profiling Reveals Genes Up-Regulated by Growth on Fucose in the Human Gut Bacterium *"Roseburia inulinivorans"*. *J. Bacteriol.* **2006**, *188*, 4340–4349. [CrossRef]

 https://doi.org/10.3390/molecules31020333

208. Ferguson, G.P.; Booth, I.R. Importance of Glutathione for Growth and Survival of *Escherichia Coli* Cells: Detoxification of Methylglyoxal and Maintenance of Intracellular K^+. *J. Bacteriol.* **1998**, *180*, 4314–4318. [CrossRef] [PubMed]
209. Furlan Sandrini, D.M.; Morgado, D.L.; De Oliveira, A.J.A.; De Moraes, D.A.; Varanda, L.C.; Frollini, E. Cellulose Esters: Synthesis for Further Formation of Films with Magnetite Nanoparticles Incorporated. *Int. J. Biol. Macromol.* **2024**, *264*, 130594. [CrossRef]

 https://doi.org/10.3390/molecules31020333

molecules

MDPI

Review

Probiotics and Fermented Foods in Human Nutrition

Irene Dini

Department of Pharmacy, University of Naples Federico II, 80131 Napoli, Italy; irdini@unina.it

Abstract

Fermented foods and probiotics represent complementary yet distinct components of human nutrition. Fermented foods are shaped by biochemical transformations driven by microbial metabolism, whereas probiotics are live microorganisms that may confer health benefits to the host. In both cases, bacteria, yeasts, and filamentous fungi mediate key metabolic activities that generate bioactive compounds and modulate host–microbiota interactions. During fermentation, microbial communities synthesize organic acids, peptides, exopolysaccharides, vitamins, and other metabolites that enhance food safety, sensory attributes, and potential health-promoting properties. Several microbial products, such as bacteriocins, reuterin, hydroxylated fatty acids, and exopolysaccharides, exhibit antimicrobial, immunomodulatory, antioxidant, and cholesterol-lowering activities. Advancing our understanding of microbial metabolism in fermented foods is essential for developing next-generation functional foods and nutraceuticals that leverage microbial biotransformations to support human health. Nonetheless, multiple challenges limit the translation of these advances into commercial products. Inadequately controlled fermentation may introduce microbiological or chemical hazards, regulatory frameworks governing microbial use in foods remain insufficiently defined, and standardized procedures for microbial strain handling and characterization are still lacking. This narrative review integrates current evidence on the nutraceutical properties of fermented foods and probiotics, while also examining the associated safety considerations and the technological factors that influence fermentation processes.

Keywords: bioactive metabolites; nutraceutical development; microbial metabolism; postbiotics; food safety; food risk assessment

1. Introduction

Although commonly used interchangeably, probiotics, fermented foods, and fermented beverages represent distinct categories within contemporary food science and regulatory frameworks (Figure 1).

Probiotics, fermented foods, and fermented beverages are often conflated, yet they constitute distinct categories within food science and regulatory frameworks. Probiotics are live, clinically validated microorganisms by human studies that confer health benefits when consumed in adequate amounts, with effects that are strain-specific [1]. In contrast, fermented foods and beverages result from microbial transformations that enhance sensory, nutritional, and functional properties, but the viability and characterization of their microbial communities vary widely, and most strains have not undergone clinical evaluation [2].

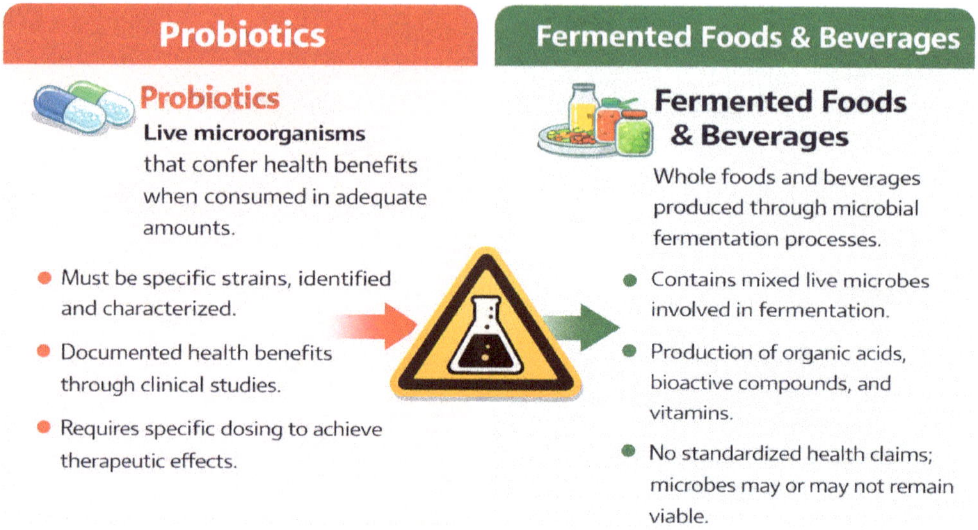

Figure 1. "Probiotics" vs. "fermented food or beverage": key differences and overlaps. Image generated by Microsoft Copilot (April 2026).

This narrative review synthesizes current evidence on the nutraceutical properties of fermented foods and probiotics and examines safety considerations and technological factors influencing fermentation processes. To improve methodological transparency, the literature search for this narrative review followed a structured approach. Databases consulted included PubMed/MEDLINE, Scopus, Web of Science, and Google Scholar. Searches were conducted between January 2010 and February 2026 using predefined keyword combinations ("fermented foods", "microbial metabolism", "probiotics", "postbiotics", "lactic acid bacteria", "food safety", "microbial detoxification"). Inclusion criteria prioritized peer-reviewed studies addressing microbial metabolism, functional metabolites, safety aspects, and nutritional implications. Exclusion criteria included non-scientific sources, papers lacking methodological detail, and studies focused solely on industrial processing without biological relevance. Although not intended as a systematic review, this structured strategy reduces selection bias and enhances reproducibility.

2. Fermented Food and Beverage

Fermentation is one of the earliest biotechnological processes developed by human societies. It is obtained by the metabolic activities of bacteria, yeasts, and filamentous fungi that convert raw substrates into products with improved sensory, nutritional, and functional qualities [3,4]. Beyond enhancing preservation through mechanisms such as acidification, alcohol production, and microbial competition [5], fermentation can generate a wide range of bioactive metabolites with potential health benefits [6]. In recent years, the focus on fermented foods and beverages has shifted beyond the culinary domain toward their capacity to generate postbiotic compounds and support the development of functional foods. According to the ISAPP consensus, postbiotics are preparations of inanimate microorganisms and/or their components that confer a health benefit to the host. This definition includes fermentation-derived metabolites, cell-free supernatants, exopolysaccharides, peptides, and structurally modified phytochemicals, provided that their biological activity is demonstrated independently of microbial viability [7].

 https://doi.org/10.3390/molecules31081353

2.1. Microbial Ecology of Fermentation

Microbial fermentation is an anaerobic metabolic process in which microorganisms convert carbohydrate-rich substrates, including simple sugars, cereal grains, and other plant-derived materials, into alcohols, organic acids, and gaseous by-products [8]. The composition of microbial communities involved in fermentation varies substantially depending on the substrate, environmental conditions, and traditional or regional processing practices [9] (Table 1). Fermentation pathways can be categorized into lactic acid, alcoholic, acetic acid, and mold-based enzymatic fermentations.

Lactic acid fermentation converts carbohydrates into lactic acid via homofermentative or heterofermentative pathways, reducing pH and improving microbial safety [10]. Alcoholic fermentation relies on the glycolytic conversion of sugars into ethanol and carbon dioxide, accompanied by the formation of esters, higher alcohols, and other volatile compounds that shape the sensory profile of fermented beverages [11].

Acetic acid fermentation, via membrane-bound dehydrogenases, involves the oxidation of ethanol to acetic acid, which contributes to acidity and preservative effects [12].

Alkaline fermentation is characterized by extensive proteolysis, during which proteins are degraded into amino acids and peptides with the release of ammonia. The resulting increase in pH to around 8–9 inhibits spoilage microorganisms and imparts the characteristic umami aroma to foods such as nattō, traditionally fermented African legumes, and eggs. This process is primarily driven by *Bacillus* species and coagulase-negative *Staphylococcus*, which secrete extracellular proteinases responsible for proteolysis [13].

Mold-based fermentations produce foods with pronounced umami notes and complex textures, owing to extracellular proteases, amylases, and lipases that release amino acids, peptides, sugars, and aromatic compounds [14].

Table 1. Microbial ecology and metabolite profiles of major fermented food substrates.

Substrate	Examples	Dominant Microbes	Key Metabolites	Reference
Cereals	Sourdough, boza, and kvass.	*Aspergillus, Paecilomyces, Cladosporium, Penicillium, Fusarium*, and *Trichothecium*.	Organic acids, CO_2, ethanol	[15–17]
Dairy	Kefir, yogurt.	Lactic acid bacteria, yeasts.	Lactic acid, peptides, exopolysaccharides	[18–20]
Legumes	Miso, tempeh, natto.	*Aspergillus oryzae, Mucor* spp. *Fusarium* spp., *Rhizopus, Bacillus subtilis.*	Amino acids, glutamate, isoflavone aglycones, and peptides	[21]
Meat/Fish	Fish sauce and salami.	Lactic acid bacteria, *Staphylococcus*, and halophiles.	Amino acids, peptides, organic acids, volatile esters, saturated alcohols	[22,23]
Vegetables	Sauerkrau, kimchi, pickles.	Lactic acid bacteria.	Lactic acid, vitamins, phenolic derivatives (flavonoids, polyphenols), glucosinolate breakdown products, and their derivatives.	[24]
Beverages	Cider, wine, beer, kombucha.	Yeasts, acetic acid bacteria, and lactic acid bacteria.	Ethanol, acetic acid, esters	[25,26]

2.2. Fermentative Microorganisms Contributing to Food Transformation

The diversity of microbial ecosystems involved in fermentation underpins the broad spectrum of biochemical processes responsible for the functional and sensory transformations observed in fermented foods.

https://doi.org/10.3390/molecules31081353

2.2.1. Lactic Acid Bacteria

Lactic acid bacteria (LAB), including *Lactobacillus, Streptococcus, Leuconostoc,* and *Pediococcus*, predominate in dairy, vegetable, and cereal fermentations, where they produce lactic acid and contribute to texture, flavor, and microbial safety.

These Gram-positive microorganisms can be non-spore-forming, aerotolerant, or facultatively anaerobic. Homofermentative LAB produce primarily lactic acid. Heterofermentative LAB generate lactic acid along with ethanol, CO_2, acetic acid, and other metabolites [27].

According to the EFSA BIOHAZ Panel, food-associated LAB, including some antibiotic-resistant strains, rarely pose clinical risks, although they may serve as reservoirs of transmissible resistance genes within the food chain [28]. Most LAB species are classified as Generally Recognized as Safe (GRAS) due to their long history of safe use [29], and several taxa have received Qualified Presumption of Safety (QPS) status from EFSA [30].

Current market analyses project that the LAB sector will reach USD 2.22 billion by 2034 [31].

LAB comprise diverse genera and species used for specific technological functions such as enhancing microbial safety, extending shelf life, and improving sensory attributes in fermented foods [32,33].

They can produce proteolytic enzymes and a wide range of antimicrobial metabolites, including lactic and acetic acids, hydrogen peroxide, reuterin, exopolysaccharides (EPS), hydroxylated fatty acids, and bacteriocins that inhibit spoilage organisms and pathogens [34].

Organic acid production lowers pH, inhibits fungal growth, and contributes to mycotoxin mitigation [35]. Diacetyl disrupts microbial metabolism through reactions with amino groups, whereas hydrogen peroxide generates reactive oxygen species that damage fungal membranes and nucleic acids.

Reuterin, a multifunctional antimicrobial system produced by *Limosilactobacillus reuteri* during glycerol metabolism, exerts broad-spectrum inhibitory activity through multiple mechanisms. Reuterin alkylates thiol groups in microbial proteins and enzymes, disrupting redox homeostasis, impairing essential metabolic pathways, and inducing oxidative and electrophilic stress. This thiol-reactive activity leads to the inactivation of key enzymes involved in DNA synthesis, amino acid metabolism, and energy production, ultimately compromising cell viability. Although its synergistic action with lactic acid has been confirmed against bacteria, whether this interaction enhances antifungal efficacy remains unclear [36].

Fermentation by lactic acid bacteria reduces acrylamide levels by enabling its binding to cell wall components, thereby decreasing its bioavailability and overall concentration in food products [37].

Some *Lactiplantibacillus plantarum* strains can hydroxylate unsaturated fatty acids to form hydroxy derivatives, such as 3-hydroxytetradecanoic and 3-hydroxydodecanoic acids, able to disrupt fungal membranes and exert antifungal activity under the acidic conditions typical of fermented foods [29].

The EPS produced by LAB can have functional activities [38,39], support beneficial gut colonization [40], enhance probiotic persistence in the gastrointestinal tract [41], and support gut protection and cholesterol reduction [42].

Moreover, LAB can produce bacteriocins that act as antibacterial, antifungal, and antimycotoxigenic agents [43]. Due to their physicochemical stability and minimal impact on sensory attributes, bacteriocins are regarded as promising biopreservatives. Their antimicrobial activity is mediated through well-defined mechanisms, including pore formation in target cell membranes, disruption of the proton motive force, inhibition of cell wall biosynthesis, and

 https://doi.org/10.3390/molecules31081353

interference with essential enzymatic pathways. Despite these advantages, their application in food systems may be limited by proteolytic degradation, uneven distribution within complex matrices, and interactions with food components that reduce bioavailability. These limitations can be mitigated through micro- and nanocapsulation strategies, which enhance stability and controlled release, or through their combination with complementary preservation technologies, such as mild heat treatments, high-pressure processing, or organic acid supplementation, to achieve synergistic antimicrobial effects [35]. The only bacteriocin approved for food use is Nisin, produced by *Lactococcus lactis*. It acts by binding lipid II and inducing pore formation in bacterial membranes [44]. It can limit Gram-positive bacteria and, indirectly, fungal growth by suppressing the growth of supportive bacteria [45].

LAB contribute to organoleptic development primarily through glycolysis, proteolysis, and, to a lesser extent, lipolysis [34]. Proteolysis is particularly relevant for flavor formation, generating amino acid-derived compounds such as alcohols, acids, esters, sulfur compounds, and aldehydes [46]. Some *Pediococcus* species, however, are associated with food spoilage [47].

2.2.2. Acetic Acid Bacteria

Acetic acid bacteria (AAB), including *Acetobacter* and *Gluconobacter*, are Gram-negative bacteria that oxidize ethanol to acetic acid in aerobic conditions [12].

AAB are most active at 40–50 °C [48]. They are considered safe in traditional food fermentations; however, they are not included in the EFSA Qualified Presumption of Safety (QPS) list, cannot be automatically approved for use in novel foods or starter cultures, and therefore require strain-specific toxicological and genomic assessment [49]. AABs have significant industrial relevance, particularly in the production of vinegar, fermented beverages, and other oxidative fermentations, including sour beer, cocoa and coffee fermentations, kefir, kombucha, and sourdough [4]. AAB are obligately aerobic microorganisms capable of incompletely oxidizing sugars and alcohols into organic acids. Their metabolic activities also yield valuable compounds, including bacteriocins, extracellular polysaccharides, bacterial cellulose, aromatic metabolites, and essential coenzymes [26]. AAB can interact with co-occurring microorganisms to produce metabolites and drive the distinctive sensory profiles of fermented foods. Because these interactions vary with strain traits, nutrient availability, ecological conditions, and fermentation time, understanding them is essential for optimizing AAB performance in food processes [50].

2.2.3. Yeasts

Yeast metabolism is crucial for establishing the thermal and biochemical conditions required for efficient fermentation. Through alcoholic fermentation, yeasts generate ethanol, CO_2, and a diverse array of volatile metabolites, including higher alcohols, aldehydes, ketones, glycerol, and organic acids. The associated heat production, together with ethanol oxidation by acetic acid bacteria, maintains temperatures above 40 °C, a threshold essential for embryo inactivation, flavor-precursor formation, and proper color development [48]. Yeasts also degrade pectin in the pulp, improving aeration and supporting AAB growth [51]. Key yeast species include *Saccharomyces cerevisiae*, *Candida milleri*, *Pichia kudriavzevii*, and *Hanseniaspora opuntiae* [52].

Saccharomyces cerevisiae is highly fermentative and widely used as a starter culture to enhance fermentation efficiency and sensory quality of beer, wine, bread, sake, cider, and kefir.

Candida milleri contributes to the fermentation of kombucha, sourdough, and kefir [4].

 https://doi.org/10.3390/molecules31081353

Pichia kudriavzevii contributes aromatic compounds and tolerates high ethanol levels, while *Hanseniaspora opuntiae* produces esters associated with fruity notes [53].

Debaryomyces hansenii has been reported to limit or prevent the formation of biogenic amines by reducing the activity of decarboxylase-positive microorganisms [54].

Several yeasts have been evaluated as biological control agents against pathogenic bacteria. *Candida* spp., *Geotrichum candidum*, and *Pichia* spp. have exhibited inhibitory activity, particularly against *Listeria monocytogenes*.

Unfortunately, most evidence for yeast-mediated antibacterial effects derives from studies conducted in commercial media rather than in dry-cured fermented food matrices, so it is not known how they affect the technological or sensory properties of foods [55].

2.2.4. Molds

Filamentous fungi such as *Aspergillus*, *Sordariaceae*, *Paecilomyces*, *Penicillium*, *Cladosporium*, and *Rhizopus* are central to the fermentation of many traditional foods, especially across Asia.

Molds can hydrolyze complex carbohydrates, proteins, and lipids through their powerful enzyme systems, thereby enhancing food flavor, texture, and nutrient bioavailability [56].

Rhizopus-derived phytases degrade phytates, thereby enhancing the bioavailability of minerals (zinc, iron, and calcium) [57]. The reduction in oligosaccharides such as stachyose and raffinose diminishes the flatulence-inducing properties of soy [58]. β-Glucosidases convert glycosylated polyphenols into their aglycones, which exhibit superior bioavailability and biological activity [59].

Mold fermentations can also generate beneficial metabolites, such as GABA, associated with relaxation and reduced anxiety [60], as well as the biosynthesis and release of polyphenols with antioxidant properties [17,61]. Examples of filamentous fungi that play a key role in traditional mold-based fermentations are *Aspergillus oryzae*, *Rhizopus oligosporus*, *Neurospora crassa*, *Penicillium roqueforti*, *Penicillium camemberti*, and *Monascus purpureus*.

Aspergillus oryzae drives the fermentation of soybeans and cereal substrates in traditional products such as soy sauce, miso, and sake [62,63]. *Rhizopus oligosporus* catalyzes the transformation of soy proteins into peptides and free amino acids during tempeh production [64], while *Neurospora crassa* contributes to specific soy-based fermentations [65]. In dairy systems, *Penicillium roqueforti* and *Penicillium camemberti* are responsible for the characteristic textures, flavors, and aromatic profiles of blue and soft-ripened cheeses [66]. Additionally, *Monascus purpureus* is employed in the production of red yeast rice, generating pigments and bioactive metabolites of technological and nutritional relevance [67].

2.2.5. Mixed Fungal–Bacterial Consortia

Microbial consortia, structured communities of bacteria, yeasts, and filamentous fungi, exhibit synergistic interactions that markedly enhance plant-based fermentation. These mixed cultures efficiently depolymerize complex plant macromolecules, improve flavor and texture, and generate a wide array of bioactive metabolites, including polyphenols, peptides, carotenoids, and bacteriocins [68] (Figure 2). They also reduce antinutritional factors such as alkaloids, oxalates, and tannins [69].

Selected fungal strains can be employed as starter cultures to standardize production and accelerate fermentation. Traditional mixed fermentations, such as kefir and sourdough, rely on stable microbial consortia whose coordinated metabolic activities yield complex and distinctive metabolite profiles [70].

https://doi.org/10.3390/molecules31081353

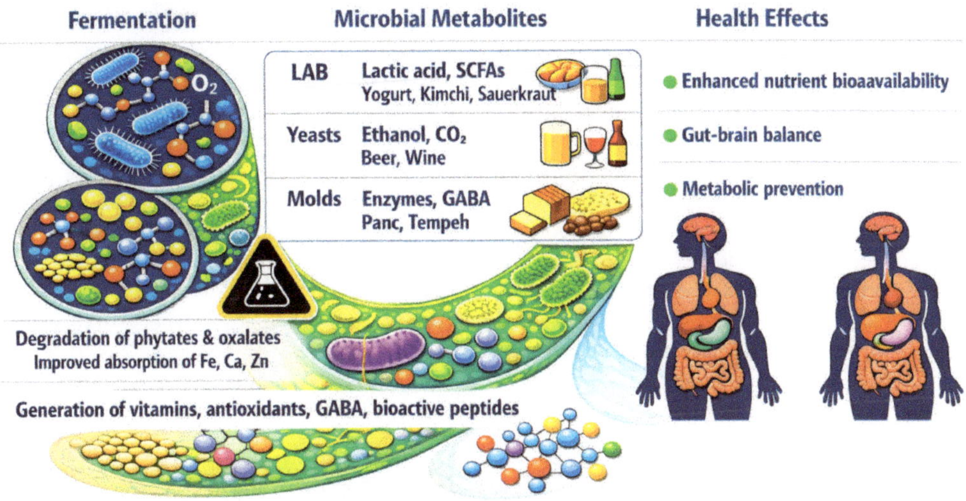

Figure 2. The fermentation frontier: how microbes transform foods in nutraceuticals. Image generated by Microsoft Copilot (April 2026).

Co-culturing lactic acid bacteria with yeasts has been shown to extend shelf life and enhance sensory quality, while fungal partners contribute additional enzymatic functions that increase nutrient bioavailability [71].

Emerging strategies, including synthetic microbial communities, enable the rational design of simplified consortia that improve safety, limit the formation of undesirable compounds, and tailor the functional properties of fermented foods [71].

Advances in multi-omics technologies have further accelerated the characterization of these communities, supporting the selection of strains that enhance flavor, texture, and nutrient accessibility in plant-based matrices [72,73]. Moreover, genetic engineering and optimized bioreactor systems are driving the development of more efficient, scalable, and sustainable fermentation processes [74].

3. Nutritional and Functional Implications

Microorganisms can enhance nutrient bioavailability, detoxify foods contaminated with microbial toxins, mycotoxins, pesticides, heavy metals, and naturally occurring antinutritional factors. Moreover, they can generate a broad spectrum of bioactive compounds and modulate the host gut microbiota [6–42]. Health effects attributed to probiotics and microorganisms present in fermented foods are highly strain-specific and cannot be generalized at the species or genus level. Even closely related strains may differ substantially in their genomic profiles, metabolic outputs, and functional properties, including the production of organic acids, exopolysaccharides, bioactive peptides, and detoxifying enzymes. In addition, microbial behavior is strongly influenced by the food matrix: the same strain may exhibit different growth dynamics, enzymatic activities, and metabolite profiles in dairy, cereal, legume, or vegetable substrates. These matrix-dependent interactions further limit the extrapolation of findings across different fermented foods. To avoid overgeneralization, the revised manuscript explicitly distinguishes mechanistic observations from clinically validated outcomes and highlights the need for strain-resolved characterization when interpreting functional effects.

 https://doi.org/10.3390/molecules31081353

3.1. Improvements in Mineral and Protein Bioaccessibility Through Microbial Fermentation

Fermentation enhances protein digestibility by hydrolyzing proteins into peptides and free amino acids. This process is particularly relevant for legumes such as soybeans, which, despite their high protein content, contain antinutritional inhibitors that impair protein absorption. Fermentation with *Lactobacillus* species has been shown to enhance the digestibility and bioavailability of soybean proteins [75]. Comparable improvements have been reported in in vitro studies on fermented lentils, where fermentation significantly increased the proportion of digested protein [76]. Moreover, fermentation can improve mineral bioavailability. The limited bioavailability of minerals in plant-derived foods is largely attributed to their association with non-digestible compounds that form stable complexes resistant to human digestion. Fermentation can mitigate this limitation by degrading such complexes and releasing bound minerals, thereby improving their accessibility [77]. In vitro studies have reported increased levels of zinc, calcium, iron, and magnesium in fermented products, primarily due to the reduction in phytate content during microbial activity [78]. However, the apparent increase in mineral concentrations may also reflect dry-matter losses resulting from microbial degradation of proteins and carbohydrates during fermentation [79]. Additionally, in vitro evidence indicates that oxalate breakdown enhances the bioavailability of minerals, including iron, phosphorus, and calcium, by preventing the formation of insoluble complexes [80].

3.2. Mitigation of Microbial and Chemical Food Contaminants Through Microbial Fermentation

LAB can detoxify foodborne hazards through a combination of microbiological and chemical mechanisms that target both biological toxins and chemical contaminants.

Microbiological detoxification is largely mediated by the antimicrobial activity of metabolites produced by lactic acid bacteria, including lactic acid, short-chain fatty acids, hydrogen peroxide, carbonyl compounds, and bacteriocins. These metabolites inhibit or inactivate toxigenic microorganisms such as Escherichia coli, Listeria monocytogenes, Salmonella, and Clostridium botulinum by disrupting membrane integrity, collapsing proton gradients, interfering with metabolic pathways, and suppressing the expression of virulence genes. Surface-associated proteins and competitive adhesion mechanisms further limit pathogen colonization [81].

Detoxification of major mycotoxins by LAB occurs through two principal mechanisms cell-wall adsorption and enzymatic or metabolic biotransformation.

Adsorption is a rapid and reversible process in which mycotoxins bind non-covalently to structural components of the microbial cell wall. In LAB, these include thick peptidoglycan layers enriched in polysaccharides, teichoic acids, and S-layer proteins, whose negatively charged functional groups enhance binding affinity. In yeasts such as Saccharomyces, β(1,3)D-glucans and mannoproteins play a central role in toxin sequestration, with adsorption efficiency determined by cell wall architecture and glucan organization. This mechanism is strain-dependent and influenced by toxin structure and environmental conditions [82].

Biodegradation involves the enzymatic conversion of mycotoxins into less toxic metabolites. Although less extensively characterized, in vitro studies indicate that several probiotic strains, including *Pediococcus pentosaceus*, *Bacillus licheniformis*, and *Saccharomyces pastorianus*, are capable of degrading aflatoxin B1, zearalenone, deoxynivalenol, T-2 toxin, HT-2 toxin, and other trichothecenes through enzymatic pathways, often mediated by extracellular enzymes that are sensitive to heat and proteolysis [83–86].

Lactic acid bacteria can also detoxify pesticides through two principal mechanisms: enzymatic biodegradation and cell-surface-mediated biosorption.

https://doi.org/10.3390/molecules31081353

Hydrolytic enzymes such as carboxylesterases, organophosphate hydrolases, phosphotriesterases, and phosphatases to cleave and metabolize organochlorine, organophosphate, neonicotinoid, and pyrethroid pesticides into less toxic derivatives [87]. Several strains, including *Leuconostoc mesenteroides*, *Lactiplantibacillus plantarum*, *Lactobacillus sakei*, and *Levilactobacillus brevis*, can utilize pesticides as carbon and energy sources, with degradation efficiencies often associated with specific genes such as *opdB*, which encodes an esterase containing the conserved 'Gly-X-Ser-X-Gly' motif characteristic of organophosphate-degrading enzymes [88].

LAB can mitigate heavy metal toxicity primarily through: biosorption (rapid, metabolism-independent) and bioaccumulation (slower, metabolism-dependent).

Biosorption is a rapid, metabolism-independent process in which metal cations bind to negatively charged functional groups, carboxyl, phosphate, and hydroxyl, located on peptidoglycan, teichoic acids, S-layer proteins, and exopolysaccharides of the LAB cell wall [89]. Bioaccumulation is slower and requires metabolic activity, allowing metals to cross the cell membrane and accumulate intracellularly once surface binding sites become saturated. Both mechanisms are strongly strain-specific and are influenced by environmental pH [90].

3.3. *Mitigation of Antinutritional Compounds Through Microbial Fermentation*

Microbial fermentation offers an effective strategy to mitigate the impact of antinutritional factors, including phytic acid, tannins, lectins, saponins, and protease inhibitors, plant-derived compounds that hinder nutrient digestion and absorption [91] (Table 2).

Table 2. Main classes of antinutrients and their mechanisms of action.

Antinutrient	Mechanism of Action	Physiological/Nutritional Effects	References
Tannins	Water-soluble phenolics form insoluble complexes with proteins and metal ions	Reduced protein digestibility and mineral bioavailability	[92]
Oxalates	Present as soluble oxalic acid and insoluble calcium oxalate; chelate minerals	Reduced mineral absorption; soluble oxalates may crystallize in the kidneys (nephrolithiasis)	[93]
Phytates (phytic acid)	Chelate divalent and trivalent cations (Fe^{2+}, Fe^{3+}, Cu^{2+}, Mg^{2+}, Ca^{2+}, Zn^{2+}); bind proteins and starches	Marked reduction in mineral, protein, and starch bioavailability	[94]
Protease inhibitors (PPIs)	Inhibit digestive enzymes (trypsin, chymotrypsin)	Impaired protein digestion; reduced availability of sulfur amino acids (Met, Cys); pancreatic hypertrophy/hyperplasia	[95,96]
Lectins	Carbohydrate-binding glycoproteins resistant to digestion; bind epithelial cells	Gastrointestinal irritation; immune activation; associated with IBS, Crohn's disease, leaky gut; insulin mimicry; NLRP3 activation	[97]
Cyanogenic glycosides	Hydrolyzed by β-glucosidases → release hydrogen cyanide (HCN), inhibitor of cellular respiration	Acute/chronic cyanide toxicity; metabolic acidosis; growth impairment	[98–101]
Goitrogenic glycosides (glucosinolates)	Interfere with iodine uptake and thyroid hormone synthesis	Potential risk of hypothyroidism; evidence is inconsistent in balanced diets	[102,103]
Saponins	Amphiphilic glycosides form complexes with membrane sterols (cholesterol)	Hemolysis increase membrane permeability, hemoglobin leakage); effects vary by aglycone and sugar composition	[104,105]

https://doi.org/10.3390/molecules31081353

Antinutritional factors can be classified according to their thermal sensitivity (heat-stable vs. heat-labile), chemical structure (proteins, glycosides, phenolics), and functional effects (anti-mineral, anti-vitamin, anti-enzyme) (Figure 3). From a thermal standpoint, they include heat-stable compounds, such as phytic acid, tannins, alkaloids, saponins, and non-protein amino acids, as well as heat-labile compounds, including lectins, cyanogenic glycosides, and protease inhibitors. A complementary classification considers their chemical structure (e.g., proteins, glycosides, phenolic compounds) and their functional effects, distinguishing factors with anti-mineral, anti-vitamin, or anti-enzyme activities [106].

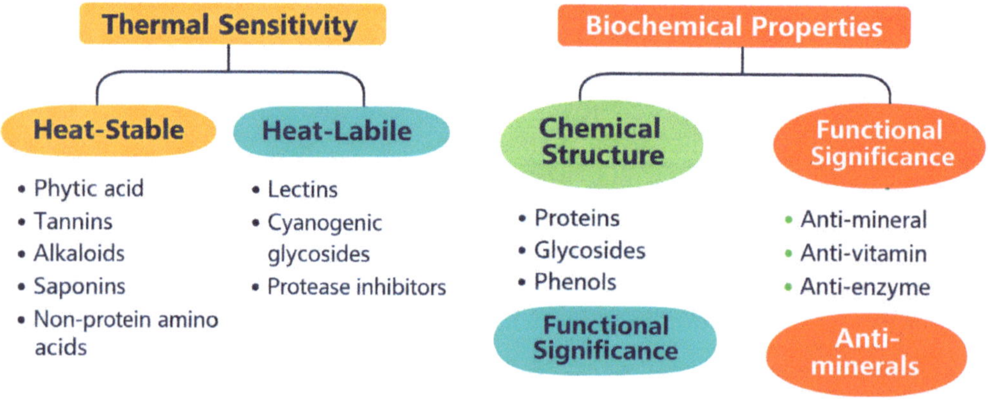

Figure 3. Classification of antinutritional factors. Image generated by Microsoft Copilot (April 2026).

During fermentation, enzymes such as phytases, tannases, oxalate decarboxylases, and β-glucosidases may be released, enabling the degradation of antinutritional factors. Phytases hydrolyze phytic acid into inorganic phosphate and lower myo-inositol esters, thereby improving the availability of minerals such as Ca^{2+} and Mg^{2+} [107]. Tannases cleave ester and depside bonds in tannins, generating gallic acid, ellagic acid, and glucose. The gallate decarboxylase generated by *Lactiplantibacillus plantarum* can convert gallic acid into pyrogallol [108]. Oxalate decarboxylases produced by *Lactobacillus plantarum* and *Lactobacillus acidophilus* can catalyze the decarboxylation of oxalate to CO_2, reducing oxalate content in fermented matrices [108]. β-Glucosidases hydrolyze saponin glycosides, decreasing saponin solubility and toxicity, and can also promote cyanide release from cyanogenic glycosides through cleavage of β-glycosidic bonds [97].

3.4. Bioactive Metabolites Derived from Fermentation and Their Physiological Roles

Fermented foods can act not only as nutrient sources but also as functional carriers of microbiota-driven bioactivity, reinforcing their relevance in personalized nutrition and functional food development.

Lactic acid bacteria, yeasts, and other probiotic strains can interact with the host gut microbiota, modulating its composition and metabolic activity [109,110]. Through the bioconversion of complex carbohydrates, proteins, lipids, and other dietary substrates, the resident and introduced microbes generate a wide spectrum of bioactive compounds, including short-chain fatty acids (SCFAs), bioactive peptides, EPS, and vitamins [109,111–113].

SCFAs, in particular, play a central role in gut–immune crosstalk modulating immune cell function and promoting anti-inflammatory cytokine production [114]. Butyrate also supports intestinal barrier integrity by serving as a primary energy source for colonocytes and reducing gut permeability [115]. Beyond local effects, SCFAs and other micro-

 https://doi.org/10.3390/molecules31081353

bial metabolites influence systemic metabolic pathways, oxidative stress responses, and gut–brain axis signaling, contributing to improved cognitive function, stress resilience, and mental well-being [116–118].

EPS can exhibit multiple functional effects, including antioxidant, immunomodulatory, antibiofilm, and cholesterol-lowering activities [38,39]. They also promote beneficial gut colonization and probiotic persistence by enhancing biofilm formation, limiting pathogen adhesion, and protecting epithelial cells [40,41]. Additionally, exopolysaccharides (EPS) can contribute to gut health by binding dietary toxins and reducing their bioavailability, and they may also lower cholesterol levels by enhancing bile acid secretion [42].

4. Safety and Quality Considerations

Fermentation entails intrinsic risks when not properly managed. These risks primarily involve microbiological and chemical hazards that may compromise product safety. Health effects attributed to probiotics are strain-specific and cannot be generalized at the species level. Functional properties of fermented foods often derive from matrix-dependent interactions rather than microbial viability alone. Many bioactive metabolites (e.g., EPS, bacteriocins, peptides) have demonstrated activity primarily in vitro or in animal models, and clinically validated effects remain limited. This clarification prevents overgeneralization of mechanistic findings.

4.1. Microbiological Hazards

A major safety concern in fermented foods is contamination by pathogenic microorganisms. Although fermentation relies on beneficial microbes, such as LAB, to inhibit pathogens, including *Listeria monocytogenes*, *Escherichia coli*, *Salmonella*, and *Clostridium botulinum*, they may persist if fermentation parameters are not adequately controlled [119]. Documented cases include *Listeria* contamination in soft cheeses [120] and *Clostridium botulinum* outbreaks associated with improperly fermented fish or vegetable products [121,122]. Traditional mitigation requires strict hygiene practices, maintenance of appropriate temperature, acidity, anaerobic conditions, the use of starter cultures with established safety profiles, thermal pasteurization, and synthetic preservatives [123]. Conventional approaches, although effective in ensuring safety, often disrupt beneficial microbiota and compromise sensory and regional characteristics of fermented products [124–129]. Emerging technologies, including cold plasma, plasma-activated water, in-package plasma systems, electromagnetic wave–based treatments (photodynamic inactivation, pulsed light, catalytic infrared radiation, microwave, and radio-frequency heating), and natural essential oils, offer targeted antimicrobial effects [24–40].

Cold plasma and its derivatives, such as plasma-activated water, plasma-activated interfaces, and in-package plasma, selectively inhibit Gram-negative and nitrite-producing bacteria through RONS (molecules generated during cold plasma treatments, including reactive oxygen species (ROS) such as O_2^-, OH·, H_2O_2, and reactive nitrogen species (RNS) such as NO·, NO_2^-, and NO_3^-), membrane disruption, and differences in antioxidant defenses, thereby improving nitrite control and enhancing sensory quality during fermentation [130–132].

Cold plasma exhibits varying levels of selective inactivation depending on the vegetable's surface morphology. Smooth root and stem vegetables allow microorganisms to remain fully exposed, making plasma action more efficient, whereas leafy vegetables with deep folds create protective niches that limit plasma penetration. As a result, treatment parameters must be specifically adjusted for each type of vegetable to ensure effective microbial control while preserving desirable fermentative microbiota [123].

https://doi.org/10.3390/molecules31081353

Photodynamic inactivation (PDI) and pulsed light (PLS) exploit ROS generation and photophysical effects to inactivate pathogens with minimal thermal damage. PDI technology provides non-thermal antimicrobial action, leaves no chemical residues, and requires low energy input [133]. PDI can inhibit or eliminate Gram-positive (*Listeria monocytogenes)*, Gram-negative (*Escherichia coli, Pseudomonas aeruginosa*), mold (*Aspergillus flavus, Fusarium avenaceum, Trichotecium roseum, Rhizopus oryzae*), yeast (*Staphilococcus cerevisiae, Rhodotorula mucilaginosa, Kloeckera javanica)*, and filamentous fungi (*Alternaria alternata, S. cerevisiae*) [134].

PLS combines photothermal and photochemical actions that selectively damage microorganisms, but its UV-C component, responsible for most of the inactivation, penetrates only shallowly into solid matrices, allowing microbes hidden in surface folds to survive. PLS exhibits a broad antimicrobial spectrum, effectively targeting Gram-negative bacteria (*Salmonella enterica, Escherichia coli*), Gram-positive bacteria (*Listeria monocytogenes, Bacillus cereus*), yeasts (*Saccharomyces cerevisiae*), and molds (*Aspergillus niger, Penicillium roqueforti*) [135,136].

Catalytic infrared radiation (CIR), microwave heating, and radio-frequency treatments all provide selective thermal or thermo-electromagnetic inactivation by exploiting differences among microbes in moisture content and electromagnetic sensitivity. CIR uses mid-infrared radiation (3.3–3.8 μm) generated by Pt–Pd catalytic combustion to induce protein denaturation, DNA melting, membrane phase transitions, and peptidoglycan cleavage, achieving efficient and clean microbial reduction while sparing low-moisture vegetable tissues; this selectivity is enhanced by the strong water-gradient effect, which concentrates infrared energy in high-moisture microbial cells [137–139].

Microwave sterilization combines rapid volumetric heating with non-thermal electromagnetic effects that alter membrane potential and ion permeability, producing differential susceptibility among microbial groups and enabling the inactivation of resistant genera such as *Bacillus*, *Clostridium*, *Penicillium*, and *Aspergillus* under optimized frequency and field-strength conditions [140,141].

Radio-frequency heating (3 kHz–300 MHz) enables deep, uniform heat penetration through dipole rotation and ion migration. This treatment disrupts membrane homeostasis, interferes with cell wall biosynthesis, and impairs energy metabolism, effectively targeting both Gram-positive and Gram-negative bacteria. Owing to its rapid heating rates, volumetric energy transfer, and compatibility with continuous-flow systems, radio-frequency heating is well suited for industrial-scale applications [142].

Moreover, several non-pathogenic members of fermentation microbiota (e.g., lactic acid bacteria, coagulase-negative *Staphylococci, Enterococci*) can carry intrinsic or acquired resistance traits, and horizontal gene transfer (HGT) may be facilitated by the high cell density, close intercellular proximity, biofilm formation, and abundance of mobile genetic elements characteristic of fermentation ecosystems [143,144]. Meta-analyses and culture- and metagenomic-based investigations have identified ARGs and viable resistant isolates in selected retail products (e.g., kimchi, artisan cheeses) and reported that diets enriched in fermented foods may transiently elevate gut resistome levels in healthy adults. However, the clinical implications of these shifts remain uncertain and appear to be context dependent [145]. Conversely, substrate-anchored metagenomic analyses across a wide range of fermentations (including kombucha and water kefir) reveal substantial variability among products, with several plant-based ferments exhibiting low or even undetectable ARG levels under the conditions evaluated [146]. The LAB's ability to acquire and transfer mobile AR determinants was documented, with several studies reporting horizontal gene transfer (HGT) events both among *Lactobacillus* species and between *Lactobacillus* and pathogenic bacteria. Notably,

 https://doi.org/10.3390/molecules31081353

in vitro and in vivo studies showed that vancomycin resistance can be transferred from Enterococci of human and animal origin to *Lactobacillus acidophilus* in mice [147]. Given the high frequency of such events in animal models, similar mechanisms are likely to occur in the human gastrointestinal tract, which is recognized as a major reservoir for ARGs [148].

Enterococci, frequently detected in animal-derived fermented products, combine intrinsic and acquired antimicrobial resistance with the ability to withstand low pH, high salt concentrations, and elevated temperatures, enabling their persistence throughout fermentation. Horizontal transfer of resistance determinants has been documented, including the transmission of the tetracycline resistance gene *tet(M)* from *Enterococcus faecium* isolated from dry sausage to clinical isolates of *Enterococcus faecalis* and *Enterococcus faecium*, mediated by a type-1 integron [143,149]. The potential dissemination of antimicrobial resistance genes (ARGs) by non-pathogenic members of the fermentation microbiota can be mitigated through a combination of strain selection, process control, and targeted monitoring. The use of well-characterized starter cultures with established safety profiles and lacking mobile genetic elements represents an important safeguard [150]. Advances in whole-genome sequencing enable the exclusion of strains carrying plasmids, transposons, or integrative conjugative elements associated with transmissible resistance, thereby reducing the likelihood of HGT [151].

Maintaining strict hygiene and environmental controls throughout production further prevents the introduction of resistant contaminants such as *Enterococcus* spp., which are known to persist under fermentation conditions and to transfer resistance determinants (e.g., *tet(M)*) to other bacteria [152].

Additionally, controlled fermentation parameters, including pH, salinity, temperature, and anaerobiosis, limit the proliferation of opportunistic microorganisms capable of participating in HGT and reduce the survival of pathogens such as *Listeria monocytogenes*, *Escherichia coli*, *Salmonella*, and *Clostridium botulinum* [4].

Finally, routine surveillance programs integrating metagenomics, qPCR-based ARG detection, and periodic verification of starter culture integrity can be used [143].

4.2. Chemical Hazards in Fermented Foods

Chemical hazards in fermented foods arise primarily from the accumulation of biogenic amines (BAs), such as histamine, tyramine, cadaverine, and putrescine, produced through microbial decarboxylation of amino acids [153]. Elevated BA levels can cause adverse reactions, including headaches, hypertension, nausea, and allergic responses, especially in individuals with reduced metabolic capacity for these compounds [119].

Histamine poisoning is frequently linked to fermented fish products, where inadequate fermentation or storage can result in toxic concentrations [154].

Tyramine, commonly present in fermented cheeses and certain alcoholic beverages, may trigger hypertensive crises in individuals taking monoamine oxidase inhibitors [155]. Regulatory agencies have established maximum allowable BA levels in high-risk fermented foods. Risk management strategies include controlling fermentation parameters (temperature, pH, and duration), selecting microbial strains with low decarboxylase activity, and ensuring proper post-fermentation storage [14].

5. Traditional vs. Industrial Fermentation

Both traditional and modern fermentation extend the shelf life of perishable raw materials [156]. Traditional fermentations are generally spontaneous processes driven by the indigenous microbiota naturally present in raw materials and the surrounding environment. These conditions give rise to products characterized by high microbial diversity and distinctive regional traits. Several authors suggest that fermentation practices and their derived products

https://doi.org/10.3390/molecules31081353

originated approximately 10,000 years ago, primarily as strategies to extend food shelf life during periods of scarcity while also enhancing sensory qualities [13,157]. In ancient societies, fermentation occurred without deliberate inoculation, relying instead on naturally occurring microorganisms associated with the substrate and its environment [158]. These complex microbial consortia, comprising bacteria, yeasts, and filamentous fungi, transform substrates and contribute to the characteristic flavors, aromas, and textures of traditional fermented products [159].

By contrast, modern approaches employ selected starter cultures, engineered microbial consortia, and genetically modified or synthetically designed strains to achieve predictable acidification, flavor formation, antioxidant production, enzyme production, pigment production, vitamin production, and inhibition of spoilage organisms [160–163]. Advances in genetic engineering and synthetic biology now enable microorganisms to efficiently synthesize targeted metabolites, including amino acids, organic acids, enzymes, and functional proteins, such as recombinant caseins used for animal-free cheese production [160,164].

Precision fermentation boosts the development of next-generation animal-free foods. Several companies produce milk, egg, and meat analogs through engineered yeasts and filamentous fungi that replicate the techno-functional properties of animal proteins [165]. Artificial intelligence has further accelerated ingredient design and fermentation optimization, exemplified by platforms that match plant matrices to the molecular profiles of animal products [166].

Recent research has used precision fermentation to produce sauces, powders, and pastes with sensory properties comparable to those of conventional seafood-based products, starting from edible insects [167,168].

The global precision fermentation market is projected to grow from USD 2.8 billion in 2023 to over USD 36 billion by 2030 [169].

5.1. Modern Fermentation and Its Contribution to Food Sustainability

Modern fermentation improves sustainability by supporting circularity, reducing waste by converting food waste into novel bioactive ingredients [170], enhancing resource efficiency, and enabling the development of novel, nutrient-dense food products [171]. Precision fermentation and microbial biomass production require substantially less energy and water than conventional livestock or crop-based protein systems, as microorganisms efficiently convert substrates into high-value nutrients with minimal spatial demands [172]. These processes are also associated with lower greenhouse gas emissions [173], offering a viable alternative to ruminant agriculture and other emission-intensive production models. Advances in bioreactor engineering, metabolic optimization, and process control have further improved the energy efficiency of fermentation-based bioprocesses [174]. Fermentation plays a key role in the valorization of agricultural and food waste by transforming low-value side streams, such as whey, fruit pomace, and cereal bran, into functional ingredients, organic acids, bioactive peptides, and microbial biomass, thereby supporting circular bioeconomy strategies and reducing the volume of organic waste requiring disposal [175]. Controlled fermentation strengthens food security and resilience by supporting decentralized, scalable production systems that are less dependent on climate-sensitive agriculture, while rapid microbial growth cycles help stabilize supply chains [69]. Diversifying protein sources through fermentation reduces vulnerability to environmental and geopolitical disruptions [176]. Additional sustainability benefits include reduced antibiotic use, as fermentation-derived enzymes and probiotics can replace antimicrobials in animal feed [177]. Finally, fermentation serves as a versatile platform technology for producing bio-

https://doi.org/10.3390/molecules31081353

based materials such as bioplastics and biopolymers, indirectly contributing to sustainable packaging solutions and reducing dependence on petrochemical resources [178].

Persistent research and regulatory gaps continue to limit the integration of microbiome-derived innovations into food systems (Figure 4). A major barrier is the lack of standardized methods for isolating, characterizing, and applying microbial strains and consortia, which undermines reproducibility, complicates cross-study comparisons, and slows both clinical validation and regulatory approval [166,179]. Harmonized international guidelines are needed to define essential parameters, including accurate taxonomic and strain-level identification, assessment of cell viability and in vivo bioactivity, rigorous randomized clinical trial design, evaluation of long-term stability during processing and storage, and comprehensive toxicological and genomic safety testing [180].

Figure 4. Modern fermentation challenge for effective application. The figure was generated using NotebookLM (Gemini 1.5 model), which was consulted in February 2026.

5.2. Consumer Perceptions

Consumer acceptance of fermentation technologies is shaped by a wide range of interconnected factors, with perceptions of naturalness emerging as a central determinant. Traditional fermentation is generally viewed as familiar, historically rooted, and therefore more natural and safer, whereas precision fermentation, despite its sustainability potential, may be perceived as less natural or unfamiliar [181,182]. These perceptions contribute to broader acceptance challenges, including concerns about technological complexity, genetic modification, food safety, and allergenicity [183,184]. Technology neophobia further affects both traditional and precision fermentation: while some consumers express apprehension toward live microbes and contamination risks in traditional ferments [185,186], precision fermentation elicits uncertainty due to its novelty, biotechnological basis, and unfamiliar

 https://doi.org/10.3390/molecules31081353

production processes [161,187]. Concerns about genetic modification remain particularly salient, as precision fermentation often relies on engineered microorganisms, and skepticism toward GMOs, rooted in perceived health and environmental risks, can extend to these products [188]. Although precision-fermented ingredients may be compositionally identical to traditional ones, consumers may still question their safety, long-term effects, and ecological implications [189].

Ethical and societal considerations also play a role, including worries about the displacement of traditional farming, job losses, and the broader implications of high-tech food systems. Cultural and emotional attachments to traditional food practices, along with concerns about transparency, trust, and regulatory oversight, further shape consumer responses. Economic factors add another layer of complexity: perceptions of naturalness strongly influence purchasing behavior and market viability [190], while fears of industry monopolization, pricing, and accessibility contribute to skepticism [191,192].

Moreover, the current high production costs of precision-fermented ingredients pose challenges for affordability and widespread adoption.

Addressing these multifaceted concerns requires transparent communication, inclusive public engagement, robust regulatory frameworks, and evidence-based information to support informed consumer understanding and foster confidence in precision fermentation.

6. Fermented Foods and Beverage

Although documentation is limited, more than 5000 fermented products are known worldwide [193]. Cultural and geographical factors influenced the substrates used for traditional fermented foods. In the Middle East, Europe, and India, fermentations use mainly milk as a substrate, whereas in Asian regions, where animal husbandry is historically more limited, rice, grains, vegetables, fish, and soybeans are used. In Africa, millet, sorghum, maize, and wheat are the predominant substrates for fermentation [194].

6.1. Fermented Foods and Beverages from Africa

Africa has a long tradition of fermenting plant and animal-based foods, typically produced at the household level under rudimentary, often unhygienic conditions. This knowledge is largely transmitted orally and preserved by elderly rural women. Sour milk is the most widely described fermented food in Africa. These foods fall into three broad categories: foodstuffs (cereals, tubers, vegetables), relishes/condiments/sauces, and beverages (alcoholic and non-alcoholic) (Table 3). The lack of a written tradition complicates historical reconstruction, yet fermented foods are central to African cultural identity and gastronomy [193].

6.1.1. Soubala

Soumbala is produced through the alkaline fermentation of *Parkia biglobosa* seeds and is widely consumed across West Africa for its nutritional and functional properties [195]. Fermentation enhances the bioavailability of essential amino acids and fatty acids, making soumbala a nutrient-dense food [196]. This process also reduces antinutritional factors and their associated allergenic effects, improving digestibility and overall safety [196,197]. Due to its low sugar content, soumbala is recommended for individuals with diabetes [197]. Its richness in unsaturated fatty acids and phenolic compounds contributes to the prevention of cardiovascular diseases and other oxidative-stress-related disorders [198]. Additionally, it provides dietary fiber and minerals that may help combat nutritional deficiencies [199]. Regular consumption has been associated with reduced risks of hypertension, diabetes, and obesity [200,201]. Although most studies focus on raw soumbala, it is commonly consumed after cooking, which can degrade heat-sensitive nutrients, phenolics, flavonoids,

 https://doi.org/10.3390/molecules31081353

and microbial enzymes, thereby diminishing its health benefits [199,202]. To preserve its nutritional value, raw soumbala should be encouraged, yet its strong sensory characteristics limit its acceptance [196]. A promising strategy is the development of a soumbala-based seasoning enriched with locally used spices, such as onion, pepper, parsley, and green anise, which can improve flavor while providing additional anti-inflammatory, antimicrobial, antioxidant, and anticancer properties [183,203–205].

6.1.2. Bikalga

Bikalga is a traditional, alkaline-fermented condiment made from *Hibiscus sabdariffa* seeds, widely consumed in West and Central Africa. Its processing involves prolonged cooking with alkalinizing ash leachate followed by spontaneous fermentation dominated by *Bacillus* species, particularly *B. subtilis* [206]. Fermentation induces extensive proteolysis, lipolysis, and carbohydrate degradation, thereby improving digestibility and enhancing nutrient bioavailability. Notably, Bikalga retains high protein and lipid levels [207]. The process also leads to a substantial increase in mineral content, especially potassium, sodium, calcium, and sulfur, due to both microbial metabolism and the mineral-rich ash leachate [207]. Beyond its nutritional benefits, Bikalga is traditionally valued for its medicinal uses, including the management of high blood pressure, diarrhea, and infections, suggesting potential antimicrobial effects due to *Bacillus* spp. [207].

6.1.3. Dawadawa

Dawadawa is a condiment produced by alkaline fermentation of *Parkia biglobosa* seeds. It is a culturally and nutritionally important food across West Africa. Fermentation profoundly enhances its nutraceutical value through proteolysis, lipid modification, the release of bioactive phenolic compounds, and increased antioxidant activity. Nutritionally, dawadawa provides 35–43% protein, a complete essential amino acid profile that meets FAO/WHO requirements, and substantial levels of calcium, iron, zinc, magnesium, and phosphorus. These features underscore its relevance as a functional food that supports metabolic health, immune function, and nutrient adequacy in populations with limited access to animal protein. Functional properties such as favorable water and oil absorption capacities further support its incorporation into diverse food formulations. Local processing practices influence its nutraceutical quality [208].

6.1.4. Okpehe

Okpehe is a traditional fermented condiment made from *Prosopis africana* seeds consumed across several West African regions. Its nutraceutical relevance stems from the biochemical transformations that occur during alkaline fermentation by autochthonous *Bacillus* species. Proteolysis hydrolyzes seed proteins into peptides, amino acids, and ammonia, thereby enhancing digestibility and contributing to the product's characteristic aroma [209]. Amylolytic activity increases the availability of soluble sugars, and selected strains of *Bacillus subtilis* synthesize polyglutamic acid (PGA), a compound associated with improved texture, viscosity, and bioactive functions [210]. Okpehe also contains bioactive peptides and volatile compounds that may exert antioxidant, antimicrobial, and gut-modulating effects [210]. The use of carefully selected starter cultures, particularly *Bacillus subtilis* strains BFE 5301 and BFE 5372, enhances both safety and quality. Notably, strain BFE 5301 produces subtilisin, a heat-stable bacteriocin active at acidic pH values and capable of inhibiting *Bacillus cereus*, thereby improving microbiological safety during fermentation [211].

 https://doi.org/10.3390/molecules31081353

6.1.5. Ugba

The African "Ugba" or "Ukpaka" oil obtained from *Pentaclethra macrophylla (P. macrophylla) Benth* bean seed, is traditionally used in the preparation of the popular African salad. The seeds require extensive cooking followed by fermentation to ensure palatability and enhance nutrient bioavailability. Fermented *Pentaclethra macrophylla* seeds have been associated with several health-promoting properties, as demonstrated by in vivo studies in animal models, including anti-anemic effects [212], anti-hyperlipidemic activity [213], and blood-pressure-lowering potential [214]. They also serve as a valuable source of phytochemicals and essential nutritional elements [215]. Phytochemical investigations have identified bioactive secondary metabolites such as the diterpene secopentaclethrolide [16α,17-dihydroxy-6,7-seco-ent-kaurane-19,6β-olide] and the 1,3-diglyceride 1-arachidonyl-3-linoleoyl-sn-glycerol, further underscoring the nutraceutical potential of this traditionally fermented legume. Finally, fermented *P. macrophylla* seeds offer significant protection against diabetes-associated neurological disorders through combined antioxidant, anti-inflammatory, and enzyme-modulating mechanisms. Specifically, its use significantly reduces the activities of acetylcholinesterase, butyrylcholinesterase, angiotensin-converting enzyme, and arginase, key biochemical markers implicated in neurodegeneration and metabolic dysfunction. Finally, as evidenced by in vivo studies in diabetic rats, ugba markedly enhances brain antioxidant levels, elevates thiol concentrations, increases superoxide dismutase and catalase activities, and reduces malondialdehyde equivalents [216].

6.1.6. Iru

Iru is a traditional fermented condiment produced from the seeds of the African locust bean (*Parkia biglobosa*). Widely consumed across West Africa and particularly appreciated among the Yoruba people of Nigeria, it is commonly incorporated into traditional dishes such as melon soup, okra soup, and various vegetable-based preparations [217], *Bacillus* spp. (*B. encimensis, B. safensis*), *Enterococcus* (*E. dispar*) and *Lysinibacillus* spp. (*L. fusiformis, L. macroides*), represent the key bacteria responsible for the fermentation of *iru* [218]. Over the course of fermentation, crude fat, ash, and fiber contents increase [219]. Processing methods markedly shape both the nutritional profile and sensory attributes of *iru*. Chemical catalysts negatively affect the nutritive quality of iru. Comparative analyses of traditional and no-potash iru show that *iru* produced without chemical additives and no potash methods retains the highest levels of essential nutrients. Sensory characteristics are strongly method-dependent: steaming preserves the creamy color of the cotyledons, minimizes off-odors by preventing direct water contact with the testa, yields a slightly marshy but clean-tasting product, and permits superior hygiene. In contrast, water-based cooking promotes higher dehulling efficiency but increases absorption of undesirable flavors and alters texture through greater hydration of the cotyledons. Overall, steaming offers the most favorable balance of nutritional retention, sensory quality, and processing hygiene, making it a recommended method for high-quality, scalable *iru* production [220]. Recent findings suggest that fermenting *P. biglobosa* seeds with selected bacterial strains may constrain dyslipidemia by reducing total cholesterol, LDL cholesterol, and triglycerides. Notably, several formulations demonstrated lipid-lowering effects that surpassed those of conventional statin therapy, as evidenced by in vivo studies conducted in Wistar rats [221].

6.1.7. Kawal

Kawal is a traditional fermented food condiment produced through natural alkaline fermentation due to the spore-forming *Bacillus* of the leaves of *Senna obtusifolia* (also known as *Cassia obtusifolia*). The process begins with harvesting mature leaves at the end of the

 https://doi.org/10.3390/molecules31081353

rainy season, then cleaning and pounding them into a paste, which is placed in earthenware jars and buried for 15 days to allow fermentation. Kawal provides antimicrobial compounds (surfactins, iturins, fengycins, and kurstakins) with potential applications in food bioprotection and agriculture [222].

6.1.8. Sigda

Sigda is a traditional Sudanese fermented food produced from sesame (*Sesamum indicum* L.) oilseed cake, a process that markedly enhances its palatability and nutritional value. The preparation involves pounding the seedcake with warm water, allowing the mixture to ferment for 3–7 days at about 30 °C in sealed earthenware or metal vessels, sometimes supplemented with kambo (an alkaline ash extract), and shaping the mixture into small balls for long-term storage. The fermentation is driven primarily by a homofermentative lactic acid bacterium (*Streptococcus* sp.), which dominates the microflora and produces substantial amounts of lactic acid, up to "50 g kg^{-1} during fermentation", resulting in a rapid decline in pH to approximately 4.0 within three days. This acidification contributes to microbial safety and improved digestibility. Nutritionally, Sigda retains the high protein content of sesame seedcake (approximately 440–460 g kg^{-1}) and preserves its rich sulfur amino acid profile. Fermentation increases the levels of leucine, isoleucine, and alanine [223].

6.1.9. Furundu

Furundu is a traditional Sudanese fermented food made from crushed seeds of Hibiscus sabdariffa (karkade) through spontaneous fermentation, primarily driven by *Bacillus* species. This fermentation transforms the raw seeds, which are naturally rich in sulfur-containing amino acids but otherwise unpalatable, into a nutritionally valuable, shelf-stable food. Fermentation preserves and enhances the biological quality of the seed proteins, improves digestibility, reduces antinutritional factors, and generates a meat-like flavor highly appreciated by local populations [223].

6.1.10. Umqombothi

Umqombothi is a nutrient-dense traditional opaque sorghum beer produced through spontaneous fermentation of a mixture of sorghum malt, maize meal, and maize malt, with promising functional properties. Its preparation involves steeping and malting sorghum, mixing it with maize meal, cooking the mixture, and allowing it to undergo natural fermentation driven by lactic acid bacteria and yeasts, a process that shapes its characteristic flavor, low alcohol content (2–3.5% v/v), and nutrient profile. Umqombothi contains dietary fiber, essential minerals (Na, Mg, K, P, Fe, Zn, Ca, S), amino acids such as glutamic acid and leucine, B-group vitamins (including nicotinic acid and folate), and soluble sugars such as maltose and glucose. Controlled fermentation has been shown to improve its nutritional value, increasing crude protein, total amino acids, nicotinic acid, ash content, and bioactive metabolites, while reducing antinutritional factors such as phytates. Moreover, sorghum fermentation produces phenolic acids, flavonoids, phytosterols, and antioxidant compounds [224].

6.1.11. Burukutu

Burukutu is a traditional Nigerian fermented beverage made from sorghum or millet enriched with gari, a fermented cassava product. Its production involves steeping and germinating sorghum grains, drying and milling the malt, boiling the flour–water mixture, adding gari, and then spontaneous fermentation for 2–3 days. Fermentation increases crude protein content while reducing sugars and ascorbic acid, and it enhances amino acid availability, particularly lysine, methionine, and serine, although heat-sensitive

 https://doi.org/10.3390/molecules31081353

amino acids such as cystine and tryptophan are lost. The final beverage contains simple sugars (glucose, fructose, sucrose, maltose) and reaches an alcohol content of about 4%. Overall, burukutu provides easily digestible nutrients that support energy and growth. Fermentation improves flavor, digestibility, and protein quality [225].

6.1.12. Dégué

Dégué is a traditional West African fermented cereal product typically prepared from pearl millet, which is cleaned, soaked, germinated, dried, milled, and then mixed with milk or dairy components before undergoing spontaneous or controlled fermentation. Fermentation of millet reduces antinutritional factors such as phytic acid (−78%) and polyphenols (−46%), while enhancing protein content, protein digestibility, starch digestibility, and mineral bioavailability, particularly calcium, iron, magnesium, and zinc. The process also increases free phosphorus and decreases phytate-bound phosphorus, further improving mineral absorption [226].

6.1.13. Ogi

Ogi is a traditional West African fermented cereal porridge made from maize, sorghum, or millet through a sequence of steeping, wet milling, sieving, and spontaneous lactic fermentation. The grains are typically soaked for 24–72 h to allow natural microbial activity, then wet-milled and sieved to remove bran and fiber, after which the slurry is left to ferment for 1–3 days, yielding a mildly acidic, smooth paste. The use of *Lactobacillus* spp. and *Saccharomyces* spp. enhances fermentation efficiency, improves safety, and increases nutritional quality by accelerating acidification, reducing pathogenic load, and promoting the degradation of antinutritional factors such as phytates and polyphenols. Controlled fermentation also increases the bioavailability of essential minerals (Ca, Fe, Zn), improves protein digestibility, and enriches the product with B-group vitamins and beneficial metabolites. As a result, ogi prepared with defined starter cultures offers superior nutritional value, enhanced microbial safety, and more consistent sensory quality compared with spontaneously fermented ogi [227].

6.1.14. Gari

Gari is a fermented, free-flowing granular product derived from cassava and widely consumed across Nigeria, where it serves as a major carbohydrate source and is commonly reconstituted with hot water to form a stiff dough known as *Eba*. The process involves peeling and washing cassava roots, grating them into a mash, allowing spontaneous lactic fermentation to occur, then dewatering, sieving, and toasting to obtain the final granular product. Fermentation plays a crucial role in improving the safety and quality of gari by reducing the potential toxicity of cassava cyanogenic compounds and modifying the physicochemical characteristics of the starch matrix. Fermentation, together with the incorporation of fermented maize residues, alters water absorption, swelling power, solubility, and pasting behavior, largely due to insoluble dietary fiber and partially degraded starch structures. These biochemical modifications also influence nutritional attributes, increasing protein, fat, fiber, and mineral content while reducing carbohydrate concentration and glycemic index [228].

6.1.15. Togwa

Togwa is a traditional fermented beverage consumed in Tanzanian households, commonly used as a weaning food or as a refreshing drink. It is produced from the short fermentation of the finger millet and maize flour. Its preparation begins with cleaning and steeping maize kernels in water for 1–3 days to allow natural fermentation, followed by wet milling and sieving to separate the starchy slurry from the fibrous residue. The slurry is

https://doi.org/10.3390/molecules31081353

then allowed to sediment, after which the supernatant is decanted, and the fermented starch is collected [229]. This spontaneous lactic acid fermentation improves the product's microbial safety, enhances nutrient bioavailability, and contributes desirable sensory attributes. Because Togwa relies on spontaneous fermentation, product quality can vary considerably. Fermentation reduces antinutritional factors such as phytates and polyphenols, increases mineral accessibility, and improves protein digestibility, while the dominance of lactic acid bacteria (e.g., *Lactobacillus plantarum*, *Lactobacillus fermentum*, *Pediococcus pentosaceus*) contributes probiotic potential and extended shelf-life [230].

6.1.16. Bushera

Bushera is the most produced traditional non-alcoholic beverage in Uganda. *Lactococcus*, *Lactobacillus*, *Leuconostoc*, *Enterococcus*, and *Weissella* are LAB employed in *bushera* production. It is typically prepared at the household level by low-income women, using sorghum and millet, both for home consumption and for small-scale sale. Its fermentation is frequently accelerated by back-slopping, resulting in a more pronounced acidity. In local practice, Bushera at different fermentation stages serves distinct consumer groups: the mildly fermented Day-1 product is commonly given to young children, whereas adults prefer the fully fermented beverage. Bushera, as a natural probiotic source, may support digestion, enhance nutrient uptake, and promote gut health [230].

6.1.17. Shamita

Shamita is a traditional Ethiopian cereal-based fermented beverage produced primarily from roasted barley flour mixed with water to form a slurry, which is then subjected to spontaneous lactic acid fermentation. The process typically involves mixing the flour with lukewarm water, adding small quantities of spices, and allowing the mixture to ferment for 12–24 h at ambient temperature, during which naturally occurring LAB dominate the microbial community. This fermentation reduces pH, enhances microbial safety, and contributes to the beverage's characteristic sour flavor. Nutritionally, Shamita provides carbohydrates, dietary fiber, and moderate amounts of protein, while fermentation increases the bioavailability of micronutrients by reducing antinutritional factors such as phytates. Moreover, the proliferation of LAB, particularly species of *Lactobacillus*, *Pediococcus*, and *Leuconostoc*, confers potential probiotic benefits, including improved gut microbial balance, production of short-chain fatty acids, and modulation of immune responses [231].

6.1.18. Munkoyo

Munkoyo is a traditional Zambian non-alcoholic fermented beverage produced from maize porridge and Rhynchosia root extracts, used as both an enzymatic source and a microbial inoculum. Production begins by mixing maize flour with water and cooking the mixture for approximately 30 min to gelatinize the starch; after cooling, *Rhynchosia* roots are added to hydrolyze complex carbohydrates and initiate spontaneous fermentation, which proceeds for about 48 h at ambient temperature in vessels such as calabashes or metal drums. The fermentation is driven predominantly by lactic acid bacteria (*Lactobacillus*, *Streptococcus*, *Lactococcus*, *Klebsiella*, *Enterobacter*, and *Acetobacter*), typically four to ten species, which acidify the matrix and contribute to product safety and stability. Nutritionally, Munkoyo provides carbohydrates, small amounts of protein, and B-vitamins. Fermentation enhances micronutrient bioavailability by reducing antinutritional factors such as phytates. The dominance of LAB also confers potential probiotic benefits, including modulation of the gut microbiota, production of organic acids, and possible contributions to immune and metabolic health [232].

https://doi.org/10.3390/molecules31081353

6.1.19. Mahewu

Mahewu is a traditional Southern African non-alcoholic fermented beverage produced from maize porridge that undergoes spontaneous or inoculated lactic acid fermentation. Its preparation typically involves cooking maize meal into a thin porridge, cooling it to lukewarm, and introducing either natural back-slopping material or specific inocula, such as *Lactobacillus* spp., to initiate fermentation. The mixture is then allowed to ferment for 24–48 h at ambient temperature, driving acidification, starch hydrolysis, and flavor development. The fermentation process significantly enhances the availability of protein and carbohydrates. Mahewu contains essential amino acids such as leucine and arginine, as well as non-essential amino acids such as aspartic acid, glutamic acid, and alanine. Potassium is the most abundant mineral; other minerals detected are sodium, phosphorus, magnesium, calcium, manganese, copper, iron, and zinc [233].

Table 3. Several fermented foods from Africa.

Product	Description	Microorganisms Identified	Region of Origin	References
Soumbala	Fermented locust bean néré (*Parkia biglobosa*),	*Bacillus subtilis*, *Bacillus. pumilus*, *Priestia megaterium*, *Bacillus licheniformis*	Burkina Faso	[183]
Bikalga	Fermented roselle seeds (*Hibiscus sabdariffa*)	*Bacillus subtilis*, *Bacillus licheniformis*, *Bacillus cereus*, *Bacillus pumilus*, *Pseudobacillus badius*, *Brevibacillus porteri*, *Lysinibacillus sphaericus*, *Lysinibacillus fusiformis*	Burkina Faso	[195]
Dawadawa	Fermented locust bean	*Bacillus* spp.	Ghana & Nigeria	[234]
Okpehe	Fermented *Prosopis africana* seeds	*Bacillus* spp.	Nigeria	[211,216]
Ugba	Fermented African oil bean	*Bacillus* spp.	Nigeria	[223,224]
Iru	Fermented locust bean	*Bacillus* spp., *Enterococcus* and *Lysinibacillus* spp.	Nigeria, Benin	[228]
Kawal	Fermented *Cassia* leaves	*Bacillus subtilis*, *Propionibacterium* spp.	Sudan	[222]
Sigda	Fermented sesame seed	*Candida* and *Saccharomyces* sp. yeasts and *Pediococcus* sp. (eliminated after the second day of fermentation) and *Streptococcus* sp. (that remain throughout the rest of the fermentation) bacterial species	Kordofan and Darfur provinces of western Sudan	[223]
Furundu	Fermented *karkadè* seeds	*Bacillus* sp.	Sudan	[223]
Umqombothi	Maize–sorghum beer	*Lactobacillus* spp., *Saccharomyces cerevisiae*	Southern Africa	[235]
Burukutu	Alcoholic beverage from sorghum and millet	*Escherichia*, *Staphylococcus*, *Bacillus*, *Lactobacillus*, *Leuconostoc*, *Acetobacter*	Nigeria, Benin, Ghana	[236]
Dégué	Fermented millet dough	*Escherichia*, *Bacillus*, *Lactobacillus*, *Enterococcus*	Burkina Faso	[237]
Ogi	Fermented cereal pudding (maize, sorghum, millet)	*Candida krusei*, *Lactiplantibacillus plantarum*, *Limosilactobacillus fermentum*, *Saccharomyces cerevisiae*, *Acetobacter* spp., *Corynebacterium* spp.	West Africa	[238]
Gari	Fermented cassava–cereal derivative (often combined with maize)	*Leuconostoc mesenteroides*, *Lactobaillus plantarum*, *Bacillus subtilis*, *Candida krusei*	West Africa	[239]
Togwa	Cereal-based lactic beverage (cassava, maize, sorghum, millet)	*Lactobacillus* spp., *Pediococcus pentosaceus*, *Weissella confusa*, *Issatchenkia orientalis*, *Saccharomyces cerevisiae*, *Candida pelliculosa*, *Candida tropicalis*	Eastern Africa	[240]
Bushera	Fermented sorghum beverage	*Lactobacillus*, *Streptococcus*, *Leuconostoc*, *Pediococcus*, *Weissella* spp.	Eastern Africa	[241]
Shamita	Fermented barley beverage	*Lactobacillus* spp.	North-East Africa	[242]

https://doi.org/10.3390/molecules31081353

Table 3. *Cont.*

Product	Description	Microorganisms Identified	Region of Origin	References
Munkoyo	Fermented maize beverage with root extract	*Weissella* spp., *Lactobacillus* spp.	Southern Africa	[243]
Mahewu	Fermented maize or sorghum beverage	*Lactobacillus plantarum*, *Lactobacillus fermentum*	Southern Africa	[244]

6.2. Fermented Foods and Beverages from Asia

Across East, South, and Southeast Asia, fermentation is applied to cereals, legumes, vegetables, dairy, and fish, often through spontaneous microbial activity or the use of defined starter cultures [245] (Table 4). Cereal-based fermentations such as Japanese *koji* products, Korean makgeolli, Chinese huangjiu, and Indian kanji use molds (*Aspergillus*, *Rhizopus*), yeasts (*Saccharomyces*, *Candida*), and LAB to hydrolyze starches and generate organic acids, ethanol, and bioactive metabolites [246]. Legume fermentations, including Indonesian tempeh, Japanese natto, and Korean doenjang, enhance protein digestibility, reduce antinutritional factors, and increase B-vitamin content [247,248]. Fermentations in kimchi and paocai are dominated by LAB that produce lactic acid, bacteriocins, and health-promoting metabolites [249]. Dairy fermentations, including Indian lassi and Central Asian kumis, improve lactose digestibility and contribute probiotic microorganisms [250]. Beyond their sensory diversity, Asian fermented fish products provide valuable nutrients and bioactive compounds generated during fermentation. A key metabolite is γ-aminobutyric acid (GABA), produced primarily by lactic acid bacteria, which may contribute to potential antihypertensive effects through mechanisms such as vasodilation and modulation of sympathetic activity. Fermentation also enhances protein digestibility, liberates peptides with antioxidant or antihypertensive potential, and may improve micronutrient availability [251]. Halophilic bacteria and endogenous enzymes in hai *pla-ra* and Filipino *bagoong* can generate amino acids and flavor compounds [252].

6.2.1. Koji

Koji is a traditional food developed in ancient China and later disseminated to Japan and Korea, where it was produced by *Aspergillus* spp. (*A. oryzae*, *A. sojae*, *A. luchuensis* mut. *kawachii*, *A. tamarii*) are cultivated on steamed cereals and legumes [253]. Unlike the mixed amylolytic starters used across Southeast Asia, koji represents a highly specialized microbial system characterized by non-aflatoxigenic strains recognized as safe. Its functional relevance stems from the enzymes (proteases, lipases, and pectinases) that drive efficient saccharification, proteolysis, and flavor formation [254]. In addition to enzymatic activities, koji contains diverse bioactive metabolites, including phenolic acids, hydroquinones, soyasaponins, flavonoids, lysophospholipids, and dipeptides, which contribute to nutritional enhancement, antioxidant potential, and sensory complexity [236–255].

6.2.2. Miso

Miso is fermented by *Aspergillus oryzae* and associated microbial consortia. From a macronutrient perspective, miso provides protein (≈9 g/100 g), complex carbohydrates (≈37 g/100 g), and a modest amount of lipids (≈2 g/100 g), largely in the form of polyunsaturated fatty acids (≈1.85 g/100 g), including linoleic and α-linolenic acids. Its micronutrient profile includes potassium, calcium, iron, and B vitamins, while its sodium content is naturally high due to salt-dependent fermentation [256]. Fermentation enhances the nutritional value of soybeans by generating bioactive peptides, increasing the bioavailability of

https://doi.org/10.3390/molecules31081353

isoflavones, and improving the digestibility of proteins and carbohydrates. These biochemical transformations contribute to miso's recognized antioxidant capacity, anti-inflammatory potential, and metabolic benefits, including improved lipid handling, reduced oxidative stress, and modulation of inflammatory markers such as IL-6, CRP, and IL-18, as documented in clinical and experimental studies [256].

6.2.3. Tempe

Tempe is a traditional Indonesian fermented soybean product developed through the long-standing ethno-microbiological practices of Javanese communities, relying primarily on the domesticated mold *Rhizopus oligosporus*. The production involves soaking, dehulling, cooking, and incubating soybeans for 24–48 h, during which *Rhizopus* mycelia bind the cotyledons into a compact, white cake. Although several *Rhizopus* species (*R. oligosporus, R. delemar, R. oryzae, R. stolonifer, R. arrhizus, and R. chinensis*), *R. oligosporus* confers superior sensory and nutritional qualities [257]. Fermentation process produces degradation of mono, di-, and oligosaccharides and polyol compounds that create flatulence [258], enrichment of vitamin B12, attributed mainly to *Saccharomyces cerevisiae*, phytoestrogen production due to *Klebsiella pneumoniae*, including daidzein, genistein, genistin, daidzein, daidzin, and glycitein [259], and immunomodulatory peptides generated predominantly through the enzymatic activity of *Rhizopus oligosporus* [260]. The combination of these events contributes to improved gastrointestinal tolerance [261], antimicrobial activity [262], modulation of the gut microbiota, and potential neurophysiological and cardiometabolic benefits [257].

6.2.4. Natto

Natto, a traditional fermented soybean product produced by *Bacillus subtilis* var. *natto*, is a nutrient-dense food with significant medicinal value. During fermentation, soybean proteins and polysaccharides undergo extensive enzymatic hydrolysis, generating short-chain peptides, free amino acids, and oligosaccharides that are more readily absorbed [263]. Fermentation markedly increases the levels of proteins, cellulose, minerals (calcium, potassium, iron), and vitamin B2 compared with raw soybeans. Fermentation also solubilizes otherwise insoluble soy proteins and promotes the de novo synthesis of digestive enzymes produced by natto bacteria, thereby enhancing gastrointestinal absorption [264]. A key functional metabolite is nattokinase, a fibrinolytic enzyme responsible for natto's thrombolytic activity [265]. Nattokinase can degrade fibrin, inhibit platelet aggregation, and stimulate endogenous fibrinolytic pathways [266]. Natto also exhibits a broad spectrum of functional activities, including neurological benefits [267], supported by its capacity to degrade amyloid-β peptides, a mechanism relevant to Alzheimer's disease [268], as well as antibacterial [269] and antitumor properties [270]. Its high vitamin K2 content contributes to bone metabolism and anti-osteoporotic activity [271], while oligosaccharides and γ-polyglutamic acid promote gut health [272]. Moreover, natto can have immunomodulatory and anti-inflammatory activities [273] and reduce the risk of dysglycemia and cardiovascular disease by decreasing the postprandial glucose response in the early phase [266,274]. Finally, the natto's antioxidant [275], antihypertensive [276], and promotion of gut health via oligosaccharides and γ-polyglutamic acid [272] were confirmed by in vivo studies using animal models.

6.2.5. Doenjang

Doenjang is a traditional Korean fermented soybean paste produced by fermenting *Meju* with *Bacillus subtilis* and molds such as *Mucor*, *Rhizopus*, and *Aspergillus oryzae*. Soybean blocks are formed after soaking, cooking, crushing, and shaping the soybeans, then brining them for several months [277]. The quality, flavor, and nutritional value of Doenjang depend

https://doi.org/10.3390/molecules31081353

strongly on fermentation conditions, raw materials, and the associated microbial community. Incorporating new ingredients or selecting superior soybean varieties can further enhance its sensory and functional properties [278]. Doenjang is known for its antioxidant, anti-inflammatory, anticancer, and antimutagenic activities [279] as well as its hypocholesterolemic properties [277]. γ-polyglutamic acid produced by *Bacillus* species in doenjang exhibits notable antioxidant and antihypertensive activities in both in vitro and in vivo models. Doenjang is also enriched in isoflavones such as genistein and daidzein, which have demonstrated estrogenic and anticancer effects in cell-based assays and limited human studies. Furthermore, bioactive peptides generated during soybean fermentation show ACE-inhibitory activity, supporting their potential role in blood pressure regulation [280,281].

6.2.6. Kimchi

Kimchi is a traditional Korean fermented vegetable food characterized by a complex matrix of probiotics, prebiotics, postbiotics, and fermentation-derived bioactive compounds [24]. Its fermentation is driven primarily by heterofermentative lactic acid bacteria, including *Leuconostoc*, *Weissella*, and *Lactobacillus*, which contribute to flavor development and functional properties. Kimchi contains diverse phytochemicals, such as isothiocyanates, indole-3-carbinol, allyl sulfides, carotenoids, flavonoids, tocopherols, and polyunsaturated fatty acids, along with vitamins, minerals, and dietary fiber. These components collectively confer antioxidant and immune-enhancing effects [282,283], antiaging [284], anticancer [285,286], and anti-obesity properties [287]. Experimental and clinical studies have demonstrated that kimchi and its LAB strains modulate lipid metabolism and reduce adiposity [288], improve gut microbiota composition by modulating cytokines and inflammatory factors [289], and exert antimutagenic and anticancer activities [290].

6.2.7. Yogurt

Yogurt is a fermented dairy product obtained through the controlled action of *Streptococcus thermophilus* and *Lactobacillus delbrueckii* subsp. *bulgaricus*, which converts lactose into lactic acid and gradually decreases the pH of milk, leading to the formation of a soft, hydrated casein network that entraps whey proteins and water [291,292]. Heat treatment prior to fermentation denatures 90–99% of whey proteins, promoting their interaction with casein micelles and contributing to the characteristic gel structure of yogurt [293,294]. This slow acidification process prevents the abrupt coagulation typical of gastric acid exposure and results in a homogeneous, mesh-like matrix with distinct physical properties compared with raw milk [295]. Nutritionally, yogurt retains the high-quality protein profile of milk, which is rich in essential amino acids and composed of casein and whey in an approximately 8:2 ratio. At the same time, fermentation partially hydrolyzes these proteins, generating bioactive peptides and enhancing digestibility [296–298]. Functionally, yogurt provides not only readily absorbable protein but also probiotic and prebiotic components derived from viable and non-viable bacterial cells, metabolites, and fermentation-derived compounds. These include short-chain fatty acids, polysaccharides, and branched-chain hydroxy acids, which contribute to improved gut health, modulation of inflammation, and metabolic benefits such as enhanced glucose regulation [299].

6.2.8. Paocai

Paocai is a traditional Chinese fermented vegetable product prepared from cabbage or radish and seasoned with spices. Paocai possesses high nutritional value, providing abundant vitamins, minerals, dietary fiber, and other functional constituents. These components not only meet essential nutritional requirements but also modulate intestinal microbiota, reduce

 https://doi.org/10.3390/molecules31081353

serum cholesterol, and exert antioxidant, weight-management, and other health-promoting effects [300]. Its spontaneous high-salt fermentation is driven mainly by native LAB, which convert sugars into volatile and non-volatile metabolites that determine flavor and quality. The reliance on natural microbiota leads to variability in sensory properties. To achieve controlled fermentation, selected LAB starters, such as *Lactiplantibacillus brevis*, *Lactiplantibacillus pentosus*, *Leuconostoc mesenteroides*, and *Weissella cibaria*, are increasingly used [301]. Among them, *Levilactobacillus brevis* L., an obligatory heterofermentative species, exhibits probiotic traits, acid tolerance, strong cell-surface hydrophobicity, and substantial auto-aggregation capacity, supporting its survivability and colonization potential. Its cell-free ethyl acetate extract shows broad-spectrum antibacterial and antifungal activity, accompanied by marked inhibition of biofilm formation and reductions in violacein, prodigiosin, and pyocyanin. Metabolomics analyses (FTIR, HPTLC, HR-LCMS) have revealed a diverse array of secondary metabolites, peptides, fatty acids, and phenolic compounds [302].

6.2.9. Lassi

Lassi is a traditional Punjabi fermented dairy beverage, typically made from buffalo milk. It may be prepared in sweet or salted forms and enriched with fruits, spices, or herbs. Although chemical acidification is sometimes used, the best sensory quality is obtained using reconstituted milk fermented with lactic cultures and supplemented with water and sugar [303]. Variants such as bhang lassi and herb- or spice-based (antro, cumin, or black pepper) are common across the Indian subcontinent [304]. Regular consumption of lassi provides multiple nutritional and functional benefits. It supports probiotic intake, digestion, immune function, weight management, bone health, and reduces bloating. Its high digestibility and efficient nutrient absorption make it an excellent vehicle for fortification. Lassi supplies essential and non-essential amino acids, moderate amounts of calcium and phosphorus, and is less acidic than fruit juices [250]. Seven *Lactobacillus* strains (MS001–MS007) were isolated and identified through morphological, biochemical, and PCR-based molecular characterization from Lassi. All isolates showed no hemolysis and were sensitive to most tested antibiotics. They exhibited strong tolerance to gastrointestinal stressors, pH 3.0, bile salts (0.3%), pancreatin (0.5%), and lysozyme, and showed phytase activity, antioxidative capacity, and the ability to form biofilms. Among them, strain MS007 displayed the highest bile salt hydrolase activity, while MS005 showed superior phytate degradation [305]. Probiotic cultures contribute to intestinal health, lactose digestion, and reductions in blood pressure and cholesterol. Its functionality can be further enhanced by adding probiotic strains or herbal extracts. Fiber enrichment helps lower LDL cholesterol, while fruits such as mango can increase vitamin A content and consumer acceptance. Ingredients such as turmeric, ginger, and carrot can boost antioxidant capacity [250].

6.2.10. Pla-ra

Pla-ra is a traditional Thai salt-fermented fish paste. It is produced from freshwater fish mixed with 17–20% salt and a small proportion of roasted rice or rice bran fermented for 6–24 months with *Lactobacillus*, *Pediococcus*, *Weissella*, *Enterococcus*, *Aerococcus*, *Tetragenococcus*, and *Leuconostoc* spp. During this extended maturation, endogenous fish enzymes and halotolerant microorganisms drive extensive proteolysis, generating free amino acids and bioactive peptides that enhance both nutritional value and sensory properties [306]. The addition of roasted rice or rice bran provides fermentable carbohydrates that support mild lactic fermentation, contributing to flavor development and improved digestibility [307]. Pla-ra is a rich source of umami-enhancing amino acids, particularly glutamic and aspartic acids, along

 https://doi.org/10.3390/molecules31081353

with essential amino acids [308] and bioactive compounds that exhibit antihypertensive (via angiotensin-I-converting enzyme inhibition) and antioxidant activities [309].

6.2.11. Bagoong

Bagoong is traditionally prepared in the Philippines through a controlled-salt fermentation process in which small fish (e.g., *Stolephorus* spp., *Sardinella* spp., and *Decapterus* spp. herring, anchovy) or *Atya* spp. or their roe are mixed with 25% salt and allowed to undergo long-term enzymatic and microbial degradation [154]. The early stages of bagoong fermentation are characterized by the presence of Gram-negative rods originating from the raw fish and handling environment. As salting progresses and osmotic pressure increases, these microorganisms rapidly decline due to the high sensitivity of marine bacteria to hypertonic conditions. Protein hydrolysis during fermentation is driven by both endogenous intestinal enzymes and microbial proteases, particularly those produced by *Bacillus subtilis* and *Bacillus coagulans*. These enzymes catalyze deamination and decarboxylation reactions, generating low fatty acids, amides, and other volatile compounds that contribute to the characteristic aroma and flavor of the final product [310]. Throughout fermentation, *Bacillus pumilus* is the dominant species. In the early fermentation phase, *Bacillus megaterium, Bacillus coagulans,* and *Bacillus subtilis* also contribute to substrate degradation. As fermentation progresses, the microbial community shifts toward *Bacillus licheniformis, Micrococcus* spp. (*Micrococcus colpogenes, Micrococcus varians, Micrococcus roseus)*, and *Staphylococcus* species, which participate in the later stages of flavor development and stabilization of the fermented matrix [310]. Bagoong shows a moderate nutrient density, with 15.25% protein, 1.90% fat, and relatively low mineral levels (potassium 80.39 mg/L, magnesium 176.01 mg/L, calcium 3.64 mg/L, sodium 3.21 mg/L, phosphorus 280.96 mg/L). Its lipid fraction contains 2.6% α-linolenic acid (n-3), 0.2% docosahexaenoic acid (n-3), and 4.5% linoleic acid (n-6) [311].

Table 4. Several fermented foods from Asia.

Product	Description	Microorganisms Identified	Region of Origin	References
Koji	Fermented rice, barley, or soybeans	*Aspergillus oryzae, Aspergillus sojae, Aspergillus luchuensis* mut. *kawachii, Aspergillus tamarii*	China, Japan, and Korea	[253]
Miso	Mold-fermented soybean paste	*Aspergillus oryzae, Clavispora lusitaniae, Meyerozyma guilliermondii; Bacillus velezensis, Bacillus subtilis, Enterococcus durans, Rothia kristinae, Lactiplantibacillus plantarum, Leuconostoc* spp., *Pediococcus* spp., *Weissella* spp.	Japan	[256]
Tempe	Fermented soybean	*Rhizopus microsporus* variety *oligosporus* (*Rhizopus oligosporus*), *Rhizopus oryzae, Rhizopus delemar, Rhizopus arrhizus, Rhizopus stolonifer, and Rhizopus chinensis,* along with *Mucor* spp. and other deuteromycete fungi and yeasts	Indonesia	[257]
Natto	Fermented soybean	*Bacillus subtilis*	China	[24,266]
Kimchi	Fermented Napa cabbage, radish, red chilies, garlic, fish, and salt	*Lactobacillus, Leuconostoc*, and *Weissella*	Corea	[287]
Poacai	Fermented cabbage or radish	*Levilactobacillus brevis, Lactiplantibacillus plantarum, Lactiplantibacillus pentosus, Leuconostoc mesenteroides*, and *Weissella cibaria*		[301]

https://doi.org/10.3390/molecules31081353

Table 4. *Cont.*

Product	Description	Microorganisms Identified	Region of Origin	References
Yogurt	Fermented milk	*Streptococcus thermophilus* and *Lactobacillus delbrueckii* subsp. *bulgaricus*	Mesopotamia, Anatolia, Iran, and Central Asia	[312]
Lassi	buffalo milk	*Limosilactobacillus fermentum* MS005, *Lactiplantibacillus plantarum* MS007	Punjabi	[305]
Pla-ra	Channa striata(Striped snake head fish, Chon), Trichogastertrichopterus (Gourami, Kra-dee), T. leeri (Kra-dee-nang), Cyclocheilichthys repasson (SilverCarp, Soi), Puntius gonionotus (Barb, Ta-pien).	*Lactobacillus, Weissella, Pediococcus, Enterococcus, Tetragenococcus, Aerococcus,* and *Leuconostoc* spp.	Thailand	[313]
Bagoon	*Stolephorus* spp., *Sardinella* spp., and *Decapterus* spp. herring, anchovy) or *Atya* spp. or their roe are mixed with 25% salt and allowed to undergo long-term enzymatic and microbial degradation	*Bacillus megaterium, Bacillus coagulans, Bacillus subtilis, Bacillus licheniformis, Micrococcus colpogenes, Micrococcus varians, Micrococcus roseus,* and *Staphylococcus*	Philippines	[165]

6.3. Fermented Foods and Beverages from Latin America

Latin American fermented foods and beverages rely on region-specific raw materials, including maize, cassava, tubers, fruits, vegetables, and animal products (Table 5). Although many countries prepare comparable products, differences in names, processing steps, and seasonings reflect the continent's cultural and ecological diversity [314]. Because knowledge was transmitted orally, and because ingredient availability varies by geography and season, multiple recipe variants have emerged over time [315]. Solid fermented foods made from native ingredients remain central to Indigenous culinary identity, whereas other fermented items, such as cheeses and cured sausages, were introduced later through European immigration and are not considered traditional [316]. Fermented beverages, both alcoholic and non-alcoholic, also hold cultural importance. Alcoholic drinks are used in ritual or spiritual functions in Indigenous groups, while recreational consumption is more common in industrialized settings. Many traditional beverages are obtained from LAB, which often coexist symbiotically with yeasts. Yeast metabolites stimulate LAB growth, and bacterial acidification enhances yeast activity [317,318]. A wide range of beverages is produced from cassava (*Manihot esculenta*) and maize (*Zea mays*), with fermentation time and alcohol content varying by region. *Chicha* exemplifies this diversity, as it encompasses numerous local variants differing in raw materials, alcohol levels, and preparation methods [319]. In some Brazilian tribes, beverages such as caxiri ferment for up to five days to increase ethanol content. Dominant bacterial genera in the production of caium, calugi, and chicha include *Lacticaseibacillus, Lactobacillus, Lactiplantibacillus, Limosilactobacillus, Streptococcus, Leuconostoc*, and *Weissella* [315]. In several Indigenous groups, chewed sweet potato is used as an inoculum; its endogenous amylases and associated microbiota convert starch into fermentable sugars, enabling the production of beverages such as calugi and caium [314].

6.3.1. Chicha

Chicha is a traditional fermented beverage produced by Indigenous communities across South America. It typically contains 2–12% (v/v) alcohol and is most commonly

https://doi.org/10.3390/molecules31081353

prepared from maize, although rice, peanut, cassava, banana, cottonseed, and carob may also be used. It is consumed during religious, agricultural, and social events. In Peru, chicha de jora is produced from local maize varieties and has a low alcoholic content (1–3%). Traditional fermentation historically relied on saliva as a source of amylases, though this practice has increasingly been replaced by alternative enzymatic approaches [320]. Studies on commercial chichas from Peru revealed diverse microbial communities dominated by *Lactobacillaceae*, with *Leuconostocaceae, Acetobacteraceae, and Streptococcaceae* prevailing in specific formulations [321]. In non-commercial maize chichas from Argentina, bacterial communities differ between production sites: *Lactiplantibacillus plantarum, Weissella viridescens, Leuconostoc lactis*, and *Furfurilactobacillus rossiae* predominate in Maimará, whereas *Enterococcus faecium, Enterococcus hirae, Leuconostoc mesenteroides*, and *Weissella confusa* dominate in Tumbaya. Yeast communities are predominantly composed of *Saccharomyces cerevisiae*, with *Fusarium* spp. [315]. In the chica produced by Guarani-Kaiowá in Brazil, *W. confusa*, and *Candida bohaiensis* were identified as the predominant bacterial and yeast species, respectively [322]. Other regional variants include rice chicha from the Umutina people, dominated by *Lacticaseibacillus casei*, and bacaba chicha, produced from *Oenocarpus bacaba*, where *Enterococcus durans* and *Pichia caribbica* prevail [323]. Cassava-based chicha (*masato*) is common among Amazonian groups and involves boiling, mashing, and inoculating cassava with chewed root before fermenting for 1–3 days [324]. LAB typically dominates fermented cassava chicha, while high-throughput sequencing studies show a shift from *Bacillus* in unfermented material to *Lactobacillus* and *Acetobacter* in fermented samples, with community differences observed between villages, suggesting long-term microbial domestication linked to communal consumption practices [319].

6.3.2. Calugi

Calugi is a traditional non-alcoholic fermented porridge produced by the Javaé Indigenous people of Tocantins, Brazil. It is typically made from corn, cassava, and sweet potato, with rice occasionally used as an additional substrate [314]. Cassava is peeled, washed, grated, and pressed, while corn is soaked and macerated with wooden tools to obtain a coarse flour. The corn paste is mixed with cassava, diluted with water, and cooked for about two hours [325]. Once cooled, the mixture is inoculated with chewed sweet potato, a practice in which saliva provides the fermentative microbiota [326], often dominated by *Lactobacillus* spp. The mixture ferments in open containers at ambient temperature for 24–48 h, producing a creamy, mildly acidic beverage with very low alcohol content. The microbiota of calugi includes lactic acid bacteria, acetic acid bacteria, mesophilic aerobic bacteria, enterobacteria, and yeasts, with *Bacillus* spp. frequently predominant. Fermentation produces a wide range of organic acids, including lactic, fumaric, acetic, citric, malic, succinic, propionic, tartaric, and oxalic acids, along with ethanol and carbon dioxide, which collectively lower the pH to around 4.0 and shape the flavor profile. Additional compounds, including terpenes, butyric acid, and diacetyl, contribute to its characteristic sensory attributes [327].

6.3.3. Cauim

Cauim is a traditional non-alcoholic fermented beverage produced by the Tapirapé Indigenous people of Mato Grosso, Brazil, using substrates such as cassava, rice, maize, peanut, banana, cottonseed, and pumpkin. The raw materials are cooked for about 2 h, cooled, and then inoculated with juice obtained from sweet potatoes chewed by women of the tribe, which provides saliva-borne microorganisms and amylases that initiate starch hydrolysis. Fermentation proceeds for 24–48 h [328]. Studies on non-commercial cauim

 https://doi.org/10.3390/molecules31081353

have shown a diverse yeast community, with *Candida tropicalis* and *Candida intermedia* as predominant species, alongside *Saccharomyces cerevisiae* and *Kluyveromyces lactis*. These yeasts contribute to starch degradation, flavor formation, and increases in vitamin B and free amino acids. Bacterial diversity is dominated by *Lactiplantibacillus plantarum* and *Limosilactobacillus fermentum* in the later stages of fermentation [315].

6.3.4. Pozol

Pozol, also known as *pochotl* or *pozolli* in Nahuatl, is a traditional fermented beverage originating in Tabasco, Mexico, with roots in pre-Hispanic Mayan culture. It is valued for its nutritional properties and consumed widely across different population groups [329]. The process entails nixtamalizing whole corn kernels with calcium hydroxide, grinding the resulting nixtamal into a dough, forming the dough into spherical units, enclosing them in banana leaves, and allowing them to ferment at ambient temperature for 1–4 days. The fermented dough is then suspended in water to produce the final beverage [330]. This drink contains LAB, including *Lactococcus, Enterococcus, Leuconostoc, Weissella,* and *Streptococcus,* particularly *Streptococcus infantarius* and *Streptococcus infantarius* 25,124 strains [331]. The nutritional composition of pozol varies according to corn genotype and processing conditions [332].

Table 5. Several fermented foods and beverages from Latin America.

Product	Description	Microorganisms Identified	Region of Origin	References
Chicha	Fermented rice, peanut, cassava, banana, cottonseed, and carob beverage	*Enterococcus, Lactococcus, Streptococcus, Weissella, Leuconostoc, Lactobacillus, Saccharomyces.*	Argentina	[331,333]
Calugi	Fermented corn, cassava, and sweet potato	*Lactobacillus mesenteroides, Lactobacillus lactis; Lactobacillus plantarum, Lactobacillus brevis). Bacillus* sp., *enterobacteria,* and *mesophilic aerobic bacteria* *Fructilactobacillus Rossiae, Weissella viridescens, Weissella confusa, Enterococcus casseliflavus, Enterococcus faecium, Enterococcus mundtti, Enterococcus durans, Enterococcus. hirae, Pediococcus acidilactici,* and *Streptococcus.*	Brazil	[326]
Caium	Fermented cassava, rice, maize, peanut, banana, cottonseed, and pumpkin	*Lactiplantibacillus plantarum* and *Limosilactobacillus fermentum*	Brazil	[315]
Pozol	Fermented maize–cocoa drink	LAB, *Saccharomyces, Candida, Aspergillus, Penicillium, Rhizopus*	Mexico	[334,335]

6.4. Fermented Foods and Beverages from Europe

European fermented foods and beverages include dairy products such as yogurt, kefir, and a wide range of cheeses, which rely primarily on lactic acid bacteria (*Lactobacillus, Lactococcus, Streptococcus*) (Table 6). In kefir, there are mixed consortia of yeasts and acetic acid bacteria [336]. Cereal-based fermentations, including sourdough breads and kvass, are driven by stable LAB–yeast associations that contribute to characteristic acidity, texture, and flavor [331]. Vegetable fermentations, such as sauerkraut and pickled vegetables, are dominated by spontaneous LAB communities (*Leuconostoc, Lactiplantibacillus, Lactobacil-*

 https://doi.org/10.3390/molecules31081353

lus) [337], whereas fermented meats and fish (e.g., rakfisk) involve LAB, coagulase-negative staphylococci, and salt-tolerant microorganisms [338].

6.4.1. Kefir

Kefir is an ancient fermented beverage originating from the Caucasus region, traditionally produced and consumed for centuries on the northeastern slopes of the Georgian and Caucasus mountains, where kefir grains were historically regarded as a form of family wealth and used to prepare a slightly acidic, low-alcohol drink with recognized health benefits [339–341]. Kefir is obtained by inoculating milk with kefir grains, a complex symbiotic consortium of LAB such as *Lactobacillus, Lactococcus, Leuconostoc, and Acetobacter*, as well as yeasts such as *Saccharomyces, Kluyveromyces, Pichia*, and Candida, embedded in a polysaccharide matrix known as kefiran [342–346]. Traditional milk kefir is produced by adding 2.5–10% grains to pasteurized and cooled milk, followed by 18–24 h fermentation at 20–25 °C, after which grains are removed and reused [347], while water kefir is prepared using sucrose solutions or fruit juices and water kefir grains [348,349]. Fermentation generates a wide spectrum of metabolites, including lactic and acetic acids, ethanol, CO_2, bioactive peptides, exopolysaccharides, vitamins, and organic acids, which shape kefir's sensory, nutritional, and functional properties Nutritionally, kefir provides high-quality proteins (≈2.7–3.1%), partially hydrolyzed caseins and whey proteins, essential amino acids, B-vitamins (B1, B2, B5, B6, B12), vitamins A, K, and C, and minerals such as potassium, calcium, phosphorus, magnesium, iron, zinc, and manganese [350]. Nutritional and lactose content are reduced during fermentation, and β-galactosidase activity increases, making kefir suitable for lactose-intolerant individuals [351]. Kefir exhibits multiple nutraceutical properties, including probiotic activity, with LAB and yeasts surviving gastrointestinal transit and inhibiting pathogens through bacteriocins, organic acids, and antimicrobial peptides [352,353], immunomodulatory effects [354], with reductions in inflammatory markers, antimicrobial, hypocholesterolemic, anti-diabetic, and antioxidant activities [355], and antihemorrhagic potential [356]. Kefir also supports gut health by enhancing the production of short-chain fatty acids, an effect demonstrated in vivo in animal models. [357,358] and improving intestinal barrier function [359]. Furthermore, fortification with plant extracts, juices, or functional ingredients (e.g., flaxseed, tea extracts, pomegranate juice) can enhance antioxidant capacity, microbial viability, and sensory attributes [360]. An in vivo study showed that microorganisms from homemade kefir are safe [18].

6.4.2. Sourdough

Sourdough is traditionally produced by mixing ground cereals, legumes, or pseudocereals with water, allowing spontaneous fermentation to occur and gradually establishing a stable microbial ecosystem composed of LAB and yeasts [361]. The dominant microorganisms include *Lactiplantibacillus plantarum, Levilactobacillus brevis, Limosilactobacillus reuteri, Companilactobacillus* spp., and yeasts such as *Saccharomyces cerevisiae, Kluyveromyces marxianus, K. lactis, and K. aestuarii*, which contribute to acidification, proteolysis, phytase activity, and dough maturation [361]. Their metabolic activity enhances the nutritional and functional attributes of sourdough bread by reducing phytic acid and thereby improving mineral bioavailability; modulating starch digestibility and lowering the glycemic index; increasing protein digestibility through extensive proteolysis; and improving both the technological and physiological properties of dietary fiber [362].

 https://doi.org/10.3390/molecules31081353

6.4.3. Kwass

Kvass is traditionally produced through a two-stage fermentation process in which rye-based raw materials (such as rye flour, dried rye bread, or rye malt) are mixed with water, sugar, and yeast, undergoing lactic acid fermentation first, followed by alcoholic fermentation. The final beverage typically contains ≤1.5% alcohol, although prolonged fermentation may raise this to 2.5% or more [363–366]. The microbial community responsible for kvass production includes *Lacticaseibacillus paracasei, Acetobacter pasteurianus,* and *Saccharomyces cerevisiae,* which together produce organic acids, mild carbonation, and characteristic aromatic compounds [367]. Lactic acid acts as a natural preservative, eliminating the need for pasteurization and contributing to kvass's stability. Nutritionally, kvass contains organic acids, amino acids, carbohydrates, B vitamins, and folic acid, supporting digestive function and metabolic regulation. Recent developments include enriching kvass with polyphenol-rich botanicals, such as black chokeberry (*Aronia melanocarpa*), whose anthocyanins, proanthocyanidins, and hydroxycinnamic acids enhance antioxidant capacity and functional value [368].

6.4.4. Rakfisk

Rakfisk is a traditional Norwegian fermented fish. The fish is dry-salted and stacked belly-up under pressure in sealed containers. Some producers add small amounts of sugar or a LAB-based starter culture. During maturation, the fish remains fully submerged in brine with a final salt concentration of 4–7% (*w*/*w*). Storage at 3–8 °C for 3–12 months enables a slow fermentation/ripening process that shapes the product's sensory and microbial profile [369].

Table 6. Several fermented foods and beverages from Europe.

Product	Description	Microorganisms Identified	Region of Origin	References
Kefir	Fermentation of milk or plant-based substrates	*Lactobacillus, Lactococcus, Leuconostoc,* and *Acetobacter, Saccharomyces,* and *Kluyveromyces.*	Caucas	[355]
Sourdough Breads	Fermentation of cereals, legumes, or pseudocereals	*Lactiplantibacillus plantarum, Levilactobacillus brevis, Limosilactobacillus reuteri, Companilactobacillus* spp., and yeasts such as *Saccharomyces cerevisiae, Kluyveromyces marxianus, Kluyveromyces lactis,* and *Kluyveromyces aestuarii.*	Mediterranean, the Middle East, and Europe,	[362,370]
Kvass	Fermentation of rye flour, dried rye bread, or rye malt	*Lacticaseibacillus paracasei, Acetobacter pasteurianus,* and *Saccharomyces cerevisiae*	Eastern and Central Europe	[368]
Rakfisk	salmon	*Latilactobacillus sakei*	Norwegian	[369]

7. Discussion

The findings summarized in this review underscore the broad functional potential of fermented foods and probiotics; however, the current evidence base remains fragmented and characterized by substantial methodological heterogeneity. Differences in microbial strains, fermentation parameters, analytical techniques, and outcome measures complicate direct comparisons across studies and limit the ability to draw robust, generalizable conclusions. A major challenge arises from the intrinsic strain-specificity of microbial functions.

https://doi.org/10.3390/molecules31081353

Even within the same species, genomic variability leads to divergent metabolic outputs, including differences in organic acid production, enzymatic activities, detoxification capacity, and postbiotic profiles. This variability is further amplified by the influence of the food matrix. Identical strains may exhibit distinct growth kinetics, metabolite production, and functional properties when inoculated into dairy, cereal, legume, or vegetable substrates. Matrix-dependent buffering capacity, nutrient composition, and endogenous enzymatic activities contribute to these discrepancies, making it difficult to extrapolate findings from one fermented food system to another. Another limitation concerns the predominance of in vitro and small-scale experimental studies. While these models provide valuable mechanistic insights, they often fail to replicate the complexity of human gastrointestinal physiology, host–microbiota interactions, and interindividual variability. Reported outcomes such as antioxidant capacity, immunomodulation, or toxin degradation frequently rely on non-standardized assays, further hindering cross-study comparability. Human clinical trials remain limited in number, heterogeneous in design, and often underpowered, restricting the ability to establish causal relationships or define effective doses and exposure durations. Advances in high-throughput sequencing have improved the characterization of microbial communities in fermented foods, yet inconsistencies persist in sequencing platforms, reference databases, and bioinformatic pipelines. Many studies do not provide strain-level genomic data, limiting the capacity to link specific genetic determinants to functional traits or safety profiles. This gap is particularly relevant for assessing the presence of antibiotic resistance genes, virulence factors, or undesirable metabolic pathways, which remain insufficiently monitored in traditional and artisanal fermentations. Safety considerations also warrant further attention. Although most microorganisms associated with fermented foods have a long history of safe use, the potential transfer of antibiotic resistance genes, the formation of biogenic amines, and the presence of opportunistic pathogens in spontaneous fermentations represent ongoing concerns. Regulatory frameworks differ substantially across regions, and the absence of harmonized guidelines for microbial strain selection, fermentation control, and product labeling complicates risk assessment and consumer protection. The growing interest in postbiotics and fermentation-derived metabolites highlights the need to distinguish the effects of live microorganisms from those of their metabolic products. This distinction is essential for understanding mechanisms of action, ensuring reproducibility, and developing next-generation functional foods with predictable health outcomes. However, the lack of standardized definitions, quantification methods, and biological assays currently limits progress in this area. These considerations emphasize the need for integrated, multidisciplinary approaches combining microbial ecology, food science, metabolomics, genomics, and clinical research. Harmonized methodologies, strain-resolved characterization, and rigorously designed human studies will be essential to strengthen the evidence base, improve reproducibility, and support the development of safe, effective, and scientifically substantiated fermented foods and probiotics.

8. Review Limitation

This review is subject to several limitations. The available literature is characterized by substantial heterogeneity in microbial strains, fermentation conditions, and analytical methods, limiting comparability across studies. Much of the evidence derives from in vitro and animal models, which provide mechanistic insights but limited translational relevance. Human studies remain few, heterogeneous, and often underpowered. Strain-level characterization is frequently incomplete, complicating attribution of specific effects. Finally, although supported by a structured search strategy, the narrative nature of this review does not eliminate the possibility of selection bias.

https://doi.org/10.3390/molecules31081353

9. Conclusions

The available evidence indicates that fermented foods and probiotics can enhance nutrient bioaccessibility, generate bioactive metabolites, and contribute to the detoxification of chemical and microbial contaminants. However, the strength of these findings is limited by substantial heterogeneity in microbial strains, fermentation conditions, analytical methods, and study designs. Most functional claims are supported primarily by in vitro or small-scale experimental models, while well-controlled human studies remain scarce and often underpowered. Safety considerations, including the presence of antibiotic resistance determinants, biogenic amines, and variability in spontaneous fermentations, require more systematic evaluation. To advance the field, standardized methodologies, strain-level characterization, and integrated multi-omics approaches are needed, together with rigorously designed clinical trials. These efforts will be essential to develop fermented foods and probiotics with reproducible, evidence-based functional properties and clearly defined safety profiles.

Funding: This research received no external funding.

Institutional Review Board Statement: Not applicable.

Informed Consent Statement: Not applicable.

Data Availability Statement: No new data were created or analyzed in this study.

Acknowledgments: During the preparation of this manuscript/study, the author used Grammarly.com to improve clarity and readability of the English text, and Microsoft Copilot (to assist in generating illustrative figures. The author has reviewed and edited the output and take full responsibility for the content of this publication.

Conflicts of Interest: The authors declare no conflicts of interest.

References

1. International Scientific Association for Probiotics and Prebiotics (ISAPP). Available online: https://isappscience.org/topic/probiotics (accessed on 19 March 2026).
2. International Scientific Association for Probiotics and Prebiotics (ISAPP). Available online: https://isappscience.org/topic/fermented-foods (accessed on 19 March 2026).
3. Jung, S.-J.; Chae, S.-W.; Shin, D.-H. Fermented Foods of Korea and Their Functionalities. *Fermentation* **2022**, *8*, 645. [CrossRef]
4. Praveen, M.; Brogi, S. Microbial Fermentation in Food and Beverage Industries: Innovations, Challenges, and Opportunities. *Foods* **2025**, *14*, 114. [CrossRef] [PubMed]
5. Zapaśnik, A.; Sokołowska, B.; Bryła, M. Role of Lactic Acid Bacteria in Food Preservation and Safety. *Foods* **2022**, *11*, 1283. [CrossRef]
6. Deveci, G.; Çelik, E.; Ağagündüz, D.; Bartkiene, E.; Rocha, J.M.F.; Özogul, F. Certain Fermented Foods and Their Possible Health Effects with a Focus on Bioactive Compounds and Microorganisms. *Fermentation* **2023**, *9*, 923. [CrossRef]
7. Ricci, A.; Allende, A.; Bolton, D.; Chemaly, M.; Davies, R.; Girones, R.; Herman, L.; Koutsoumanis, K.; Lindqvist, R.; Nørrung, B.; et al. Scientific Opinion on the Update of the List of QPS-recommended Biological Agents Intentionally Added to Food or Feed as Notified to EFSA. *EFSA J.* **2017**, *15*, e04664.
8. Tamang, J.P.; Watanabe, K.; Holzapfel, W.H. Review: Diversity of Microorganisms in Global Fermented Foods and Beverages. *Front. Microbiol.* **2016**, *7*, 377. [CrossRef]
9. Singh, P.; Kulkarni, R.; Shetty, K.; Rale, V. Lactic Acid Bacteria and Biotechnology of Cereal and Legume-Based Fermented Foods from India. In *Fermentation Biotechnology for Functional Foods*; Shetty, K., Ranilla, L.G., Eds.; CRC Press: Boca Raton, FL, USA, 2025; pp. 175–191.
10. Huang, Y.; Wang, Y.; Shang, N.; Li, P. Microbial Fermentation Processes of Lactic Acid: Challenges, Solutions, and Future Prospects. *Foods* **2023**, *12*, 2311. [CrossRef]
11. Sun, W.; Shahrajabian, M.H.; Lin, M. Research Progress of Fermented Functional Foods and Protein Factory-Microbial Fermentation Technology. *Fermentation* **2022**, *8*, 688. [CrossRef]

https://doi.org/10.3390/molecules31081353

12. Mota, J.; Vilela, A. Aged to Perfection: The Scientific Symphony behind Port Wine, Vinegar, and Acetic Acid Bacteria. *Fermentation* **2024**, *10*, 200. [CrossRef]
13. Gomathy, M.; Sabarinathan, K.G.; Rajakumar, D.; Manoj, M.; Bhimani, H.D.; John, P.; Wijayaratne, T.; Ekanayake, S.; Maheswari, U.; Rajan, A.S.; et al. Fermented Foods of South Asia. In *Microbial Fermentations in Nature and as Designed Processes*; Hurst, C.J., Ed.; Wiley: Hoboken, NJ, USA, 2023; pp. 323–351, ISBN 9781119849971.
14. Sawant, S.S.; Park, H.-Y.; Sim, E.-Y.; Kim, H.-S.; Choi, H.-S. Microbial Fermentation in Food: Impact on Functional Properties and Nutritional Enhancement—A Review of Recent Developments. *Fermentation* **2025**, *11*, 15. [CrossRef]
15. Xu, D.; Zhang, H.; Xi, J.; Jin, Y.; Chen, Y.; Guo, L.; Jin, Z.; Xu, X. Improving bread aroma using low-temperature sourdough fermentation. *Food Biosci.* **2020**, *37*, 100704. [CrossRef]
16. Zugravu, C.-A.; Constantin, C. Eastern European Fermented Foods: Nutritional Value, Functional Potential, and Cultural Heritage. *Foods* **2026**, *15*, 28. [CrossRef]
17. Zhang, J.; Liu, M.; Zhao, Y.; Zhu, Y.; Bai, J.; Fan, S.; Zhu, L.; Song, C.; Xiao, X. Recent Developments in Fermented Cereals on Nutritional Constituents and Potential Health Benefits. *Foods* **2022**, *11*, 2243. [CrossRef] [PubMed]
18. Maione, A.; Imparato, M.; Buonanno, A.; Salvatore, M.M.; Carraturo, F.; de Alteriis, E.; Guida, M.; Galdiero, E. Evaluation of Potential Probiotic Properties and In Vivo Safety of Lactic Acid Bacteria and Yeast Strains Isolated from Traditional Home-Made Kefir. *Foods* **2024**, *13*, 1013. [CrossRef] [PubMed]
19. Kanak, E.K.; Öztürk Yılmaz, S. Determination of the Probiotic and Functional Properties of Yeasts Isolated from Different Dairy Products. *Fermentation* **2025**, *11*, 104. [CrossRef]
20. Biçer, Y.; Telli, A.E.; Turkal, G.; Telli, N.; Uçar, G. From Raw to Fermented: Uncovering the Microbial Wealth of Dairy. *Fermentation* **2025**, *11*, 552. [CrossRef]
21. do Prado, F.G.; Pagnoncelli, M.G.B.; de Melo Pereira, G.V.; Karp, S.G.; Soccol, C.R. Fermented Soy Products and Their Potential Health Benefits: A Review. *Microorganisms* **2022**, *10*, 1606. [CrossRef]
22. Li, C.; Li, L.; Chen, S.; Zhao, Y.; Wu, Y. Volatile Flavor Improvement and Spoilage Microorganism Inhibition in Low-Salt Fish Sauce (Yulu) by Salt-Tolerant *Bacillus subtilis*. *Fermentation* **2023**, *9*, 515. [CrossRef]
23. Woldemariam, K.Y.; Cai, M.; Jiao, Y.S.; Tang, W.S.; Liu, Y.L.; Wang, J. Impact of Lactic Acid Bacteria on the Flavor Profile of Salami and Role of Microbial Diversity and Functional Gene. *J. Food Biochem.* **2024**, *2024*, 6073261. [CrossRef]
24. Fijan, S.; Fijan, P.; Wei, L.; Marco, M.L. Health Benefits of Kimchi, Sauerkraut, and Other Fermented Foods of the Genus Brassica. *Appl. Microbiol.* **2024**, *4*, 1165–1176. [CrossRef]
25. Tran, T.; Grandvalet, C.; Verdier, F.; Martin, A.; Alexandre, H.; Tourdot-Maréchal, R. Microbial Dynamics between Yeasts and Acetic Acid Bacteria in Kombucha: Impacts on the Chemical Composition of the Beverage. *Foods* **2020**, *9*, 963. [CrossRef] [PubMed]
26. Han, D.; Yang, Y.; Guo, Z.; Dai, S.; Jiang, M.; Zhu, Y.; Wang, Y.; Yu, Z.; Wang, K.; Rong, C.; et al. A Review on the Interaction of Acetic Acid Bacteria and Microbes in Food Fermentation: A Microbial Ecology Perspective. *Foods* **2024**, *13*, 2534. [CrossRef]
27. Nicolescu, C.M.; Bumbac, M.; Buruleanu, C.L.; Popescu, E.C.; Stanescu, S.G.; Georgescu, A.A.; Toma, S.M. Biopolymers Produced by Lactic Acid Bacteria: Characterization and Food Application. *Polymers* **2023**, *15*, 1539. [CrossRef] [PubMed]
28. Koutsoumanis, K.; Allende, A.; Alvarez-Ordóñez, A.; Bolton, D.; Bover-Cid, S.; Chemaly, M.; Davies, R.; De Cesare, A.; Hilbert, F.; Lindqvist, R.; et al. Update of the List of QPS-Recommended Microbiological Agents Intentionally Added to Food or Feed as Notified to EFSA 16: Suitability of Taxonomic Units Notified to EFSA until March 2022. *EFSA J.* **2022**, *20*, 7408.
29. Nasrollahzadeh, A.; Mokhtari, S.; Khomeiri, M.; Saris, P.E.J. Antifungal Preservation of Food by Lactic Acid Bacteria. *Foods* **2022**, *11*, 395. [CrossRef] [PubMed]
30. Hazards, E.P.O.B.; Allende, A.; Alvarez-Ordóñez, A.; Bortolaia, V.; Bover-Cid, S.; De Cesare, A.; Dohmen, W.; Guillier, L.; Jacxsens, L.; Nauta, M. Update of the list of qualified presumption of safety (QPS) recommended microbiological agents intentionally added to food or feed as notified to EFSA 21: Suitability of taxonomic units notified to EFSA until September 2024. *EFSA J.* **2026**, *24*, e9824.
31. Grand View Research. Lactic Acid Bacteria Market Size, Share & Trends Analysis Report, 2024–2034. Available online: https://www.grandviewresearch.com (accessed on 5 February 2026).
32. Yang, S.; Bai, M.; Liu, W.; Li, W.; Zhong, Z.; Kwok, L.-Y.; Dong, G.; Sun, Z. Predicting *Lactobacillus* delbrueckii subsp. bulgaricus-*Streptococcus* thermophilus interactions based on a highly accurate semi-supervised learning method. *Sci. China Life Sci.* **2025**, *68*, 558–574. [CrossRef]
33. Abdul Hakim, B.N.; Xuan, N.J.; Oslan, S.N.H. A Comprehensive Review of Bioactive Compounds from Lactic Acid Bacteria: Potential Functions as Functional Food in Dietetics and the Food Industry. *Foods* **2023**, *12*, 2850. [CrossRef]
34. Anumudu, C.K.; Miri, T.; Onyeaka, H. Multifunctional Applications of Lactic Acid Bacteria: Enhancing Safety, Quality, and Nutritional Value in Foods and Fermented Beverages. *Foods* **2024**, *13*, 3714. [CrossRef]

35. Rahman, M.S.; Soltani, S.; LaPointe, G.; Karboune, S.; Fliss, I. Lactic acid bacteria: Beyond fermentation to bio-protection against fungal spoilage and mycotoxins in food systems. *Front. Microbiol.* **2025**, *16*, 1580670. [CrossRef]
36. Soltani, S.; Biron, E.; Said, L.B.; Subirade, M.; Fliss, I. Bacteriocin-based synergetic consortia: A promising strategy to enhance antimicrobial activity and broaden the spectrum of inhibition. *Microbiol. Spectr.* **2022**, *10*, e00406–e00421. [CrossRef] [PubMed]
37. Petka, K.; Sroka, P.; Tarko, T.; Duda-Chodak, A. The Acrylamide Degradation by Probiotic Strain *Lactobacillus acidophilus* LA-5. *Foods* **2022**, *11*, 365. [CrossRef]
38. Jurášková, D.; Ribeiro, S.C.; Silva, C.C.G. Exopolysaccharides Produced by Lactic Acid Bacteria: From Biosynthesis to Health-Promoting Properties. *Foods* **2022**, *11*, 156. [CrossRef]
39. Sharma, S.; Padhi, S.; Kumari, M.; Rai, A.K.; Sahoo, D. *Bioactive Compounds in Fermented Foods. Health Aspects*; CRC Press: Boca Raton, FL, USA, 2021; p. 48.
40. Zhou, Y.; Cui, Y.; Qu, X. Exopolysaccharides of lactic acid bacteria: Structure, bioactivity and associations: A review. *Carbohydr. Polym.* **2019**, *207*, 317–332. [CrossRef]
41. Oerlemans, M.M.P.; Akkerman, R.; Ferrari, M.; Walvoort, M.T.C.; de Vos, P. Benefits of bacteria-derived exopolysaccharides on gastrointestinal microbiota, immunity and health. *J. Funct. Foods* **2021**, *76*, 104289. [CrossRef]
42. Abdalla, A.K.; Ayyash, M.M.; Olaimat, A.N.; Osaili, T.M.; Al-Nabulsi, A.A.; Shah, N.P.; Holley, R. Exopolysaccharides as antimicrobial agents: Mechanism and spectrum of activity. *Front. Microbiol.* **2021**, *12*, 664395. [CrossRef]
43. Souza, L.V.; da Silva, R.R.; Falqueto, A.; Fusieger, A.; Martins, E.; Caggia, C.; Randazzo, C.L.; de Carvalho, A.F. Evaluation of antifungal activity of lactic acid bacteria against fungi in simulated cheese matrix. *LWT—Food Sci. Technol.* **2023**, *182*, 114773. [CrossRef]
44. Field, D.; De Ullivarri, M.F.; Ross, R.P.; Field, D. After a century of Nisin research—Where are we now? *FEMS Microbiol. Rev.* **2023**, *47*, fuad023. [CrossRef]
45. Gut, I.M.; Blanke, S.R.; Van der Donk, W.A. Mechanism of inhibition of *Bacillus anthracis* spore outgrowth by the lantibiotic nisin. *ACS Chem. Biol.* **2011**, *6*, 744–752. [CrossRef] [PubMed]
46. Wang, Y.; Han, J.; Wang, D.; Gao, F.; Zhang, K.; Tian, J.; Jin, Y. Research update on the impact of lactic acid bacteria on the substance metabolism, flavor, and quality characteristics of fermented meat products. *Foods* **2022**, *11*, 2090. [CrossRef] [PubMed]
47. Mokoena, M.P.; Omatola, C.A.; Olaniran, A.O. Applications of Lactic Acid Bacteria and Their Bacteriocins against Food Spoilage Microorganisms and Foodborne Pathogens. *Molecules* **2021**, *26*, 7055. [CrossRef] [PubMed]
48. Van de Voorde, D.; Díaz-Muñoz, C.; Hernandez, C.E.; Weckx, S.; De Vuyst, L. Yeast Strains Do Have an Impact on the Production of Cured Cocoa Beans, as Assessed with Costa Rican Trinitario Cocoa Fermentation Processes and Chocolates Thereof. *Front. Microbiol.* **2023**, *14*, 1232323. [CrossRef]
49. EFSA BIOHAZ Panel (EFSA Panel on Biological Hazards). The Qualified Presumption of Safety (QPS) approach for the safety assessment of microorganisms intentionally added to food or feed. *EFSA J.* **2023**, *21*, 7747.
50. Campos, S.d.M.; Martínez-Burgos, W.J.; dos Reis, G.A.; Ocán-Torres, D.Y.; dos Santos Costa, G.; Rosas Vega, F.; Alvarez Badel, B.; Sotelo Coronado, L.; Lima Serra, J.; Soccol, C.R. The Role of Microbial Dynamics, Sensorial Compounds, and Producing Regions in Cocoa Fermentation. *Microbiol. Res.* **2025**, *16*, 75. [CrossRef]
51. Morales-Rodriguez, W.J.; Morante-Carriel, J.; Herrera-Feijoo, R.J.; Ayuso-Yuste, M.C.; Bernalte-García, M.J. Effect of Addition of Yeasts and Enzymes during Fermentation on Physicochemical Quality of Fine Aroma Cocoa Beans. *J. Agric. Food Res.* **2024**, *16*, 101126. [CrossRef]
52. Pokharel, B. Cocoa Bean Fermentation: Impact on Chocolate Flavor and Quality. *Int. J. Sci. Res. (IJSR)* **2023**, *12*, 1668–1674. [CrossRef]
53. Gutiérrez-Ríos, H.G.; Suárez-Quiroz, M.L.; Hernández-Estrada, Z.J.; Castellanos-Onorio, O.P.; Alonso-Villegas, R.; Rayas-Duarte, P.; Cano-Sarmiento, C.; Figueroa-Hernández, C.Y.; González-Rios, O. Yeasts as Producers of Flavor Precursors during Cocoa Bean Fermentation and Their Relevance as Starter Cultures: A Review. *Fermentation* **2022**, *8*, 33. [CrossRef]
54. Petka, K.; Walczycka, M. Microbiological Aspects of Meat Fermentation: From Traditional Methods to Advanced Microflora Control Techniques—A Systematic Review. *Appl. Sci.* **2026**, *16*, 641. [CrossRef]
55. Delgado, J.; Álvarez, M.; Cebrián, E.; Martín, I.; Roncero, E.; Rodríguez, M. Biocontrol of Pathogen Microorganisms in Ripened Foods of Animal Origin. *Microorganisms* **2023**, *11*, 1578. [CrossRef] [PubMed]
56. Adebo, J.A.; Njobeh, P.B.; Gbashi, S.; Oyedeji, A.B.; Ogundele, O.M.; Oyeyinka, S.A.; Adebo, O.A. Fermentation of Cereals and Legumes: Impact on Nutritional Constituents and Nutrient Bioavailability. *Fermentation* **2022**, *8*, 63. [CrossRef]
57. Toor, B.S.; Kaur, A.; Kaur, J. Fermentation of legumes with Rhizopus oligosporus: Effect on physicochemical, functional and microstructural properties. *Int. J. Food Sci. Technol.* **2022**, *57*, 1763–1772. [CrossRef]
58. Astuti, M.; Meliala, A.; Dalais, F.S.; Wahlqvist, M.L. Tempe, a Nutritious and Healthy Food from Indonesia. *Asia Pac. J. Clin. Nutr.* **2000**, *9*, 322–325. [CrossRef]

59. Rizzo, G. Soy-Based Tempeh as a Functional Food: Evidence for Human Health and Future Perspective. *Front. Biosci.* **2024**, *16*, 3. [CrossRef]
60. Liwinski, T.; Lang, U.E.; Brühl, A.B.; Schneider, E. Exploring the Therapeutic Potential of Gamma-Aminobutyric Acid in Stress and Depressive Disorders through the Gut–Brain Axis. *Biomedicines* **2023**, *11*, 3128. [CrossRef]
61. Górska, K.; Pejcz, E.; Harasym, J. Tempeh and Fermentation—Innovative Substrates in a Classical Microbial Process. *Appl. Sci.* **2025**, *15*, 8888. [CrossRef]
62. Singh, A.; Kumar, S. Exploring the Functionality of Microbes in Fermented Foods: Technological Advancements and Future Directions. *Fermentation* **2025**, *11*, 300. [CrossRef]
63. Zhao, G.; Ding, L.L.; Yao, Y.; Cao, Y.; Pan, Z.H.; Kong, D.H. Extracellular Proteome Analysis and Flavor Formation During Soy Sauce Fermentation. *Front. Microbiol.* **2018**, *9*, 1872. [CrossRef]
64. Vong, W.C.; Hua, X.Y.; Liu, S.-Q. Solid-state fermentation with *Rhizopus* oligosporus and Yarrowia lipolytica improved nutritional and flavour properties of okara. *LWT-Food Sci. Technol.* **2018**, *90*, 316–322. [CrossRef]
65. Huang, W.; Deng, Z.; Lu, L.; Ouyang, Y.; Zhong, S.; Luo, T.; Fan, Y.; Zheng, L. Polysaccharides from soybean residue fermented by Neurospora crassa alleviate DSS-induced gut barrier damage and microbiota disturbance in mice. *Food Funct.* **2022**, *13*, 5739–5751. [CrossRef] [PubMed]
66. Amiri, S.; Aghamirzaei, M.; Mostashari, P.; Sarbazi, M.; Tizchang, S.; Madahi, H. The Impact of Biotechnology on the Dairy Industry. In *Microbial Biotechnology in Food and Health*; Academic Press: Cambridge, MA, USA, 2021; pp. 53–79.
67. Chen, D.; Xu, Y.; Li, H.; Zhu, X. Optimization of *Monascus purpureus* Culture Conditions in Rice Bran for Enhanced *Monascus* Pigment Biosynthesis. *Fermentation* **2025**, *11*, 505. [CrossRef]
68. Jin, R.; Song, J.; Liu, C.; Lin, R.; Liang, D.; Aweya, J.J.; Weng, W.; Zhu, L.; Shang, J.; Yang, S. Synthetic microbial communities: Novel strategies to enhance the quality of traditional fermented foods. *Compr. Rev. Food Sci. Food Saf.* **2024**, *23*, e13388. [CrossRef]
69. Dhiman, S.; Kaur, S.; Thakur, B.; Singh, P.; Tripathi, M. Nutritional Enhancement of Plant-Based Fermented Foods: Microbial Innovations for a Sustainable Future. *Fermentation* **2025**, *11*, 346. [CrossRef]
70. Muhammed, Y.M.R.; Minervini, F.; Cavoski, I. From Ancient Fermentations to Modern Biotechnology: Historical Evolution, Microbial Mechanisms, and the Role of Natural and Commercial Starter Cultures in Shaping Organic and Sustainable Food Systems. *Foods* **2025**, *14*, 4240. [CrossRef]
71. Siroli, L.; Giordani, B.; Rossi, S.; Gottardi, D.; McMahon, H.; Augustyniak, A.; Menon, A.; Vannini, L.; Vitali, B.; Patrignani, F.; et al. Antioxidant and Functional Features of Pre-Fermented Ingredients Obtained by the Fermentation of Milling By-Products. *Fermentation* **2022**, *8*, 722. [CrossRef]
72. Mannaa, M.; Mannaa, M.; Han, G.; Seo, Y.-S.; Park, I. Evolution of Food Fermentation Processes and the Use of Multi-Omics in Deciphering the Roles of the Microbiota. *Foods* **2021**, *10*, 2861. [CrossRef]
73. Bergsveinson, J.; Kajala, I.; Ziola, B.; Ziola, B. Next-generation sequencing approaches for improvement of lactic acid bacteria-fermented plant-based beverages. *AIMS Microbiol.* **2017**, *3*, 8–24. [CrossRef]
74. Janavi, R.; Punia, D.; Soumya, C. Fermentation technology. In *Futuristic Trends in Biotechnology Volume 3 Book 4*; Iterative International Publishers: Chikmagalur, India, 2024; pp. 61–73.
75. Pérez–Velasco, R.; Gómez-Gil, B.; Martínez-Montaño, E.; González-Córdova, A.F.; Hernández, C. Nutritional Attributes and Microbial Metagenomic Profile during Solid-state Fermentation of Soybean Meal Inoculated with *Lactobacillus* acidophilus under Non-sterile Conditions. *J. Sci. Food Agric.* **2024**, *104*, 8219–8229. [CrossRef]
76. Asensio-Grau, A.; Calvo-Lerma, J.; Heredia, A.; Andrés, A. Enhancing the Nutritional Profile and Digestibility of Lentil Flour by Solid State Fermentation with Pleurotus ostreatus. *Food Funct.* **2020**, *11*, 7905–7912. [CrossRef]
77. Nkhata, S.G.; Ayua, E.; Kamau, E.H.; Shingiro, J. Fermentation and Germination Improve Nutritional Value of Cereals and Legumes through Activation of Endogenous Enzymes. *Food Sci. Nutr.* **2018**, *6*, 2446–2458. [CrossRef]
78. Ahmed, M.I.; Xu, X.; Sulieman, A.A.; Na, Y.; Mahdi, A.A. The Effect of Fermentation Time on in Vitro Bioavailability of Iron, Zinc, and Calcium of Kisra Bread Produced from Koreeb (Dactyloctenium aegyptium) Seeds Flour. *Microchem. J.* **2020**, *154*, 104644. [CrossRef]
79. Day, C.N.; Morawicki, R.O. Effects of Fermentation by Yeast and Amylolytic Lactic Acid Bacteria on Grain Sorghum Protein Content and Digestibility. *J. Food Qual.* **2018**, *2018*, 3964392. [CrossRef]
80. Hajimohammadi, A.; Mottaghitalab, M.; Hashemi, M. Effects of Microbial Fermented Sesame Meal and Enzyme Supplementation on the Intestinal Morphology, Microbiota, pH, Tibia Bone and Blood Parameters of Broiler Chicks. *Ital. J. Anim. Sci.* **2020**, *19*, 457–467. [CrossRef]
81. Petrova, P.; Arsov, A.; Tsvetanova, F.; Parvanova-Mancheva, T.; Vasileva, E.; Tsigoriyna, L.; Petrov, K. The Complex Role of Lactic Acid Bacteria in Food Detoxification. *Nutrients* **2022**, *14*, 2038. [CrossRef]

 https://doi.org/10.3390/molecules31081353

82. Liu, L.; Xie, M.; Wei, D. Biological Detoxification of Mycotoxins: Current Status and Future Advances. *Int. J. Mol. Sci.* **2022**, *23*, 1064. [CrossRef]
83. Harkai, P.; Szabó, I.; Cserháti, M.; Krifaton, C.; Risa, A.; Radó, J.; Balázs, A.; Berta, K.; Kriszt, B. Biodegradation of aflatoxin-B1, and zearalenone by Streptomyces sp. collection. *Int. Biodeterior. Biodegrad.* **2016**, *108*, 48–56. [CrossRef]
84. Rao, K.R.; Vipin, A.V.; Hariprasad, P.; Appaiah, K.A.A.; Venkateswaran, G. Biological detoxification of Aflatoxin B1, by Bacillus licheniformis CFR1. *Food Control* **2017**, *71*, 234–241.
85. Juodeikiene, G.; Bartkiene, E.; Cernauskas, D.; Cizeikiene, D.; Zadeike, D.; Lele, V.; Bartkevics, V. Antifungal activity of lactic acid bacteria and their application for Fusarium mycotoxin reduction in malting wheat grains. *LWT Food Sci. Technol.* **2018**, *89*, 307–314. [CrossRef]
86. Nathanail, A.V.; Gibson, B.; Han, L.; Peltonen, K.; Ollilainen, V.; Jestoi, M.; Laitila, A. The lager yeast *Saccharomyces pastorianus* removes and transforms *Fusarium* trichothecene mycotoxins during fermentation of brewer's wort. *Food Chem.* **2016**, *203*, 448–455. [CrossRef]
87. Ghosh, S.; Sarkar, B. Microbial enzymes for biodegradation and detoxification of pesticides. In *Current Developments in Biotechnology and Bioengineering*; Elsevier: Amsterdam, The Netherlands, 2023; pp. 321–356.
88. Islam, S.M.A.; Math, R.K.; Cho, K.M.; Lim, W.J.; Hong, S.Y.; Kim, J.M.; Yun, M.G.; Cho, J.J.; Yun, H.D. Organophosphorus hydrolase (*OpdB*) of *Lactobacillus brevis* WCP902 from kimchi is able to degrade organophosphorus pesticides. *J. Agric. Food Chem.* **2010**, *58*, 5380–5386. [CrossRef]
89. Chen, W.; Zhai, Q. Applications of Lactic Acid Bacteria in Heavy Metal Pollution Environment. In *Lactic Acid Bacteria in Foodborne Hazards Reduction*, 1st ed.; Chen, W., Narbad, A., Eds.; Springer: Singapore, 2018; pp. 213–248.
90. Wang, Y.; Han, J.; Ren, Q.; Liu, Z.; Zhang, X.; Wu, Z. The Involvement of Lactic Acid Bacteria and Their Exopolysaccharides in the Biosorption and Detoxication of Heavy Metals in the Gut. *Biol. Trace Elem. Res.* **2024**, *202*, 671–684. [CrossRef]
91. Salim, R.; Nehvi, I.B.; Mir, R.A.; Tyagi, A.; Ali, S.; Bhat, O.M. A review on antinutritional factors: Unraveling the natural gateways to human health. *Front. Nutr.* **2023**, *10*, 1215873. [CrossRef]
92. Ribeiro, M.M.S.; Viganó, J.; Meireles, M.A.A.; Veggi, P.C. Recent advances in the recovery of tannins from natural matter. *Stud. Nat. Prod. Chem.* **2022**, *75*, 289–328.
93. Álvarez, A.; Rodríguez, A.; Chaparro, S.; Borrás, L.M.; Rache, L.Y.; Brijaldo, M.H.; Martínez, J.J. Solid-State Fermentation as a Biotechnological Tool to Reduce Antinutrients and Increase Nutritional Content in Legumes and Cereals for Animal Feed. *Fermentation* **2025**, *11*, 359. [CrossRef]
94. Cheng, S.; Langrish, T.A.G. A Review of the Treatments to Reduce Anti-Nutritional Factors and Fluidized Bed Drying of Pulses. *Foods* **2025**, *14*, 681. [CrossRef]
95. Abbas, G. An Augmented Review about Antinutrients and Toxins of Feed Stuff and their Control Strategies, a Step toward Sustainable Resource Utilization. *Int. J. Agric. Innov. Res.* **2020**, *8*, 361–377.
96. Cid-Gallegos, M.; Corzo-Ríos, L.; Jiménez-Martinez, C.; Sánchez-Chino, X. Protease inhibitors from plants as therapeutic agents—A review. *Plant Foods Hum. Nutr.* **2022**, *77*, 20–29. [CrossRef]
97. Arsov, A.; Tsigoriyna, L.; Batovska, D.; Armenova, N.; Mu, W.; Zhang, W.; Petrov, K.; Petrova, P. Bacterial Degradation of Antinutrients in Foods: The Genomic Insight. *Foods* **2024**, *13*, 2408. [CrossRef]
98. Martinez, M.; Diaz, I. Plant Cyanogenic-Derived Metabolites and Herbivore Counter-Defences. *Plants* **2024**, *13*, 1239. [CrossRef]
99. Lachowicz, J.I.; Alexander, J.; Aaseth, J.O. Cyanide and Cyanogenic Compounds—Toxicity, Molecular Targets, and Therapeutic Agents. *Biomolecules* **2024**, *14*, 1420. [CrossRef]
100. Deen, A.U.; Kumari, V.; Sharma, A.N.; Mondal, G.; Singh, G.P. Understanding cyanogenic glycoside toxicity in livestock: A review. *Int. J. Chem. Stud.* **2018**, *6*, 1559–1561.
101. Urugo, M.M.; Tringo, T.T. Naturally Occurring Plant Food Toxicants and the Role of Food Processing Methods in Their Detoxification. *Int. J. Food Sci.* **2023**, *2023*, 9947841. [CrossRef]
102. Babiker, A.; Alawi, A.; Al Atawi, M.; Al Alwan, I. The role of micronutrients in thyroid dysfunction. *Sudan. J. Paediatr.* **2020**, *20*, 13–19. [CrossRef]
103. Galanty, A.; Grudzińska, M.; Paździora, W.; Służały, P.; Paśko, P. Do Brassica Vegetables Affect Thyroid Function?—A Comprehensive Systematic Review. *Int. J. Mol. Sci.* **2024**, *25*, 3988. [CrossRef]
104. Dini, I.; Tenore, G.C.; Dini, A. Oleanane Saponins in "Kancolla", a Sweet Variety of Chenopodium quinoa. *J. Nat. Prod.* **2002**, *65*, 1023–1026. [CrossRef]
105. Cao, S.; Liu, M.; Han, Y.; Li, S.; Zhu, X.; Li, D.; Shi, Y.; Liu, B. Effects of Saponins on Lipid Metabolism: The Gut–Liver Axis Plays a Key Role. *Nutrients* **2024**, *16*, 1514. [CrossRef]
106. Imam, Y.T.; Irondi, E.A.; Awoyale, W.; Ajani, E.O.; Alamu, E.O. Application of legumes in the formulation of gluten-free foods: Functional, nutritional and nutraceutical importance. *Front. Sustain. Food Syst.* **2024**, *8*, 1251760. [CrossRef]

 https://doi.org/10.3390/molecules31081353

107. Liang, J.; Han, B.-Z.; Nout, M.J.R.; Hamer, R.J. Effects of Soaking, Germination and Fermentation on Phytic Acid, Total and in Vitro Soluble Zinc in Brown Rice. *Food Chem.* **2008**, *110*, 821–828. [CrossRef]
108. Murru, N.; Blaiotta, G.; Peruzy, M.F.; Santonicola, S.; Mercogliano, R.; Aponte, M. Screening of Oxalate Degrading Lactic Acid Bacteria of Food Origin. *Ital. J. Food Saf.* **2017**, *6*, 6345. [CrossRef]
109. Terpou, A.; Rai, A.K. 2—Microbial transformation for improving food functionality. In *Current Developments in Biotechnology and Bioengineering*; Rai, A.K., Singh, S.P., Pandey, A., Larroche, C., Soccol, C.R., Eds.; Elsevier: Amsterdam, The Netherlands, 2022; pp. 31–45.
110. Terpou, A.; Papadaki, A.; Lappa, I.K.; Kachrimanidou, V.; Bosnea, L.A.; Kopsahelis, N. Probiotics in food systems: Significance and emerging strategies towards improved viability and delivery of enhanced beneficial value. *Nutrients* **2019**, *11*, 1591. [CrossRef]
111. Dahiya, D.; Nigam, P.S. Use of Characterized Microorganisms in Fermentation of Non-Dairy-Based Substrates to Produce Probiotic Food for Gut-Health and Nutrition. *Fermentation* **2023**, *9*, 1. [CrossRef]
112. Kamada, N.; Seo, S.U.; Chen, G.Y.; Nunez, G. Role of the gut microbiota in immunity and inflammatory disease. *Nat. Rev. Immunol.* **2013**, *13*, 321–335. [CrossRef]
113. Kovtonyuk, L.V.; McCoy, K.D. Microbial metabolites and immunotherapy: Basic rationale and clinical indications. *Semin. Immunol.* **2023**, *67*, 101755. [CrossRef]
114. Yang, W.; Cong, Y. Gut microbiota-derived metabolites in the regulation of host immune responses and immune-related inflammatory diseases. *Cell. Mol. Immunol.* **2021**, *18*, 866–877. [CrossRef]
115. Garavaglia, B.; Vallino, L.; Amoruso, A.; Pane, M.; Ferraresi, A.; Isidoro, C. The role of gut microbiota, immune system, and autophagy in the pathogenesis of inflammatory bowel disease: Molecular mechanisms and therapeutic approaches. *Asp. Mol. Med.* **2024**, *4*, 100056. [CrossRef]
116. Dimidi, E.; Cox, S.R.; Rossi, M.; Whelan, K. Fermented Foods: Definitions and Characteristics, Impact on the Gut Microbiota and Effects on Gastrointestinal Health and Disease. *Nutrients* **2019**, *11*, 1806. [CrossRef]
117. Balasubramanian, R.; Schneider, E.; Gunnigle, E.; Cotter, P.D.; Cryan, J.F. Fermented foods: Harnessing their potential to modulate the microbiota-gut-brain axis for mental health. *Neurosci. Biobehav. Rev.* **2024**, *158*, 105562. [CrossRef]
118. Hernández-Velázquez, R.; Flörl, L.; Lavrinienko, A.; Sebechlebská, Z.; Merk, L.; Greppi, A.; Bokulich, N.A. The future is fermented: Microbial biodiversity of fermented foods is a critical resource for food innovation and human health. *Trends Food Sci. Technol.* **2024**, *150*, 104569. [CrossRef]
119. Todorovic, S.; Akpinar, A.; Assunção, R.; Bär, C.; Bavaro, S.L.; Berkel Kasikci, M.; Domínguez-Soberanes, J.; Capozzi, V.; Cotter, P.D.; Doo, E.-H.; et al. Health Benefits and Risks of Fermented Foods—The PIMENTO Initiative. *Front. Nutr.* **2024**, *11*, 1458536. [CrossRef]
120. Ewida, R.M.; Hasan, W.S.; Elfaruk, M.S.; Alayouni, R.R.; Hammam, A.R.A.; Kamel, D.G. Occurrence of *Listeria* spp. in Soft Cheese and Ice Cream: Effect of Probiotic *Bifidobacterium* spp. on Survival of *Listeria monocytogenes* in Soft Cheese. *Foods* **2022**, *11*, 3443. [CrossRef]
121. Uzal, F.A.; Henderson, E.; Asin, J. Botulism in fish: A review. *J. Vet. Diagn. Investig.* **2024**, *36*, 312–318. [CrossRef]
122. Allwood, J.G.; Wakeling, L.T.; Post, L.S.; Bean, D.C. Food Safety Considerations in the Production of Traditional Fermented Products: Japanese Rice Koji and Miso. *J. Food Saf.* **2023**, *43*, e13048. [CrossRef]
123. Zhu, L.; Cheng, M.; Xu, C.; Wang, R.; Zhang, M.; Tao, Y.; Qi, S.; Wei, W. Selective Inactivation Strategies for Vegetable Raw Materials: Regulating Microbial Communities to Ensure the Safety and Quality of Fermented Vegetables. *Foods* **2025**, *14*, 3291. [CrossRef]
124. Behera, S.S.; El Sheikha, A.F.; Hammami, R.; Kumar, A. Traditionally fermented pickles: How the microbial diversity associated with their nutritional and health benefits? *J. Funct. Foods* **2020**, *70*, 103971. [CrossRef]
125. Gao, Q.; Wang, Y.; Li, Y.; Hou, J.; Liang, Y.; Zhang, Z. Investigation of the formation of furfural compounds in apple products treated with pasteurization and high pressure processing. *Food Res. Int.* **2024**, *190*, 114546. [CrossRef]
126. Singh, H.; Ramaswamy, H.S. Thermal Processing of Acidified Vegetables: Effect on Process Time-Temperature, Color and Texture. *Processes* **2023**, *11*, 1272. [CrossRef]
127. Fujiyoshi, T.; Ikami, T.; Kikukawa, K.; Kobayashi, M.; Takai, R.; Kozaki, D.; Yamamoto, A. Direct quantitation of the preservatives benzoic and sorbic acid in processed foods using derivative spectrophotometry combined with micro dialysis. *Food Chem.* **2018**, *240*, 386–390. [CrossRef]
128. De Jesus, J.H.F.; Szilágyi, I.M.; Regdon, G.; Cavalheiro, E.T.G. Thermal behavior of food preservative sorbic acid and its derivates. *Food Chem.* **2021**, *337*, 127770. [CrossRef]
129. Dong, W.; Shen, H.; Liu, H.; Song, F.; Li, P.; Peng, N.; Liang, Y.; Zhao, S. Unraveling the microbial community and succession during zha-chili fermentation and their relationships with flavor formation. *Food Res. Int.* **2022**, *157*, 111239. [CrossRef]

 https://doi.org/10.3390/molecules31081353

130. Zhu, L.; Cheng, M.; Wang, Y.; Hu, X.; Wang, R.; Yang, B.; Xiao, X.; Wei, W. A Novel Strategy for Controlling Nitrite Accumulation in Fermented Cabbage by Regulating Microbial Community Structure via Cold Plasma. *Food Control* **2025**, *180*, 111628. [CrossRef]
131. Wei, W.; Yang, S.; Yang, F.; Hu, X.; Wang, Y.; Guo, W.; Yang, B.; Xiao, X.; Zhu, L. Cold plasma controls nitrite hazards by modulating microbial communities in pickled radish. *Foods* **2023**, *12*, 2550. [CrossRef]
132. Zhao, N.; Ge, L.; Huang, Y.; Wang, Y.; Wang, Y.; Lai, H.; Wang, Y.; Zhu, Y.; Zhang, J. Impact of cold plasma processing on quality parameters of packaged fermented vegetable (radish paocai) in comparison with pasteurization processing: Insight into safety and storage stability of products. *Innov. Food Sci. Emerg. Technol.* **2020**, *60*, 102300. [CrossRef]
133. Zhou, R.; Rezaeimotlagh, A.; Zhou, R.; Zhang, T.; Wang, P.; Hong, J.; Soltani, B.; Mai-Prochnow, A.; Liao, X.; Ding, T. In-package plasma: From reactive chemistry to innovative food preservation technologies. *Trends Food Sci. Technol.* **2022**, *120*, 59–74. [CrossRef]
134. Ghate, V.S.; Zhou, W.; Yuk, H.G. Perspectives and Trends in the Application of Photodynamic Inactivation for Microbiological Food Safety. *Compr. Rev. Food Sci. Food Saf.* **2019**, *18*, 402–424. [CrossRef]
135. Huang, G.; Sun, W.; Dai, C.; Sun, L.; Tang, Y.; He, R.; Ma, H. Sterilization of Bacillus tequilensis isolated from aerogenic vinegar by intense pulsed light. *LWT* **2020**, *118*, 108811. [CrossRef]
136. Cassar, J.R.; Mills, E.W.; Demirci, A. Characterization of Pulsed Light for Microbial Inactivation. *J. Food Eng.* **2022**, *334*, 111152. [CrossRef]
137. Ren, M.; Ren, Z.; Chen, L.; Zhou, C.; Okonkwo, C.E.; Mujumdar, A.S. Comparison of ultrasound and ethanol pretreatments before catalytic infrared drying on physicochemical properties, drying, and contamination of Chinese ginger (*Zingiber officinale* Roscoe). *Food Chem.* **2022**, *386*, 132759. [CrossRef]
138. Wang, Y.; Li, T.; Pan, Z.; Ye, X.; Ma, H. Effectiveness of combined catalytic infrared radiation and holding time for decontamination *Aspergillus* niger on dried shiitake mushrooms (*Lentinus edodes*) with different moisture contents. *LWT* **2023**, *176*, 114503. [CrossRef]
139. Wu, B.; Ma, Y.; Guo, Y.; Zielinska, M.; Gao, K.; Song, C.; Bouhile, Y.; Qiu, C.; Pan, Z.; Ma, H. Research progress in the application of catalytic infrared technology in fruit and vegetable processing. *Compr. Rev. Food Sci. Food Saf.* **2024**, *23*, e13291. [CrossRef]
140. Li, M.; Zhou, C.; Wang, B.; Zeng, S.; Mu, R.; Li, G.; Li, B.; Lv, W. Research progress and application of ultrasonic- and microwave-assisted food processing technology. *Compr. Rev. Food Sci. Food Saf.* **2023**, *22*, 3707–3731. [CrossRef]
141. Canumir, J.A.; Celis, J.E.; de Bruijn, J.; Vidal, L.V. Pasteurisation of apple juice by using microwaves. *LWT-Food Sci. Technol.* **2002**, *35*, 389–392. [CrossRef]
142. Espitia, J.; Verheyen, D.; Kozak, D.S.; Van Impe, J.F. Influence of Microbial Cell Morphology and Composition on Radio Frequency Heating of Simple Media at Different Frequencies. *Sci. Rep.* **2023**, *13*, 10839. [CrossRef]
143. Vinayamohan, P.G.; Salihu, S.; Kalia, V.C. Fermented Foods as a Potential Vehicle of Antimicrobial-Resistant Bacteria and Genes. *Fermentation* **2023**, *9*, 688. [CrossRef]
144. Wolfe, B.E. Are Fermented Foods an Overlooked Reservoir of Antimicrobial Resistance? *Curr. Opin. Food Sci.* **2023**, *51*, 101018. [CrossRef]
145. Li, Y.; Fu, S.; Klein, M.S.; Wang, H. High Prevalence of Antibiotic Resistance in Traditionally Fermented Foods as a Critical Risk Factor for Host Gut Antibiotic Resistome. *Microorganisms* **2024**, *12*, 1433. [CrossRef]
146. Leech, J.; Cabrera-Rubio, R.; Walsh, A.M.; Macori, G.; Walsh, C.J.; Barton, W.; Finnegan, L.; Crispie, F.; O'Sullivan, O.; Claesson, M.J.; et al. Fermented-Food Metagenomics Reveals Substrate-Associated Differences in Taxonomy and Health-Associated and Antibiotic Resistance Determinants. *mSystems* **2020**, *5*, e00522-20. [CrossRef]
147. Mater, D.D.G.; Langella, P.; Corthier, G.; Flores, M.-J. A Probiotic Lactobacillus Strain Can Acquire Vancomycin Resistance during Digestive Transit in Mice. *J. Mol. Microbiol. Biotechnol.* **2008**, *14*, 123–127. [CrossRef]
148. Broaders, E.; Gahan, C.G.; Marchesi, J.R. Gut Microbes Mobile Genetic Elements of the Human Gastrointestinal Tract Potential for Spread of Antibiotic Resistance Genes. *Gut Microbes* **2013**, *4*, 271–280. [CrossRef]
149. Jahan, M.; Zhanel, G.G.; Sparling, R.; Holley, R.A. Horizontal Transfer of Antibiotic Resistance from Enterococcus faecium of Fermented Meat Origin to Clinical Isolates of E. Faecium and Enterococcus faecalis. *Int. J. Food Microbiol.* **2015**, *199*, 78–85. [CrossRef]
150. Magalhães, K.T. Probiotic-Fermented Foods and Antimicrobial Stewardship: Mechanisms, Evidence, and Translational Pathways Against AMR. *Fermentation* **2025**, *11*, 684. [CrossRef]
151. Gomes, E.; Araújo, D.; Nogueira, T.; Oliveira, R.; Silva, S.; Oliveira, L.V.N.; Azevedo, N.F.; Almeida, C.; Castro, J. Advances in Whole Genome Sequencing for Foodborne Pathogens: Implications for Clinical Infectious Disease Surveillance and Public Health. *Front. Cell. Infect. Microbiol.* **2025**, *15*, 1593219. [CrossRef]
152. EFSA Panel on Biological Hazards (BIOHAZ); Koutsoumanis, K.; Allende, A.; Bolton, D.; Bover-Cid, S.; Chemaly, M.; De Cesare, A.; Herman, L.; Hilbert, F.; Lindqvist, R.; et al. Persistence of Microbiological Hazards in Food and Feed Production and Processing Environments. *EFSA J.* **2024**, *22*, e8521. [CrossRef]

153. Saha Turna, N.; Chung, R.; McIntyre, L. A Review of Biogenic Amines in Fermented Foods: Occurrence and Health Effects. *Heliyon* **2024**, *10*, e24501. [CrossRef]
154. Chan, S.X.Y.; Fitri, N.; Mio Asni, N.S.; Sayuti, N.H.; Azlan, U.K.; Qadi, W.S.M.; Dawoud, E.A.D.; Kamal, N.; Sarian, M.N.; Mohd Lazaldin, M.A.; et al. A Comprehensive Review with Future Insights on the Processing and Safety of Fermented Fish and the Associated Changes. *Foods* **2023**, *12*, 558. [CrossRef]
155. Costa, V.M.; Grando, L.G.R.; Milandri, E.; Nardi, J.; Teixeira, P.; Mladěnka, P.; Remião, F.; on behalf of The OEMONOM. Natural Sympathomimetic Drugs: From Pharmacology to Toxicology. *Biomolecules* **2022**, *12*, 1793. [CrossRef]
156. Salas-Millán, J.Á.; Aguayo, E. Fermentation for Revalorisation of Fruit and Vegetable By-Products: A Sustainable Approach Towards Minimising Food Loss and Waste. *Foods* **2024**, *13*, 3680. [CrossRef]
157. Bell, V.; Ferrão, J.; Fernandes, T. Nutritional guidelines and fermented food frameworks. *Foods* **2017**, *6*, 65. [CrossRef]
158. Tamang, J.P. History and culture of Indian ethnic fermented foods and beverages. In *Ethnic Fermented Foods and Beverages of India: Science, History and Culture*; Tamang, J.P., Ed.; Springer: Singapore, 2020; pp. 1–40.
159. Voidarou, C.; Antoniadou, M.; Rozos, G.; Tzora, A.; Skoufos, I.; Varzakas, T.; Lagiou, A.; Bezirtzoglou, E. Fermentative Foods: Microbiology, Biochemistry, Potential Human Health Benefits and Public Health Issues. *Foods* **2020**, *10*, 69. [CrossRef] [PubMed]
160. Nielsen, M.B.; Meyer, A.S.; Arnau, J. The next food revolution is here: Recombinant microbial production of milk and egg proteins by precision fermentation. *Annu. Rev. Food Sci. Technol.* **2024**, *15*, 173–187. [CrossRef]
161. Augustin, M.A.; Hartley, C.J.; Maloney, G.; Tyndall, S. Innovation in precision fermentation for food ingredients. *Crit. Rev. Food Sci. Nutr.* **2023**, *64*, 6218–6238. [CrossRef] [PubMed]
162. Leroy, F.; De Vuyst, L. Lactic acid bacteria as functional starter cultures for the food fermentation industry. *Trends Food Sci. Technol.* **2004**, *15*, 67–78. [CrossRef]
163. Santiago-Díaz, P.; Rico, M.; Rivero, A.; Santana-Casiano, M. Bioactive Metabolites of Microalgae from Canary Islands for Functional Food and Feed Uses. *Chem. Biodivers.* **2022**, *19*, e202200230. [CrossRef]
164. Obayomi, O.V.; Malomo, A.A.; Olaniran, A.F.; Osemwegie, O.O.; Olojede, A.O.; Beyioku, O.E.; Adeyemi, O.S. Transition from fermentation to precision fermentation: Role in sustainable food system. *Biotech Rep* **2026**, e00952. [CrossRef]
165. Eastham, J.L.; Leman, A.R. Precision fermentation for food proteins: Ingredient innovations, bioprocess considerations, and outlook. *Curr. Opin. Food Sci.* **2024**, *58*, 101194. [CrossRef]
166. Kuhl, E. IA para alimentos: Acelerando e democratizando a descoberta e a inovação. *NPJ Sci. Food* **2025**, *9*, 82. [CrossRef] [PubMed]
167. Khurana, R.K.; Jain, A.; Jain, A.; Sharma, T.; Singh, B.; Kesharwani, P. Administration of antioxidants in cancer: Debate of the decade. *Drug Discov. Today* **2018**, *23*, 763–770. [CrossRef]
168. Mouritsen, O.G.; Duelund, L.; Calleja, G.; Frøst, M.B. Flavour of fermented fish, insect, game, and pea sauces: Garum revisited. *Int. J. Gastron. Food Sci.* **2017**, *9*, 16–28. [CrossRef]
169. **169.** MarketsandMarkets. Precision Fermentation Ingredients Market by Ingredient (Egg, Dairy Proteins, Collagen, Heme), Microbe (Yeast, Algae, Bacteria, Fungi), Application (Dairy Alternatives, Meat Alternatives, Egg Alternatives), and Region—Global Forecast to 2030. MarketsandMarkets^{TM}. 2025. Available online: https://www.marketsandmarkets.com/Market-Reports/precision-fermentation-market-147520320.html (accessed on 25 May 2025).
170. Yang, R.; Chen, Z.; Hu, P.; Zhang, S.; Luo, G. Two-stage fermentation enhanced single-cell protein production by *Yarrowia lipolytica* from food waste. *Bioresour. Technol.* **2022**, *361*, 127677. [CrossRef]
171. Yaekob, A.T.; Gebreslassie, G.; Mesele, E.; Semere, T. Current advancements of probiotic foods and their role in sustainable food security. *Food Bioeng.* **2026**, e70046. [CrossRef]
172. Adeyeye, S.A.O.; Babu, A.S.; Subudhi, A.; Adeyeye, B.R. Precision fermentation processes for producing novel foods and its sustainable applications. *J. Basic Microbiol.* **2026**, *66*, e70160. [CrossRef]
173. Kossmann, H.; Karslioglu, Ö.Ö.; Breunig, P. From Moo to Microbes: Pathways for precision fermentation in recombinant protein production. *J. Agric. Food Res.* **2025**, *22*, 102056. [CrossRef]
174. Yadav, H.; Singh, S.; Sinha, R. Fermentation Technology for Microbial Products and Their Process Optimization. In *Industrial Microbiology and Biotechnology: A New Horizon of the Microbial World*; Springer: Berlin/Heidelberg, Germany, 2024; pp. 35–64.
175. Kanwal, N.; Zhang, M.; Khan, S.; Batool, U. AI-Driven Precision Fermentation: From Restaurant Food Waste to Sustainable Protein Production. *Food Bioprocess Technol.* **2025**, *18*, 8354–8379. [CrossRef]
176. Ajayeoba, T.A.; Ijabadeniyi, O.A. Transforming Food for the Future: Precision Fermentation as a Key to Sustainability, Nutrition, and Health. Available online: https://papers.ssrn.com/sol3/papers.cfm?abstract_id=5119686 (accessed on 19 March 2025).
177. Niyigaba, T.; Küçükgöz, K.; Kołożyn-Krajewska, D.; Królikowski, T.; Trząskowska, M. Advances in Fermentation Technology: A Focus on Health and Safety. *Appl. Sci.* **2025**, *15*, 3001. [CrossRef]

178. Ewing, T.A.; Nouse, N.; van Lint, M.; van Haveren, J.; Hugenholtz, J.; van Es, D.S. Fermentation for the Production of Biobased Chemicals in a Circular Economy: A Perspective for the Period 2022–2050. *Green Chem.* **2022**, *24*, 6373–6405. [CrossRef]
179. Marco, M.L.; Sanders, M.E.; Gänzle, M.; Arrieta, M.C.; Cotter, P.D.; De Vuyst, L.; Hill, C.; Holzapfel, W.; Lebeer, S.; Merenstein, D.; et al. Declaração de consenso da Associação Científica Internacional para Probióticos e Prebióticos (ISAPP) sobre alimentos fermentados. *Nat. Rev. Gastroenterol. Hepatol.* **2021**, *18*, 196–208. [CrossRef] [PubMed]
180. Nascimento, A.P.S.; Barros, A.N. Sustainable Innovations in Food Microbiology: Fermentation, Biocontrol, and Functional Foods. *Foods* **2025**, *14*, 2320. [CrossRef]
181. Banovic, M.; Grunert, K.G. Consumer Acceptance of Precision Fermentation Technology: A Cross-Cultural Study. *Innov. Food Sci. Emerg. Technol.* **2023**, *88*, 103435. [CrossRef]
182. Inbar, Y.; Phelps, J.; Rozin, P. Recency negativity: Newer food crops are evaluated less favorably. *Appetite* **2020**, *154*, 104754. [CrossRef]
183. Siegrist, M.; Hartmann, C. Consumer acceptance of novel food technologies. *Nat. Food* **2020**, *1*, 343–350. [CrossRef] [PubMed]
184. Frewer, L.J.; Bergmann, K.; Brennan, M.; Lion, R.; Meertens, R.; Rowe, G.; Siegrist, M.; Vereijken, C. Consumer response to novel agri-food technologies: Implications for predicting consumer acceptance of emerging food technologies. *Trends Food Sci. Technol.* **2011**, *22*, 442–456. [CrossRef]
185. Rezac, S.; Kok, C.R.; Heermann, M.; Hutkins, R. Fermented foods as a dietary source of live organisms. *Front. Microbiol.* **2018**, *9*, 1785. [CrossRef]
186. Skowron, K.; Budzyńska, A.; Grudlewska-Buda, K.; Wiktorczyk-Kapischke, N.; Andrzejewska, M.; Wałecka-Zacharska, E.; Gospodarek-Komkowska, E. Two faces of fermented foods—The benefits and threats of its consumption. *Front. Microbiol.* **2022**, *13*, 845166. [CrossRef]
187. Perkins, C. 'Test Tube Milk' More Likely to Win Brits than Lab-Grown Meat. The Grocer. 2018. Available online: https://www.thegrocer.co.uk/sourcing/test-tube-milk-more-likely-to-win-brits-than-lab-grown-meat/571598.article (accessed on 19 March 2026).
188. Boukid, F.; Ganeshan, S.; Wang, Y.; Tülbek, M.Ç.; Nickerson, M.T. Bioengineered Enzymes and Precision Fermentation in the Food Industry. *Int. J. Mol. Sci.* **2023**, *24*, 10156. [CrossRef] [PubMed]
189. Broad, G.M.; Zollman Thomas, O.; Dillard, C.; Bowman, D.; Le Roy, B. Framing the Futures of Animal-Free Dairy: Using Focus Groups to Explore Early-Adopter Perceptions of the Precision Fermentation Process. *Front. Nutr.* **2022**, *9*, 997632. [CrossRef]
190. Etale, A.; Siegrist, M. Food Processing and Perceived Naturalness: Is It More Natural or Just More Traditional? *Food Qual. Prefer.* **2021**, *94*, 104323. [CrossRef]
191. Bogueva, D.; Apostolova, M.; Danova, S. Food, Nutrition and Health in Bulgaria. In *Nutritional and Health Aspects of Food in the Balkans*; Gostin, A., Bogueva, D., Kakurinov, V., Eds.; Elsevier: Amsterdam, The Netherlands, 2021; pp. 67–89.
192. Sogari, G.; Bogueva, D.; Marinova, D. Australian Consumers' Response to Insects as Food. *Agriculture* **2019**, *9*, 108. [CrossRef]
193. Bell, V.; Guina, J.; Fernandes, T.H. African Fermented Foods and Beverages: Potential Impact on Health. In *Microbial Fermentations in Nature and as Designed Processes*; Wiley: Hoboken, NJ, USA, 2023; pp. 293–322.
194. Tamang, J.P.; Cotter, P.D.; Endo, A.; Han, N.S.; Kort, R.; Liu, S.Q.; Mayo, B.; Westerik, N.; Hutkins, R. Fermented foods in a global age: East meets West. *Compr. Rev. Food Sci. Food Saf.* **2020**, *19*, 184–217. [CrossRef]
195. Parkouda, C.; Ba/Hama, F.; Ouattara/Songre, L.; Tano-Debrah, K.; Diawara, B. Biochemical changes associated with the fermentation of baobab seeds in *Maari*: An alkaline fermented seeds condiment from western Africa. *J. Ethn. Foods* **2015**, *2*, 58–63. [CrossRef]
196. Shahidah, A.A.; Farouq, A.A.; Magashi, M.A.; Sokoto, A.M. Comparative amino acid and volatile flavor profile of *dawadawa* produced from the seeds of *P. biglobosa*, *G. max* and *H. sabdariffa*. *J. Adv. Microbiol.* **2019**, *14*, 1–13. [CrossRef]
197. Kouamé, K.A.; Bouatenin, K.M.J.-P.; Coulibaly, W.H.; Marcellin, D.K. Biochemical and microbiological characterization of "Soumbara" from African locust bean (*Parkia biglobosa*) seeds consumed in Abidjan (Côte d'Ivoire). *N. Afr. J. Food Nutr. Res.* **2021**, *5*, 35–42. [CrossRef]
198. Esse, M.Y.; Guehi, T.S.; Lebrun, M.; Morel, G.; Grabulos, J.; Mestres, C.; Achir, N. Kinetic study of some flavor and bioactive compounds during fermentation of *Parkia biglobosa*. *J. Food Process. Preserv.* **2022**, *46*, e16888. [CrossRef]
199. Ahon-Gnamien, M.; Théodore, A.K.; Losséni, B.; Ernest, A.K. Manufacturing technique and nutritive value of *Soumbara* (*Parkia biglobosa* seeds) and dark seed of Korhogo (Côte d'Ivoire). *Open J. Appl. Sci.* **2022**, *12*, 1673–1686. [CrossRef]
200. Laya, A.; Wangso, H.; Fernandes, I.; Djakba, R.; Oliveira, J.; Carvalho, E. Bioactive ingredients in traditional fermented food condiments: Emerging products for prevention and treatment of obesity and type 2 diabetes. *J. Food Qual.* **2023**, *2023*, 5236509. [CrossRef]
201. Palazzuoli, A.; Iacoviello, M. Diabetes leading to heart failure and heart failure leading to diabetes: Epidemiological and clinical evidence. *Heart Fail. Rev.* **2023**, *28*, 585–596. [CrossRef]

202. ElGamal, R.; Song, C.; Rayan, A.M.; Liu, C.; Al-Rejaie, S.; ElMasry, G. Thermal degradation of bioactive compounds during drying process of horticultural and agronomic products: A comprehensive overview. *Agronomy* **2023**, *13*, 1580. [CrossRef]
203. Singletary, K.W. Anise: Potential health benefits. *Nutr. Today* **2022**, *57*, 96–109. [CrossRef]
204. Vojnović, Đ.; Maksimović, I.; Tepić Horecki, A.; Karadžić Banjac, M.; Kovačević, S.; Daničić, T.; Podunavac-Kuzmanović, S.; Ilin, Ž. Onion (*Allium cepa* L.) yield and quality depending on biostimulants and nitrogen fertilization—A chemometric perspective. *Processes* **2023**, *11*, 684. [CrossRef]
205. Woldemariam, H.W.; Admassu Emire, S.; Getachew Teshome, P.; Toepfl, S.; Aganovic, K. Physicochemical, functional, oxidative stability and rheological properties of red pepper (*Capsicum annuum* L.) powder and paste. *Int. J. Food Prop.* **2021**, *24*, 1416–1437. [CrossRef]
206. Ouoba, L.I.I.; Parkouda, C.; Diawara, B.; Scotti, C.; Varnam, A.H. Identification of *Bacillus* spp. from Bikalga, fermented seeds of *Hibiscus sabdariffa*: Phenotypic and genotypic characterization. *J. Appl. Microbiol.* **2007**, *104*, 122–131. [CrossRef] [PubMed]
207. Parkouda, C.; Diawara, B.; Ouoba, L.I.I. Technology and physico-chemical characteristics of Bikalga, alkaline fermented seeds of *Hibiscus sabdariffa*. *Afr. J. Biotechnol.* **2008**, *7*, 916–922.
208. Zakari, D.A.; Adefila, A.M. Antioxidant activities, nutritional composition, and functional properties of traditionally fermented African locust bean condiment (Dawadawa) from Kogi State, Nigeria. *Diet. Suppl. Nutraceuticals* **2025**, *4*, 98–111. [CrossRef]
209. Odunfa, S.A. Biochemical changes during "iru" fermentation. *J. Food Technol.* **1985**, *20*, 295–303. [CrossRef]
210. Ouoba, L.I.I.; Diawara, B.; Annan, N.T.; Poll, L.; Jakobsen, M. Volatile compounds of soumbala, a fermented African locust bean condiment. *J. Appl. Microbiol.* **2005**, *99*, 1413–1421. [CrossRef]
211. Oguntoyinbo, F.A.; Sanni, A.I.; Franz, C.M.A.P.; Holzapfel, W.H. In vitro fermentation studies for selection and evaluation of *Bacillus* strains as starter cultures for the production of okpehe, a traditional African fermented condiment. *Int. J. Food Microbiol.* **2007**, *113*, 208–218. [CrossRef]
212. Nwanjo, H.; Iroagba, I.; Nnatuanya, I.; Eze, N. Is fermented *Pentaclethra macrophylla* nutritional or antinutritional? Response from haematological studies in protein-malnourished guinea pigs. *Internet J. Nutr. Wellness* **2006**, *4*, 1–5. [CrossRef]
213. Anike, I.C. Effect of fermented *Pentaclethra macrophylla* Bent (African oil bean) seed extract on plasma lipid profile in a healthy rat model—A preliminary study. *S. Asian Res. J. Nat. Prod.* **2019**, *2*, 43–51.
214. Okwuonu, G.C.; Ojimadu, N.E.; Ozoemelam, K.; Esione, A.; Iwe, I.S.; Chimezie, B.O.; Onwuchekwa, C.J. Case report: The potential antihypertensive effects of the African oil bean seed (*Pentaclethra macrophylla*). *Bentham Pioneer Med. J.* **2013**, *3*, 1–11.
215. Igwenyi, I.O.; Isiguzo, O.E.; Aja, P.M.; Ugwu Okechukwu, P.C.; Ezeani, N.N.; Uraku, A.J. Proximate composition, mineral content and phytochemical analysis of the African oil bean (*Pentaclethra macrophylla*) seed. *Am.-Eurasian J. Agric. Environ. Sci.* **2015**, *15*, 1873–1875.
216. Agunloye, O.M.; Oboh, G. Fermented seeds of *Pentaclethra macrophylla* mitigate memory deficit and restore altered enzymatic activity in the brain of streptozotocin-diabetic rats. *Metab. Brain Dis.* **2023**, *38*, 973–981. [CrossRef]
217. Aderibigbe, E.Y.; Visessanguan, W.; Sumpavapol, P.; Kongtong, K. Sourcing starter cultures for *Parkia biglobosa* fermentation I: Phylogenic grouping of *Bacillus* species from commercial "iru" samples. *Int. J. Biotechnol. Mol. Biol. Res.* **2011**, *2*, 121–127.
218. Ademola, O.M.; Adeyemi, T.E.; Ezeokoli, O.T.; Ayeni, K.I.; Obadina, A.O.; Somorin, Y.M.; Omemu, A.M.; Adeleke, R.A.; Nwangburuka, C.C.; Oluwafemi, F.; et al. Phylogenetic analyses of bacteria associated with the processing of iru and ogiri condiments. *Lett. Appl. Microbiol.* **2018**, *67*, 354–362. [CrossRef]
219. Azokpota, P.; Hounhouigan, D.J.; Nago, M.C. Microbiological and chemical changes during the fermentation of African locust bean (*Parkia biglobosa*) to produce afitin, iru and sonru, three traditional condiments produced in Benin. *Int. J. Food Microbiol.* **2006**, *107*, 304–309. [CrossRef]
220. Sadiku, O.A. Processing methods influence the quality of fermented African locust bean (iru/ogiri/dadawa) *Parkia biglobosa*. *J. Appl. Sci. Res.* **2010**, *6*, 1656–1661.
221. Atere, A.V.; Adedeji, A.; Akinmoladun, A.C.; Oyetayo, V.O.; Akinyosoye, F.A. Local Condiment, Iru, Obtained from the Fermentation of *Parkia biglobosa* Seed Substantially Reduced the Serum Cholesterol Level of Wistar Rats. *Prev. Nutr. Food Sci.* **2020**, *25*, 153–157. [CrossRef]
222. Waongo, B.; Pupin, M.; Duban, M.; Chataigne, G.; Zongo, O.; Cisse, H.; Savadogo, A.; Leclère, V. *Kawal*: A fermented food as a source of *Bacillus* strain producing antimicrobial peptides. *Sci. Afr.* **2023**, *20*, e01714. [CrossRef]
223. Elfaki, A.E.; Dirar, H.A.; Collins, M.A.; Harper, D.B. Biochemical and microbiological investigations of sigda—A Sudanese fermented food derived from sesame oilseed cake. *J. Sci. Food Agric.* **1991**, *57*, 351–365. [CrossRef]
224. Nyathi, L.A.; Njobeh, P.B.; Dlamini, B.; Tesfamariam, K.; De Saeger, S.; Valerio, F.; Avantaggiato, G.; Akinmoladun, O.F.; Akanni, G.B.; Adebo, O.A. Nutritional benefits, technological innovations and safety of fermented cereal products: Insights from umqombothi. *Food Biosci.* **2025**, *74*, 107982. [CrossRef]

225. Odetokun, S.M. Chemical changes and nutritive values of burukutu (a Nigerian beverage). *Nahrung/Food* **1997**, *41*, 375–377. [CrossRef]
226. Ganguly, S.; Sabikhi, L.; Singh, A.K. Effect of probiotic fermentation on physico-chemical and nutritional parameters of milk-cereal based composite substrate. *J. Food Sci. Technol.* **2022**, *59*, 3073–3085. [CrossRef] [PubMed]
227. Olakunle, D.T.; Holzapfel, W.H.; Odunfa, S.A. Selection, Use and the Influence of Starter Cultures in the Nutrition and Processing Improvement of Ogi. *J. Food Sci. Nutr. Res.* **2023**, *6*, 61–70. [CrossRef]
228. Friday, O.; Kiin-Kabari, D.B.; Akusu, M.; Achinewhu, S.C. Nutritional, Functional and Sensory Properties of Gari Enhanced with Fermented Maize Residues Flour. *Int. J. Food Sci. Biotechnol.* **2021**, *6*, 59–65. [CrossRef]
229. Oi, Y.; Kitabatake, N. Chemical Composition of an East African Traditional Beverage, Togwa. *J. Agric. Food Chem.* **2003**, *51*, 7024–7028. [CrossRef]
230. Hawaz, H.; Bottari, B.; Scazzina, F.; Carini, E. Eastern African traditional fermented foods and beverages: Advancements, challenges, and perspectives on food technology, nutrition, and safety. *Compr. Rev. Food Sci. Food Saf.* **2025**, *24*, 3727–3756. [CrossRef] [PubMed]
231. Alemu, T.T.; Kuyu, C.G. A Review of the Production, Quality, and Safety of Traditionally Fermented Cereal-Based Alcoholic Beverages in Ethiopia. *Food Sci. Nutr.* **2024**, *12*, 3125–3136. [CrossRef]
232. Chileshe, J.; van den Heuvel, J.; Handema, R.; Zwaan, B.J.; Talsma, E.F.; Schoustra, S. Nutritional Composition and Microbial Communities of Two Non-alcoholic Traditional Fermented Beverages from Zambia: A Study of *Mabisi* and *Munkoyo*. *Nutrients* **2020**, *12*, 1628. [CrossRef]
233. Daji, G.A.; Green, E.; Adebo, O.A. Nutritional and Phytochemical Composition of *Mahewu* (a Southern African Fermented Food Product) Derived from White and Yellow Maize (*Zea mays*) with Different Inocula. *Fermentation* **2023**, *9*, 58. [CrossRef]
234. Ajayi, O.A.; Akinrinde, I.M.; Akinwunmi, O.O. Towards the development of shelf stable "iru" (*Parkia biglobosa*) condiment bouillon cubes using corn, cassava and potato starch extracts as binders. *Niger. Food. J.* **2015**, *33*, 67–72. [CrossRef]
235. Hlangwani, E.; Adebiyi, J.; Doorsamy, W.; Adebo, O. Processing, characteristics and composition of *Umqombothi* (a South African traditional beer). *Processes* **2020**, *8*, 1451. [CrossRef]
236. Atter, A.; Obiri-Danso, K.; Amoa-Awua, W.K. Microbiological and chemical processes associated with the production of *Burukutu*, a traditional beer in Ghana. *Int. Food Res. J.* **2014**, *21*, 1769–1776.
237. Angelov, A.; Petrova, G.; Angelov, A.; Stefanova, P.; Bokossa, I.Y.; Tchekessi, C.; Marco, M.L.; Gotcheva, V. Molecular identification of yeasts and lactic acid bacteria involved in the production of Beninese fermented food *Degue*. *Open Biotechnol. J.* **2017**, *11*, 94–104. [CrossRef]
238. Afolayan, A.O.; Ayeni, F.A.; Ruppitsch, W. Antagonistic and quantitative assessment of indigenous lactic acid bacteria in different varieties of ogi against gastrointestinal pathogens. *Pan Afr. Med. J.* **2017**, *27*, 22. [CrossRef]
239. Mokoena, M.P.; Mutanda, T.; Olaniran, A.O. Perspectives on the probiotic potential of lactic acid bacteria from African traditional fermented foods and beverages. *Food Nutr. Res.* **2016**, *60*, 29630. [CrossRef] [PubMed]
240. Johansen, P.G.; Owusu-Kwarteng, J.; Parkouda, C.; Padonou, S.W.; Jespersen, L. Occurrence and importance of yeasts in indigenous fermented foods and beverages produced in sub-Saharan Africa. *Front. Microbiol.* **2019**, *10*, 1789. [CrossRef] [PubMed]
241. Muyanja, C.M.B.K.; Narvhus, J.A.; Treimo, J.; Langsrud, T. Isolation, characterisation and identification of lactic acid bacteria from bushera: A Ugandan traditional fermented beverage. *Int.J. Food Microbiol.* **2003**, *80*, 201–210. [CrossRef]
242. Wassie, M.; Wassie, T. Isolation, characterization and identification of lactic acid bacteria from ready-to-consume *Shamita*: Ethiopian traditional fermented beverage. *Int. J. Life Sci. Technol.* **2016**, *9*, 51–55.
243. Phiri, S.; Schoustra, S.E.; Van Den Heuvel, J.; Smid, E.J.; Shindano, J.; Linnemann, A. Fermented cereal-based *Munkoyo* beverage: Processing practices, microbial diversity and aroma compounds. *PLoS ONE* **2019**, *14*, e0223501. [CrossRef]
244. Pswarayi, F.; Gänzle, M.G. Composition and origin of the fermentation microbiota of *Mahewu*, a Zimbabwean fermented cereal beverage. *Appl. Environ. Microbiol.* **2019**, *85*, e03130-18. [CrossRef] [PubMed]
245. Ashaolu, T.J.; Reale, A. A Holistic Review on Euro-Asian Lactic Acid Bacteria Fermented Cereals and Vegetables. *Microorganisms* **2020**, *8*, 1176. [CrossRef] [PubMed]
246. Chen, W.; Lv, X.; Tran, V.-T.; Maruyama, J.; Han, K.-H.; Yu, J.-H. From Traditional to Modern: Progress of Molds and Yeasts in Fermented-Food Production. *Front. Microbiol.* **2022**, *13*, 876872.
247. Di, D.; He, S.; Zhang, R.; Gao, K.; Qiu, M.; Li, X.; Sun, H.; Xue, S.; Shi, J. Exploring the Dual Role of Anti-Nutritional Factors in Soybeans: A Comprehensive Analysis of Health Risks and Benefits. *Crit. Rev. Food Sci. Nutr.* **2025**, *65*, 5772–5789. [CrossRef]
248. Harahap, I.A.; Suliburska, J.; Karaca, A.C.; Capanoglu, E.; Esatbeyoglu, T. Fermented soy products: A review of bioactives for health from fermentation to functionality. *Compr. Rev. Food Sci. Food Saf.* **2025**, *24*, e70080. [CrossRef]
249. Patra, J.K.; Das, G.; Paramithiotis, S.; Shin, H.-S. Kimchi and Other Widely Consumed Traditional Fermented Foods of Korea: A Review. *Front. Microbiol.* **2016**, *7*, 1493. [CrossRef]

250. Rani, P.; Kanetkar, P.; Bakshi, S.; Binchhar, B.; Yadav, S.; Paswan, V. Exploring Some Traditional Fermented Dairy Beverages of Global Importance: A Review. *Asian J. Dairy Food Res.* **2025**. [CrossRef]
251. Utama, G.L.; Inayah, D.N.; Nurmilah, S.; Rialita, T.; Andriana, B.B.; Balia, R.L. Fish-Based Traditional Fermented Products in Asia: A Functional Food Rich in GABA for Hypertension Control. *Cogent Food Agric.* **2026**, *12*, 2620172. [CrossRef]
252. Phuwapraisirisan, P.; Phewpan, A.; Ngamchuachit, P.; Lopetcharat, K.; Phraephaisarn, C.; Dawid, C.; Keeratipibul, S. Kokumi Substances in Thai-Fermented Freshwater Fish "Pla-ra". In *Kokumi Substance as an Enhancer of Koku: Biochemistry, Physiology, and Food Science*; Springer Nature: Singapore, 2024; pp. 83–110.
253. Daba, G.M.; Mostafa, F.A.; Elkhateeb, W.A. The Ancient Koji Mold (*Aspergillus oryzae*) as a Modern Biotechnological Tool. *Bioresour. Bioprocess.* **2021**, *8*, 52. [CrossRef] [PubMed]
254. Kusumoto, K.-I.; Yamagata, Y.; Tazawa, R.; Kitagawa, M.; Kato, T.; Isobe, K.; Kashiwagi, Y. Japanese Traditional *Miso* and *Koji* Making. *J. Fungi* **2021**, *7*, 579. [CrossRef] [PubMed]
255. Lee, H.; Lee, S.; Kyung, S.; Ryu, J.; Kang, S.; Park, M.; Lee, C. Metabolite Profiling and Anti-Aging Activity of Rice *Koji* Fermented with *Aspergillus oryzae* and *Aspergillus cristatus*: A Comparative Study. *Metabolites* **2021**, *11*, 524. [CrossRef]
256. Papagianni, O.I.; Dimou, C.; Koutelidakis, A.E. Biofunctional Miso-Type Sauce Enhanced with Biocarotenoids: How Does Its Habitual Consumption Affect Lipidemic, Glycemic, and Oxidative Stress Markers? A Pilot Cross-Over Clinical Study. *Appl. Sci.* **2025**, *15*, 5962. [CrossRef]
257. Tamang, J.P.; Anupma, A.; Shangpliang, H.N.J. Ethno-microbiology of Tempe, an Indonesian fungal-fermented soybean food and Koji, a Japanese fungal starter culture. *Curr. Opin. Food Sci.* **2022**, *48*, 100912. [CrossRef]
258. Huang, L.; Wang, C.; Zhang, Y.; Chen, X.; Huang, Z.; Xing, G.; Dong, M. Degradation of anti-nutritional factors and reduction of immunoreactivity of tempeh by co-fermentation with *Rhizopus oligosporus* RT-3 and *Actinomucor elegans* DCY-1. *Int. J. Food Sci. Technol.* **2019**, *54*, 1836–1848. [CrossRef]
259. Athaillah, Z.A.; Muzdalifah, D.; Lestari, A.; Fitria, A.; Udin, L.Z.; Artanti, N.; Lioe, H.N. Phenolic Compound Profile and Functionality of Aqueous Overripe Tempe Extracts. *Curr. Res. Nutr. Food Sci.* **2019**, *7*, 382–392. [CrossRef]
260. Aoki, H.; Chuma, S.; Iba, Y.; Tashiro, H.; Watanabe, N.; Oyama, H. Comparison of Bioactive Components in Tempeh Produced by Three Different *Rhizopus* Starters and Immunomodulatory Effect of Tempeh on Atopic Dermatitis Mice. *Food Sci. Technol. Res.* **2020**, *26*, 665–672. [CrossRef]
261. Nyyssölä, A.; Ellilä, S.; Nordlund, E.; Poutanen, K. Reduction of FODMAP content by bioprocessing. *Trends Food Sci. Technol.* **2020**, *99*, 257–272. [CrossRef]
262. Chen, Y.-C.; Hsieh, S.-L.; Hu, C.-Y. Effects of Red-Bean Tempeh with Various Strains of *Rhizopus* on GABA Content and Cortisol Level in Zebrafish. *Microorganisms* **2020**, *8*, 1330. [CrossRef]
263. Liu, Y.; Su, H.; Song, H.L. Comparison of four extraction methods, SPME, DHS, SAFE, versus SDE, for analysis of flavor compounds in Natto. *Food Anal. Methods* **2018**, *11*, 343–354. [CrossRef]
264. Kanghae, A.; Eungwanichayapant, P.D.; Chukeatirote, E. Fatty acid profiles of fermented soybean prepared by *Bacillus subtilis* and *Rhizopus oligosporus*. *Environ. Exp. Biol.* **2017**, *15*, 173–176.
265. Hu, Y.; Ge, C.; Yuan, W.; Zhu, R.; Zhang, W.; Du, L.; Xue, J. Characterization of fermented black soybean natto inoculated with Bacillus natto during fermentation. *J. Sci. Food Agric.* **2010**, *90*, 1194–1202. [CrossRef] [PubMed]
266. Wang, C.; Chen, J.; Tian, W.; Han, Y.; Xu, X.; Ren, T.; Tian, C.; Chen, C. Natto: A medicinal and edible food with health function. *Chin. Herb. Med.* **2023**, *15*, 349–359. [CrossRef]
267. Borrego-Ruiz, A.; González-Domenech, C.M.; Borrego, J.J. The Role of Fermented Vegetables as a Sustainable and Health-Promoting Nutritional Resource. *Appl. Sci.* **2024**, *14*, 10853. [CrossRef]
268. Hsu, R.-L.; Lee, K.-T.; Wang, J.-H.; Lee, L.Y.L.; Chen, R.P.Y. Amyloid-Degrading Ability of Nattokinase from Bacillus subtilis Natto. *J. Agric. Food Chem.* **2009**, *57*, 503–508. [CrossRef]
269. Zhang, J.; Bilal, M.; Liu, S.; Zhang, J.; Lu, H.; Luo, H.; Luo, C.; Shi, H.; Iqbal, H.M.N.; Zhao, Y. Isolation, Identification and Antimicrobial Evaluation of Bactericides Secreting *Bacillus subtilis* Natto as a Biocontrol Agent. *Processes* **2020**, *8*, 259. [CrossRef]
270. Yan, Y.; Wang, Y.; Qian, J.; Wu, S.; Ji, Y.; Liu, Y.; Zeng, J.; Gong, A. Nattokinase Crude Extract Inhibits Hepatocellular Carcinoma Growth in Mice. *J. Microbiol. Biotechnol.* **2019**, *29*, 1281–1287. [CrossRef]
271. Afzaal, M.; Saeed, F.; Islam, F.; Ateeq, H.; Asghar, A.; Shah, Y.A.; Ofoedu, C.E.; Chacha, J.S. Nutritional Health Perspective of Natto: A Critical Review. *Biochem. Res. Int.* **2022**, *2022*, 5863887. [CrossRef]
272. Raphel, S.; Halami, P.M. Bioactive compounds from food-grade Bacillus. *J. Sci. Food Agric.* **2025**, *105*, 4085–4095. [CrossRef]
273. Araki, R.; Yamada, T.; Maruo, K.; Araki, A.; Miyakawa, R.; Suzuki, H.; Hashimoto, K. Gamma-Polyglutamic Acid-Rich Natto Suppresses Postprandial Blood Glucose Response in the Early Phase after Meals: A Randomized Crossover Study. *Nutrients* **2020**, *12*, 2374. [CrossRef]

 https://doi.org/10.3390/molecules31081353

274. Iwai, K.; Nakaya, N.; Kawasaki, Y.; Matsue, H. Antioxidative Functions of Natto, A Kind of Fermented Soybeans: Effect on LDL Oxidation and Lipid Metabolism in Cholesterol-Fed Rats. *J. Agric. Food Chem.* **2002**, *50*, 3597–3601. [CrossRef] [PubMed]
275. Fujita, M.; Ohnishi, K.; Takaoka, S.; Ogasawara, K.; Fukuyama, R.; Nakamuta, H. Antihypertensive Effects of Continuous Oral Administration of Nattokinase and Its Fragments in Spontaneously Hypertensive Rats. *Biol. Pharm. Bull.* **2011**, *34*, 1696–1701. [CrossRef] [PubMed]
276. Tamura, M.; Watanabe, J.; Hori, S.; Inose, A.; Kubo, Y.; Noguchi, T.; Kobori, M. Effects of a high-γ-polyglutamic acid-containing natto diet on liver lipids and cecal microbiota of adult female mice. *Biosci. Microbiota Food Health* **2021**, *40*, 176–185. [CrossRef]
277. Han, A.L.; Ryu, M.S.; Yang, H.-J.; Jeong, D.-Y.; Choi, K.H. Evaluation of Menopausal Syndrome Relief and Anti-Obesity Efficacy of the Korean Fermented Food Doenjang: A Randomized, Double-Blind Clinical Trial. *Nutrients* **2024**, *16*, 1194. [CrossRef]
278. Bahuguna, A.; Shukla, S.; Lee, J.S.; Bajpai, V.K.; Kim, S.Y.; Huh, Y.S.; Han, Y.K.; Kim, M. Garlic augments the functional and nutritional behavior of Doenjang, a traditional Korean fermented soybean paste. *Sci. Rep.* **2019**, *9*, 5436. [CrossRef]
279. Lee, S.; Lee, S.; Singh, D.; Oh, J.Y.; Jeon, E.J.; Ryu, H.S.K.; Lee, D.W.; Kim, B.S.; Lee, C.H. Comparative evaluation of microbial diversity and metabolite profiles in doenjang, a fermented soybean paste, during the two different industrial manufacturing processes. *Food Chem.* **2017**, *221*, 1578–1586. [CrossRef] [PubMed]
280. Jung, J.Y.; Han, S.-S.; Kim, Z.-H.; Kim, M.H.; Kang, H.K.; Jin, H.M.; Lee, M.H. In-Vitro Characterization of Growth Inhibition against the Gut Pathogen of Potentially Probiotic Lactic Acid Bacteria Strains Isolated from Fermented Products. *Microorganisms* **2021**, *9*, 2141. [CrossRef]
281. Song, E.J.; Lee, E.S.; Park, S.L.; Choi, H.J.; Roh, S.W.; Nam, Y. Do Bacterial community analysis in three types of the fermented seafood, jeotgal, produced in South Korea. *Biosci. Biotechnol. Biochem.* **2018**, *82*, 1444–1454. [CrossRef] [PubMed]
282. Nugroho, D.; Thinthasit, A.; Surya, E.; Hartati; Oh, J.-S.; Jang, J.-G.; Benchawattananon, R.; Surya, R. Immunoenhancing and Antioxidant Potentials of Kimchi, an Ethnic Food from Korea, as a Probiotic and Postbiotic Food. *J. Ethn. Foods* **2024**, *11*, 12. [CrossRef]
283. Kim, M.E.; Lee, J.S. The Potential of Korean Bioactive Substances and Functional Foods for Immune Enhancement. *Int. J. Mol. Sci.* **2024**, *25*, 1334. [CrossRef] [PubMed]
284. Das, G.; Kameswaran, S.; Ramesh, B.; Bangeppagari, M.; Nath, R.; Das Talukdar, A.; Shin, H.-S.; Patra, J.K. Anti-Aging Effect of Traditional Plant-Based Food: An Overview. *Foods* **2024**, *13*, 3785. [CrossRef]
285. Ko, J.; Kim, W.; Park, H. Effects of microencapsulated Allyl isothiocyanate (AITC) on the extension of the shelf-life of Kimchi. *Int. J. Food Microbiol.* **2012**, *153*, 92–98. [CrossRef]
286. Jang, M.; Kim, G.-H. Glucosinolate and isothiocyabate contents according to processing of Kimchi cabbage (*Brassica rapa* L. ssp. pekinensis). *Korean J. Food Preserv. (Food Sci. Preserv.)* **2017**, *24*, 367–373. [CrossRef]
287. Park, K.Y.; Hong, G.H.; Lee, S.Y.; Lee, Y.J. Kimchi and Its Antiobesity and Anticancer Functions. *J. Ethn. Foods* **2024**, *11*, 37. [CrossRef]
288. Baek, Y.-H.; Kwak, J.-R.; Kim, S.-J.; Han, S.-S.; Song, Y.-O. Effects of kimchi supplementation and/or exercise training on body composition and plasma lipids in obese middle school girls. *J. Korean Soc. Food Sci. Nutr.* **2001**, *30*, 906–912.
289. Lee, N.-K.; Lee, Y.; Shin, D.-S.; Choi, Y.-M.; Lee, J.; Park, E.; Paik, H.-D. Effect of *Lactiplantibacillus plantarum* DSW3805 Isolated from Kimchi for Gut Health Attenuating Colonic Inflammation in a Dextran Sulfate Sodium-Induced Mouse Model. *Nutrients* **2025**, *17*, 1259. [CrossRef]
290. Lee, Y.-J.; Pan, Y.; Kwack, K.-B.; Chung, J.H.; Park, K.-Y. Increased Anticancer Activity of Organic Kimchi with Starters Demonstrated in HT-29 Cancer Cells. *Appl. Sci.* **2023**, *13*, 6654. [CrossRef]
291. *CXS 243-2003*; WHO Standard for Fermented Milks. Food and Agriculture Organization of the United Nations: Rome, Italy, 2018.
292. FAO. *Food and Agriculture Organization of the United Nations: Food Energy—Methods of Analysis and Conversion Factors*; FAO Food and Nutrition Paper 77; Food and Agriculture Organization of the United Nations: Rome, Italy, 2003.
293. Ichimura, T.; Osada, T.; Yonekura, K.; Horiuchi, H. A new method for producing superior set yogurt, focusing on heat treatment and homogenization. *J. Dairy Sci.* **2022**, *105*, 2978–2987. [CrossRef]
294. Dannenberg, F.; Kessler, H.G. Effect of denaturation of beta-lactoglobulin on texture properties of set-style nonfat yoghurt. 2. Firmness and flow properties. *Milchwiss. Milk Sci. Int.* **1988**, *43*, 700–704.
295. Rasic, J.L.; Kurmann, J.A. *Yoghurt. Scientific Grounds, Technology, Manufacture and Preparations*; Technical Dairy Publishing House: Copenhagen, Denmark, 1978; 466p.
296. Nguyen, H.T.H.; Gathercole, J.L.; Day, L.; Dalziel, J.E. Differences in peptide generation following in vitro gastrointestinal digestion of yogurt and milk from cow, sheep and goat. *Food Chem.* **2020**, *317*, 126419. [CrossRef]
297. Savijoki, K.; Ingmer, H.; Varmanen, P. Proteolytic systems of lactic acid bacteria. *Appl. Microbiol. Biotechnol.* **2006**, *71*, 394–406. [CrossRef]

298. Tzvetkova, I.; Dalgalarrondo, M.; Danova, S.; Iliev, I.; Ivanova, I.; Chobert, J.-M.; Haertlé, T. Hydrolysis of Major Dairy Proteins by Lactic Acid Bacteria from Bulgarian Yogurts. *J. Food Biochem.* **2007**, *31*, 680–702. [CrossRef]
299. Sumi, K.; Tagawa, R.; Yamazaki, K.; Nakayama, K.; Ichimura, T.; Sanbongi, C.; Nakazato, K. Nutritional Value of Yogurt as a Protein Source: Digestibility/Absorbability and Effects on Skeletal Muscle. *Nutrients* **2023**, *15*, 4366. [CrossRef]
300. Ding, W.; Shi, C.; Chen, M. Screening for lactic acid bacteria in traditional fermented Tibetan yak milk and evaluating their probiotic and cholesterol-lowering potentials in rats fed a high-cholesterol diet. *J. Funct. Foods* **2017**, *32*, 324–332. [CrossRef]
301. Zhang, X.; Li, Y.; Zhao, Y.; Guan, H.; Jin, C.; Gong, H.; Sun, X. Effect of *Levilactobacillus brevis* as a starter on the flavor quality of radish paocai. *Food Res. Int.* **2023**, *168*, 112780. [CrossRef] [PubMed]
302. Ibrahim, N.A.; Patel, M.; Trivedi, D.; Janiyani, K.; Sulieman, A.M.E.; Tahir, H.E.; Mir, R.H.; Fatima, S.B.; Adnan, M. Metabolomics-Based Functional Characterization of *Levilactobacillus brevis*: Insights into Probiotic and Antimicrobial Potential. *Mol. Nutr. Food Res.* **2026**, *70*, e70366. [CrossRef]
303. Zohra, F.T.; Islam, M.N.; Siddiki, M.S.R.; Habib, M.R.; Afrin, S. Effect of Different Levels of Water on the Qualitative Characteristics of Lassi Prepared from Reconstituted Milk. *Bangladesh J. Anim. Sci.* **2016**, *45*, 46–51. [CrossRef]
304. Saha, A.; Das, M.; Das, A.; Mandal, S. Dietetic Benefits of Yogurt-Based Beverage (Lassi) Becoming a Fact-Finding Probe in Research. *Int. J. Pharm. Res.* **2021**, *13*, 1150–1156.
305. Samtiya, M.; Bhushan, B.; Sari, T.P.; Badgujar, P.C.; Chandratre, G.A.; Singh, P.; Dhewa, T. Characterization of Indigenous Lactobacilli from Dairy Fermented Foods of Haryana as Potential Probiotics Utilizing Multiple Attribute Decision-Making Approach. *Food Prod. Process. Nutr.* **2024**, *6*, 81. [CrossRef]
306. Settachaimongkon, S.; Det-udom, R.; Suriya, A.; Suphamityotin, P.; Thanthithum, C.; Manowan, P.; Prakitchaiwattana, C. Active Dried Co-Autochthonous Culture as Accelerators and Safety Enhancers in Thai-Salty Fish (Pla-ra) Fermentation: Laboratory and Large-Scale Studies. *LWT* **2025**, *228*, 118093. [CrossRef]
307. Chotechuang, N. Taste Active Components in Thai Foods: A Review of Thai Traditional Seasonings. *J. Nutr. Food Sci.* **2012**, *S10*, 004. [CrossRef]
308. Rodpai, R.; Sanpool, O.; Thanchomnang, T.; Wangwiwatsin, A.; Sadaow, L.; Phupiewkham, W.; Boonroumkaew, P.; Intapan, P.M.; Maleewong, W. Investigating the Microbiota of Fermented Fish Products (Pla-ra) from Different Communities of Northeastern Thailand. *PLoS ONE* **2021**, *16*, e0245227. [CrossRef]
309. Cha, Y.-J.; Yu, D. Health Benefits and Functions of Salt-Fermented Fish. *J. Ethn. Food* **2024**, *11*, 34. [CrossRef]
310. Priscilla, S.H.A. Microorganisms and Technology of Philippine Fermented Foods. *J. Jpn. Lact. Acid Bact.* **1999**, *10*, 19–28.
311. Julmohammad, N.; Atun, M.; Roslan, J.; Shariff, A.H.M.; Adzitey, F.; Sukoso; Huda, N. Compositional Analysis and Nutritional Profiling of Southeast Asian Fermented Fish Products: Insights into Macronutrients, Micronutrients, and Bioactive Compounds. *Discov. Food* **2025**, *5*, 75. [CrossRef]
312. Hersh, J. *Yoghurt: A Global History*; Reaktion Books: London, UK, 2021.
313. Sangjindavong, M.; Chuapoehuk, P.; Runglerdkriangkrai, J.; Klaypradit, W.; Vareevanich, D. Fermented Fish Product (Pla-ra) from Marine Fish and Preservation. *Kasetsart J. Nat. Sci.* **2008**, *42*, 129–136.
314. Ramos, C.L.; Schwan, R.F. Technological and nutritional aspects of indigenous Latin America fermented foods. *Curr. Opin. Food Sci.* **2017**, *13*, 97–102. [CrossRef]
315. Jimenez, M.E.; O'Donovan, C.M.; de Ullivarri, M.F.; Cotter, P.D. Microorganisms present in artisanal fermented food from South America. *Front. Microbiol.* **2022**, *13*, 941866. [CrossRef] [PubMed]
316. Carboni, A.D.; Martins, G.N.; Gómez-Zavaglia, A.; Castilho, P.C. Lactic Acid Bacteria in the Production of Traditional Fermented Foods and Beverages of Latin America. *Fermentation* **2023**, *9*, 315. [CrossRef]
317. Chacón Mayorga, G.A.; Arias Palma, G.B.; Sandoval-Cañas, G.J.; Ordoñez-Araque, R.H. Ancestral fermented indigenous beverages from South America made from cassava (Manihot esculenta). *Food Sci. Technol.* **2021**, *41*, 360–367. [CrossRef]
318. Faria-Oliveira, F.; Diniz, R.H.S.; Godoy-Santos, F.; Piló, F.B.; Mezadri, H.; Castro, I.M.; Brandão, R.L. The role of yeast and lactic acid bacteria in the production of fermented beverages in South America. In *Food Production and Industry*; Eissa, A., Ed.; IntechOpen: London, UK, 2015; pp. 107–135.
319. Colehour, A.M.; Meadow, J.F.; Liebert, M.A.; Cepon-Robins, T.J.; Gildner, T.E.; Urlacher, S.S.; Bohannan, B.J.M.; Snodgrass, J.J.; Sugiyama, L.S. Local domestication of lactic acid bacteria via cassava beer fermentation. *PeerJ* **2014**, *2*, e479. [CrossRef]
320. Freire, A.L.; Zapata, S.; Mosquera, J.; Mejia, M.L.; Trueba, G. Bacteria associated with human saliva are major microbial components of Ecuadorian indigenous beers (chicha). *PeerJ* **2016**, *4*, e1962. [CrossRef]
321. Bassi, D.; Orrù, L.; Vasquez, J.C.; Cocconcelli, P.S.; Fontana, C. Peruvian *chicha*: A focus on the microbial populations of this ancient maize-based fermented beverage. *Microorganisms* **2020**, *8*, 93. [CrossRef]

322. Resende, L.V.; Pinheiro, L.K.; Miguel, M.G.d.C.P.; Ramos, C.L.; Vilela, D.M.; Schwan, R.F. Microbial community and physico-chemical dynamics during the production of 'Chicha', a traditional beverage of Indigenous people of Brazil. *World J. Microbiol. Biotechnol.* **2018**, *34*, 46. [CrossRef] [PubMed]
323. Puerari, C.; Magalhães-Guedes, K.T.; Schwan, R.F. Physicochemical and microbiological characterization of chicha, a rice-based fermented beverage produced by Umutina Brazilian Amerindians. *Food Microbiol.* **2015**, *46*, 210–217. [CrossRef]
324. Penna, A.L.B.; Nero, L.A.; Todorov, S.D. Fermented Foods of Latin America: From Traditional Knowledge to Innovative Applications. In *Food Biology Series*; CRC Press: Boca Raton, FL, USA, 2017; p. 338. ISBN 9781498738118.
325. Rosane, F.S.; Claudia, C.A.D.A.S.; Marianna, R.R.M.S.; Whasley, F.D.; Schwan, R.F.; Santos, C.C.A.D.A.; Duarte, W.F. Bacterial dynamics and chemical changes during the spontaneous production of the fermented porridge (Calugi) from cassava and corn. *Afr. J. Microbiol. Res.* **2014**, *8*, 839–849. [CrossRef]
326. Almeida, E.G.; Rachid, C.C.; Schwan, R.F. Microbial population present in fermented beverage 'cauim' produced by Brazilian Amerindians. *Int. J. Food Microbiol.* **2007**, *120*, 146–151. [CrossRef]
327. Lima, T.T.M.; Hosken, B.D.O.; Venturim, B.C.; Lopes, I.L.; Martin, J.G.P. Traditional Brazilian Fermented Foods: Cultural and Technological Aspects. *J. Ethn. Food* **2022**, *9*, 35. [CrossRef]
328. Schwan, R.F.; Almeida, E.G.; Souza-Dias, M.A.G.; Jespersen, L. Yeast diversity in rice–cassava fermentations produced by the indigenous Tapirapé people of Brazil. *FEMS Yeast Res.* **2007**, *7*, 966–972. [CrossRef] [PubMed]
329. Kljak, K.; Duvnjak, M.; Bedeković, D.; Kiš, G.; Janječić, Z.; Grbeša, D. Commercial Corn Hybrids as a Single Source of Dietary Carotenoids: Effect on Egg Yolk Carotenoid Profile and Pigmentation. *Sustainability* **2021**, *13*, 12287. [CrossRef]
330. Rizo, J.; Rogel, M.A.; Guillén, D.; Wacher, C.; Martinez-Romero, E.; Encarnación, S.; Sánchez, S.; Rodríguez-Sanoja, R. Nitrogen Fixation in Pozol, a Traditional Fermented Beverage. *Appl. Environ. Microbiol.* **2020**, *86*, e00588-20. [CrossRef]
331. Reyes-Pavón, D.; Soto-Sigala, K.S.; Cano-Sampedro, E.; Méndez-Trujillo, V.; Navarro-Ibarra, M.J.; Pérez-Pasten-Borja, R.; Olvera-Sandoval, C.; Torres-Maravilla, E. Pigmented Native Maize: Unlocking the Potential of Anthocyanins and Bioactive Compounds from Traditional to Functional Beverages. *Beverages* **2024**, *10*, 69. [CrossRef]
332. López-Reynoso, M.; Martínez-Medina, G.A.; Londoño-Hernández, L.; Aguilar-Zarate, P.; Hernández-Beltrán, J.U.; Hernández-Almanza, A.Y. Corn-Based Fermented Beverages: Nutritional Value, Microbial Dynamics, and Functional Potential—An Overview. *Foods* **2026**, *15*, 27. [CrossRef]
333. Elizaquível, P.; Pérez-Cataluña, A.; Yépez, A.; Aristimuño, C.; Jiménez, E.; Cocconcelli, P.S.; Vignolo, G.; Aznar, R. Pyrosequencing vs. culture-dependent approaches to analyze lactic acid bacteria associated to chicha, a traditional maize-based fermented beverage from Northwestern Argentina. *Int. J. Food Microbiol.* **2015**, *198*, 9–18. [CrossRef]
334. Romero-Luna, H.E.; Hernández-Sánchez, H.; Dávila-Ortiz, G. Traditional fermented beverages from Mexico as a potential probiotic source. *Ann. Microbiol.* **2017**, *67*, 577–586. [CrossRef]
335. Velázquez-López, A.; Covatzin-Jirón, D.; Toledo-Meza, M.D.; Vela-Gutiérrez, G. Bebida fermentada elaborada con bacterias ácido lácticas aisladas del pozol tradicional chiapaneco. *CienciaUAT* **2018**, *13*, 165. [CrossRef]
336. Baschali, A.; Tsakalidou, E.; Kyriacou, A.; Karavasiloglou, N.; Matalas, A.L. Traditional low-alcoholic and non-alcoholic fermented beverages consumed in European countries: A neglected food group. *Nutr. Res. Rev.* **2017**, *30*, 1–24. [CrossRef]
337. Xu, J.; Peng, S.; Xiong, Y.; Zheng, Z.; Liu, M.; Xu, J.; Chen, W.; Liu, M.; Kong, J.; Wang, C.; et al. A review on fermented vegetables: Microbial community and potential upgrading strategy via inoculated fermentation. *Compr. Rev. Food Sci. Food Saf.* **2024**, *23*, e13362. [CrossRef]
338. Axelsson, L.; Bjerke, G.A.; McLeod, A.; Berget, I.; Holck, A.L. Growth Behavior of *Listeria monocytogenes* in a Traditional Norwegian Fermented Fish Product (*Rakfisk*), and Its Inhibition through Bacteriophage Addition. *Foods* **2020**, *9*, 119. [CrossRef]
339. Terpou, A. Ethnic Selected Fermented Foods of Greece. In *Fermented Food Products*, 1st ed.; Sankaranarayanan, N., Dhanasekaran, D., Eds.; CRC Press: London, UK; Taylor & Francis Group: New York, NY, USA, 2020; p. 10.
340. Terpou, A.; Mantzourani, I. Vinegars Made with Kefir. In *Advances in Vinegar Production*, 1st ed.; Bekatorou, A., Ed.; CRC Press: London, UK; Taylor & Francis Group: New York, NY, USA, 2019; p. 16.
341. Guzel-Seydim, Z.B.; Gokirmaklı, C.; Greene, A.K. A Comparison of Milk Kefir and Water Kefir: Physical, Chemical, Microbiological and Functional Properties. *Trends Food Sci. Technol.* **2021**, *113*, 42–53. [CrossRef]
342. Walsh, A.M.; Crispie, F.; Kilcawley, K.; O'Sullivan, O.; O'Sullivan, M.G.; Claesson, M.J.; Cotter, P.D.; Dutton, R.J. Microbial Succession and Flavor Production in the Fermented Dairy Beverage Kefir. *mSystems* **2016**, *1*, e00052-16. [CrossRef]
343. Bengoa, A.A.; Iraporda, C.; Garrote, G.L.; Abraham, A.G. Kefir Microorganisms: Their Role in Grain Assembly and Health Properties of Fermented Milk. *J. Appl. Microbiol.* **2019**, *126*, 686–700. [CrossRef] [PubMed]
344. Salari, A.; Hashemi, M.; Afshari, A. Functional Properties of Kefiran in Medical Field and Food Industry. *Curr. Pharm. Biotechnol.* **2021**, *23*, 388–395. [CrossRef]

https://doi.org/10.3390/molecules31081353

345. Tzavaras, D.; Papadelli, M.; Ntaikou, I. From Milk Kefir to Water Kefir: Assessment of Fermentation Processes, Microbial Changes and Evaluation of the Produced Beverages. *Fermentation* **2022**, *8*, 135. [CrossRef]
346. Rosa, D.D.; Dias, M.M.; Grześkowiak, Ł.M.; Reis, S.A.; Conceição, L.L.; Maria do Carmo, G.P. Milk Kefir: Nutritional, Microbiological and Health Benefits. *Nutr. Res. Rev.* **2017**, *30*, 82–96. [CrossRef] [PubMed]
347. Karagozlu, C.; Kavas, G. Alkollü Fermente Süt İçecekleri: Kefir ve Kimizin Özellikleri ve İnsan Beslenmesindeki Önemi. *Gida* **2000**, *6*, 86–93.
348. Açik, M.; Çakiroğlu, F.P.; Altan, M.; Baybo, T. Alternative Source of Probiotics for Lactose Intolerance and Vegan Individuals: Sugary Kefir. *Food Sci. Technol.* **2020**, *40*, 523–531. [CrossRef]
349. Laureys, D.; Aerts, M.; Vandamme, P.; De Vuyst, L. Oxygen and Diverse Nutrients Influence the Water Kefir Fermentation Process. *Food Microbiol.* **2018**, *73*, 351–361. [CrossRef]
350. Saleem, K.; Ikram, A.; Saeed, F.; Afzaal, M.; Ateeq, H.; Hussain, M.; Raza, A.; Rasheed, A.; Asghar, A.; Asif Shah, M. Nutritional and Functional Properties of Kefir. *Int. J. Food Prop.* **2023**, *26*, 3261–3274. [CrossRef]
351. Arslan, S. A Review: Chemical, Microbiological and Nutritional Characteristics of Kefir. *CyTA-J. Food* **2015**, *13*, 340–345. [CrossRef]
352. Mathipa, M.G.; Thantsha, M.S. Probiotic Engineering: Towards Development of Robust Probiotic Strains with Enhanced Functional Properties and for Targeted Control of Enteric Pathogens. *Gut Pathog.* **2017**, *9*, 28. [CrossRef]
353. Bourrie, B.C.T.; Willing, B.P.; Cotter, P.D. The Microbiota and Health Promoting Characteristics of the Fermented Beverage Kefir. *Front. Microbiol.* **2016**, *7*, 647. [CrossRef] [PubMed]
354. Flynn, J.; Ryan, A.; Hudson, S.P. Pre-Formulation and Delivery Strategies for the Development of Bacteriocins as Next Generation Antibiotics. *Eur. J. Pharm. Biopharm.* **2021**, *165*, 149–163. [CrossRef]
355. Qaisrani, Z.N.; Lin, W.P.; Lay, B.B.; Phyo, K.Y.; San, M.M.; Awaeloh, N.; Aunsorn, S.; Pattanayaiying, R.; Na Ayudthaya, S.P.; Hongkulsup, C.; et al. The Impact of Kefir Consumption on Inflammation, Oxidative Stress Status, and Metabolic-Syndrome-Related Parameters in Animal Models: A Systematic Review and Meta-Analysis. *Foods* **2025**, *14*, 2077. [CrossRef] [PubMed]
356. Oryan, A.; Alemzadeh, E.; Eskandari, M.H. Kefir Accelerates Burn Wound Healing Through Inducing Fibroblast Cell Migration in vitro and Modulating the Expression of IL-1ß, TGF-ß1, and bFGF genes in vivo. *Probiotics Antimicrob. Proteins* **2018**, *11*, 874–886. [CrossRef]
357. Yang, L.; Xie, X.; Li, Y.; Wu, L.; Fan, C.; Liang, T.; Xi, Y.; Yang, S.; Li, H.; Zhang, J.; et al. Evaluation of the cholesterol-lowering mechanism of Enterococcus faecium strain 132 and *Lactobacillus paracasei* strain 201 in hypercholesterolemia rats. *Nutrients* **2021**, *13*, 1982. [CrossRef]
358. Bernini, L.J.; Simão, A.N.C.; de Souza, C.H.B.; Alfieri, D.F.; Segura, L.G.; Costa, G.N.; Dichi, I. Effect of Bifidobacterium lactis HN019 on inflammatory markers and oxidative stress in subjects with and without the metabolic syndrome. *Br. J. Nutr.* **2018**, *120*, 645–652. [CrossRef]
359. Akar, F.; Sumlu, E.; Alçığır, M.E.; Bostancı, A.; Sadi, G. Potential mechanistic pathways underlying intestinal and hepatic effects of kefir in high-fructose-fed rats. *Food Res. Int.* **2021**, *143*, 110287. [CrossRef] [PubMed]
360. Lau, S.W.; Chong, A.Q.; Chin, N.L.; Talib, R.A.; Basha, R.K. Sourdough microbiome comparison and benefits. *Microorganisms* **2021**, *9*, 1355. [CrossRef]
361. Fekri, A.; Torbati, M.; Khosrowshahi, A.Y.; Shamloo, H.B.; Azadmard-Damirchi, S. Functional effects of phytate-degrading, probiotic lactic acid bacteria and yeast strains isolated from Iranian traditional sourdough on the technological and nutritional properties of whole wheat bread. *Food Chem.* **2020**, *306*, 125620. [CrossRef]
362. Alkay, Z.; Falah, F.; Cankurt, H.; Dertli, E. Exploring the Nutritional Impact of Sourdough Fermentation: Its Mechanisms and Functional Potential. *Foods* **2024**, *13*, 1732. [CrossRef]
363. Xiong, Y.; Zhang, P.; Warner, R.D.; Shen, S.; Fang, Z. Cereal grain-based functional beverages: From cereal grain bioactive phytochemicals to beverage processing technologies, health benefits and product features. *Crit. Rev. Food Sci. Nutr.* **2020**, *62*, 2404–2431. [CrossRef] [PubMed]
364. Dlusskaya, E.; Jänsch, A.; Schwab, C.; Gänzle, M.G. Microbial and chemical analysis of a kvass fermentation. *Eur. Food Res. Technol.* **2008**, *227*, 261–266. [CrossRef]
365. Gambuś, H.; Mickowska, B.; Bartoń, H.; Augustyn, G.; Zięć, G.; Litwinek, D.; Szary-Sworst, K.; Berski, W. Health benefits of kvass manufactured from rye wholemeal bread. *J. Microbiol. Biotechnol. Food Sci.* **2015**, *4*, 34–39. [CrossRef]
366. Lidums, I.; Karklina, D.; Kirse, A. Quality parameters of fermented kvass extract. *Chem. Technol.* **2016**, *1*, 67. [CrossRef]
367. Wang, P.; Wu, J.; Wang, T.; Zhang, Y.; Yao, X.; Li, J.; Wang, X.; Lü, X. Fermentation process optimization, chemical analysis, and storage stability evaluation of a probiotic barley malt kvass. *Bioprocess. Biosyst. Eng.* **2022**, *45*, 1175–1188. [CrossRef]
368. Kaszuba, J.; Jańczak-Pieniążek, M.; Migut, D.; Kapusta, I.; Buczek, J. Comparison of the Antioxidant and Sensorial Properties of Kvass Produced from Mountain Rye Bread with the Addition of Selected Plant Raw Materials. *Foods* **2024**, *13*, 357. [CrossRef]

369. Bjerke, G.A.; Rudi, K.; Avershina, E.; Moen, B.; Blom, H.; Axelsson, L. Exploring the Brine Microbiota of a Traditional Norwegian Fermented Fish Product (*Rakfisk*) from Six Different Producers during Two Consecutive Seasonal Productions. *Foods* **2019**, *8*, 72. [CrossRef] [PubMed]
370. Kotzekidou, P.; Tsakalidou, E. Fermentation Biotechnology of Plant-Based Traditional Foods of the Middle East and Mediterranean Region. In *Food Biotechnology*; CRC Press: Boca Raton, FL, USA, 2005; pp. 1822–1855.

https://doi.org/10.3390/molecules31081353

molecules

MDPI

Article

Evaluation of Physicochemical Properties, Bioactive Composition, and Antioxidant Activity of *Prunus armeniaca* L. Cultivars for Functional Food and Nutraceutical Development

Ceren Birinci [1], Anna Kurek-Górecka [2,*], Elsevar Asadov [3], Zenon P. Czuba [2] and Sevgi Kolaylı [1,3,*]

[1] Department of Chemistry, Faculty of Science, Karadeniz Technical University, 61080 Trabzon, Türkiye; cerendidar.birinci@gmail.com
[2] Department of Microbiology and Immunology, Faculty of Medical Sciences in Zabrze, Medical University of Silesia in Katowice, Jordana 19, 41-808 Zabrze, Poland; zczuba@sum.edu.pl
[3] Nakhchivan State University, Nakhchivan AZ7012, Azerbaijan; asadoves1974@gmail.com
* Correspondence: akurekgorecka@sum.edu.pl (A.K.-G.); skolayli@ktu.edu.tr (S.K.)

Abstract

Advancing the characterization of bioactive compounds in *Prunus armeniaca* L. is critical for identifying high-value cultivars with enhanced nutritional and functional potential. In this study, selected apricot varieties were evaluated, with particular emphasis on their sugar composition, phenolic profile, and antioxidant capacity. Sugar and phenolic compounds were analyzed using RP-HPLC-RID and RP-HPLC-PDA methods, respectively. Total phenolic content (TPC), total flavonoid content (TFC), and ascorbic acid levels were determined, and antioxidant activities were assessed using FRAP and DPPH assays. Distinct varietal differences were observed among the apricot cultivars. Sucrose, glucose, and fructose were identified as the dominant sugars contributing to the nutritional quality of the fruits. The phenolic composition was characterized by high levels of flavonoids and phenolic acids, which were strongly associated with antioxidant activity. Among the studied cultivars, Amasya (A-7) and Nakhchivan Adi Şalax (A-2) exhibited markedly higher total phenolic and flavonoid contents, as well as superior antioxidant capacities compared with other varieties. These findings demonstrate substantial biochemical variability among apricot cultivars and highlight A-7 and A-2 as promising candidates for functional food and nutraceutical applications due to their rich bioactive compound content and strong antioxidant potential.

Keywords: apricot; antioxidant; phenolic; sugar; Malatya; Nakhchivan

1. Introduction

Apricot (*Prunus armeniaca* L.), a member of the Rosaceae family, is a widely cultivated fruit of considerable nutritional and economic importance. Although native to Central Asia, apricot cultivation has expanded globally, demonstrating adaptability to diverse climates, particularly temperate regions [1,2]. Apricot trees are adapted to growing in regions characterized by hot summers, cool winters, and high sunshine hours; however, climate conditions play a decisive role in yield and fruit quality, particularly due to their high sensitivity to late spring frosts. Differences in temperature, light intensity, and growing environment can significantly affect the accumulation of sugars, phenolic compounds, and other nutritional components in apricot fruits. Apricots are a widely cultivated fruit in Europe, the Middle East, and the

Mediterranean basin. Today, Türkiye, Spain, Italy, Iran, China, and Azerbaijan are among the leading producing countries [3,4]. Among these countries, Türkiye holds a leading position globally, particularly in the production and export of dried apricots, thanks to its favorable agroclimatic conditions. The Malatya region, considered the center of apricot cultivation in Türkiye, plays a decisive role in the international market with its high-quality dried apricot production. The fruit is consumed fresh, dried, or processed into jams, juices, and oils, and is highly valued for its beneficial effects on intestinal health due to its rich fiber content and prebiotic sugars [5]. Owing to these nutritional attributes, along with its bioactive compound profile, apricot is also considered an important fruit with significant potential as a functional food. The diversity of apricot cultivars is notable, as they vary according to region and intended use. Prominent varieties include Hacıhaliloğlu, Şekerpare, Patterson, Moorpark, and Tilton. These cultivars exhibit varying attributes in terms of taste, texture, size, and durability, catering to diverse consumer preferences and industrial needs [1,3]. Malatya apricot is a globally renowned product with geographical indication; Malatya is often referred to as the "apricot capital of the world." The region predominantly cultivates varieties such as Soğancı, Kabaaşı, Hasanbey, Hacıhaliloğlu, Çataloğlu, and Hacıkız. These cultivars are recognized for their exceptional quality and distinct flavor profiles, contributing significantly to Malatya's esteemed position in the global apricot industry [5,6]. The Nakhchivan region is one of Azerbaijan's main apricot growing centers due to its favorable climate and long-standing fruit-growing tradition. The region, dominated by a continental climate, experiences hot, dry summers and cold winters, significantly affecting apricot development and fruit quality. Local apricot genotypes are reported to be rich in sugars and phenolic compounds thanks to high sunshine duration and soil characteristics. However, scientific data on the physicochemical and biochemical properties of apricot varieties grown in Nakhchivan are still limited [7].

In terms of nutritional value, apricots are a low-calorie, high-fiber fruit that is highly beneficial for those following calorie-controlled diets. A 100 g serving of apricots provides approximately 45–50 kcal of energy and contains about 2 g of dietary fiber. This fiber content is reported to contribute to regulating energy intake and reducing the risk of overeating by increasing feelings of fullness [8,9]. Additionally, apricots are rich in essential nutrients such as potassium and vitamins A and C, which support overall health and metabolic function. Dried apricots maintain a similar nutritional profile, offering high levels of fiber, potassium, and phenolic compounds. Fresh and dried apricots are rich sources of vitamins A and C, potassium, iron, and dietary fiber. These nutrients contribute significantly to cardiovascular health and support the optimal functioning of the digestive and circulatory system [3,9].

Apricot is an economically and nutritionally important fruit, and, in Türkiye, Malatya and, in Azerbaijan, Nakhchivan are among the main apricot growing centers with long-standing production traditions. While the biochemical characteristics of Malatya apricot varieties have been studied in detail, physicochemical and nutritional data on apricots grown in Nakhchivan are limited. Furthermore, there are very few studies comparing these two neighboring production regions. This study aims to address this knowledge gap by comparatively evaluating the physical properties, sugar and phenolic compositions, and antioxidant capacities of apricot varieties from both regions [3,5,8].

2. Results and Discussion

2.1. Physicochemical Properties of Apricot Extracts

This study investigated the physical, chemical, and nutritional attributes of several apricot cultivars collected from geographically proximal regions located within two neighboring countries. The physicochemical properties of the fruits, including their mass, dimensional characteristics, and color parameters, are summarized in Table 1. The results

 https://doi.org/10.3390/molecules31060988

indicated that the mean fruit weight varied between 42.03 and 74.10 g, while the average fruit length ranged from 41.33 to 57.00 mm. Furthermore, the mean seed weight of the examined cultivars was determined to be within the range of 2.23 to 3.39 g.

Table 1. Physicochemical characteristics of the apricot fruits.

Apricot Fruits	A-1	A-2	A-3	A-4	A-5	A-6	A-7
Weight (g)	74.10 ± 6.65 ^{d}	58.97 ± 1.80 $^{b.c}$	60.33 ± 2.40 $^{b.c}$	42.03 ± 3.38 ^{a}	55.34 ± 4.26 $^{a.b}$	42.56 ± 2.50 ^{a}	70.40 ± 8.42 $^{c.d}$
Seedless weight (g)	72.96 ± 5.00 ^{c}	54.89 ± 6.30 ^{b}	55.44 ± 2.28 ^{b}	40.21 ± 1.95 ^{a}	49.96 ± 3.37 $^{a.b}$	40.58 ± 0.94 ^{a}	71.56 ± 3.53 ^{c}
Length (mm)	57.00 ± 6.25 ^{c}	50.67 ± 2.52 $^{b.c}$	53.00 ± 2.65 $^{b.c}$	41.33 ± 2.52 ^{a}	51.33 ± 2.52 $^{b.c}$	47.00 ± 1.00 $^{a.b}$	51.67 ± 5.46 $^{b.c}$
Width (mm)	50.33 ± 3.22 $^{a.c}$	45.67 ± 0.58 $^{a.b.c}$	48.00 ± 2.00 $^{b.c}$	42.67 ± 1.53 $^{a.b}$	44.00 ± 2.00 $^{a.b.c}$	41.00 ± 2.65 ^{a}	50.33 ± 3.51 $^{a.c}$
Seed weight (g)	2.81 ± 0.30 $^{a.b}$	2.23 ± 0.18 ^{a}	2.74 ± 0.24 ^{a}	3.39 ± 0.17 ^{b}	2.39 ± 0.22 ^{a}	2.84 ± 0.10 $^{a.b}$	2.76 ± 0.31 ^{a}
Soluble solid %	18.52 ± 0.20 ^{c}	19.80 ± 0.35 ^{d}	19.44 ± 0.23 $^{c.d}$	18.72 ± 0.24 ^{c}	13.47 ± 0.20 ^{a}	20.86 ± 0.77 ^{e}	16.15 ± 0.31 ^{b}
Titratable acidity (Malic acid%)	0.41 ± 0.03 ^{b}	0.46 ± 0.02 ^{b}	0.55 ± 0.02 ^{c}	0.43 ± 0.02 ^{b}	0.46 ± 0.02 ^{b}	0.27 ± 0.02 ^{a}	0.90 ± 0.07 ^{d}
(CIE Lab*) L	67.82 ± 3.86 $^{a.b}$	72.82 ± 1.27 $^{b.c}$	72.04 ± 3.24 $^{b.c}$	72.58 ± 1.52 $^{b.c}$	75.47 ± 1.28 ^{c}	71.15 ± 2.32 $^{b.c}$	61.82 ± 3.03 ^{a}
(CIE Lab*) a	23.68 ± 2.03 $^{b.c}$	21.78 ± 0.88 ^{b}	22.04 ± 2.41 ^{b}	13.83 ± 1.66 ^{a}	21.00 ± 1.23 ^{b}	19.65 ± 2.20 ^{b}	26.91 ± 0.80 ^{c}
(CIE Lab*) b	47.76 ± 2.30 ^{a}	50.27 ± 3.09 ^{a}	47.66 ± 3.60 ^{a}	52.58 ± 1.70 ^{a}	47.96 ± 2.49 ^{a}	49.52 ± 3.84 ^{a}	45.63 ± 3.84 ^{a}

The values in the table are given as average values. Letters in each cell indicate differences at $p < 0.05$ (one-way ANOVA test) level.

Minor variations in length, weight, and width were observed among the apricot samples. Although apricot fruit size is influenced by species, deriving definitive morphological characteristics from these measurements alone is not feasible, as they are also impacted by factors such as tree age, cultivation practices, and pruning and thinning techniques [1].

Nevertheless, it can be concluded that the A-1 and A-7 apricot cultivars are characterized by larger fruit sizes. Indeed, a study on Malatya apricots reported that the Hasanbey variety exhibited the highest average fruit weight compared to other cultivars. In the same study, the average weight of apricots from the Iğdır region was recorded as 32.33 ± 1.47 g. When compared with the values measured in the present study, significant differences were observed among the apricot cultivars [3]. A study conducted on 27 apricot genotypes from the Budapest region reported that fruit weights ranged from 22 to 78 g [8]. Similarly, variability was observed in kernel weight, with the A-4 apricot cultivar exhibiting the highest kernel weight, while the A-6 cultivar had the lowest. Apricot kernels, which are the seeds inside the hard pits of apricots, are highly valued for their nutritional and functional properties. Rich in oils, proteins, and bioactive compounds, they are used in various industries, including food, cosmetics, and traditional medicine. However, their consumption should be monitored due to the presence of amygdalin, which can release toxic cyanide when consumed in excessive amounts [10–13]. Dried apricot kernels, traditionally consumed as a snack after being sun-dried, are a popular food product in many cultures.

The soluble solid (SS) content of the apricot samples was found to range from 13.47% to 20.86%. The SS in apricots refers to the total amount of dissolved substances in the fruit, including sugars, organic acids, vitamins, and other soluble compounds. Soluble solid is also a crucial parameter for determining the sweetness and ripeness of apricots and is commonly used as an indicator of the fruit's quality, influencing consumer preferences. Among the studied varieties, the highest SS value was found in the Hacıoğlu variety from the Malatya region (A-6), while the lowest value was observed in the Iğdır region apricot (A-5). No significant differences were observed among the apricot varieties from the Nakhchivan region apricots (from A-1 to A-4).

https://doi.org/10.3390/molecules31060988

The external skin color of the fruits was quantified using CIE Lab color coordinates, revealing a color spectrum ranging from yellow to orange. The L* value of color ranged from 61 to 75, with no significant differences observed among the varieties, except for the A-7 variety. Although all apricot cultivars exhibit a light skin color, a comparison indicates that the A-7 cultivar differs significantly from the others. The CIE a* value (−a/+a) reflects the greenness and redness of the fruit, with our findings ranging from 13 to 26. Upon evaluation, it can be concluded that the A-4 cultivar shows the strongest greenish tone, while the A-7 cultivar reveals the most pronounced reddish hue. The b value (−b/+b) indicates the blueness and yellowness, with measurements ranging from 44 to 52. Very slight differences were observed between these color values. In a study conducted on 10 different apricot varieties, the L value ranged from 49 to 61, the a value varied between 9 and 15, and the b value ranged from 39 to 47. The CIE Lab* color values of the fruits were found to exhibit similar characteristics to those of apricots from the Kayseri region [14]. In this study, the sensory quality attributes of apricot samples were evaluated using easily measurable physical parameters, including external skin color, fruit weight, and size, together with sweetness-related indicators. However, comprehensive sensory attributes such as taste and aroma, as well as textural properties including fruit firmness, were not included within the scope of this study. These limitations are acknowledged, and the incorporation of detailed sensory analysis and texture-related parameters is recommended for future studies to achieve a more complete assessment of apricot sensory quality. Skin color is a key quality indicator closely linked to fruit maturity, consumer acceptance, and the accumulation of bioactive compounds.

The acidity of apricot samples was quantified through titratable acidity, expressed as malic acid equivalent. Acidity is a key quality attribute in fruits, as it plays an important role in sensory perception, fruit maturity, product stability, and the bioavailability of bioactive compounds [14,15]. The results demonstrated a wide range of titratable acidity (TA) values, spanning from 0.27% to 0.90%. The lowest acidity was recorded in the Malatya (A-6), while the highest was observed in the Amasya (A-7). Among the Nakhchivan varieties, no significant differences in TA were detected. Comparable findings were reported in a study on apricots from the Pakistan region, where titratable acidity ranged between 0.31% and 0.34% as malic acid [3]. In a study conducted on different apricots using similar analytical methods, a variation ranging from 0.90% to 2.28% was reported [15]. In a study, titratable acidity was calculated in terms of citric acid, and it was reported that the Malatya apricot variety showed the lowest TA value among the samples [3]. Similarly, in our study, the Malatya variety was found to have a notably low TA value, consistent with previous findings.

This indicates consistency across different geographical regions and highlights the variability of TA levels due to genetic and environmental factors. Titratable acidity is a critical quality parameter for apricots as it significantly influences the fruit's sensory properties, including taste and aroma. Higher TA levels often result in a tangy flavor, while lower levels are associated with a sweeter taste profile. Furthermore, TA is widely used as an indicator of fruit maturity and ripeness, providing insights into the optimal harvesting period. Organic acids, such as ascorbic acid, citric acid, malic acid, succinic, oxalic, and phenolic acids collectively define the total acidity of fruits [16–20]. These compounds play an integral role in shaping the flavor, aroma, and overall quality of apricots, underscoring their significance in both consumer preferences and commercial applications [3,4].

2.2. *Ascorbic Acid and Sugar Profiles of Apricot Extracts*

Ascorbic acid is one of the major organic acids present in fruits, alongside citric acid, and phenolic acids, which are among the key organic compounds contributing to the fruit's chemical composition and quality. In this study, the analysis focused on ascorbic acid

 https://doi.org/10.3390/molecules31060988

and specific phenolic acids (Figure 1). The ascorbic acid content was quantified using a titrimetric method, ensuring precise and reliable measurement. This approach allowed for the simultaneous evaluation of key organic compounds that contribute to the antioxidant properties and overall quality of the fruit. The ascorbic acid content in the apricot samples was found to range from 2.82 to 9.31 mg/100 g. Among the varieties analyzed, the A-1 and A-3 varieties exhibited the highest ascorbic acid concentrations. Studies on apricots have reported that the major organic acids are citric acid and malic acid, while the ascorbic acid content is generally lower than that of these two organic acids [3,4]. In Hungary, apricot samples from different genotypes had ascorbic acid levels ranging from 3.04 to 16.17 mg/100 g FW [11].

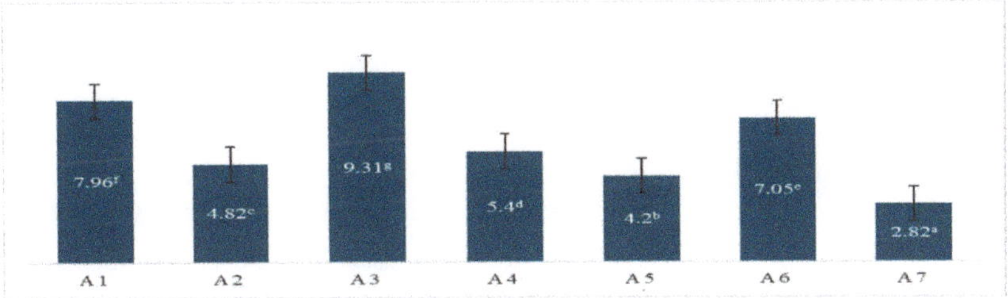

Figure 1. Ascorbic acid contents of the apricot samples (mg/100 g). Different letters (a–g) in the same lines are significantly different at the 5% level ($p < 0.05$).

The values in the table are given as average values. Letters in each cell indicate differences at the $p < 0.05$ (one-way ANOVA test) level.

The sugar profile was determined by HPLC-RID through the quantification of eight individual sugars, and the corresponding results are presented in Table 2. Sucrose emerged as the predominant sugar in the fruits, followed by glucose, melibiose, fructose, and trehalose. Notably, ribose, maltose and melezitose were not detected in any fruits.

Table 2. Sugar profiles and sweetness index (SI) of the apricot fruit samples.

(g/100 g FW)	A-1	A-2	A-3	A-4	A-5	A-6	A-7
Ribose	-	-	-	-	-	-	-
Fructose	0.67 ± 0.02^{c}	0.66 ± 0.03^{c}	1.08 ± 0.06^{d}	0.03 ± 0.01^{a}	-	0.26 ± 0.03^{b}	-
Glucose	4.34 ± 0.03^{c}	5.39 ± 0.04^{e}	4.29 ± 0.03^{c}	4.12 ± 0.04^{b}	4.75 ± 0.03^{d}	8.77 ± 0.04^{f}	1.10 ± 0.01^{a}
Sucrose	12.81 ± 0.19^{b}	12.44 ± 0.21^{b}	12.17 ± 0.54^{b}	19.30 ± 0.27^{c}	8.48 ± 0.32^{a}	9.31 ± 0.21^{a}	12.66 ± 0.48^{b}
Maltose	-	-	-	-	-	-	-
Trehalose	$0.14 \pm 0.02^{a,c}$	$0.03 \pm 0.03^{a,b}$	$0.07 \pm 0.02^{a,b,c}$	$0.10 \pm 0.04^{b,c}$	0.01 ± 0.01^{a}	0.48 ± 0.05^{d}	$0.03 \pm 0.01^{a,b}$
Melibiose	1.00 ± 0.05^{c}	0.92 ± 0.04^{c}	1.71 ± 0.06^{e}	0.54 ± 0.05^{a}	3.05 ± 0.06^{f}	0.73 ± 0.06^{b}	1.30 ± 0.07^{d}
Melezitose	-	-	-	-	-	-	-
Total sugar	18.99 ± 0.03^{c}	19.45 ± 0.03^{d}	19.31 ± 0.14^{d}	23.91 ± 0.04^{e}	16.36 ± 0.05^{b}	19.51 ± 0.04^{d}	14.99 ± 0.13^{a}
Sweetness index	27.99 ± 0.13^{c}	29.88 ± 0.08^{d}	27.37 ± 0.79^{c}	35.32 ± 0.18^{f}	22.46 ± 0.15^{b}	32.95 ± 0.10^{e}	19.49 ± 0.74^{a}

(-) not detected. The values in the table are given as average values. Letters in each cell indicate differences at $p < 0.05$ (one-way ANOVA test) level.

Fructose, commonly referred to as fruit sugar, is a primary sugar naturally found in most fruits. However, unlike many other fruits such as figs [20], grapes [21], dates [22] and apples [22], apricots do not typically exhibit higher levels of fructose compared to sucrose

 https://doi.org/10.3390/molecules31060988

and glucose [16]. This deviation is a characteristic feature of apricots, distinguishing their sugar composition profile from that of other fruits. Similarly, other studies have reported that fructose levels in apricots are lower than those of sucrose and glucose. For instance, a study on Malatya apricots reported sucrose concentrations ranging from approximately 22.96 to 56.83 mg/100 g, glucose concentrations from 9.47 to 23.67 mg/100 g, and fructose concentrations from 6.34 to 15.68 mg/100 g [3]. A study comparing local and foreign apricot varieties reported that sugar concentrations follow the descending order of sucrose, glucose, fructose, and maltose. In this study, it was shown that sucrose exhibited the greatest variation in its concentration [18]. However, the different studies also highlighted the presence of significant amounts of sorbitol in apricots, with sorbitol concentrations varying between 4.26 and 8.16 mg/g depending on the apricot variety [17]. The higher content of melibiose compared to fructose in apricots is considered a notable finding. Melibiose, a disaccharide, is formed by the α-1,6-glycosidic bond between galactose and glucose molecules. This sugar, which is typically found in high concentrations in honeydew honey, cannot be directly metabolized by the human digestive system; however, it is fermented by gut microorganisms, providing a prebiotic effect [23,24]. Another disaccharide present in apricots in relatively low amounts is trehalose, with values ranging from 0.01 to 0.48 g/100 g. The highest concentration of this sugar was observed in Malatya apricots. Trehalose, like melibiose, is considered to be sugar with prebiotic properties. This disaccharide, formed by an α,α-1,1-glycosidic bond between two glucose molecules, cannot be directly metabolized in humans due to the absence of the enzyme trehalase. Instead, trehalose is fermented and broken down by gut microorganisms, a process that provides prebiotic effects and supports gut health [25–27].

The differences in sugar composition among the analyzed *P. armeniaca* varieties can be attributed to environmental and agroclimatic conditions affecting carbohydrate metabolism during fruit development and ripening. Temperature, sunshine, water availability, and soil characteristics influence photosynthetic activity and carbon distribution, determining sugar synthesis and accumulation [17,19]. And, also, high glucose and fructose levels in some varieties can be explained by increased solar radiation and high temperatures enhancing photosynthesis and sucrose hydrolysis, while a lower sugar content may be attributed to restricted carbohydrate transport due to cool climate conditions, limited light, or water stress [22,23].

Different types of sugars are important for health because they affect glycemic response, energy metabolism, and nutritional quality. In fruits, the ratios of glucose, fructose, and sucrose determine taste perception and blood sugar response, while their presence along with fiber and bioactive compounds contributes to their potential as functional foods [24]. Fructose, a fruit sugar, is one of the sweetest sugars. Although the fructose content is relatively low in apricot fruits, it appears that the fruit's sweetness primarily results from other sugars. When calculating the sweetness index (SI) of the fruit, based on the combined function of fructose, glucose, and sucrose, the highest SI was found in A-4 and A-6 apricots. It was determined that the apricot with the lowest sweetness value was the A-7 variety. While organic acids (such as citric acid, ascorbic acid, malic acid) and organic alcohols (such as sorbitol) in the fruit give the fruit its sourness, fructose, sucrose and some oligosaccharides give it its sweetness [1,17]. Indeed, when the relationship between acidity and sweetness index was examined using Pearson correlation, a negative correlation was found between them ($r = -0.712$, $p < 0.01$). A similar relationship has also been demonstrated in other studies [17,19]. The sugar composition determines the SI, which is a measure of the sweetness value of apricot fruit. The relationship between the measured sugar values and SI indices is provided in Table 3, using Pearson correlation.

https://doi.org/10.3390/molecules31060988

Table 3. Pearson correlation coefficients among individual sugar components of the apricot fruits.

	Fru	Glu	Suc	Tre	Mel	TS	SI
Fru	1	0.150	−0.074	−0.038	−0.111	0.152	0.162
Glu		1	−0.373	0.777 **	−0.183	0.358	0.634 **
Suc			1	−0.237	−0.590 **	0.703 **	0.475 *
Tre				1	−0.449 *	0.240	0.519 *
Mel					1	−0.599 **	−0.668 **
TS						1	0.932 **
SI							1

**: $p < 0.01$, *: $p < 0.05$, Fru: fructose, Glu: glucose, Suc: sucrose, Tre: trehalose, Mel: melibiose, TS: Total Sugar, SI: sweetness index.

2.3. Antioxidant Activity of Apricot Extracts

The total phenolic content (TPC) and antioxidant capacities of methanolic apricot extracts are given in Table 4. The TPC values were found to range widely, from 10.04 mg GAE/100 g FW to 85.88 mg GAE/100 g FW. Similarly, the total flavonoid content varied between 0.21 mg QE/100 g FW and 3.36 mg QE/100 g FW. The highest total phenolic content was observed in the A-2 and A-7 varieties, whereas it was lower in the A-3 and A-5 apricot varieties. Similarly, the total flavonoid content was also found to be highest in A-2 and A-7, consistent with the trends observed for total phenolic content.

Table 4. Total phenolic and flavonoid content, and antioxidant capacities of the apricot fruits.

	A-1	A-2	A-3	A-4	A-5	A-6	A-7
Total Phenolic Content (mg GAE/100 g FW)	25.48 ± 0.69 ^{c}	85.88 ± 3.96 ^{d}	10.28 ± 2.26 ^{a}	27.83 ± 1.68 ^{c}	10.04 ± 0.34 ^{a}	16.59 ± 0.33 ^{b}	85.32 ± 1.38 ^{d}
Total Flavonoid Content (mg QE/100 g FW)	1.75 ± 0.08 ^{e}	3.36 ± 0.13 ^{f}	1.10 ± 0.03 ^{c}	0.64 ± 0.10 ^{b}	0.21 ± 0.02 ^{a}	1.37 ± 0.08 ^{d}	1.97 ± 0.13 ^{e}
FRAP (μmol $FeSO_4.7H_2O$/100 g FW)	19.53 ± 0.10 ^{d}	77.91 ± 1.14 ^{e}	11.29 ± 0.65 ^{b}	10.12 ± 0.97 ^{a,b}	9.08 ± 0.17 ^{a}	16.35 ± 0.30 ^{c}	81.84 ± 0.28 ^{f}
$DPPH^{\bullet}$ radical scavenging activity SC_{50} (mg/mL)	25.06 ± 0.53 ^{b}	7.18 ± 0.37 ^{a}	56.28 ± 1.58 ^{d}	74.45 ± 1.91 ^{e}	77.02 ± 2.00 ^{e}	34.72 ± 1.74 ^{c}	6.32 ± 0.19 ^{a}

Different letters (a–f) in the same lines are significantly different at the 5% level ($p < 0.05$).

Table 5 presents some phenolic properties reported in the literature for different apricot varieties. Previous studies have shown that apricots are rich in various phenolic compounds, primarily phenolic acids and flavonoids, but the quantity and distribution of these compounds vary significantly depending on the variety [25]. These differences are reported to stem from factors such as genetic structure, growing conditions, and maturity level. In particular, compounds such as chlorogenic acid, caffeic acid, and rutin stand out as dominant phenolics in apricots and make a significant contribution to their antioxidant potential. A study of 27 apricot genotypes reported that total phenolic content ranged from 82.0 to 2891.6 mg GAE/L and 168.5 to 3499.6 mg GAE/L in 2006 and 2007, respectively [8]. Another apricot study reported that total phenolic amounts ranged from 29 to 69 mg GA/100 g. These findings demonstrate that apricot represents a nutritionally valuable fruit with significant potential health benefits [28].

The antioxidant activity of the methanolic extracts was assessed using two distinct methods (Table 4). The Ferric Reducing Antioxidant Power (FRAP) assay, a widely recognized and straightforward method, measures antioxidant capacity based on the reduction of the Fe^{3+}- TPTZ complex to its ferrous form (Fe^{2+}) in the presence of antioxidants. Higher

 https://doi.org/10.3390/molecules31060988

FRAP values correspond to a greater antioxidant potential. The FRAP values of the samples exhibited a wide range, from 9.08 mg TE/100 g FW to 81.84 mg TE/100 g FW, with the A-7 and A-2 samples demonstrating the highest antioxidant capacity. Similarly, the DPPH• radical scavenging activity assay, which evaluates the ability of antioxidants to neutralize free radicals, revealed a significant relationship between SC_{50} values and radical inhibition efficacy. Lower SC_{50} values indicate a stronger radical scavenging effect. Among the samples analyzed, A-7 and A-2 displayed the highest radical scavenging activity. A robust negative correlation was observed between the FRAP values and DPPH• radical scavenging activities ($r = -0.832$, $p < 0.05$), underscoring the consistency of these methods in assessing antioxidant potential. The DPPH assay was chosen as a widely used and well-established method for assessing the free radical scavenging capacity of natural products. Its simplicity and reproducibility allow reliable comparison with previous studies on fruit- and plant-derived extracts. Together with ferric reducing power, the DPPH assay was considered adequate to represent the antioxidant potential of the apricot extracts within the scope of this study [29,30].

Table 5. Biochemical properties of selected apricot cultivars reported in the literature.

References	Total Phenolic Content (TPC)	Total Flavonoid Content (TFC)	Total Antioxidant Capacity (FRAP)	Major Phenolics	Cited
32 cultivars of different origin	63.5–1277.30 mg GAE/100 g DW	0–153.10 mg CE/100 g DW	483.4–2348.40 mg TE/100 g DW	-	[31]
11 apricot genotypes	24.87–41.31 (GAE/g)	-	-	-	[32]
14 apricot genotypes	92.20–162.10 (mg GAE/100 g)	-	154.10–243.60 FeSO4.7H_2O µg/mL	-	[33]
	41–170 mg GAE/100 g FW	-	12.00–102.90 mg TE/100 g FW	chlorogenic acid, salicylic acid, quercetin, rutin	[34]
27 apricot cultivars and hybrids of diverse origins	82.0–2891.60 mg GA/L	-	0.48 to 14 mmol AA/L	-	[11]
10 apricot genotypes	69.3–80.80 mg GAE/100 g	9.20–15.10 mg CE/100 g	-	-	[13]
14 apricot varieties	-	24.2–72.50 mg/100 g	12.2–36.10 mg/100 g	*m*-coumaric acid, chlorogenic acid	[35]

Antioxidant activity was evaluated to determine the capacity of the fruit to counteract oxidative stress through its bioactive compounds. This parameter is particularly important for functional foods, as it reflects potential health-promoting effects beyond basic nutrition and supports the biological relevance of the obtained results. The antioxidant capacity of apricots is primarily attributed to their content of phenolic compounds and vitamins, including ascorbic acid and β-carotene [3,32]. The strong correlations observed between total phenolic content (TPC), total flavonoid content (TFC) and the results of FRAP and DPPH• assays further substantiate this relationship (Table 5). These findings highlight the significant contribution of phenolic substances and vitamins to the overall antioxidant activity of apricots. A study examining 27 distinct apricot species identified genotypes as the primary determinant of antioxidant capacity. Researchers have identified that genotypes are the most significant determining factor for variations in antioxidant capacity, although the harvest year also exerts a notable effect on total phenolic content [11,32].

 https://doi.org/10.3390/molecules31060988

2.4. Phenolic Compounds of Apricot Extracts

Chromatographic profiles of phenolic components in apricot methanol extracts were determined using the RP-HPLC-PDA method, and the results of analyses performed according to 25 phenolic standards are presented in Table 6. The findings revealed that all apricot varieties examined generally exhibited similar profiles in terms of phenolic composition, but there were significant differences between varieties in the concentrations of individual phenolic compounds. The phenolic standards used for calibration were selected to represent the major polyphenolic compounds commonly reported in natural products. Care was taken to include compounds that showed good chromatographic separation and no peak overlap or interference in HPLC-PDA analysis, thereby ensuring reliable and reproducible quantification [36]. Among the phenolic standards used in the study, only four compounds were consistently detected in all apricot samples: chrysanthemum, pinosembrin, galangin, and *t*-cinnamic acid. Other phenolic components were not evaluated as they fell below the LOD and LOQ values (Table S1, Figure S1). The common presence of these compounds in all samples suggests that this quadruple phenolic compound pattern could be considered a distinctive phenolic profile or a potential "phenolic fingerprint" for apricot samples.

Table 6. Phenolics profiles of the apricot samples from 25 phenolic compounds from HPLC-PDA.

	A-1	A-2	A-3	A-4	A-5	A-6	A-7
	Phenolic acids (μg/100 g FW)						
t-Cinnamic acid	48.60	56.60	19.80	13.80	42.50	13.50	5.50
	Flavonoids (μg/100 g FW)						
Chyrisin	68.10	61.20	39.90	52.10	34.80	46.4	49.2
Pinocembrin	67.20	46.60	38.70	43.90	28.4	38.2	35.0
Galangin	64.50	54.20	38.30	44.70	35.50	40.80	50.80

Gallic acid, protocatechuic acid, p-hydroxybenzoic acid, chlorogenic acid, caffeic acid, syringic acid, vanillic acid, p-coumaric acid, ferulic acid, ellagic acid, epicatechin, rutin, myricetin, daidzein, luteolin, quercetin, resveratrol, apigenin, hesperetin, rhamnetin, and CAPE were all below the limit of detection (LOD) in the analyzed apricot samples.

A study conducted on domestic and foreign apricot varieties in Türkiye detected the presence of chlorogenic acid, caffeic acid, syringic acid, *p*-coumaric acid, ferulic acid, protocatechuic acid, vanillic acid, and gallic acid. However, in the current analysis, these phenolic compounds were found below detectable levels in the samples [19]. A study like ours has also demonstrated that cinnamic acid, chrysin, and pinocembrin are among the major phenolic compounds found in strawberries [29]. HPLC analysis of fourteen apricot varieties originating from North India revealed a phenolic profile that differs from the findings of the present study [28]. Similarly, another investigation conducted on local apricot cultivars reported chlorogenic acid, protocatechuic acid, caffeic acid, ferulic acid, and gallic acid as the predominant phenolic compounds [18]. These discrepancies in phenolic composition may be attributed to inter-cultivar variability as well as to methodological differences, particularly variations in the extraction procedures employed, which are known to significantly influence the qualitative and quantitative recovery of phenolic compounds. Phenolic compounds are of interest in functional foods due to their strong antioxidant, anti-inflammatory properties, as well as potential protective effects against oxidative stress-related diseases. Their presence enhances the health-promoting value of food beyond basic nutrition [29,37]. A total of 25 phenolic standards were screened using HPLC-PDA analysis to determine the phenolic composition; however, within the detection and quantification limits of the method, only chrysin, pinocembrin, galangin,

 https://doi.org/10.3390/molecules31060988

and cinnamic acid could be detected. The limited number of compounds in the phenolic profile may be related to the lower sensitivity of the HPLC-PDA method compared to more advanced analytical techniques with higher sensitivity and selectivity, such as LC-MS/MS or GC-MS. Therefore, the phenolic compounds identified in this study reflect those that could be determined under the applied analytical conditions.

2.5. Principal Component Analysis (PCA)

Before applying PCA, all variables were standardized using autoscaling (mean centering and division by standard deviation) to prevent bias that might arise from different units of measurement. The clustering observed in the PCA score graph was found to be directly related to differences in the biochemical profiles of the varieties, particularly changes in phenolic compound content and antioxidant-related parameters. Variance explanation ratios exceeding 70% are generally considered sufficient to ensure the reliability and interpretability of PCA models [38]. In the present study, the first two principal components (PC1 and PC2) together accounted for 86.059% of the total variance, with PC1 contributing 45.945% and PC2 contributing 40.115% (Figure 2). This high cumulative variance indicates that the multidimensional structure of the dataset is effectively captured within a two-component model.

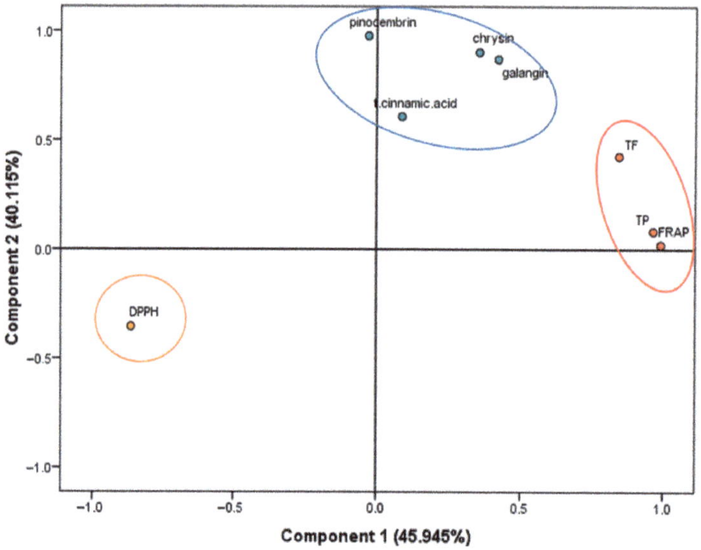

Figure 2. PCA score plot is based on the TPC, TFC, DPPH•, FRAP and phenolic acid parameters.

In accordance with Kaiser's criterion, which designates components with eigenvalues greater than 1.0 as meaningful descriptors of data structure [38,39], both PC1 (4.932) and PC2 (1.952) exceeded this threshold. The magnitude of these eigenvalues suggests that each component encapsulates a substantial portion of the dataset's informational content and contributes significantly to the overall discriminatory power of the model.

Collectively, these results confirm the robustness and statistical adequacy of the PCA. The combination of a high cumulative explained variance and eigenvalues surpassing the Kaiser threshold indicates that PC1 and PC2 provide a comprehensive and reliable representation of the underlying relationships among the variables. Thus, the principal components

 https://doi.org/10.3390/molecules31060988

derived in this analysis can be considered effective in summarizing the dominant patterns and driving sources of variability within the dataset.

3. Materials and Methods

3.1. Apricot Samples and Extraction

In June 2024, a minimum of 1 kg of ripe apricot samples was collected from gardens (Figure S3). The samples were placed in polyethylene bags and transported to the laboratory under cold-chain conditions. The average length and weight of each apricot variety were determined using measurements from at least 20 individual apricots. Fresh samples were stored at 4 °C until analysis, with a maximum storage period of five days. Depending on the specific analysis parameters, either aqueous or methanolic extraction methods were employed. Table 7 shows the codes, genotypes, and growing regions of the apricot varieties used in the study.

Table 7. Apricot (*P. armeniaca*) samples and their corresponding production regions.

A1	A2	A3	A4	A5	A6	A7
Nakhchivan Şeftali şalax	Nakhchivan Adi şalax	Nakhchivan Badami	Nakhchivan Balyarim	Türkiye Iğdır	Türkiye Malatya (Hacıoğlu)	Türkiye Amasya

Aqueous apricot extracts were prepared using distilled water as the solvent. Each apricot sample was homogenized with distilled water and subsequently filtered. The obtained filtration was then centrifuged at 1000 rpm for 15 min to remove particulate matter and obtain a clear extract. The methanolic apricot extract was prepared using 80% methanol as the extraction solvent. A total of 100 g of apricot mixture was homogenized with 250 mL of 80% methanol and subjected to ultrasonic-assisted extraction for 2 h, followed by agitation in a shaker for 6 h. The resulting extract was filtered and stored in aliquots at deep-freeze temperatures until further analysis [37].

3.2. Determination of Physicochemical Characteristics

The dry matter content was determined in accordance with AOAC protocols [10]; soluble solid content, expressed as a percentage (%), was measured in the juice of each sample using an Abbe-3 L refractometer (Tokyo, Japan) at 20 °C.

Following centrifugation of the aqueous apricot extract, the supernatant was filtered and subsequently diluted at a ratio of 1:20 with bidistilled water. An aliquot of the diluted filtrate was then titrated with 0.01 M NaOH using phenolphthalein as an indicator. The titratable acidity was calculated and expressed as percentage of malic acid equivalents, in accordance with the method described by Wani et al. [40]. The aqueous extracts were prepared for the determination of titratable acidity (TA). TA was analyzed by titrating 5 mL of juice with 0.1 M NaOH, and the results were reported as a percentage of malic acid, following the guidelines outlined in AOAC [10]. All analyses were performed in triplicate to ensure the reliability and reproducibility of the results. The color of fresh apricot samples was measured using the classical method for food analysis, based on the CIE Lab* color space model (Konica Minolta, CR-400 Plus, Tokyo, Japan). In this model: L* represents lightness, with values ranging from 0 (black) to 100 (white); a* denotes the color axis from green (−a) to red (+a); and b* represents the color axis from blue (−b) to yellow (+b).

https://doi.org/10.3390/molecules31060988

3.3. Determination of Ascorbic Acid Amount

The amount of ascorbic acid in aqueous solution was measured through classical titrimetric analysis based on the redox reaction between Potassium iodate (KIO_3) and ascorbic acid. It uses the principle of oxidation of ascorbic acid by potassium iodate, a strong reducing agent [41]. The experiment utilized the reaction of iodate ions with iodide ions in an acidic solution to generate iodine. For this purpose, KI (1 g), HCl (1 M, 5 mL), and starch solution were added to the sample, which had been diluted at a ratio of 1:50 to contain vitamin C (Sigma-Aldrich, Merck, Darmstadt, Germany). The mixture was then titrated with 5×10^{-5} M KIO_3. Calculations based on the reaction were performed to determine the ascorbic acid content, expressed as mg/100 g FW.

3.4. Determination of Sugar Profiles

For the sugar profile analysis, an Elite LaChrom Hitachi HPLC system equipped with a refractive index detector (RID) was utilized. The analysis employed eight sugar standards: ribose, fructose, glucose, sucrose, maltose, trehalose, melibiose, and melezitose. All analytical standards (≥98% purity) were purchased from Sigma-Aldrich, Merck (Darmstadt, Germany). Standard solutions at varying concentrations (40, 20, 10, 5, 2.5, and 1.25 mg/mL) were prepared, filtered through a 0.45 μm membrane, transferred to vials, and loaded onto the HPLC autosampler. Calibration curves were constructed from the resulting data prior to analysis.

For sample preparation, 1 g of homogenized apricot was mixed with 10 mL of 80% ethanol and filtered through filter paper, and the filtrate was evaporated to dryness. The residue was reconstituted in ultrapure water, filtered through a 0.45 μm membrane, and injected into the HPLC. A reverse-phase -NH_2 column (EC 250/4 Nucleosil, Macherey-Nagel, Düren, Germany) was employed for separation, using an isocratic mobile phase of 80% acetonitrile and 20% ultrapure water. The injection volume for both standards and samples were optimized at 20 μL, with a mobile phase flow rate of 1.5 mL/min and a column temperature of 30 °C [42].

3.5. Determination of Sweetness Index (SI)

The sweetness index (SI) of the fruit, which serves as an estimate of the overall sweetness perception, was calculated by considering the relative amounts and sweetness properties of individual carbohydrates [1]. The contribution of each carbohydrate to the overall sweetness was determined based on the relative sweetness values: fructose is 2.3 times sweeter than glucose; sucrose is 1.35 times sweeter; and sorbitol is 0.81 times sweeter than glucose. Consequently, the SI was calculated using Formula (1).

$$SI = (1.00 \times \text{glucose}) + (1.35 \times \text{sucrose}) + (2.3 \times \text{fructose}), \quad (1)$$

This method enables the quantification of sweetness based on the contribution of each sugar, providing a more comprehensive measure of the fruit's overall sweetness profile; however, since sorbitol was not included in the analysis, its contribution to the sweetness was not considered.

3.6. Determination of Total Phenolic Contents (TPC)

The total phenolic content was determined spectrophotometrically using the Folin–Ciocalteu method [43]. In this method, methanolic apricot extracts were used. A reaction mixture was prepared by combining 50 μL of the extract with 400 μL of 0.2 N Folin–Ciocalteu reagent, followed by dilution with 680 μL of distilled water. After a 3 min

https://doi.org/10.3390/molecules31060988

incubation at room temperature, 400 μL of 10% Na_2CO_3 was added, and the mixture was incubated at 25 °C for 2 h. The absorbance of the reaction was measured at 760 nm using a UV-Vis spectrophotometer (Thermo Scientific Evolution TM 201, Madison, WI, USA). A calibration curve was generated using six gallic acid standards ranging from 0.0625 to 1.00 mg GAE/100 mL. The total phenolic content was expressed as mg GAE/100 g FW, based on the standard curve.

3.7. Determination of Total Flavonoid Contents (TFC)

The total flavonoid content was determined in the methanolic extract using a modified method based on Fukumoto and Mazza [44]. In this modification, $AlNO_3$ was used in place of $AlCl_3$ due to its lower solubility in water. To begin, 50 μL of ethanolic extract was combined with 50 μL of 10% $Al(NO_3)_3$ and 50 μL of 1.0 M NH_4CH_3COO. The mixture was then diluted to a final volume of 3.0 mL with 99% ethanol and incubated at 25 °C for 45 min. After incubation, the absorbance was measured at 415 nm. A calibration curve was constructed using six different quercetin standards, ranging from 0.031 to 0.50 mg QE/100 mL. The total flavonoid content was expressed as mg QE/100 g FW, based on the standard curve.

3.8. Determination of Ferric Reducing Antioxidant Power (FRAP)

The total antioxidant activity of the methanolic extract was assessed using the Ferric-Reducing Antioxidant Power (FRAP) assay [45]. To prepare the FRAP reagent, ferric tripyridyl-triazine (Fe-III-TPTZ) (Sigma-Aldrich, Darmstadt, Germany), acetate buffer in 40 mM HCl, and a 2.5 mL solution of mM $FeCl_3.6H_2O$ were combined in a test tube. Subsequently, 3 mL of the FRAP reagent was mixed with 100 μL of the extract and incubated for 4 min at 37 °C. The absorbance was measured at 595 nm. A standard calibration curve was generated using different concentrations of $FeSO_4.7H_2O$ (ranging from 1000 to 31.25 μmol/mL). The results were reported as μmol $FeSO_4.7H_2O$/100 g FW.

3.9. Determination of DPPH• Radical Scavenging Capacity

The DPPH• radical scavenging activity was evaluated using a spectrophotometric method [46]. In this assay, 1.0 mL of 0.04 mg/mL DPPH• radical (Sigma-Aldrich, Darmstadt, Germany) solution was mixed with 1.0 mL of the extract. The mixture was then kept in the dark for 45 min at 25 °C, after which the absorbance was measured at 517 nm. To determine the SC_{50} value, six different dilutions of the ethanolic extract were prepared and treated with the DPPH• solution, followed by absorbance measurement at 517 nm. The SC_{50} value, representing the concentration of extract required to scavenge 50% of the DPPH• radical, was obtained from the resulting graph.

3.10. Determination of Phenolic Composition Using HPLC-PDA

Prior to the phenolic component analysis of methanolic extracts via HPLC–PDA, liquid–liquid extraction was conducted for enrichment. A 10 mL aliquot of the extract was evaporated at 40 °C using a rotary evaporator. The residue was dissolved in 10 mL of distilled water, and the pH was adjusted to 2 using concentrated HCl. The organic phases were combined after three successive extractions with diethyl ether and ethyl acetate. Following solvent removal, the residue was dissolved in 2 mL of methanol, filtered through a 0.45 μm RC membrane, and subjected to phenolic analysis.

The phenolic composition of the mixture was analyzed using an HPLC-PDA system equipped with a photodiode array (PDA) detector (Shimadzu Liquid Corporation LC 20AT, Kyoto, Japan) and a C18 column (250 mm × 4.6 mm, 5 μm; GL Sciences, Tokyo, Japan). A

https://doi.org/10.3390/molecules31060988

calibration curve was constructed using 25 phenolic standards. The mobile phase consisted of (A) 2% acetic acid in water and (B) a mixture of acetonitrile and water (70:30). Sample and standard injection volumes were set at 20 µL, with a column temperature of 30 °C and a flow rate of 1.0 mL/min [36]. In addition, analytical validation parameters for the phenolic compounds, including the limits of detection (LOD) and limits of quantification (LOQ), as well as linearity ranges and correlation coefficients, were thoroughly established and reported in our previous study [36].

3.11. Statistical Analysis

All experiments were conducted in triplicate, with results expressed as mean ± standard deviation. Data analysis was performed using SPSS software (version 21; IBM Corp., Armonk, NY, USA, 2012). One-way ANOVA test was used to find differences. The correlations between sugar components and antioxidant parameters were analyzed by Pearson correlation. Principal component analysis was also performed using antioxidants and phenolic components. $p < 0.01$ and $p < 0.05$ were used as statistical significance levels.

4. Conclusions

The present study comprehensively evaluated the physicochemical, biochemical, and nutritional characteristics of apricot genotypes cultivated in the Nakhchivan region of Azerbaijan and in Türkiye, revealing substantial inter-genotypic and inter-regional variability. Fruit weight, color parameters (Hunter L* and a*), soluble solid content, and titratable acidity (TA) exhibited wide ranges, reflecting significant differences in pomological and quality-related traits. According to the CIE L*a*b* color system, the apricot samples exhibited L* (lightness) values ranging from 61.82 to 72.58, a* (greenness) values from 13.83 to 26.91, and b* (yellowness) values from 45.63 to 52.58. Although variations were observed in the numerical color parameters, all samples were visually characterized by a light-yellow appearance, indicating a generally similar overall color profile despite measurable differences. The ascorbic acid concentrations in the analyzed samples were found to range from 2.82 to 9.31 mg/100 g, indicating a significant difference in content between the samples. In the analyzed samples, sucrose and glucose were identified as the dominant sugars, while fructose levels were found to be low. The total phenolic content (TPC) in the analyzed samples ranged from 10.04 to 85.88 mg GAE/g. The highest TPC values were detected in the Nakhchivan Adi shalax (A2) and Amasya (A7) samples, respectively. In addition, a positive correlation was observed between TPC and antioxidant capacity measurements (FRAP and DPPH), supporting the contribution of phenolic compounds to antioxidant activity. Apricot samples were found to contain varying concentrations of the same polyphenolic compounds, including chrysin, pinocembrin, galangin, and *trans*-cinnamic acid, indicating qualitative consistency but quantitative variability in their phenolic profiles.

Overall, the observed genotype–environment interactions appear to shape the biochemical profile of apricots, and certain cultivars demonstrated more favorable compositional characteristics, suggesting their potential suitability for functional food and nutraceutical applications. Nevertheless, further in vivo and clinical investigations are required to substantiate specific health-related claims.

Supplementary Materials: The following supporting information can be downloaded at: https://www.mdpi.com/article/10.3390/molecules31060988/s1.

Author Contributions: Conceptualization: S.K.; Methodology: S.K. and C.B.; All experiments: C.B.; Formal analysis: C.B. and E.A.; Data curation: C.B.; Writing—original draft preparation: C.B.;

 https://doi.org/10.3390/molecules31060988

Writing—review and editing: S.K. and A.K.-G.; Supervision: S.K. and Z.P.C. All authors have read and agreed to the published version of the manuscript.

Funding: This research received no external funding.

Institutional Review Board Statement: Not applicable.

Informed Consent Statement: Not applicable.

Data Availability Statement: Data will be made available upon reasonable request.

Conflicts of Interest: The authors declare no conflicts of interest.

Abbreviations

The following abbreviations are used in this manuscript:

ANNOVA	One-way analysis of variance
DPPH$^{\bullet}$	2,2-diphenyl-1-picrylhydrazyl
FRAP	Ferric reducing antioxidant power
FW	Fresh weight
GAE	Gallic acid equivalents
HPLC	High-pressure liquid chromatography
PDA	Photodiode array
QE	Quercetin equivalent
RID	Refractive index detector
SI	Sweetness index
SS	Soluble solid
TA	Titratable acidity
TFC	Total flavonoid content
TPC	Total phenolic content

References

1. Roussos, P.A.; Sefferou, V.; Denaxa, N.K.; Tsantili, E.; Stathis, V. Apricot (*Prunus armeniaca* L.) fruit quality attributes and phytochemicals under different crop load. *Sci. Hortic.* **2011**, *129*, 472–478. [CrossRef]
2. Zhebentyayeva, T.; Ledbetter, C.; Burgos, L.; Llácer, G. Apricot. In *Fruit Breeding*; Badenes, M.L., Byrne, D.H., Eds.; Springer: New York, NY, USA, 2012; pp. 415–458. [CrossRef]
3. Akin, E.B.; Karabulut, I.; Topcu, A. Some compositional properties of main Malatya apricot (*Prunus armeniaca* L.) varieties. *Food Chem.* **2008**, *107*, 939–948. [CrossRef]
4. Ali, S.; Masud, T.; Abbasi, K.S.; Mahmood, T.; Hussain, A. Apricot: Nutritional potentials and health benefits—A review. *Ann. Food Sci. Technol.* **2015**, *16*, 175–184.
5. Alpaslan, M.; Hayta, M. Apricot kernel: Physical and chemical properties. *J. Am. Oil Chem. Soc.* **2006**, *83*, 469–471. [CrossRef]
6. Ercisli, S. Apricot culture in Türkiye. *Sci. Res. Essays* **2009**, *4*, 715–719.
7. Rakida, A. Analysis of morphological and pomological features of apricot in the Nakhchivan Autonomous Republic of Azerbaijan. *Turk. J. Agric. For.* **2023**, *47*, 23–30. [CrossRef]
8. Reale, A.; Di Renzo, T.; Russo, A.; Niro, S.; Ottombrino, A.; Pellicano, M.P. Production of low-calorie apricot nectar sweetened with stevia: Impact on qualitative, sensory, and nutritional profiles. *Food Sci. Nutr.* **2020**, *8*, 1837–1847. [CrossRef]
9. Gupta, S.; Chhajed, M.; Arora, S.; Thakur, G.; Gupta, R. Medicinal Value of Apricot: A Review. *Indian J. Pharm. Sci.* **2018**, *80*, 790–794. [CrossRef]
10. Cunniff, P. *Official Methods of Analysis of AOAC International*; AOAC: Gaithersburg, MD, USA, 1995; Volume 1–2.
11. Hegedus, A.; Engel, R.; Abrankó, L.; Balogh, E.; Blázovics, A.; Hermán, R.; Halász, J.; Ercisli, S.; Pedryc, A.; Stefanovits-Bányai, É. Antioxidant and antiradical capacities in apricot (*Prunus armeniaca* L.) fruits: Variations from genotypes, years, and analytical methods. *J. Food Sci.* **2010**, *75*, C722–C730. [CrossRef] [PubMed]
12. Akhone, M.A.; Bains, A.; Tosif, M.M.; Chawla, P.; Fogarasi, M.; Fogarasi, S. Apricot kernel: Bioactivity, characterization, applications, and health attributes. *Foods* **2022**, *11*, 2184. [CrossRef] [PubMed]
13. Karatas, N. Evaluation of nutritional content in wild apricot fruits for sustainable apricot production. *Sustainability* **2022**, *14*, 1063. [CrossRef]

 https://doi.org/10.3390/molecules31060988

14. Yaman, M.; Turan, S. Determination of fruit and leaf characteristics of some apricot varieties in Kayseri ecology. *Erzincan Univ. J. Sci. Technol.* **2021**, *14*, 962–969. [CrossRef]
15. Lobit, P.; Soing, P.; Génard, M.; Habib, R. Theoretical analysis of relationships between composition, pH, and titratable acidity of peach fruit. *J. Plant Nutr.* **2002**, *25*, 2775–2792. [CrossRef]
16. Ayour, J.; Alahyane, A.; Harrak, H.; Neffa, M.; Taourirte, M.; Benichou, M. Assessment of nutritional, technological, and commercial apricot quality criteria of the Moroccan cultivar "Maoui" compared to introduced Spanish cultivars "Canino" and "Delpatriarca". *J. Food Qual.* **2021**, *2021*, 6679128. [CrossRef]
17. Su, C.; Zheng, X.; Zhang, D.; Chen, Y.; Xiao, J.; He, Y.; Wang, B.; Shi, X. Investigation of sugars, organic acids, phenolic compounds, antioxidant activity, and the aroma fingerprint of small white apricots grown in Xinjiang. *J. Food Sci.* **2020**, *85*, 4300–4311. [CrossRef] [PubMed]
18. Etienne, A.; Génard, M.; Lobit, P.; Mbeguié-A-Mbéguié, D.; Bugaud, C. What controls fleshy fruit acidity? A review of malate and citrate accumulation in fruit cells. *J. Exp. Bot.* **2013**, *64*, 1451–1469. [CrossRef] [PubMed]
19. Muradoglu, F.; Kayakeser, U. Multivariate analysis of Turkish and foreign apricot cultivars based on biochemical components. *Erwerbs-Obstbau* **2022**, *64*, 639–646. [CrossRef]
20. Slatnar, A.; Klancar, U.; Stampar, F.; Veberic, R. Effect of drying of figs (*Ficus carica* L.) on the contents of sugars, organic acids, and phenolic compounds. *J. Agric. Food Chem.* **2011**, *59*, 11696–11702. [CrossRef]
21. Eyduran, S.P.; Akin, M.; Ercisli, S.; Eyduran, E.; Maghradze, D. Sugars, organic acids, and phenolic compounds of ancient grape cultivars (*Vitis vinifera* L.) from Igdir province of Eastern Türkiye. *Biol. Res.* **2015**, *48*, 2. [CrossRef]
22. Assirey, E.A.R. Nutritional composition of fruit of ten date palm (*Phoenix dactylifera* L.) cultivars grown in Saudi Arabia. *J. Taibah Univ. Sci.* **2015**, *9*, 75–79. [CrossRef]
23. Soleymani, A.; Shahrajabian, M.H. Effects of planting dates and row distance on sugar content, root yield and solar radiation absorption in sugar beet at different plant densities. *Rom. Agric. Res* **2017**, *34*, 1–11.
24. Ryu, J.H.; Yim, J.E.; Suk, W.H.; Lee, H.; Ahn, H.; Kim, Y.S.; Choue, R. Sugar composition and glycemic indices of frequently consumed fruits in Korea. *Korean J. Nutr.* **2012**, *45*, 192–200. [CrossRef]
25. Zupan, A.; Mikulic-Petkovsek, M.; Stampar, F.; Veberic, R. Sugar and phenol content in apple with or without watercore. *J. Sci. Food Agric.* **2016**, *96*, 2845–2850. [CrossRef]
26. Seeburger, V.C.; D'Alvise, P.; Shaaban, B.; Schweikert, K.; Lohaus, G.; Schroeder, A.; Hasselmann, M. The trisaccharide melezitose impacts honey bees and their intestinal microbiota. *PLoS ONE* **2020**, *15*, e0230871. [CrossRef]
27. Recklies, K.; Peukert, C.; Kölling-Speer, I.; Speer, K. Differentiation of honeydew honeys from blossom honeys and according to their botanical origin by electrical conductivity and phenolic and sugar spectra. *J. Agric. Food Chem.* **2021**, *69*, 1329–1347. [CrossRef] [PubMed]
28. Saridaş, M.; Ağçam, E.; Paydaş Kargı, S. The nutritional composition of key apricot varieties cultivated in Türkiye with a focus on health-related compounds. *Int. J. Agric. Environ. Food Sci.* **2023**, *7*, 934–939. [CrossRef]
29. Dinç, S.; Kara, M.; Takma, Ç.; Kara, Y.; Kolaylı, S. Strawberries from Konya in the Central Anatolia Region of Türkiye: Phenolic profile, antioxidant potential, and mineral composition. *Appl. Fruit Sci.* **2024**, *66*, 1229–1240. [CrossRef]
30. Kolaylı, S.; Tunca, E.; Özkök, A. The Botanic, physicochemical, phenolic and antioxidant properties of Muş Basin honeys. *J. Apitherapy Nat.* **2025**, *8*, 175–192. [CrossRef]
31. Rampáčková, E.; Göttingerová, M.; Gála, P.; Kiss, T.; Ercişli, S.; Nečas, T. Evaluation of protein and antioxidant content in apricot kernels as a sustainable additional source of nutrition. *Sustainability* **2021**, *13*, 4742. [CrossRef]
32. Wani, S.M.; Masoodi, F.A.; Wani, T.A.; Ahmad, M.; Gani, A.; Ganai, S.A. Physical characteristics, mineral analysis and antioxidant properties of some apricot varieties grown in North India. *Cogent Food Agric.* **2015**, *1*, 1118961. [CrossRef]
33. Korekar, G.; Stobdan, T.; Arora, R.; Yadav, A.; Singh, S.B. Antioxidant capacity and phenolics content of apricot (*Prunus armeniaca* L.) kernel as a function of genotype. *Plant Foods Hum. Nutr.* **2011**, *66*, 376–383. [CrossRef] [PubMed]
34. Sochor, J.; Skutkova, H.; Babula, P.; Zitka, O.; Cernei, N.; Rop, O.; Kizek, R. Mathematical evaluation of the amino acid and polyphenol content and antioxidant activities of fruits from different apricot cultivars. *Molecules* **2011**, *16*, 7428–7457. [CrossRef]
35. Wani, S.M.; Hussain, P.R.; Masoodi, F.A.; Ahmad, M.; Wani, T.A.; Gani, A.; Rather, S.A.; Suradkar, P. Evaluation of the composition of bioactive compounds and antioxidant activity in fourteen apricot varieties of North India. *J. Agric. Sci.* **2017**, *9*, 66–78. [CrossRef]
36. Kara, Y.; Birinci, C. Usability of the phenolic profile analysis method developed in RP-HPLC-PDA in natural products. *J. Apitherapy Nat.* **2024**, *7*, 14–27. [CrossRef]
37. Kara, Y.; Aybar, M.; Kolaylı, S. Physicochemical, phenolic, and antioxidant profiles of *Vitis vinifera* L. and *Vitis labrusca* L. grapes from Artvin, Eastern Black Sea Region. *J. Food Sci.* **2025**, *90*, e70435. [CrossRef] [PubMed]

 https://doi.org/10.3390/molecules31060988

38. Larrigaudière, C.; Lentheric, I.; Puy, J.; Pintó, E. Biochemical characterisation of core browning and brown heart disorders in pear by multivariate analysis. *Postharvest Biol. Technol.* **2004**, *31*, 29–39. [CrossRef]
39. Kaiser, H.F. The application of electronic computers to factor analysis. *Educ. Psychol. Meas.* **1960**, *20*, 141–151. [CrossRef]
40. Wani, S.M.; Gull, A.; Ahad, T.; Malik, A.R.; Ganaie, T.A.; Masoodi, F.A.; Gani, A. Effect of gum Arabic, xanthan, and carrageenan coatings containing antimicrobial agent on postharvest quality of strawberry: Assessing physicochemical, enzyme activity, and bioactive properties. *Int. J. Biol. Macromol.* **2021**, *183*, 2100–2108. [CrossRef]
41. Silva, C.R.; Simoni, J.A.; Collins, C.H.; Volpe, P.L.O. Ascorbic acid as a standard for iodometric titrations: An analytical experiment for general chemistry. *J. Chem. Educ.* **1999**, *76*, 1421–1422. [CrossRef]
42. Cao, S.; Yang, Z.; Zheng, Y. Sugar metabolism in relation to chilling tolerance of loquat fruit. *Food Chem.* **2013**, *136*, 139–143. [CrossRef]
43. Slinkard, K.; Singleton, V.L. Total phenol analysis: Automation and comparison with manual methods. *Am. J. Enol. Vitic.* **1977**, *28*, 49–55. [CrossRef]
44. Fukumoto, L.R.; Mazza, G. Assessing antioxidant and prooxidant activities of phenolic compounds. *J. Agric. Food Chem.* **2000**, *48*, 3597–3604. [CrossRef]
45. Benzie, I.F.F.; Strain, J.J. The ferric reducing ability of plasma (FRAP) as a measure of "antioxidant power". *Anal. Biochem.* **1996**, *239*, 70–76. [CrossRef] [PubMed]
46. Molyneux, P. The use of the stable free radical diphenylpicrylhydrazyl (DPPH) for estimating antioxidant activity. *Songklanakarin J. Sci. Technol.* **2004**, *26*, 211–219.

 https://doi.org/10.3390/molecules31060988

molecules

MDPI

Article

Identification of Active Anti-Inflammatory Compounds in Sweet Potato Storage Roots Extracted with Ethanol via LC-Q-TOF-MS

Ryohei Sakuraoka, Hirofumi Masutomi * and Katsuyuki Ishihara

Research and Development Division, Calbee, Inc., Utsunomiya 321-3231, Tochigi, Japan; r_sakuraoka@calbee.co.jp (R.S.); k_ishihara@calbee.co.jp (K.I.)

* Correspondence: h_masutomi@calbee.co.jp; Tel.: +81-70-3872-7681

Abstract

Sweet potatoes (*Ipomoea batatas* (L.) Lam.) are known for their anti-inflammatory effects, which are attributed to their phytochemical content. Our previous study revealed that ethanol extracts of sweet potato storage roots (SP-EtOH-Ex) inhibit interleukin-6 (IL-6) production in RAW264.7 cells stimulated with lipopolysaccharide (LPS). However, the causative compounds responsible for the anti-inflammatory effect have not yet been identified. This study aims to identify the compounds responsible for the anti-inflammatory effect of SP-EtOH-Ex using liquid chromatography–quadrupole time-of-flight mass spectrometry (LC-Q-TOF-MS). The unknown compounds were measured using the auto MS/MS mode (data-dependent acquisition; DDA) of LC-Q-TOF-MS, and the resulting data were analyzed using MS-DIAL and MS-FINDER and also compared with those of the corresponding reference standards in terms of retention time and fragment ions. As a result, β-sitosterol (2.527–4.850 μg/mL), campesterol (75.74–93.63 ng/mL), and lauroyl diethanolamide (4.568–9.260 ng/mL) were identified and quantified in SP-EtOH-Ex. Moreover, the anti-inflammatory effect of these three compounds against RAW264.7 cells was investigated at varying concentrations of β-sitosterol (1 μg/mL, 5 μg/mL, 10 μg/mL), campesterol (10 ng/mL, 100 ng/mL, 1000 ng/mL), and lauroyl diethanolamide (1 ng/mL, 10 ng/mL, 100 ng/mL). The phytosterols β-sitosterol and campesterol suppressed LPS-induced IL-6 production at concentrations comparable to those present in SP-EtOH-Ex. In contrast, lauroyl diethanolamide did not similarly suppress LPS-induced IL-6 production. These results suggest that β-sitosterol and campesterol in sweet potato storage roots contribute to their anti-inflammatory effects. The lack of activity in lauroyl diethanolamide further supports that phytosterols are the primary anti-inflammatory constituents. The edible portion of sweet potatoes holds promise as a promising raw material with anti-inflammatory properties.

Keywords: anti-inflammatory effect; sweet potato; phytosterols; LC-Q-TOF-MS; MS-FINDER

1. Introduction

Sweet potato (*Ipomoea batatas* (L.) Lam.) is a perennial herbaceous species of the family Convolvulaceae. It exhibits strong drought tolerance and can produce high yields even in nutrient-poor soils. Although global production is concentrated in China and regions of Africa, its storage roots are consumed worldwide and are recognized as nutrient-dense foods rich in dietary fiber, vitamins, and minerals, thereby offering potential health benefits [1]. Sweet potatoes are also notable for their abundant contents of phytochemicals, including carotenoids and anthocyanins [2,3]. Accumulating evidence has highlighted their relevance in immune modulation [4–6]. For instance, anthocyanins derived from purple-fleshed sweet potatoes suppress lipopolysaccharide (LPS)-induced production of pro-inflammatory mediators, including tumor necrosis factor-α (TNFα), interleukin-6 (IL-6), and nitric oxide (NO), in RAW264.7 macrophages and attenuate the proliferation of several cancer cell lines [4]. Likewise, the β-carotene present in orange-fleshed sweet potatoes has been reported to inhibit LPS-induced IL-6, NO, and prostaglandin E_2 (PGE_2) production in the same cell model [5]. Our previous findings demonstrated that we fractionated the ethanol extract of sweet potato storage roots using reversed-phase chromatography and found that the fraction eluting between 51 and 60 min significantly suppressed LPS-induced IL-6 production in RAW264.7 macrophages, whereas the other fractions showed little or no activity [7]. Based on these findings, the present study focused specifically on this bioactive fraction, rather than the entire extract, to identify the constituents responsible for its anti-inflammatory effects. For clarity, this fraction is referred to as SP-EtOH-Ex throughout the manuscript. Mass spectrometry has become indispensable in chemical biology research—including metabolomics, biosynthetic pathway analysis, foodomics, and exposomics—and remains one of the most powerful analytical platforms for characterizing low-molecular-weight natural products [8–11]. Numerous studies have proposed workflows for the structural elucidation of unknown compounds using high-resolution mass spectrometry [12–15]. In plant metabolite profiling, such efforts have predominantly focused on phenolic compounds, and many earlier studies relied heavily on manual interpretation of spectra with reference to the literature or spectral databases [16–19]. More recently, data-processing platforms such as MS-DIAL (RIKEN Center for Sustainable Resource Science, Kanagawa, Japan) have been developed to perform peak detection, deconvolution, and preliminary annotation of features from non-targeted LC–MS datasets in a systematic and reproducible manner. Annotation of unknown features typically involves matching experimentally acquired product-ion spectra with curated reference data. A range of computational tools—such as MS-FINDER, CSI:FingerID, CFM-ID, and MetFrag—enable in silico fragmentation and structural prediction, and their combined application has been shown to substantially improve identification accuracy [20–24].

The present study aimed to identify constituents within the ethanol extract of sweet potato storage roots that contribute to their anti-inflammatory effects using an integrated mass spectrometry-based approach.

2. Results

2.1. Annotation of Nine Compounds from the Ethanol Extract of Sweet Potato Storage Roots

All sweet potato samples used for SP-EtOH-Ex preparation were of the cultivar Beniharuka. Processing the raw LC-MS data of ethanol extracts of sweet potato storage roots (SP-EtOH-Ex) using MS-DIAL yielded a total of 18,863 detected peaks. Subsequent annotation by MS-FINDER (RIKEN Center for Sustainable Resource Science, Kanagawa, Japan) resulted in 7439 putative compounds. Among these, database filtering using FooDB and KNApSAcK narrowed the annotation candidates to gibberellin A3, cinnamic acid, N-hexadecanoylpyrrolidine, lauroyl diethanolamide, β-sitosterol, campesterol, coniferyl aldehyde, mellein, and cycloartenol.

Ranking of these candidates using MS-FINDER and MassBank further refined their likelihood. In MS-FINDER, β-sitosterol, campesterol, coniferyl aldehyde, gibberellin A3, and cinnamic acid were ranked first; lauroyl diethanolamide and N-hexadecanoylpyrrolidine ranked between first and second; mellein ranked between second and fortieth; and cycloartenol ranked between third and eighth (Table S3). In MassBank,

 https://doi.org/10.3390/molecules310304567

lauroyl diethanolamide, β-sitosterol, and campesterol ranked 1st, while coniferyl aldehyde ranked between 39th and 41st; the remaining compounds were not ranked (Table S3). Although some compounds did not rank in MassBank, their high MS-FINDER ranks supported the inclusion of these nine compounds as candidates potentially contributing to the anti-inflammatory activity of SP-EtOH-Ex (Figures 1 and S1).

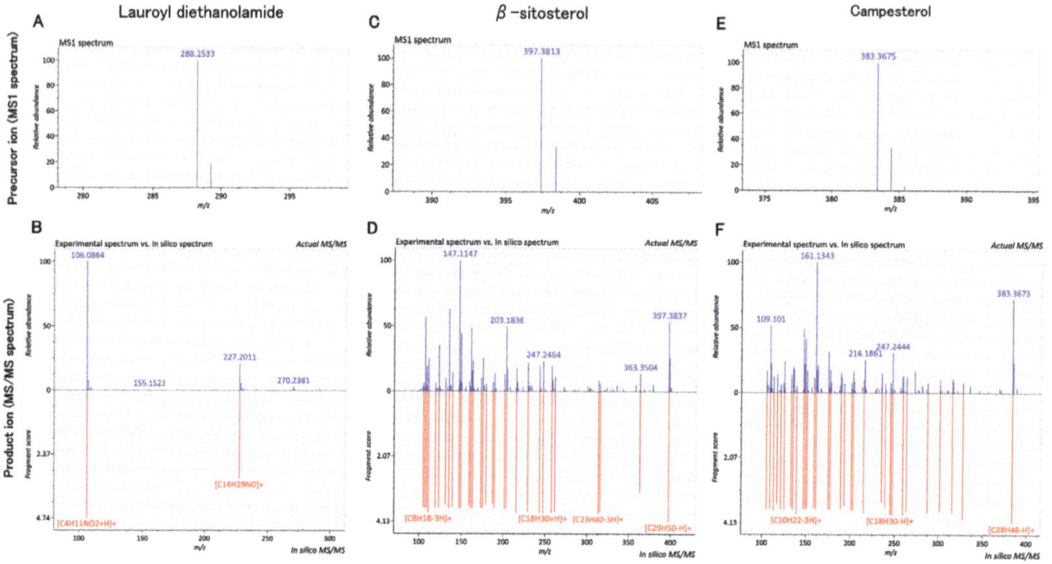

Figure 1. Precursor ions and product ions of compounds that ranked highly in both MS-FINDER and MassBank. These three compounds correspond to Level 1 according to Schymanski et al.'s criteria, indicating reliable identification. The compounds are annotated as lauroyl diethanolamide (**A**,**B**), β-sitosterol (**C**,**D**), and campesterol (**E**,**F**).

2.2. Identification and Quantification of Compounds in Sweet Potato Ethanol Extracts

The annotation confidence of the nine candidate compounds was evaluated by comparing retention times (RTs), precursor ions, and product-ion masses between SP-EtOH-Ex and authentic standards. Representative compound chromatograms (CCs) for standards and SP-EtOH-Ex are shown in Figures 2A–F and S2A–H, and product-ion spectra are shown in Figures 3A–F and S3A–H. The differences in RTs between standards and SP-EtOH-Ex were as follows: lauroyl diethanolamide, −0.04 to 0.02 min; β-sitosterol, −0.028 to 0.051 min; campesterol, −0.003 to 0.017 min; N-hexadecanoylpyrrolidine, −0.278 to 0.258 min; coniferyl aldehyde, 20.724 min; mellein, 6.215 to 6.453 min; and cycloartenol, 0.631 to 0.712 min (Tables 1 and S4). Among these, lauroyl diethanolamide, β-sitosterol, and campesterol met the retention-time criterion of ±0.25 min proposed by Pérez-Ortega et al., demonstrating sufficient RT consistency with their corresponding standards [15]. The relative mass errors were 0.69–7.98 ppm for lauroyl diethanolamide, −4.03–2.26 ppm for β-sitosterol, −1.57–6.26 ppm for campesterol, −1.29–5.48 ppm for N-hexadecanoylpyrrolidine, −464.62 ppm for coniferyl aldehyde, 19.55–37.42 ppm for mellein, and −7.82–12.21 ppm for cycloartenol. Under the same analytical conditions as those used for annotation, gibberellin A3 and cinnamic acid were not detected (signal-to-noise ratio (S/N) < 3). Absolute quantification using calibration curves for each standard revealed that SP-EtOH-Ex 1–3 contained lauroyl diethanolamide (4.568–9.260 ng/mL), β-sitosterol (2.527–4.850 μg/mL), and campesterol (75.74–93.63 ng/mL) (Table 2).

https://doi.org/10.3390/molecules310304567

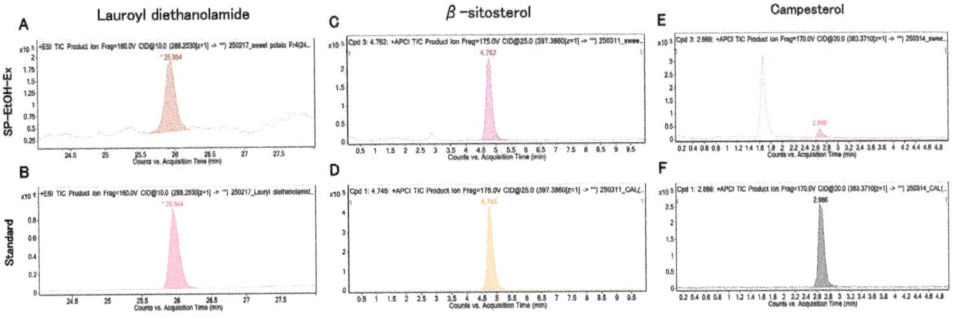

Figure 2. Compound chromatograms (CCs) of SP-EtOH-Ex corresponding to compounds meeting Level 1 of Schymanski et al.'s identification criteria based on standard reference compounds, and CCs of each standard reference compound. Lauroyl diethanolamide (**A**,**B**), β-sitosterol (**C**,**D**), campesterol (**E**,**F**). SP-EtOH-Ex: ethanol extracts of sweet potato storage roots. An asterisk (*) shown in (**A**,**B**) denotes manually integrated peaks. The double asterisk (**) shown in (**A**–**F**) indicates that the MS/MS chromatograms represent the full range of product ions obtained for the specified precursor ion.

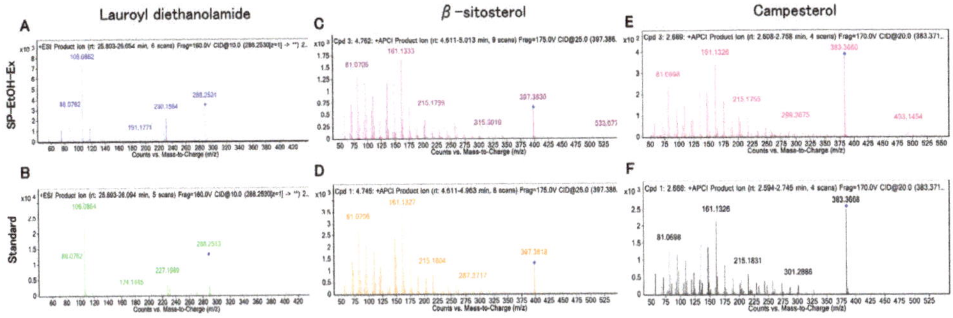

Figure 3. SP-EtOH-Ex product ions and product ions of each reference standard. Lauroyl diethanolamide (**A**,**B**), β-sitosterol (**C**,**D**), campesterol (**E**,**F**). These three compounds corresponded to Level 1 of Schymanski et al.'s identification criteria based on standard compound identification. SP-EtOH-Ex: ethanol extract of sweet potato storage roots. The blue diamond indicates the precursor ion selected for fragmentation in the MS/MS analysis. The double asterisk (**) indicates that the MS/MS chromatograms represent the full range of product ions obtained for the specified precursor ion.

Table 1. The molecular formulae, ion species, theoretical values of precursor ions, RTs, measured values of precursor ions, mass errors, and product ions for the nine candidate compounds.

Annotated Compound According to MS-FINDER	Molecular Formula	Ion Species	Theoretical (*m*/*z*)	Reference Standard or SP-EtOH-Ex	Retention Time (min)	Experimental (*m*/*z*)	Error (ppm)	Product Ions	Identification Result
Lauroyl diethanolamide	$C_{16}H_{33}NO_3$	$[M+H]^+$	288.2533	Reference standard	25.94	288.25	6.94	106.0864 (100), 288.2513 (47), 88.0762 (33)	Yes
				SP-EtOH-Ex	25.90–25.96	288.2510–288.2531	0.69–7.98	106.0861 (100), 288.2510 (47), 88.0760 (34)	
β-sitosterol	$C_{29}H_{50}O$	$[M-H_2O+H]^+$	397.3829	Reference standard	4.745	397.3818	2.77	161.1327 (100), 81.0706 (77), 95.0865 (68)	Yes
				SP-EtOH-Ex	4.717–4.796	397.3820–397.3845	−4.03–2.26	161.1333 (100), 81.0706 (78), 95.0855 (71)	

https://doi.org/10.3390/molecules310304567

Table 1. *Cont.*

Annotated Compound According to MS-FINDER	Molecular Formula	Ion Species	Theoretical (*m/z*)	Reference Standard or SP-EtOH-Ex	Retention Time (min)	Experimental (*m/z*)	Error (ppm)	Product Ions	Identification Result
Campesterol	$C_{28}H_{48}O$	$[M-H_2O+H]^+$	383.3672	Reference standard	2.672	383.3668	1.04	383.3668 (100), 161.1326 (85), 147.1173 (55)	Yes
				SP-EtOH-Ex	2.669–2.689	383.3623–383.3678	−1.57–6.26	383.3661 (100), 161.1329 (73), 147.1184 (62)	
N-Hexadecanoylpyrrolidine	$C_{20}H_{39}NO$	$[M+H]^+$	310.3104	Reference standard	31.458	310.31	1.29	310.3100 (100), 311.3130 (23), 312.3168 (3)	No
				SP-EtOH-Ex	31.18–31.20	310.3087–310.3108	−1.29–5.48	109.0649 (100), 109.1001 (81), 135.1164 (80)	
Coniferyl aldehyde	$C_{10}H_{10}O_3$	$[M+H]^+$	179.0703	Reference standard	12.711	179.0707	−2.23	119.0493 (100), 147.0441 (60), 146.0364 (35)	No
				SP-EtOH-Ex	33.435	179.1535	−464.62	119.0493 (100), 147.0440 (83), 148.0462 (11)	
Mellein	$C_{10}H_{10}O_3$	$[M+H]^+$	179.0703	Reference standard	1.069	179.0694	5.03	161.0594 (100), 179.0694 (70), 133.0642 (26)	No
				SP-EtOH-Ex	7.284–7.522	179.0636–179.0668	19.55–37.42	147.0435 (100), 148.0470 (12), 119.0493 (11)	
Gibberellin A3	$C_{19}H_{22}O_6$	$[M+H]^+$	347.1489	Reference standard	N.A.	N.A.	N.A.	N.A.	No
				SP-EtOH-Ex	–	–	–	–	
Cinnamic acid	$C_9H_8O_2$	$[M-H_2O+H]^+$	131.0491	Reference standard	N.A.	N.A.	N.A.	N.A.	No
				SP-EtOH-Ex	–	–	–	–	
Cycloartenol	$C_{30}H_{50}O$	$[M-H_2O+H]^+$	409.3829	Reference standard	3.57	409.3833	−0.98	109.1015 (100), 191.1799 (80), 121.1015 (71)	No
				SP-EtOH-Ex	4.201–4.282	409.3779–409.3861	−7.82–12.21	191.1802 (100), 409.3799 (63), 109.1012 (60)	

For SP-EtOH-Ex, RT, precursor ion measured value, and mass error. For product ions, one representative ion was selected from SP-EtOH-Ex 1, 2, or 3 and listed. The brackets for the product ions represent the relative intensity ratio, with the intensity of the strongest product ion set to 100. SP-EtOH-Ex: ethanol extract of sweet potato storage roots; RT: retention time; N.A.: not applicable.

Table 2. Quantitative results of SP-EtOH-Ex for lauroyl diethanolamide, β-sitosterol, and campesterol.

Compound	Concentration of SP-EtOH-Ex		
	1	2	3
Lauroyl diethanolamide (ng/mL)	8.113	9.260	4.568
β-sitosterol (μg/mL)	4.850	2.839	2.527
Campesterol (ng/mL)	93.63	75.74	N.A.

SP-EtOH-Ex: ethanol extract of sweet potato storage roots; N.A.: not applicable.

2.3. *β-Sitosterol and Campesterol in Sweet Potato Roots Contribute to Anti-Inflammatory Activity*

Measurement of IL-6 concentrations in culture supernatants 24 h after LPS stimulation showed that SP-EtOH-Ex significantly suppressed IL-6 secretion compared with the LPS-only control (Figure 4). SP-EtOH-Ex 1 produced the greatest reduction, followed

https://doi.org/10.3390/molecules310304567

by SP-EtOH-Ex 2 and SP-EtOH-Ex 3. β-Sitosterol suppressed IL-6 production at concentrations comparable to those present in SP-EtOH-Ex (2.527–4.850 μg/mL). Similarly, campesterol suppressed IL-6 production at concentrations comparable to those in SP-EtOH-Ex (75.74–93.63 ng/mL) (Figure 4). In contrast, lauroyl diethanolamide did not suppress IL-6 production at concentrations found in SP-EtOH-Ex (4.568–9.260 ng/mL).

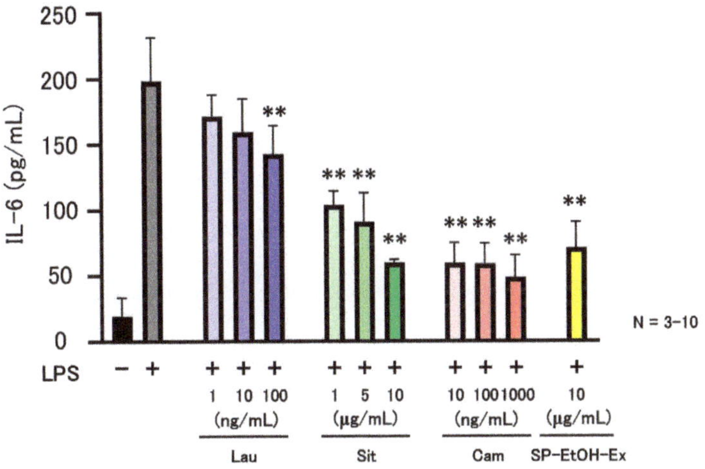

Figure 4. Measurement of the pro-inflammatory IL-6 24 h after LPS stimulation in the presence of lauroyl diethanolamide, β-sitosterol, campesterol, and SP-EtOH-Ex. Concentrations of Lau (lauroyl diethanolamide), Sit (β-sitosterol), and Cam (campesterol) are indicated on the x-axis (ng/mL or μg/mL). Mean ± standard error of the mean (SEM). SP-EtOH-Ex values indicate sample number. One-way ANOVA with Tukey's post hoc test; **: $p < 0.01$; IL-6: interleukin-6; LPS: lipopolysaccharide; Lau: lauroyl diethanolamide; Sit: β-sitosterol; Cam: campesterol; SP-EtOH-Ex: ethanol extract of sweet potato roots.

3. Discussion

In this study, we sought to annotate and identify the anti-inflammatory constituents present in an ethanol extract of sweet potato storage roots. As a result, nine compounds were annotated by MS-FINDER, among which lauroyl diethanolamide, β-sitosterol, and campesterol were successfully identified. Of these, β-sitosterol and campesterol exhibited anti-inflammatory activity at concentrations corresponding to those present in the ethanol extract of sweet potato storage roots.

The choice of mass spectrometric techniques is critical for the identification of unknown compounds. Kind et al. reported that high-resolution mass spectrometry (HRMS) used in non-target screening typically provides resolving powers of 10,000–450,000 (FWHM) and mass accuracies of <1–5 ppm [25]. Sasse et al. also emphasized that non-targeted analyses require a minimum resolving power of 10,000 and mass errors below 10 ppm [26]. The Agilent 6530 Q-TOF mass spectrometer used in the present study has a maximum resolving power of 20,000 (FWHM) and exhibits a MS1 mass accuracy better than 2 ppm. Therefore, this instrument meets the requirements for identifying unknown compounds in complex matrices.

Numerous workflows for the identification of unknown compounds have been reported [13–15,21,27–30]. MS-FINDER predicts plausible molecular structures via in silico fragmentation of candidate formulas derived from exact mass, isotopic patterns, and product-ion information. These candidate structures are ranked using multiple scoring components, including hydrogen rearrangement rules, with reference to 23 built-in databases [20]. MassBank, the first public mass spectral repository for small molecules (<3000 Da) in the life sciences, is one of the most widely used spectral databases for structural elucidation [31]. Similar to GNPS and NIST [32], MassBank enables annotation by matching experimentally obtained product ions to those in standard spectral libraries [33]. Although MS-FINDER exhibits high predictive accuracy compared to existing in silico fragmentation tools, perfect identification cannot be expected [21,34,35]. Mallmann et al. reported that MS-FINDER alone correctly identified 70% of tested compounds [34], whereas Vaniya et al. reported 53% accuracy [21]; however, integrating publicly available databases such as NIST [32], METLIN [36], and MassBank of North America [37] data increased the identification accuracy to 78%. These findings highlight the importance of using multiple tools to enhance the identification accuracy.

In this study, MS-FINDER and MassBank were used to annotate nine potential anti-inflammatory constituents in sweet potato ethanol extracts. For small molecules, Schymanski et al. proposed a five-level confidence scale for compound identification, wherein Level 1—the highest confidence—is achieved when the retention time, precursor ion, and product ions match those of an authentic standard [38]. Sasse et al. recommended targeting at least Level 2 confidence for metabolite annotation in complex samples [26], while, regarding retention time (RT), Pérez-Ortega et al. suggested an optimal window of $\pm$0.25 min [15]. Chaleckis et al. described a tiered workflow for evaluating annotation confidence, noting that the ideal RT tolerance varies with chromatographic mode; for the RP column used in this study, $\pm$0.1 min is recommended [14]. In a study screening more than 630 multi-residue food contaminants using the same Agilent 6530 Q-TOF instrument, Pérez-Ortega et al. reported relative mass errors between -5.51 and 7.94 ppm [15].

When comparing RTs and precursor ions between the annotated compounds and SP-EtOH-Ex 1–3, only lauroyl diethanolamide, β-sitosterol, and campesterol satisfied these criteria. Their product-ion spectra were also consistent with those of the corresponding standards. Jayantha et al. reported precursor ions of m/z 397.40 and 383.40 for β-sitosterol and campesterol, respectively, along with product ions at m/z 134.9 and 161.2 (β-sitosterol) and m/z 147.1 and 161.1 (campesterol) [39]. Krauss et al. reported a precursor ion of m/z 288.2528 for lauroyl diethanolamide [40], while MassBank reference spectra (Accession ID: MSBNK-CASMI_2016-SM876702) contain product ions at m/z 106.0862 and 227.2005. These reports agree with the results obtained in this study, retention time, pre-cursor ion, and product ions of lauroyl diethanolamide, β-sitosterol, and campesterol in SP-EtOH-Ex were match. (Table 1). Therefore, the annotation confidence for lauroyl diethanolamide, β-sitosterol, and campesterol in SP-EtOH-Ex corresponds to Level 1 according to the criteria of Schymanski et al., indicating highly reliable identifications. In contrast, coniferyl aldehyde, mellein, and cycloartenol did not match the RTs or precursor ions of their authentic standards. For N-hexadecanoylpyrrolidine, values were similar for RTs and precursor ions, but inconsistent for product ions (Figure S3A). Gibberellin A3 and cinnamic acid were not detected ($S/N < 3$) under the same analytical conditions used during annotation.

Phytosterols are widely reported to possess anti-inflammatory properties [41]. It has been reported that sweet potatoes also contain plant sterols such as beta-sitosterol and campesterol [42–44]. Long-term high-dose intake of total phytosterols, including campesterol and β-sitosterol, has been associated with a modest reduction in coronary artery disease (CAD) risk [28]. Studies on rheumatoid arthritis (RA) have also suggested beneficial effects of sweet potato and phytosterols on chronic inflammatory diseases [45]. Lanlan Yuan et al. reported that campesterol exhibited a stronger inhibitory effect than β-sitosterol on the proliferation of LPS-stimulated RAW264.7 macrophages [46] and in

 https://doi.org/10.3390/molecules310304567

human monocytic U937 cells, sitosterol, campesterol, and 7-keto-sitosterol decreased IL-8 secretion [47]. The mechanisms of β-sitosterol and campesterol have been reported to involve signaling pathways such as NF-κB, MAPK, and PPARγ [48–51]. In our previous study, we also reported that hydrophobic components derived from sweet potato storage roots attenuated LPS-induced inflammatory responses via activation of the Nrf2 pathway and subsequent suppression of NF-κB signaling [7]. The results of these previous studies are consistent with our findings that IL-6 production is suppressed by β-sitosterol and campesterol present in SP-EtOH-Ex (Figure 4). Beyond their cellular effects, phytosterols are well recognized for their nutritional benefits in humans. Clinical studies have demonstrated that daily intake of plant sterols lowers LDL-cholesterol by 5–15% by reducing intestinal cholesterol absorption [52,53]. Emerging evidence also suggests that phytosterols can modulate systemic inflammation and oxidative stress, and may improve glucose and lipid metabolism in individuals with metabolic syndrome, reinforcing their value as functional food ingredients [54,55]. The finding that the edible storage roots of the widely eaten yellow sweet potato can be a source of β-sitosterol and campesterol is thought to expand the future potential of sweet potatoes. Several limitations of this study should be acknowledged. First, the sweet potato cultivars, growing regions, and harvest periods analyzed were limited. Future studies are needed to explore various conditions, including those from various growing regions (e.g., Japan and Southeast Asia), various varieties, and different annual climatic conditions. Only ethanol was employed in the extraction method, and extraction conditions that might yield higher levels of phytosterols were not explored. It will be particularly important to explore efficient extraction and purification/separation conditions for industrialization. The analytical platform was restricted to LC-Q-TOF-MS; the inclusion of more advanced LC–MS/MS systems would likely enhance the confidence in compound identification. Anti-inflammatory assays were also conducted exclusively in mouse-derived macrophages, and whether similar effects occur in other cell types should be determined. In addition, the effectiveness of plant sterols has begun to be verified in humans, and the effectiveness of sweet potato-derived plant sterols needs to be verified through clinical trials.

4. Materials and Methods

4.1. Preparation of Sweet Potato Samples and Extracts

Sweet potato storage roots produced in Ibaraki or Chiba Prefecture, Japan, were washed with water and sliced into round pieces approximately 2–3 cm thick. The slices were freeze-dried using a freeze dryer (FDU-540; Tokyo Rika Kikai, Tokyo, Japan), and the dried material was pulverized to a fine powder using a food processor (MK-K62; Panasonic, Osaka, Japan). For extraction, 0.5 g of the powder was mixed with 25 mL of LC/MS-grade ethanol (Fujifilm Wako Pure Chemical, Osaka, Japan) and homogenized at 12,000 rpm for 2 min using a homogenizer (T 25 digital ULTRA-TURRAX; IKA-Werke GmbH & Co. KG, Staufen, Germany). The homogenate was centrifuged at 3000 rpm for 10 min using a KUBOTA 5922 centrifuge (Kubota Corporation, Osaka, Japan), and the resulting supernatant was collected. This extraction process was repeated four times, and all steps were performed at room temperature. The combined supernatants were evaporated to dryness with a rotary evaporator (N-1000; Tokyo Rika Kikai, Tokyo, Japan) and a diaphragm vacuum pump (DTC-41; ULVAC KIKO, Miyazaki, Japan), after which the residue was reconstituted in ethanol at a concentration of 10 mg/mL. The extract solution was then fractionated using an Agilent 1260 Infinity system (Agilent Technologies, Santa Clara, CA, USA) equipped with an ODS column (Cadenza CD-C18, 250 × 4.6 mm, particle size 3 μm; Imtakt Corporation, Kyoto, Japan). Chromatographic separation was performed

 https://doi.org/10.3390/molecules310304567

at 40 °C with an injection volume of 30 μL. Solvent A (0.4% formic acid; Fujifilm Wako Pure Chemicals, Osaka, Japan) and solvent B (100% acetonitrile; Kanto Chemical, Tokyo, Japan) were used as mobile phases at a flow rate of 1 mL/min. The gradient was programmed from the initial condition (A:B = 93:7) to A:B = 60:40 at 33 min, followed by A:B = 0:100 at 40 min, and then maintained at A:B = 0:100 for 30 min. The fraction eluted between 51 and 60 min was collected (designated as SP-EtOH-Ex). The collected fraction was evaporated to dryness using a rotary evaporator and a diaphragm vacuum pump and then reconstituted in LC/MS-grade ethanol at a concentration of 1.0 mg/mL. Three independently prepared extracts (SP-EtOH-Ex 1–3) were stored at −80 °C until use.

4.2. Reagents

LPS (*Escherichia coli* O127) was purchased from Fujifilm Wako Pure Chemical Corporation (Osaka, Japan). The reference standards used in this study were procured as follows: lauroyl diethanolamide from Combi-Blocks (San Diego, CA, USA); β-sitosterol and campesterol from Tama Biochemical Co., Ltd. (Tokyo, Japan); N-hexadecanoylpyrrolidine from Angene (London, UK); coniferyl aldehyde from ChemScene (Monmouth Junction, NJ, USA); mellein and cycloartenol from MedChemExpress (Monmouth Junction, NJ, USA); gibberellin A3 from TargetMol Chemicals (Boston, MA, USA); and cinnamic acid from BLDpharm (Shanghai, China).

4.3. Analysis of SP-EtOH-Ex by LC-Q-TOF-MS

LC analyses were performed using an Agilent 1260 series system (Agilent Technologies, Santa Clara, CA, USA) equipped with either a Waters Atlantis dC18 column (150 × 2.1 mm, 3 μm; Waters Corporation, Milford, MA, USA) or a Waters Atlantis T3 column (150 × 2.1 mm, 3 μm; Waters Corporation). Mass spectrometric detection was carried out using an Agilent 6530 Q-TOF mass spectrometer (Agilent Technologies, Santa Clara, CA, USA) operated in positive-ion mode with electrospray ionization (ESI) or atmospheric pressure chemical ionization (APCI).

For ESI, the mobile phases consisted of acetonitrile (solvent A) and 0.1% formic acid in water (solvent B). The gradient program was as follows: 0–5 min, 15% A; 5–20 min, 15–55% A; 20–27 min, 55–90% A; 27–34 min, 90% A; 34–34.1 min, 90–15% A; and 34.1–50 min, 15% A. The flow rate was set to 0.2 mL/min. The MS parameters were as follows: Vcap, 3500 V; nebulizer pressure, 35 psi; fragmentor, 150 V; skimmer, 65 V; sheath-gas temperature, 375 °C; sheath-gas flow, 11.0 L/min; collision energy, 20 eV; and mass range, m/z 100–1700.

For APCI, the mobile phases consisted of methanol or methanol/isopropanol (IPA) (50:50, v/v) (solvent A) and either 0.1% formic acid in water or 10 mM ammonium formate (pH 6.2) in water (solvent B). The gradient was programmed as follows: 0–5 min, 15% A; 5–20 min, 15–55% A; 20–27 min, 55–90% A; 27–39 min, 90% A; 39–39.1 min, 90–15% A; and 39.1–55 min, 15% A. The flow rate was 0.6 mL/min when methanol was used as solvent A and 0.4 mL/min when methanol/IPA (50:50, v/v) was used. The column temperature was maintained at 45 °C, and the injection volume was 5 μL (Table S1). The MS parameters were as follows: Vcap, 3500 V; nebulizer pressure, 60 psi; fragmentor, 150 V; skimmer, 65 V; vaporizer temperature, 350 °C; gas flow, 5.0 L/min; collision energy, 20 eV; and mass range, m/z 100–1700. Each SP-EtOH-Ex 1–3 was analyzed in triplicate to ensure reproducibility of chromatographic measurements.

Both ESI and APCI data were acquired in data-dependent acquisition (DDA) mode.

https://doi.org/10.3390/molecules310304567

4.4. Data Analysis for Annotating Unknown Compounds

Raw data were first processed using MS-DIAL (Ver. 5.3) [56]. The parameters were set as follows: MS1 tolerance, 0.01 Da; MS2 (product-ion) tolerance, 0.025 Da; minimum peak height, 1000; mass slice width, 0.1 Da; and sigma window, 0.5. The MS/MS database used for annotation was MSMS-Public_all-pos-VS19 (https://systemsomicslab.github.io/compms/msdial/main.html, accessed on 10 September 2024). Annotation of detected compounds was subsequently performed using MS-FINDER (Ver. 3.61). The parameters for MS-FINDER were set as follows: MS1 tolerance, 0.01 Da; MS2 tolerance, 0.025 Da; and mass range, m/z 0–2000. For element selection, C and H were specified as required elements and O, N, P, and S as optional elements. Annotation was conducted based on information from 23 integrated databases.

Next, FooDB [57] and KNApSAcK [58] were used to narrow down the annotated compounds and select candidate metabolites for identification with reference standards. In addition, MassBank Peak List Search (https://massbank.jp/MassBank/search, accessed on 22 January 2025) was employed to retrieve compounds whose MS2 spectra exhibited similarity to the experimentally obtained spectra.

4.5. Identification of Compounds by LC-Q-TOF-MS Using Reference Standards

The LC conditions (column, mobile phases, temperature, and injection volume) were identical to those used for compound annotation. Chromatographic separation was performed using a Waters Atlantis dC18 column for lauroyl diethanolamide and campesterol. A Waters Atlantis T3 column was used for all remaining standards. Ionization was conducted in positive-ion mode, using electrospray ionization (ESI) for lauroyl diethanolamide and N-hexadecanoylpyrrolidine and atmospheric pressure chemical ionization (APCI) for all remaining standards.

The mobile phases were as follows: acetonitrile (A) and 0.1% aqueous formic acid (B) for lauroyl diethanolamide and N-hexadecanoylpyrrolidine; methanol (A) and 0.1% aqueous formic acid (B) for coniferyl aldehyde and mellein; methanol/IPA (50:50, v/v) (A) and 0.1% aqueous formic acid (B) for gibberellin A3; methanol (A) only for campesterol; and methanol/IPA (50:50, v/v) (A) with 10 mM ammonium formate (pH 6.2) (B) for the remaining standards.

The gradient programs were as follows: for lauroyl diethanolamide and N-hexadecano ylpyrrolidine, 0–5 min: 15% A; 5–20 min: 15–55% A; 20–27 min: 55–90% A; 27–34 min: 90% A; 34–34.1 min: 90–15% A; and 34.1–50 min: 15% A. For coniferyl aldehyde, gibberellin A3, and cinnamic acid, 0–5 min: 15% A; 5–20 min: 15–55% A; 20–27 min: 55–90% A; 27–39 min: 90% A; 39–39.1 min: 90–15% A; and 39.1–55 min: 15% A. β-Sitosterol, campesterol, mellein, and cycloartenol were analyzed under isocratic conditions: 100% A from 0 to 5 min for campesterol, and 90% A from 0 to 10 min for β-sitosterol, mellein, and cycloartenol.

The flow rates were set as follows: 0.2 mL/min for lauroyl diethanolamide and N-hexadecanoylpyrrolidine; 0.4 mL/min for β-sitosterol, gibberellin A3, and cinnamic acid; 0.5 mL/min for coniferyl aldehyde, mellein, and cycloartenol; and 0.6 mL/min for campesterol (Table S2).

The mass ranges for MS and MS/MS acquisition were m/z 50–440 for lauroyl diethanolamide, m/z 50–550 for β-sitosterol and campesterol, m/z 100–1700 for gibberellin A3, m/z 50–1000 for cinnamic acid, and m/z 100–1100 for the other standards. All standards were analyzed using the targeted MS/MS (data-dependent acquisition; DDA) mode. Collision energies were set to 10 eV for lauroyl diethanolamide and mellein, 25 eV for β-sitosterol, and 20 eV for all other compounds.

 https://doi.org/10.3390/molecules310304567

4.6. Data Analysis for Confirming and Quantifying Unknown Compounds by Reference Standards

Data used for compound identification with reference standards were acquired using Agilent MassHunter Qualitative Analysis (B.07.00; Agilent Technologies, Santa Clara, CA, USA). For each target compound, the Find by Targeted MS/MS algorithm was used to extract the compound chromatogram (CC), retention time (RT), major product ions, and peak area.

4.7. Cell Culture and Stimulation Conditions

Mouse macrophage RAW264.7 cells were purchased from KAC (Kyoto, Japan). The cells were cultured in Dulbecco's Modified Eagle Medium (DMEM; Sigma–Aldrich, St. Louis, MO, USA) supplemented with 10% deactivated fetal bovine serum (FBS; Biowest, Nuaillé, France) and penicillin–streptomycin (Gibco, Grand Island, NY, USA) at 37 °C in a humidified atmosphere containing 5% CO_2.

4.8. Enzyme-Linked Immunosorbent Assay (ELISA)

RAW264.7 cells were seeded at a density of 1.0×10^4 cells/well in 96-well plates and incubated for 24 h. SP-EtOH-Ex was added to the culture medium at a final concentration of 10 µg/mL, followed by incubation for 1 h, after which lipopolysaccharide (LPS) was added at a final concentration of 50 ng/mL. Each reference standard was treated in the same manner: the compound was added to the medium, incubated for 1 h, and subsequently stimulated with LPS (50 ng/mL). SP-EtOH-Ex and all reference standards were dissolved in LC/MS-grade ethanol. The same final concentration of ethanol was added to the control group to ensure that any effects observed were not attributable to the solvent. After 24 h of incubation, the culture supernatant was collected, and the concentration of IL-6 was measured using an ELISA MAX™ Deluxe Set Mouse IL-6 kit (BioLegend, San Diego, CA, USA).

4.9. Statistical Analysis

Statistical analysis was performed with GraphPad Prism 9 version 9.0.2 (GraphPad Software Inc., San Diego, CA, USA). One-way analysis of variance (ANOVA) was performed to test three or more groups, followed by Tukey's post hoc test. The significance level was set at 5%.

5. Conclusions

In this study, β-sitosterol and campesterol were identified in the ethanol extract of sweet potato storage roots and were demonstrated to contribute to their anti-inflammatory activity. The edible portion of sweet potatoes holds promise as a raw material with anti-inflammatory properties.

Supplementary Materials: The following supporting information can be downloaded at: https://www.mdpi.com/article/10.3390/molecules310304567/s1, Figure S1. Precursor ions and product ions of compounds annotated in MS-FINDER but not meeting Schymanski et al.'s identification criteria based on reference standards; Figure S2. CC values for SP-EtOH-Ex annotated as N-hexadecanoylpyrrolidine, coniferyl aldehyde, mellein, and cycloaltenol, respectively, and CC values for corresponding standards; Figure S3. Product ions for SP-EtOH-Ex annotated as N-hexadecanoylpyrrolidine, coniferyl aldehyde, mellein, and cycloartenol, respectively, and product ions for corresponding standards; Table S1. Measurement conditions for DDA analysis; Table S2. Measurement conditions for identification using reference standards; Table S3. Ranking of candidate compounds with potential anti-inflammatory effects contained in SP-EtOH-Ex among compounds annotated by MS-FINDER in MS-FINDER and MassBank; Table S4. The molecular formulae, ion species, precursor ion theoretical values, RTs, measured values for precursor ions, mass errors, and product ions for the nine candidate compounds.

Author Contributions: Conceptualization, R.S. and H.M.; methodology, R.S. and H.M.; formal analysis, R.S.; investigation, R.S.; writing, R.S. and H.M.; review and editing: K.I.; supervision, K.I.; project administration, H.M. All authors have read and agreed to the published version of the manuscript.

Funding: The study was conducted with funding from Calbee, Inc., Tokyo, Japan.

Institutional Review Board Statement: Not applicable.

Informed Consent Statement: Not applicable.

Data Availability Statement: The data presented in this study are available on reasonable request from the corresponding author.

Conflicts of Interest: R.S., H.M., and K.I. are employed by the company Calbee, Inc.

References

1. Qin, Y.; Naumovski, N.; Ranadheera, C.S.; D'Cunha, N.M. Nutrition-related health outcomes of sweet potato (*Ipomoea batatas*) consumption: A systematic review. *Food Biosci.* **2022**, *50*, 102208. [CrossRef]
2. Islam, S.N.; Nusrat, T.; Begum, P.; Ahsan, M. Carotenoids and β-carotene in orange fleshed sweet potato: A possible solution to vitamin A deficiency. *Food Chem.* **2016**, *199*, 628–631. [CrossRef] [PubMed]
3. Chen, C.C.; Lin, C.; Chen, M.H.; Chiang, P.Y. Stability and Quality of Anthocyanin in Purple Sweet Potato Extracts. *Foods* **2019**, *8*, 393. [CrossRef] [PubMed]
4. Sugata, M.; Lin, C.Y.; Shih, Y.C. Anti-Inflammatory and Anticancer Activities of Taiwanese Purple-Fleshed Sweet Potatoes (*Ipomoea batatas* L. Lam) Extracts. *Biomed. Res. Int.* **2015**, *2015*, 768093. [CrossRef] [PubMed]
5. Bae, J.Y.; Park, W.S.; Kim, H.J.; Kim, H.S.; Kang, K.K.; Kwak, S.S.; Ahn, M.J. Protective Effect of Carotenoid Extract from Orange-Fleshed Sweet Potato on Gastric Ulcer in Mice by Inhibition of NO, IL-6 and PGE(2) Production. *Pharmaceuticals* **2021**, *14*, 1320. [CrossRef]
6. Jiang, T.; Zhou, J.; Liu, W.; Tao, W.; He, J.; Jin, W.; Guo, H.; Yang, N.; Li, Y. The anti-inflammatory potential of protein-bound anthocyanin compounds from purple sweet potato in LPS-induced RAW264.7 macrophages. *Food Res. Int.* **2020**, *137*, 109647. [CrossRef]
7. Matsumoto, Y.; Suto, M.; Umebara, I.; Masutomi, H.; Ishihara, K. Hydrophobic Components in Light-Yellow Pulp Sweet Potato (*Ipomoea batatas* (L.) Lam.) Tubers Suppress LPS-Induced Inflammatory Responses in RAW264.7 Cells via Activation of the Nrf2 Pathway. *Nutrients* **2024**, *16*, 563. [CrossRef]
8. Chen, C.J.; Lee, D.Y.; Yu, J.; Lin, Y.N.; Lin, T.M. Recent advances in LC-MS-based metabolomics for clinical biomarker discovery. *Mass. Spectrom. Rev.* **2023**, *42*, 2349–2378. [CrossRef]
9. Shen, S.; Zhan, C.; Yang, C.; Fernie, A.R.; Luo, J. Metabolomics-centered mining of plant metabolic diversity and function: Past decade and future perspectives. *Mol. Plant.* **2023**, *16*, 43–63. [CrossRef]
10. Herrero, M.; Simó, C.; García-Cañas, V.; Ibáñez, E.; Cifuentes, A. Foodomics: MS-based strategies in modern food science and nutrition. *Mass. Spectrom. Rev.* **2012**, *31*, 49–69. [CrossRef]
11. Maitre, L.; Bustamante, M.; Hernández-Ferrer, C.; Thiel, D.; Lau, C.E.; Siskos, A.P.; Vives-Usano, M.; Ruiz-Arenas, C.; Pelegrí-Sisó, D.; Robinson, O.; et al. Multi-omics signatures of the human early life exposome. *Nat. Commun.* **2022**, *13*, 7024. [CrossRef]
12. Kalogiouri, N.P.; Aalizadeh, R.; Dasenaki, M.E.; Thomaidis, N.S. Application of High Resolution Mass Spectrometric methods coupled with chemometric techniques in olive oil authenticity studies—A review. *Anal. Chim. Acta.* **2020**, *1134*, 150–173. [PubMed]
13. Bletsou, A.A.; Jeon, J.; Hollender, J.; Archontaki, E.; Thomaidis, N.S. Targeted and non-targeted liquid chromatography-mass spectrometric workflows for identification of transformation products of emerging pollutants in the aquatic environment. *TrAC Trends Anal. Chem.* **2015**, *66*, 32–44.
14. Chaleckis, R.; Meister, I.; Zhang, P.; Wheelock, C.E. Challenges, progress and promises of metabolite annotation for LC-MS-based metabolomics. *Curr. Opin. Biotechnol.* **2019**, *55*, 44–50. [PubMed]
15. Pérez-Ortega, P.; Lara-Ortega, F.J.; Gilbert-López, B.; Moreno-González, D.; García-Reyes, J.F.; Molina-Díaz, A. Screening of Over 600 Pesticides, Veterinary Drugs, Food-Packaging Contaminants, Mycotoxins, and Other Chemicals in Food by Ultra-High

 https://doi.org/10.3390/molecules310304567

Performance Liquid Chromatography Quadrupole Time-of-Flight Mass Spectrometry (UHPLC-QTOFMS). *Food Anal. Methods* **2017**, *10*, 1216–1244. [CrossRef]

16. Oniszczuk, A.; Olech, M. Optimization of ultrasound-assisted extraction and LC-ESI–MS/MS analysis of phenolic acids from *Brassica oleracea* L. var. *sabellica*. *Ind. Crops Prod.* **2016**, *83*, 359–363. [CrossRef]
17. Spínola, V.; Pinto, J.; Castilho, P.C. Identification and quantification of phenolic compounds of selected fruits from Madeira Island by HPLC-DAD-ESI-MS(n) and screening for their antioxidant activity. *Food Chem.* **2015**, *173*, 14–30.
18. Kolniak-Ostek, J. Identification and quantification of polyphenolic compounds in ten pear cultivars by UPLC-PDA-Q/TOF-MS. *J. Food Compos. Anal.* **2016**, *49*, 65–77. [CrossRef]
19. Simirgiotis, M.J.; Bórquez, J.; Schmeda-Hirschmann, G. Antioxidant capacity, polyphenolic content and tandem HPLC-DAD-ESI/MS profiling of phenolic compounds from the South American berries *Luma apiculata* and *L. chequén*. *Food Chem.* **2013**, *139*, 289–299.
20. Tsugawa, H.; Kind, T.; Nakabayashi, R.; Yukihira, D.; Tanaka, W.; Cajka, T.; Saito, K.; Fiehn, O.; Arita, M. Hydrogen Rearrangement Rules: Computational MS/MS Fragmentation and Structure Elucidation Using MS-FINDER Software. *Anal. Chem.* **2016**, *88*, 7946–7958.
21. Vaniya, A.; Samra, S.N.; Palazoglu, M.; Tsugawa, H.; Fiehn, O. Using MS-FINDER for identifying 19 natural products in the CASMI 2016 contest. *Phytochem. Lett.* **2017**, *21*, 306–312. [PubMed]
22. Dührkop, K.; Shen, H.; Meusel, M.; Rousu, J.; Böcker, S. Searching molecular structure databases with tandem mass spectra using CSI:FingerID. *Proc. Natl. Acad. Sci. USA* **2015**, *112*, 12580–12585. [CrossRef] [PubMed]
23. Allen, F.; Greiner, R.; Wishart, D. Competitive fragmentation modeling of ESI-MS/MS spectra for putative metabolite identification. *Metabolomics* **2015**, *11*, 98–110. [CrossRef]
24. Ruttkies, C.; Schymanski, E.L.; Wolf, S.; Hollender, J.; Neumann, S. MetFrag relaunched: Incorporating strategies beyond in silico fragmentation. *J. Cheminform.* **2016**, *8*, 3.
25. Kind, T.; Tsugawa, H.; Cajka, T.; Ma, Y.; Lai, Z.; Mehta, S.S.; Wohlgemuth, G.; Barupal, D.K.; Showalter, M.R.; Arita, M.; et al. Identification of small molecules using accurate mass MS/MS search. *Mass. Spectrom. Rev.* **2018**, *37*, 513–532.
26. Sasse, M.; Rainer, M. Mass Spectrometric Methods for Non-Targeted Screening of Metabolites: A Future Perspective for the Identification of Unknown Compounds in Plant Extracts. *Separations* **2022**, *9*, 415. [CrossRef]
27. Righetti, L.; Paglia, G.; Galaverna, G.; Dall'Asta, C. Recent Advances and Future Challenges in Modified Mycotoxin Analysis: Why HRMS Has Become a Key Instrument in Food Contaminant Research. *Toxins* **2016**, *8*, 361. [CrossRef]
28. Wang, Y.; Liu, B.; Hu, Y.; Sampson, L.; Manson, J.E.; Rimm, E.B.; Sun, Q. Phytosterol intake and risk of coronary artery disease: Results from 3 prospective cohort studies. *Am. J. Clin. Nutr.* **2024**, *119*, 344–353.
29. Knolhoff, A.M.; Croley, T.R. Non-targeted screening approaches for contaminants and adulterants in food using liquid chromatography hyphenated to high resolution mass spectrometry. *J. Chromatogr. A* **2016**, *1428*, 86–96. [CrossRef]
30. Ji, H.; Xu, Y.; Lu, H.; Zhang, Z. Deep MS/MS-Aided Structural-Similarity Scoring for Unknown Metabolite Identification. *Anal. Chem.* **2019**, *91*, 5629–5637.
31. Horai, H.; Arita, M.; Kanaya, S.; Nihei, Y.; Ikeda, T.; Suwa, K.; Ojima, Y.; Tanaka, K.; Tanaka, S.; Aoshima, K.; et al. MassBank: A public repository for sharing mass spectral data for life sciences. *J. Mass. Spectrom.* **2010**, *45*, 703–714. [CrossRef]
32. National Institute of Standards and Technology. NIST Mass Spectrometry Data Center. NIST 23 Mass Spectral Library (NIST/EPA/NIH), Standard Reference Database 1A. Available online: https://www.nist.gov/ (accessed on 29 December 2025).
33. Nothias, L.F.; Petras, D.; Schmid, R.; Dührkop, K.; Rainer, J.; Sarvepalli, A.; Protsyuk, I.; Ernst, M.; Tsugaywa, H.; Fleischauer, M.; et al. Feature-based molecular networking in the GNPS analysis environment. *Nat. Methods* **2020**, *17*, 905–908. [CrossRef] [PubMed]
34. Mallmann, L.P.; Alessandro, O.R.; Rodrigues, E. MS-FINDER and SIRIUS for phenolic compound identification from high-resolution mass spectrometry data. *Food Res. Int.* **2023**, *163*, 112315. [CrossRef] [PubMed]
35. Nguyen, V.; Witkowski, B.; Gierczak, T. New water-soluble, toxic tracers of wood burning identified in fine brown carbon aerosol using a non-target approach. *Atmos. Chem. Phys.* **2025**, *25*, 10965–10986. [CrossRef]
36. Smith, C.A.; O'Maille, G.; Want, E.J.; Qin, C.; Trauger, S.A.; Brandon, T.R.; Custodio, D.E.; Abagyan, R.; Siuzdak, G. METLIN: A metabolite mass spectral database. *Ther. Drug Monit.* **2005**, *27*, 747–751. [CrossRef]
37. MoNA. MassBank of North America. Available online: https://mona.fiehnlab.ucdavis.edu/ (accessed on 29 December 2025).
38. Schymanski, E.L.; Jeon, J.; Gulde, R.; Fenner, K.; Ruff, M.; Singer, H.P.; Hollender, J. Identifying small molecules via high resolution mass spectrometry: Communicating confidence. *Environ. Sci. Technol.* **2014**, *48*, 2097–2098. [CrossRef]

39. Jayantha, J.; Mashayekhy Rad, F.; Vidanarachchi, J.K.; Bergquist, J.; Kumari, A.U.S.J. A fast ultra performance supercritical fluid chromatography-tandem mass spectrometric method for profiling of targeted phytosterols. *J. Chromatogr. B Anal. Technol. Biomed. Life Sci.* **2023**, *1225*, 123737. [CrossRef]
40. Krauss, M.; Huber, C.; Schulze, T.; Bartel-Steinbach, M.; Weber, T.; Kolossa-Gehring, M.; Lermen, D. Assessing background contamination of sample tubes used in human biomonitoring by non-targeted liquid chromatography-high resolution mass spectrometry. *Environ. Int.* **2024**, *183*, 108426. [CrossRef]
41. Wang, X.; Liu, Y.; Du, F.; Zong, A.; Xu, T. Phytosterol-mediated disturbance of the lipid packing order in the plasma membrane regulates inflammatory response. *Food Funct.* **2025**, *16*, 6150–6159. [CrossRef]
42. Zhang, J.; Huang, L.; Tang, X.; Pan, M.; Ye, X.; Li, X.; Ma, H. High β-sitosterol-D-glucoside content in sweet potato varieties and its anti-breast cancer potential through multiple metastasis-associated signals. *Food Sci. Hum. Wellness* **2024**, *13*, 2779–2789. [CrossRef]
43. Xu, H.; Li, Y.; Han, B.; Li, Z.; Wang, B.; Jiang, P.; Zhang, J.; Ma, W.; Zhou, D.; Li, X.; et al. Anti-breast-Cancer Activity Exerted by β-Sitosterol-d-glucoside from Sweet Potato via Upregulation of MicroRNA-10a and via the PI3K-Akt Signaling Pathway. *J. Agric. Food Chem.* **2018**, *66*, 9704–9718. [CrossRef]
44. Alam, M.K. A comprehensive review of sweet potato (*Ipomoea batatas* [L.] Lam): Revisiting the associated health benefits. *Trends Food Sci. Technol.* **2021**, *115*, 512–529. [CrossRef]
45. Al-Okbi, S.Y. Nutraceuticals of anti-inflammatory activity as complementary therapy for rheumatoid arthritis. *Toxicol. Ind. Health* **2014**, *30*, 738–749. [CrossRef] [PubMed]
46. Yuan, L.; Zhang, F.; Shen, M.; Jia, S.; Xie, J. Phytosterols Suppress Phagocytosis and Inhibit Inflammatory Mediators via ERK Pathway on LPS-Triggered Inflammatory Responses in RAW264.7 Macrophages and the Correlation with Their Structure. *Foods* **2019**, *8*, 582. [CrossRef] [PubMed]
47. Vejux, A.; Montange, T.; Martine, L.; Zarrouk, A.; Riedinger, J.M.; Lizard, G. Absence of oxysterol-like side effects in human monocytic cells treated with phytosterols and oxyphytosterols. *J. Agric. Food Chem.* **2012**, *60*, 4060–4066. [CrossRef]
48. Zhang, Y.; Jin, X.; Xia, H.; Wu, X.; Chen, W.; Zhuang, M.; Tang, S. β-Sitosterol Ameliorates Ulcerative Colitis Through Modulation of the AMPK/MLCK Anti-Inflammatory Pathway. *J. Biochem. Mol. Toxicol.* **2025**, *39*, e70287. [CrossRef]
49. Zhang, P.; Liu, N.; Xue, M.; Zhang, M.; Liu, W.; Xu, C.; Fan, Y.; Meng, Y.; Zhang, Q.; Zhou, Y. Anti-Inflammatory and Antioxidant Properties of β-Sitosterol in Copper Sulfate-Induced Inflammation in Zebrafish (*Danio rerio*). *Antioxidants* **2023**, *12*, 391. [CrossRef]
50. Nazir, S.; Ahmad, I.; Mobashar, A.; Sharif, A.; Shabbir, A.; Chaudhary, W.A. Mechanistic evaluation of antiarthritic and anti-inflammatory effect of campesterol ester derivatives in complete Freund's adjuvant-induced arthritic rats. *Front. Pharmacol.* **2023**, *14*, 1346054. [CrossRef]
51. Nazir, S.; Chaudhary, W.; Mubashar, A.; Anjum, I.; Hameed, S.; Azhar, S. Campesterol: A Natural Phytochemical with Anti Inflammatory Properties as Potential Therapeutic Agent for Rheumatoid Arthritis: A Systematic Review: Campesterol: A Natural Phytochemical. *Pak. J. Health Sci.* **2023**, *4*, 356–363. [CrossRef]
52. Demonty, I.; Ras, R.T.; van der Knaap, H.C.; Duchateau, G.S.; Meijer, L.; Zock, P.L.; Geleijnse, J.M.; Trautwein, E.A. Continuous dose-response relationship of the LDL-cholesterol-lowering effect of phytosterol intake. *J. Nutr.* **2009**, *139*, 271–284. [CrossRef]
53. Gylling, H.; Plat, J.; Turley, S.; Ginsberg, H.N.; Ellegård, L.; Jessup, W.; Jones, P.J.; Lütjohann, D.; Maerz, W.; Masana, L.; et al. Plant sterols and plant stanols in the management of dyslipidaemia and prevention of cardiovascular disease. *Atherosclerosis* **2014**, *232*, 346–360. [CrossRef] [PubMed]
54. Vezza, T.; Canet, F.; de Marañón, A.M.; Bañuls, C.; Rocha, M.; Víctor, V.M. Phytosterols: Nutritional Health Players in the Management of Obesity and Its Related Disorders. *Antioxidants* **2020**, *9*, 1266. [CrossRef]
55. Ruscica, M.; Loh, W.J.; Sirtori, C.R.; Watts, G.F. Phytosterols and phytostanols in context: From physiology and pathophysiology to food supplementation and clinical practice. *Pharmacol. Res.* **2025**, *214*, 107681. [CrossRef]
56. Tsugawa, H.; Cajka, T.; Kind, T.; Ma, Y.; Higgins, B.; Ikeda, K.; Kanazawa, M.; VanderGheynst, J.; Fiehn, O.; Arita, M. MS-DIAL: Data-independent MS/MS deconvolution for comprehensive metabolome analysis. *Nat. Methods* **2015**, *12*, 523–526. [CrossRef]
57. The Food Database (FooDB). Available online: https://foodb.ca (accessed on 29 December 2025).
58. Afendi, F.M.; Okada, T.; Yamazaki, M.; Hirai-Morita, A.; Nakamura, Y.; Nakamura, K.; Ikeda, S.; Takahashi, H.; Altaf-Ul-Amin, M.; Darusman, L.K.; et al. KNApSAcK family databases: Integrated metabolite-plant species databases for multifaceted plant research. *Plant Cell Physiol.* **2012**, *53*, e1. [CrossRef]

 https://doi.org/10.3390/molecules310304567

molecules

MDPI

Review

Acerola and Its By-Products as Sources of Bioactive Compounds: Phytochemical Profile and Biological Effects in Experimental and Clinical Studies

Jailane de Souza Aquino [1,2,3], Alana Natalícia Vasconcelos de Araújo [1,3], Januse Míllia Dantas de Araújo [1,3], Luana Clementino Santos [2,3], Jordania Candice Costa Silva [2,3], Kamila Sabino Batista [4] and Lucas Rannier Ribeiro Antonino Carvalho [5,*]

1 Postgraduate Program in Nutrition Sciences, Federal University of Paraíba, João Pessoa 58051-900, Paraíba, Brazil; jailane.aquino@academico.ufpb.br (J.d.S.A.); allana1araujo@gmail.com (A.N.V.d.A.); januse.millia@academico.ufpb.br (J.M.D.d.A.)

2 Postgraduate Program in Food Science and Technology, Federal University of Paraíba, João Pessoa 58051-900, Paraíba, Brazil; luana.clementino@academico.ufpb.br (L.C.S.); jordania.candice@academico.ufpb.br (J.C.C.S.)

3 Experimental Nutrition Laboratory—LANEX, Department of Nutrition, Federal University of Paraíba, João Pessoa 58051-900, Paraíba, Brazil

4 Secretariat for Health and Environmental Surveillance, Ministry of Health, Brasília 70719-040, Federal District, Brazil; kamilasabino00@gmail.com

5 Department of Physiology and Pharmacology, Karolinska Institutet, Biomedicum, Solnavägen 9, SE-171 65 Solna, Sweden

* Correspondence: lucas.carvalho@ki.se; Tel.: +46-(0)8-524-800-00

Abstract

Acerola (*Malpighia emarginata* DC.) is one of the richest natural sources of vitamin C and an important source of phenolic compounds, carotenoids, and bioactive polysaccharides. Although the fruit can be consumed fresh, it is more commonly processed into juices and frozen pulp, generating substantial amounts of by-products (pomace, peels, and seeds), corresponding to approximately 20–60% of the fruit biomass, with high phytochemical content. These fractions represent underutilized sources of bioactive compounds. This narrative review, supported by a structured literature search, integrates evidence on the phytochemical composition of acerola pulp and its by-products and relates these profiles to biological effects in experimental and human studies, focusing on compound characterization, composition–function relationships, and underlying mechanisms. Key compounds, including ascorbic acid, hydroxycinnamic acids, flavonoids, and polysaccharides, are associated with the modulation of redox homeostasis, inflammatory signaling, and lipid metabolism, particularly under high-fat dietary conditions. Human evidence remains limited but suggests matrix-dependent effects on vitamin C bioavailability and selected cardiometabolic markers. Overall, the evidence is constrained by methodological heterogeneity, limited clinical data, and insufficient characterization of bioactive fractions. Future research should prioritize detailed phytochemical profiling, dose–response relationships, bioavailability assessment, and well-controlled clinical trials incorporating molecular biomarkers, supporting the development of acerola-derived matrices as functional and bioactive-rich ingredients.

Keywords: acerola; *Malpighia emarginata*; fruit by-products; phytochemical composition; animal studies; clinical trials; antioxidant activity; metabolic effects

1. Introduction

The growing demand for sustainable food systems has increased interest in the valorization of fruit by-products as alternative sources of bioactive compounds with potential health benefits [1–3]. Large volumes of by-products are generated during fruit processing, particularly in juice and pulp industries, representing both an environmental challenge and an opportunity for the recovery of high-value phytochemicals [2,4,5]. Tropical fruits are especially relevant in this context due to their chemical diversity and high concentrations of bioactive compounds [6,7].

Acerola (*Malpighia emarginata* DC.), also known as Barbados cherry or West Indian cherry, is a perennial species native to the Caribbean Basin and northern South America and is currently cultivated in tropical and subtropical regions [8]. Brazil is the leading producer, with approximately 5753 hectares under cultivation and an annual production of 60,000–61,000 tons, largely concentrated in the Northeast region [9]. This production supports an industrial chain focused on frozen pulp, juices, concentrates, and nutraceutical ingredients. During processing, a substantial proportion of the fruit biomass estimated at 20–60% is converted into by-products, mainly pomace, peels and seeds [4,9].

These by-products may contain equal or higher levels of phenolic compounds, carotenoids, and residual ascorbic acid than the edible pulp [4,5]. Acerola is also a rich source of vitamin C and contains flavonoids and hydroxycinnamic acids associated with antioxidant and anti-inflammatory effects [4,10].

Both acerola and its by-products have been increasingly explored in food and nutraceutical applications [11–14]. The pulp has been incorporated into fermented and dairy products with good sensory acceptance [11], while processing strategies, particularly drying, have improved stability and expanded its use in various food formulations [12]. Acerola by-products have also been investigated as natural antioxidant ingredients in functional foods and dietary supplements, and immature fruits are widely used as industrial sources of ascorbic acid [13,14].

Phenolic extracts from acerola by-products have been applied to the development of biodegradable films and coatings, contributing to improved preservation and extended shelf life [15–18]. Technological approaches have facilitated their stabilization and incorporation into complex matrices [12,19,20], while emerging evidence suggests their potential to enhance the gastrointestinal survival of probiotic strains [21]. More recently, acerola-derived matrices have been explored in advanced delivery systems, including naturally occurring nanovesicles capable of protecting bioactive compounds during gastrointestinal transit [22]. Together, these findings highlight the technological versatility of acerola and its by-products and their relevance for nutraceutical and functional applications.

However, evidence regarding their biological effects remains uneven and is largely derived from preclinical models, with relatively few human studies [8,14,23]. The frequent use of simplified experimental designs, crude extracts, and antioxidant-focused assays limits mechanistic insight and hampers the identification of molecular targets and dose–response relationships [1,13,14,23–28].

In parallel, most reviews on acerola have addressed composition, applications, or selected biological activities as separate topics [1,4,8,14,23], leaving the integration of phytochemical composition with biological effects across experimental settings insufficiently explored. The application of omics-based approaches also remains limited [29], and the role of matrix complexity (whole matrices versus isolated compounds) in shaping biological responses has not been comprehensively evaluated [27,30].

 https://doi.org/10.3390/molecules31111792

Thus, this narrative review examines the phytochemical composition of acerola pulp and its by-products and relates these profiles to their biological effects in experimental and clinical studies. To address the identified gap, this review provides an integrative analysis linking phytochemical composition to biological effects across different experimental settings, an approach that has not been systematically explored in previous studies. Specifically, it focuses on three complementary aspects: (i) the characterization of bioactive compounds; (ii) the integration of compositional data with reported biological effects across different experimental models; and (iii) the identification of underlying mechanisms linking specific compounds to physiological responses. By structuring the analysis in this way, the review provides an integrated perspective and outlines directions for future research, supporting the development of acerola-derived matrices as functional and bioactive-rich ingredients.

2. Literature Search and Study Selection

This study was conducted as a narrative review supported by a structured literature search. A comprehensive search was performed in the Embase, PubMed/MEDLINE, Scopus, and Web of Science Core Collection databases to identify studies on acerola, its bioactive compounds, and related biological effects. The search was performed in February 2026 using a combination of controlled vocabulary (Emtree and MeSH terms) and free-text terms, adapted to the syntax of each database.

The search strategy included terms related to acerola (e.g., "acerola", "*Malpighia emarginata*", "Barbados cherry", "West Indian cherry", and "Antilles cherry"), combined with terms associated with bioactive compounds and antioxidant properties (e.g., "antioxidant*", "phenolic*", "polyphenol*", "flavonoid*", "ascorbic acid", "vitamin C", "extract*", and "phytochemical*"), as well as terms related to biological activity and experimental models (e.g., "effect*", "activ*", "bioactiv*", "metabol*", "animal*", "in vivo", "human", and "experiment*"). Boolean operators (AND and OR) were used to combine these terms. In addition, citation searching was performed to identify relevant studies not captured in the database search.

Study selection was performed according to predefined inclusion and exclusion criteria. Only original articles published in English between January 2016 and February 2026 were considered. Reviews, editorials, conference abstracts, and studies not aligned with the scope of this review were excluded. Screening was performed based on titles and abstracts, followed by full-text assessment. A total of 33 studies met the inclusion criteria and were included in this review. The study selection process was summarized using the PRISMA 2020 flow diagram framework (Figure 1). As this study was conducted as a narrative review supported by a structured literature search, no formal risk of bias assessment was performed.

https://doi.org/10.3390/molecules31111792

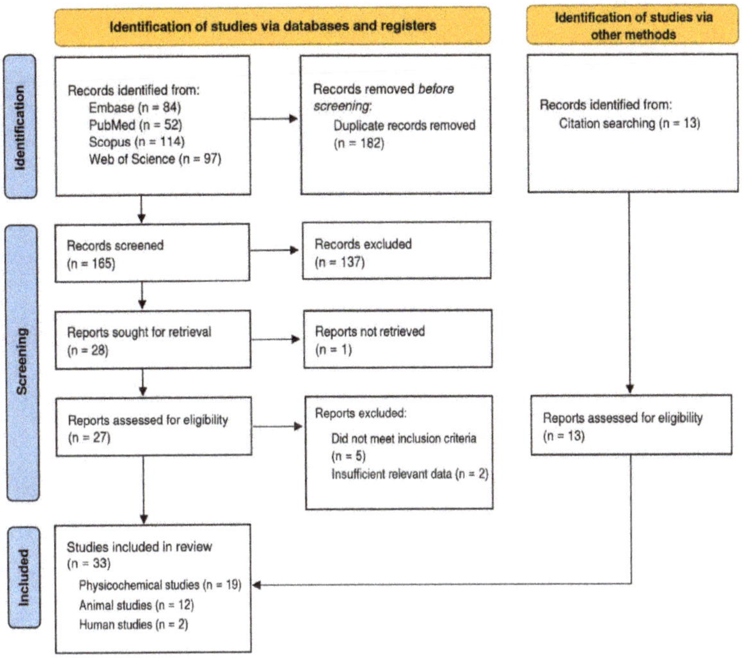

Figure 1. The PRSIMA 2020 flow diagram used to select the articles.

3. Phytochemical Profile and Biological Effects of Acerola and Its By-Products

Acerola is characterized by pronounced genotypic heterogeneity, resulting in significant variation in fruit morphology, pulp yield, soluble solids content, and, importantly, the secondary metabolites profiles [10,13,31]. Internationally, acerola cultivars are less standardized than those developed in Brazil and are often described as local or regional selections [32]. In the Caribbean, Central America, and tropical regions of the United States (such as Florida and Hawaii), common types include Barbados or West Indian, as well as Florida Sweet and Tropical Sweet. These cultivars are typically selected for traits like yield, fruit size, and lower acidity for fresh consumption. However, many are not officially registered, which limits standardization and comparability across studies, reflecting the high genetic variability of the species [31–33].

Commercial cultivars cultivated in Brazil, such as 'Okinawa', 'Sertaneja', 'Flor Branca', 'Junko', and 'Costa Rica', and internationally disseminated genotypes including 'Florida Sweet' and 'B-17', represent distinct biochemical phenotypes shaped by breeding selection and agroecological adaptation [34]. More recently, regionally adapted breeding initiatives have generated improved genotypes such as 'Rubi-PB', developed through collaboration between researchers from the Empresa Paraibana de Pesquisa, Extensão e Regularização Fundiária (EMPAER-PB) and the Federal University of Paraíba (UFPB), selected for yield stability, elevated soluble solids (°Brix), and high ascorbic acid concentration under semi-arid stress conditions [35].

Importantly, comparative analyses demonstrate that genotype and processing parameters modulate the relative abundance and structural diversity of phenolic compounds, carotenoids, and residual ascorbic acid, thereby influencing both quantitative bioactive density and qualitative phytochemical complexity [36].

 https://doi.org/10.3390/molecules31111792

Genotypic variability and processing conditions are recognized determinants of the qualitative and quantitative phytochemical composition of acerola matrices, with direct implications for biological activity [31]. An integrated framework linking genotypic variability, processing conditions, and the distribution of bioactive compounds in pulp and by-products is presented in Figure 2. Nevertheless, cultivar identity is frequently not reported in experimental animal and human studies, thereby limiting robust integration between composition and biological function [4,31]. The following sections critically synthesize the available phytochemical evidence within this framework, emphasizing genotype traceability as a key factor for strengthening translational interpretation.

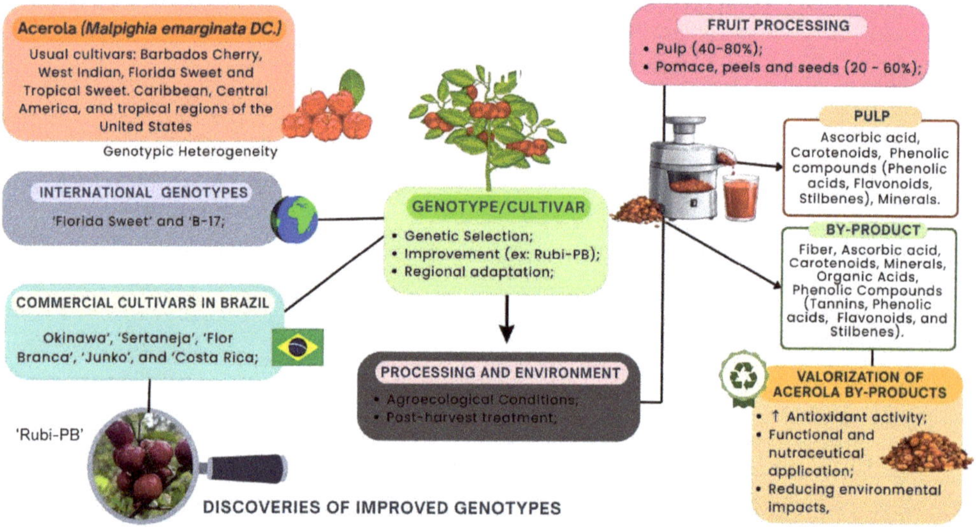

Figure 2. Processing flow of acerola (*Malpighia emarginata* DC.) and generation of value-added by-products. The diagram shows acerola processing and the resulting by-products (pomace, peels, and seeds), highlighting the phytochemical composition of pulp and by-product fractions and their potential for functional and nutraceutical applications. Arrow (↑) = indicates increase.

3.1. Proximate Composition and Phytochemical Profile of Acerola Pulp

The phytochemical profile of acerola pulp is characterized by a high content of bioactive compounds. On a dry weight basis, high concentrations of total phenolic compounds (2069.55 mg/100 g), ascorbic acid (7.94–36.36 mg/100 g), and total anthocyanins (2.24 mg/100 g) have been reported, with cyanidin-3-rhamnoside identified as the predominant pigment [37,38]. In addition, the pulp contains 3.22 µg/g of total chlorophyll and 29.71–79.2 µg/g of total carotenoids [37,38]. Previous research on acerola pulp cultivated in Brazil (cultivar not reported in the original study) identified hydroxycinnamic acids (chlorogenic and gallic acids), flavan-3-ols (catechin, procyanidin B1, and epigallocatechin gallate), and flavonols (kaempferol 3-glycoside and quercetin 3-glycoside) as the predominant quantified polyphenols [37]. The phytochemical profile and proximate composition of acerola pulp are presented in Table 1.

The quantification of ascorbic acid, total chlorophyll, flavonoids, and carotenoids in studies involving acerola pulp was performed by spectrophotometry, using specific methodologies for each class of compound based on absorbance at characteristic wavelengths [13,39]. The total phenolic content was determined using the Folin–Ciocalteu colorimetric method, which is widely employed to estimate the overall phenolic content in plant matrices [13]. Additionally, the phenolic profile was identified and quantified by high-performance liquid

 https://doi.org/10.3390/molecules31111792

chromatography coupled with a UV/Vis detector (HPLC-UV/Vis), enabling the separation, identification, and quantification of the different compounds present in the samples [37].

Table 1. Proximate composition and main bioactive compounds of acerola pulp *.

Parameter (g/100 g)	Mean Value	Reference
Moisture	93.49	[38]
	91.20	[40]
Proteins	0.04	[38]
	1.08	[40]
Lipids	0.09	[38]
	0.53	[40]
Ash	0.60	[38]
	0.31	[40]
	0.36	[37]
Carbohydrates	5.45	[38]
	6.28	[40]
Soluble fibers	0.24	[40]
Insoluble fibers	1.14	[40]
Bioactive compounds		
Ascorbic acid (mg/100 g)	36.36	[37]
	7.94	[38]
Lycopene (mg/100 g)	0.19	[37]
Anthocyanins (mg/100 g)	2.24	[37]
Total chlorophyll (μg/g)	3.22	[38]
Carotenoids (μg/g)		
Trans-β-carotene	11.70	[40]
Trans-α-carotene	1.24	
β-cryptoxanthin	1.92	
9-cis-βC	0.73	
13-cis-βC	1.44	
15-cis-βC	1.19	
Lutein	1.00	
Total carotenoids	79.20	[37]
	29.71	[38]
Phenolic compounds (mg/100 g)		
Gallic acid	2.20	[37]
Syringic acid	0.03	
Caftaric acid	0.11	
Chlorogenic acid	0.15	
Caffeic acid	0.06	
p-Coumaric acid	0.03	
Procyanidin B1	0.40	
Procyanidin B2	0.17	
Epigallocatechin gallate	0.32	
Epicatechin	0.07	
Epicatechin gallate	0.08	
Catechin	1.14	
Kaempferol 3-glicoside	1.39	
Rutin	0.18	
Quercetin 3-Glycoside	0.66	
Isorhamnetin	0.24	
Hesperidin	0.30	
Trans-resveratrol	0.10	

https://doi.org/10.3390/molecules31111792

Table 1. *Cont.*

Parameter (g/100 g)	Mean Value	Reference
Total flavonoids (mg/100 g)	12.64 3.74	[37] [38]
Total phenolic compounds (mg/100 g)	2069.55 6.12	[37] [38]
Minerals (mg/100 g)		
Iron	0.80	[41]
Zinc	0.08	
Manganese	0.24	
Copper	0.17	
Sodium	35.13	
Aluminum	0.93	
Boron	0.11	

* Values are presented as means to ensure consistency across studies, as standard deviations were not consistently reported in all the original sources.

The differences observed in bioactive compound values between studies can be attributed to multiple methodological factors, including variations in extraction procedures (e.g., solvent type, time, temperature, and processing conditions), as well as differences in the units or bases used to express the results (fresh weight basis, dry weight basis, or gallic acid equivalents). In addition, variability in the sample matrix, fruit ripening stage, cultivar/variety, and cultivation conditions may also contribute to the observed differences. Collectively, these factors limit the direct comparability between studies [17,37].

Acerola pulp also contains carotenoids, which, together with ascorbic acid and phenolic profile, may contribute to its high antioxidant activity [36,42,43]. A comparative assessment indicates that, although acerola contains relevant levels of carotenoids, it cannot be classified among the most concentrated dietary sources when compared to well-established carotenoid-rich matrices such as mango, papaya and tomato [44–46]. Nevertheless, its carotenoid, ascorbic acid and phenolic contents are comparable to, and in some cases higher than those of widely consumed fruits such as apple, guava, orange and lemon, depending on genotype, ripening stage, and processing conditions [31,37,47,48].

Chemical characterization studies have demonstrated the presence of tocopherols (vitamin E), phytosterols such as β-sitosterol, campesterol, and stigmasterol, as well as saponins and trace levels of alkaloids, compounds that contribute to the fruit's antioxidant and anti-inflammatory potential [6,23]. Volatile compounds, including terpenes with possible antimicrobial and antioxidant activity, have also been identified in the acerola matrix [49].

Several studies have demonstrated that acerola pulp (with the cultivar not reported in the original study) exhibits high antioxidant capacity, as measured by standard free radical scavenging and reducing power assays. In addition to recent data reporting antioxidant activity of approximately 6.17 μmol Trolox equivalents/g by the DPPH assay and approximately 3.19 mM TEAC/g by the ABTS assay, broader investigations have shown expressive values for ORAC and other antioxidant capacity methods [38]. In a comparative study, acerola pulp (cultivar not reported in the original study) exhibited higher antioxidant capacity, as measured by ABTS (21.99 μmol Fe^{2+}/g), DPPH (11.14 μmol TE/g), and FRAP (6.71 μmol TE/g) assays, with values higher than those observed for guava [37].

In different analyses, the ORAC method applied to acerola fruits at different ripening stages from two distinct regions in the United States (Davie and Vero Beach, Florida) showed values ranging approximately from 36 to 79 mM Trolox equivalents/kg of fresh

 https://doi.org/10.3390/molecules31111792

weight basis, depending on geographic origin and maturity stage [49]. Phenolic compounds were responsible for approximately 7.1–36.5% of the total antioxidant capacity measured by the ORAC assay, while ascorbic acid contributed between 18% and 39% of the overall antioxidant activity [49]. Regardless of geographic origin, fruits at earlier ripening stages consistently exhibited higher antioxidant capacity, as measured by the ORAC method, compared to fully ripe fruits. This trend is mainly attributed to the higher concentration of phenolic compounds in unripe or semi-ripe stages, which significantly contribute to total antioxidant activity [49]. As ripening progresses, a general reduction in phenolic content and ascorbic acid levels is observed, leading to a decrease in overall antioxidant capacity [49]. Nevertheless, these findings reflect a high oxygen radical scavenging capacity, indicating effective antioxidant activity at the molecular level [49].

Evidence has demonstrated that extracts from unripe acerola pulp (with the cultivar not reported in the original study) exhibited higher antioxidant capacity and reduced power compared to those from ripe fruit. In the current analysis, ascorbic acid contributed between 59.12% and 98.12% of the antioxidant activity, with the lowest contribution observed in ripe acerola and the highest in unripe acerola [50]. These results help explain the elevated reducing power observed in unripe samples and indicate that this method should not be used as a proxy for total phenolic content, since ascorbic acid is not a phenolic compound [50].

3.2. Proximate Composition and Phytochemical Profile of Acerola By-Products

During industrial processing of acerola mainly to produce juice and frozen pulp, substantial amounts of by-products, including pomace, peels, and seeds, are generated [4,5]. The extensive processing of acerola has raised environmental concerns, particularly regarding waste management. These by-products represent, on average, 40% of the total fruit volume, indicating a substantial loss of biomass that could otherwise be recovered and valorized in the development of value-added products [1].

The by-products generated during the processing of acerola have been characterized as nutritionally and functionally valuable matrices. According to da Silva et al. [17], acerola pulp by-product, on a dry weight basis, contains 34.2 $\pm$ 0.4 g/100 g of crude fiber and 42.3 mg/100 g of anthocyanins, expressed on a wet weight basis. Complementarily, Batista et al. [30] identified even higher concentrations of specific bioactives, reporting 5366.44 µg/g of total phenolic compounds (536.6 mg/100 g) and 2022.06 µg/g ascorbic acid. Their work also quantified individual flavonoids, including quercetin (250.80 µg/g) and rutin (22.46 µg/g), as well as total carotenoids (4.99 mg/100 g).

The studies involving acerola by-products (cultivar not reported in the original study) presented in Table 2 characterized the profile of bioactive compounds using different analytical methodologies, depending on the compound class evaluated. Anthocyanin determination was performed on a wet basis after acidified aqueous extraction, followed by quantification using UV–Vis spectrophotometry at 520 and 700 nm [17]. Chlorophyll content was also determined by spectrophotometric readings at wavelengths of 646 [50] and 663 nm [51]. Total carotenoids were also determined by UV–Vis spectrophotometry at 450, 470 and 645 nm [30,51,52]. For ascorbic acid, different approaches were employed across the studies, including organic solvent extraction followed by HPLC-DAD analysis [17], UV-Vis spectrophotometry at 518 nm [30], and the titrimetric method [52]. Total phenolic content was determined by the Folin–Ciocalteu assay at different absorbances (UV-Vis, 725 nm) [17] and (UV-Vis, 760 nm) [51,53]. Additionally, the phenolic and organic acids profile was characterized using organic solvent extraction and high-performance liquid chromatography coupled with a DAD detector (HPLC-DAD), enabling the identification and quantification of individual compounds [30].

https://doi.org/10.3390/molecules31111792

Table 2. Proximate composition and main bioactive compounds of acerola by-products.

Parameter (g/100 g)	Mean Value	By-Product	Reference
Moisture	77.0	Mixture of lumps, seeds and peels	[17]
Ash	2.8	Mixture of lumps, seeds and peels	[17]
Proteins	8.3	Mixture of lumps, seeds and peels	[17]
Lipids	2.45	Mixture of peels and seeds	[54]
Carbohydrates	52.0	Mixture of lumps, seeds and peels	[17]
Fibers	34.2	Mixture of lumps, seeds and peels	[17]
Bioactive compounds			
Anthocyanins (mg/100 g)	42.30	Mixture of lumps, seeds and peels	[17]
Chlorophyll (μg/mL)			
Chlorophyll a	5.49	Green residue ‡	[51]
Chlorophyll b	4.06		
Total chlorophyll	9.56		
β-carotene (mg/g)	5.12	Peel	[2]
	3.24	Seed	
Total carotenoids (mg/100 g)	4.99	Mixture of peels, residual pulp, and seed	[30]
	0.263 *	Mixture of peels, residual pulp, and seeds	[52]
	0.070 *	Green residue ‡	[51]
Ascorbic acid (mg/100 g)	1063.50	Mixture of lumps, seeds and peels	[17]
	2898.83	Mixture of peels, residual pulp, and seeds	[30]
	1012.19	Green residue ‡	[51]
	202.21 *	Mixture of peels, residual pulp, and seeds	[52]
Organic acids (mg/g)			
Formic acid	7.31	Mixture of peels, residual pulp, and seeds	[30]
Malic acid	6.60		
Succinic acid	15.46		
Tannins (mg/g)			
Condensed	3.86	Peel	[2]
Hydrolyzable	28.21	Peel	
Condensed	2.90	Seed	
Hydrolyzable	19.76	Seed	
Phenolic acids (μg/g)			
Ellagic acid	50.40	Mixture of peels, residual pulp, and seeds	[30]
2-Hydroxybenzoic	1374.80		
3,4,5-Trihydroxybenzoic acid	1.51		
3,5-Dimethoxy-4-hydroxybenzoic acid	101.60		
Caftaric acid	86.58		
4-Hydroxybenzoic acid	789.60		
Cinnamic acid	25.60		
3,4-Dihydroxycinnamic acid	7.35		
3-O-Caffeoylquinic acid	14.67		
4-Hydroxycinnamic acid	65.50		
4-Hydroxy-3-methoxycinnamic acid	51.97		
4-Hydroxy-3,5-dimethoxycinnamic acid	137.60		

https://doi.org/10.3390/molecules31111792

Table 2. *Cont.*

Parameter (g/100 g)	Mean Value	By-Product	Reference
Flavonoids (μg/g)			
Procyanidin A2	2.38	Mixture of peels, residual pulp, and seeds	[30]
Procyanidin B1	7.26		
Procyanidin B2	43.33		
Epigallocatechin gallate	4.48		
Epicatechin gallate	3.37		
Catechin	1622.50		
Kaempferol 3-glicoside	88.37		
Rutin	22.46		
Quercetin	250.80		
Quercetin-3-glycoside	13.39		
Myricetin	333.80		
Hesperedin	29.54		
Hesperetin	13.00		
Naringenin	48.90		
Chrysine	8.40		
Stilbenes (μg/g)			
Trans-resveratrol	6.19	Mixture of peels, residual pulp, and seeds	[30]
Cis-resveratrol	15.67		
Total phenolic compounds (mg/100 g)			
	1155.20 †	Mixture of lumps, seeds and peels	[17]
	10.680 *	Mixture of seeds and peels	[54]
	536.64 *	Mixture of peels, residual pulp, and seeds	[30]
	7276.25 * †	Green residue ‡	[51]
Minerals (mg/g)			
Potassium	58.32	Mixture of peels, residual pulp, and seeds	[30]
Calcium	31.91		
Magnesium	2.82		
Chlorine	2.25		
Phosphorus	2.43		
Sulfur	1.20		
Iron	0.25		
Zinc	0.09		
Copper	0.05		

Values are presented as means to ensure consistency across studies, as standard deviations were not consistently reported in all original sources. * Indicates data that were standardized in mg/100 g to improve comparability of the results. † Indicates data reported as mg GAE/100 g in the original study. ‡ Green residue is the term used in the original study to describe acerola processing by-products; its specific composition (e.g., peel, seed, or pulp fractions) was not detailed by the authors.

Liquid chromatographic (HPLC-DAD) profiling further identified phenolic compounds such as hesperetin, naringenin, and phenolic acids, as well as fructooligosaccharides (kestose and nystose), reinforcing the multifunctional nature of these matrices [30,54]. The phytochemical composition and physicochemical characteristics of acerola by-products are summarized in Table 2.

The substantial variability reported across characterization studies represents a key limitation for direct comparisons between matrices, as it is influenced by factors such as genotype, ripening stage, processing conditions, and the interval between harvest and

 https://doi.org/10.3390/molecules31111792

analysis. Furthermore, differences in analytical methodologies and reporting units (e.g., fresh vs. dry weight) compromise cross-study comparability and may introduce bias [55].

Importantly, this variability is compound-dependent. Ascorbic acid shows particularly high variability due to its high susceptibility to oxidation and differences in extraction conditions, whereas carotenoids are relatively more stable, although still influenced by processing, light exposure, and solvent systems [44,56]. Anthocyanins, in turn, are highly affected by pH, temperature, and extraction protocols, contributing to inconsistencies across studies. Phenolic compounds also exhibit considerable variation, largely due to differences in extraction solvents, hydrolysis steps, analytical methods, and their structural diversity [46,57]. In addition, dietary fiber content may vary depending on the type of by-product analyzed (e.g., peel, seed, or pomace) and the processing conditions applied, particularly those affecting the proportion of soluble and insoluble fractions [56].

These limitations are further exacerbated by the lack of standardization in data reporting, including differences between fresh and dry weight bases, expression per edible portion or per extract, and the use of distinct normalization criteria (e.g., per 100 g, per gram, or per serving size). Moreover, variations in moisture content and sample preparation can significantly influence reported concentrations, leading to over- or underestimation of bioactive compound levels and reducing the reliability of direct comparisons and meta-analytical interpretations [55].

Focusing on acerola, fruit by-products exhibit significant variability in the contents of phenolic compounds and dietary fiber. While grape pomace is recognized as one of the richest matrices, presenting higher levels of these compounds than acerola, by-products from banana, pineapple, and mango generally show low to moderate levels compared to acerola [2]. Acerola residues present relatively high levels of ascorbic acid, comparable to citrus and apple by-products, although generally lower than those derived from grapes. Regarding carotenoids, acerola by-products exhibit higher levels than those of fruits such as apple and strawberry, but lower than well-established sources such as mango and carrot [2,57–59].

Acerola by-products, which comprise residual pulp and peel fractions, are particularly rich in flavonoids and phenolic acids, exhibiting strong antioxidant activity in chemical and biological assays [54]. Peels have been reported to contain concentrated phenolic fractions, while seeds, although less explored, have shown promising levels of tannins and other phenolic compounds with antioxidant potential [31]. These findings reinforce the concept that acerola by-products represent underutilized matrices with relevant phytochemical and biological value.

Importantly, the antioxidant activity of acerola cannot be attributed solely to vitamin C. Several studies have demonstrated synergistic effects between ascorbic acid and polyphenols, resulting in enhanced radical scavenging capacity and improved biological activity when compared to isolated compounds [6,7]. This complex phytochemical matrix supports the classification of acerola as a functional fruit with potential applications beyond the present total phenolic contents comparable to or higher than those found in the edible pulp, highlighting its potential as an alternative source of phytochemicals [38,39].

Acerola by-products have been widely investigated for their antioxidant capacity and activity, due to the high concentration of bioactive compounds that remain after industrial processing. Studies have shown that these by-products contain substantial levels of total phenolic compounds, with concentrations that may be comparable to or even exceed those found in the pulp [1,17].

Antioxidant activity is commonly assessed using assays such as DPPH, ABTS, FRAP, and ORAC. Across studies, acerola by-products consistently exhibit high antioxidant capacity, although results vary depending on the assay and experimental conditions. ABTS values have been reported within a relatively narrow range (46.7–48.5 μmol Trolox equivalents

(TE)/g dry weight), whereas FRAP values show greater dispersion (93.0–273.8 μmol TE/g dry weight basis), indicating variability in reducing power across samples [17]. Similarly, DPPH results reported values (38.17 EC_{50} μg/mL extract) indicate strong radical scavenging capacity, while ORAC values (7.86 ± 0.06 EC_{50} μg/mL extract) confirm the ability of these matrices to neutralize peroxyl radicals [53].

Comparative evidence further supports the high antioxidant potential of acerola by-products. Acerola peels exhibited substantially higher antioxidant activity (approximately 4–7-fold) than strawberry residues across ABTS, DPPH, and FRAP assays, with consistent superiority also observed in ORAC measurements for both peels and seeds [2]. These differences are aligned with the higher concentrations of carotenoids, ascorbic acid, and phenolic compounds, including flavonoids, reported in acerola matrices [2,30,53], reinforcing the relationship between composition and antioxidant capacity. The data demonstrate that acerola peel shows promising potential as a source matrix for the extraction of phenolic compounds under different extraction conditions.

Methodological factors, particularly extraction techniques, further contribute to the variability observed across studies. Probe-type ultrasound-assisted extraction (UAE), for example, has been shown to enhance the recovery of phenolic compounds, leading to higher antioxidant activity in DPPH and FRAP assays under optimized conditions [60,61]. However, these responses are highly dependent on extraction parameters and sample characteristics, which limit direct comparisons between studies.

Despite the extensive characterization of acerola phytochemicals, important methodological limitations remain. Analytical approaches are often not standardized across studies, with variations in extraction solvents, detection techniques, and reporting units, limiting comparability and reproducibility. Furthermore, most studies report total phenolic or antioxidant capacity without detailed structural identification, hindering a precise understanding of compound-specific bioactivity. These limitations highlight the need for advanced analytical platforms and harmonized methodologies to improve the reliability and translational relevance of phytochemical data.

3.3. *Foodomics Approaches to Acerola Composition and Functional Properties*

Foodomics approaches, integrating metabolomics, transcriptomics, and proteomics, have contributed to advancing the understanding of the biochemical and molecular complexity of acerola and its by-products, providing complementary insights into the mechanisms underlying the biosynthesis of bioactive compounds associated with its functional properties [29,62,63]. Among these, metabolomics has been the most applied, allowing the characterization of low-molecular-weight compounds such as flavonoids, amino acids, organic acids, and lipids [62,63].

Untargeted analyses using advanced platforms (e.g., Ultra Performance Liquid Chromatography—Quadrupole Time-of-Flight Mass Spectrometry—UPLC-QTOF-MS) have shown clear variation in metabolite profiles across ripening stages, in agreement with the phytochemical composition described in Sections 3.1 and 3.2. In particular, the consistent association between ascorbic acid content and antioxidant capacity highlights the role of vitamin C in the functional properties of acerola [63]. These findings reinforce the characterization of acerola as a phytochemically complex matrix whose composition is dynamically modulated during fruit development.

In addition to fresh pulp, metabolomic approaches have been successfully applied to acerola by-products, including pomace and seeds [63]. These matrices retain high concentrations of phenolic compounds such as quercetin, kaempferol, and isorhamnetin, supporting their potential as sustainable sources of bioactive ingredients for nutraceutical

https://doi.org/10.3390/molecules31111792

and functional food applications [4,5,63]. Such findings are particularly relevant in the context of agro-industrial valorization and circular economy strategies [1].

Transcriptomic and genomic approaches have provided important insights into the regulation of metabolic pathways associated with fruit composition. Gene expression studies in acerola have identified key enzymes involved in the L-galactose pathway of ascorbic acid biosynthesis, with transcriptional activity closely associated with vitamin C accumulation during fruit development [64,65]. Transcriptomic analyses further demonstrate coordinated regulation of genes involved in ascorbate metabolism, ethylene signaling, and secondary metabolism, highlighting the complexity of regulatory networks involving fruit ripening [65]. These findings reinforce that the marked variability in phytochemical composition observed among acerola cultivars is strongly influenced by genetic background and gene expression dynamics [31].

Despite these advances, several limitations still restrict the application of omics approaches in acerola research. Transcriptomic analyses are restricted to fresh tissues, as RNA is unstable and prone to rapid degradation, requiring rapid processing and stabilization to preserve transcript integrity [29,31]. In fruit systems, transcriptomic analysis has therefore been predominantly conducted in fresh or minimally processed tissues, particularly in investigations of fruit development and ripening dynamics [66].

On the other hand, proteomic analyses face additional challenges, as fruit tissues typically contain low amounts of protein and high levels of sugars, organic acids, and phenolic compounds, which interfere with protein extraction and downstream analyses [67,68]. These limitations are documented in fruit proteomics, where sample preparation is considered a critical and often limiting step for reliable protein identification and quantification [69,70].

Overall, while metabolomics has been successfully applied to both acerola pulp and derived products, genomic and transcriptomic approaches remain limited in scope, and proteomics are still incipient. The lack of integrated multi-omics studies represents a critical gap in the literature. Future research combining these approaches is essential to achieve a systems-level understanding of acerola biology [29], thereby linking genetic regulation to metabolite accumulation and, ultimately, to the biological effects attributed to this species.

3.4. Biological Effects of Acerola and Its By-Products in Animal Models

3.4.1. Antioxidant and Anti-Inflammatory Effects

Preclinical evidence from animal models of metabolic stress indicates that acerola and its derived matrices exert antioxidant and anti-inflammatory effects (Table 3). These responses are consistently reflected in markers of oxidative stress, including reductions in lipid peroxidation and modulation of inflammatory mediators across different experimental conditions [26,28].

Table 3. Antioxidant and anti-inflammatory effects of acerola-derived matrices in animal models.

Matrix/Preparation	Dose, Route and Duration	Animal Model (*n*)	Experimental Condition	Main Outcomes	Reference
Acerola by-products (peels, residual pulp, and seeds; freeze-dried)	400 mg/kg, orogastric, 28 days	Wistar rats (*n* = 8/group)	HFD-induced dyslipidemia	↓ Lipid peroxidation; ↑ TAC (plasma, colon, liver)	[71]
Acerola by-products	Dietary (diet incorporation, 1%), 4 weeks	Wistar rats (*n* = 6/group)	Diet-induced obesity	↑ CAT (subcutaneous adipose tissue)	[72]
Acerola polysaccharides (isolated fraction)	200–800 mg/kg, oral, 9 weeks	C57BL/6 mice (*n* = 6/group)	HFD-induced MASLD (reported as NAFLD in the original study)	↓ TNF-α, IL-6, IL-1β; ↑ SOD, CAT; activation of Nrf2 pathway	[26]

https://doi.org/10.3390/molecules31111792

Table 3. *Cont.*

Matrix/Preparation	Dose, Route and Duration	Animal Model (*n*)	Experimental Condition	Main Outcomes	Reference
Acerola polysaccharides (arabinan-rich fraction)	0.1–1 mg/kg, intraperitoneal	Swiss mice (n = 4–8/group)	Carrageenan-induced inflammation	↓ TNF-α, IL-1β, PGE2; ↑ GSH, SOD, CAT	[23]
Lyophilized acerola bagasse extract	7–14 mg phenolics/kg, orogastric, 21 days	Wistar rats (n = 6/group)	CCl_4-induced liver injury	↓ ALT, AST; ↓ MDA; ↑ SOD, TAC	[25]

Abbreviations: ALT, alanine aminotransferase; AST, aspartate aminotransferase; CAT, catalase; CCl_4, carbon tetrachloride; GSH, reduced glutathione ratio; HFD, high-fat diet; IL-1β, interleukin-1 beta; MASLD, metabolic dysfunction-associated steatotic liver disease; MDA, malondialdehyde; Nrf2, nuclear factor (erythroid-derived-2)-like 2; SOD, superoxide dismutase; PGE_2, prostaglandin E2; TAC, total antioxidant capacity; TNF-α, tumor necrosis factor alpha; upward arrows (↑) = indicates increase; downward arrows (↓) = indicates decrease.

In rats fed a high-fat diet, administration of acerola by-products (400 mg/kg) reduced lipid peroxidation, with plasma levels decreasing from 8.19 to 3.86 nmol/L, with consistent effects also observed in other tissues such as the colon and liver [71]. These findings may be partly associated with the bioactive composition of these matrices, particularly phenolic compounds, ascorbic acid, and carotenoids, which have been previously characterized [5,39,71]. In parallel, phenolic compounds remain bioaccessible after ingestion and have been detected along the enterohepatic axis, encompassing both luminal compartments (cecal content and feces) and target tissues such as the colon and liver, supporting their physiological relevance and indicating that they reach sites of action in vivo [71]. This provides a mechanistic basis for the systemic redox response observed under dyslipidemic conditions. However, because these by-products represent a complex matrix, the relative contribution of individual constituents remains difficult to establish with certainty, especially in view of the overlapping mechanisms through which antioxidant compounds may act [73].

At the tissue level, antioxidant responses have also been observed in adipose depots following administration of acerola by-products. In obese rats fed a high-fat diet supplemented with 1% of these matrices, catalase activity increased in subcutaneous adipose tissue, accompanied by a reduction in subcutaneous fat mass, without significant changes in overall body weight gain [73]. This profile is consistent with localized modulation of oxidative status in adipose tissue, which may influence lipid handling independent of whole-body energy balance. The absence of changes in body weight further indicates that these responses are not driven by global metabolic alterations, but rather reflect tissue-specific redox adaptations.

Evidence from experimental findings using purified polysaccharide fractions extracted from acerola fruit provides mechanistic insight into the responses in high-fat diet models [26]. In mice exposed to diet-induced metabolic stress, these fractions reduced circulating and hepatic levels of pro-inflammatory cytokines (TNF-α, IL-6 and IL-1β), while improving oxidative status through decreased lipid peroxidation and restoration of antioxidant enzyme activity [26]. These responses were associated with activation of the Nrf2 signaling pathway, including increased expression of heme oxygenase-1 (HO-1) and NAD(P)H:quinone oxidoreductase-1 (NQO-1), key components of cellular antioxidant defense [26]. Improvements in cellular energy metabolism were further supported by restoration of ATP content and increased activity of respiratory chain complexes, reinforcing the link between redox and mitochondrial homeostasis [26]. However, as these findings derive from a single experimental model, their generalizability remains limited.

In acute inflammatory settings, acerola-derived polysaccharides have also been shown to exhibit anti-inflammatory and antioxidant activity. An arabinan-rich pectic polysaccharide fraction extracted from acerola pulp and peel attenuated carrageenan-induced

https://doi.org/10.3390/molecules31111792

paw edema and mechanical allodynia, reduced pro-inflammatory mediators including TNF-α, IL-1β, and prostaglandin E2 levels, and increased IL-10 levels in inflamed paw tissue, indicating a shift toward an anti-inflammatory profile [24]. These responses were accompanied by restoration of glutathione levels and antioxidant enzyme activity, along with reduced lipid peroxidation at the site of inflammation, supporting coordinated modulation of inflammatory and redox pathways [70]. These effects may be partly related to structural features of these polysaccharides, as pectic polysaccharide structures have been associated with the regulation of inflammatory mediators and immune responses [74]. This profile suggests that acerola-derived polysaccharides may act through the combined regulation of oxidative stress and inflammatory signaling.

Protective effects are also observed in chemically induced liver injury models. In Wistar rats exposed to carbon tetrachloride (CCl_4), a widely used hepatotoxic agent that induces lipid peroxidation through the generation of reactive species, administration of lyophilized acerola bagasse extract (7 and 14 mg phenolic compounds/kg, for 21 days) reduced serum aminotransferase activities and preserved hepatic histological architecture [25]. These effects were accompanied by increased superoxide dismutase activity and total antioxidant capacity, reflecting attenuation of oxidative stress associated with xenobiotic exposure [25]. This is consistent with the known mechanisms of CCl-induced hepatotoxicity. Extrapolation of these findings should be approached with caution, given the constraints of the experimental model and the specific composition of the extract used.

Available experimental evidence indicates that acerola-derived matrices modulate antioxidant and inflammatory pathways in multiple tissues, particularly under conditions of metabolic and inflammatory stress [30,48,54,72]. Reported effects include reductions in oxidative damage, modulation of inflammatory mediators and activation of endogenous antioxidant defence systems, supporting the involvement of acerola bioactives in redox-sensitive biological processes.

3.4.2. Metabolic and Hepatic Effects

Acerola and its derived matrices influence hepatic lipid homeostasis in experimental models of metabolic dysfunction (Table 4). Evidence from diet-induced steatosis and dyslipidemia indicates that these matrices modulate lipid synthesis, mitochondrial bioenergetics, and the distribution of lipids between intestinal and hepatic compartments [26,30,54]. Such effects are related to the presence of bioactive compounds, particularly phenolic constituents, ascorbic acid and dietary fiber fractions found in acerola by-products, as well as polysaccharide-rich extracts evaluated in specific experimental models, which may contribute to modulating lipid metabolism and hepatic function.

Studies conducted in high-fat diet-fed rodents, a widely used model of metabolic dysfunction-associated steatotic liver disease (MASLD), formerly termed non-alcoholic fatty liver disease, support the effects attributed to acerola-derived bioactive compounds [3,50,69]. Polysaccharides isolated from acerola fruit attenuated hepatic lipid accumulation and improved steatosis assessed by histology, with reduced expression of lipogenic enzymes such as fatty acid synthase (FAS), acetyl-CoA carboxylase (ACC) and stearoyl-CoA desaturase-1 (SCD-1) [1]. In parallel, mitochondrial adaptations were observed, including restoration of hepatic ATP content, increased activity of respiratory chain complexes I, IV and V, reduced expression of uncoupling protein 2 (UCP2), and stimulation of peroxisome proliferator-activated receptor gamma coactivator-1α (PGC-1α), supporting enhanced mitochondrial β-oxidation [1]. The extent of these changes is noteworthy, as they were obtained using a pectic polysaccharide fraction. The simultaneous modulation of lipogenesis and mitochondrial function suggests broader metabolic effects than would

https://doi.org/10.3390/molecules31111792

be expected from a single structural component, supporting a more active role of these compounds in hepatic metabolism [1,75].

Table 4. Metabolic and hepatic effects of acerola-derived matrices in experimental models.

Matrix/Preparation	Dose, Route and Duration	Animal Model (*n*)	Experimental Condition	Main Outcomes	Reference
Acerola-derived polysaccharides (isolated fraction)	200–800 mg/kg, oral, 9 weeks	C57BL/6 mice (n = 6/group)	HFD-induced MASLD (reported as NAFLD in the original study)	↓ Hepatic lipid accumulation; ↓ SREBP-1c, FAS, ACC, SCD-1; ↓ UCP2; ↑ mitochondrial complexes I, IV, V; ↑ PGC-1α	[26]
Acerola by-products (peels, residual pulp, and seeds; freeze-dried)	400 mg/kg, orogastric, 28 days	Wistar rats (n = 8/group)	HFD-induced metabolic dysfunction	↓ Liver fat accumulation; ↓ serum lipids; ↓ blood glucose; ↑ insulin tolerance; ↑ fecal bile acid excretion	[30]
Acerola by-products (peels, residual pulp, and seeds; freeze-dried)	400 mg/kg, orogastric, 28 days	Wistar rats (n = 8/group)	Diet-induced dyslipidemia	↓ Hepatic fat deposition; ↓ serum lipids; ↓ visceral fat; ↑ fecal fat excretion; ↑ organic acids	[54]
Acerola by-products (peels, residual pulp, and seeds; freeze-dried)	400 mg/kg, orogastric, 28 days	Wistar rats (n = 8/group)	Healthy condition	↑ Hepatic retinol deposition; ↓ serum lipids	[52]

Abbreviations: FAS, fatty acid synthase; HFD, high-fat diet; NAFLD, non-alcoholic fatty liver disease; PGC-1α, peroxisome proliferator-activated receptor gamma coactivator-1α; SCD-1, stearoyl-CoA desaturase-1; SREBP-1c, sterol regulatory element-binding protein-1c; UCP2, uncoupling protein 2; upward arrows (↑) = indicates increase; downward arrows (↓) = indicates decrease.

Complementary evidence from whole acerola matrices further supports these metabolic effects. Orogastric administration of acerola by-products (400 mg/kg for 28 days) in high-fat diet-fed rats reduced body weight gain, serum lipid concentrations, blood glucose and hepatic fat accumulation, while improving insulin tolerance and increasing fecal bile acid excretion [30], consistent with coordinated changes in intestinal and hepatic lipid metabolism under diet-induced metabolic stress [30,49]. From a physiological perspective, the use of a complex matrix reflects the physiological context of food consumption, and the observed responses likely involve the integration of intestinal and hepatic processes, rather than being attributable to a single pathway, particularly given the overlap between lipid metabolism and bile acid turnover [54,71]. This pattern is consistent with the heterogeneous composition of acerola by-products, including fiber-rich fractions and oligosaccharides, along with a range of micronutrients and bioactive compounds such as carotenoids, ascorbic acid, organic acids, and phenolic constituents, which may collectively contribute to their modulatory effects on lipid metabolism, as characterized in the original study [30].

Comparable responses were observed in a dyslipidemia-driven model, where supplementation with an acerola processing by-product (400 mg/kg, orogastric administration for 28 days) reduced visceral and hepatic fat deposition and improved circulating lipid parameters in female Wistar rats [54]. Increased fecal fat excretion and alterations in organic acid concentrations suggest that intestinal processes contribute to reduced hepatic lipid burden. Under the same conditions, by-products from cashew and guava were also evaluated, although the effects were more pronounced in the acerola group, likely reflecting differences in matrix composition, indicating that these responses may depend on the specific bioactive profile of each matrix rather than representing a generalizable effect across fruit by-products [54]. Histological evaluation further supported attenuation of diet-induced hepatic fat accumulation [54].

https://doi.org/10.3390/molecules31111792

Acerola processing residues also influenced basal lipid homeostasis under non-pathological conditions. In healthy Wistar rats, supplementation (400 mg/kg, orogastric administration for 28 days) was associated with reduced serum lipid concentrations and increased hepatic retinol deposition without adverse effects on body composition [52]. By-products from cashew and guava were also evaluated, with partially overlapping but distinct metabolic responses across three matrices. While acerola and guava by-products were associated with reductions in serum lipid levels, cashew by-products showed a more pronounced effect on blood glucose [52]. The persistence of these responses in healthy animals indicates that the effects of acerola-derived materials are not restricted to dyslipidemic states, but may also reflect modulation of basal lipid metabolism and vitamin A homeostasis. These differences may be partly explained by variations in the compositional profiles of the matrices [39,54], particularly the higher ascorbic acid content observed in acerola by-products, alongside comparable carotenoid levels across the three fruits, which may contribute to differences in their metabolic effects.

Acerola-derived matrices modulate hepatic lipid metabolism in experimental models, with reported effects on hepatic lipid accumulation, circulating lipid parameters and pathways related to mitochondrial and enterohepatic regulation. Available data indicate a consistent influence on mechanisms involved in lipid metabolism and metabolic homeostasis under conditions of metabolic dysfunction [30,54,71,72].

3.4.3. Other Biological Effects

Acerola and its derived matrices are also explored in experimental contexts beyond antioxidant and metabolic responses. Preclinical studies describe gastroprotective effects, attenuation of genotoxic damage, and improvements in fatigue-related performance, alongside modulation of neuroenergetic parameters under dietary or physical stress conditions [75–77]. These effects are observed across diverse experimental settings; however, variations in study design and outcome measures limit direct comparisons between studies (Table 5).

Table 5. Other biological effects of acerola-derived matrices in experimental models.

Matrix/Preparation	Dose, Route and Duration	Animal Model (*n*)	Experimental Condition	Main Outcomes	Reference
Acerola-derived polysaccharides (pectic fraction)	50–200 mg/kg, orogastric, 28 days	Swiss mice (*n* = 6/group)	Weight-loaded swimming model	↑ Time to exhaustion; ↑ mitochondrial respiration (skeletal muscle); ↑ hippocampal GSH	[75]
Polysaccharides from acerola by-products (peels, residual pulp, and seeds)	1 mg/kg, orogastric, 1 day	Wistar rats (*n* = 6/group)	Ethanol-induced gastric injury	↑ Gastric GSH; ↓ MDA (stomach); preservation of gastric mucosa	[77]
Acerola juice (unripe and ripe)	0.1 mL/10 g, orogastric, 1 day	CF1 mice (*n* = 10/group)	Genotoxicity induced by iron	↓ micronucleated erythrocytes (bone marrow)	[78]
Acerola juice (unripe, ripe, industrial)	0.1 mL/10 g, orogastric, 4 weeks	Swiss mice (*n* = 6/group)	Cafeteria diet-induced obesity	Restoration of citrate synthase activity; partial recovery of mitochondrial complex I activity (hypothalamus)	[76]

Abbreviations: GSH, reduced glutathione; MDA, malondialdehyde; upward arrows (↑) = indicates increase; downward arrows (↓) = indicates decrease.

Gastrointestinal effects have been observed in mice models of ethanol-induced gastric injury. Water-soluble polysaccharides extracted from acerola processing residues preserved

 https://doi.org/10.3390/molecules31111792

gastric mucosal architecture and attenuated oxidative damage in vivo [77]. Pretreatment (1 mg/kg, orogastric administration) prevented ethanol-induced depletion of gastric glutathione and reduced malondialdehyde levels in stomach tissue, indicating decreased lipid peroxidation and maintenance of mucosal redox balance. The lack of cytotoxicity effects in intestinal epithelial cells further supports the tolerability of these fractions within the experimental setting evaluated [77]. Comparable gastroprotective responses have also been reported for polysaccharide fractions derived from other fruit by-products [3,77], including cashew and pineapple, whereas similar effects were not observed for mango and passion fruit, highlighting variability in bioactivity among matrices. These differences are consistent with variations in monosaccharide composition, with acerola polysaccharides enriched in arabinose, while mango and passion fruit fractions are predominantly glucose-based [77], which may influence their structural properties and interaction with the gastric mucosa [79]. Given the acute nature of the model, responses likely reflect short-term protection against chemically induced stress.

Genoprotective responses further extend the range of biological effects attributed to acerola [78,80,81]. Exposure to dietary metals may include both essential and toxic elements capable of inducing DNA damage, either directly or through the generation of reactive oxygen species: a process associated with the development of several diseases [80]. Acerola contains a range of antioxidant compounds that may contribute to mitigating metal-induced oxidative stress and subsequent DNA damage [81]. In mice exposed to iron-induced genotoxicity, pre-treatment with acerola juice, regardless of maturation stage, reduced the frequency of micronuclei in bone marrow cells, indicating antimutagenic activity as assessed by the micronucleus assay [78]. This response is consistent with a protective effect against oxidative damage to the genetic material, suggesting that acerola-derived antioxidants may interfere with early events involved in metal-induced genotoxicity [80]. The absence of marked differences between unripe and ripe juice also indicates that this effect may be preserved across maturation stages, despite expected variations in fruit composition. Although these findings support a protective role against oxidative DNA damage, the use of a single genotoxic agent and acute exposure conditions limits the generalization of these effects across different types of genomic stress.

Neuroenergetic alterations also emerge in the context of diet-related metabolic disturbances. Using a cafeteria diet model of obesity, acerola juice intake partially restored key enzymes involved in cerebral energy metabolism, including citrate synthase and mitochondrial respiratory complexes, across metabolically relevant regions such as the hypothalamus and prefrontal cortex [76]. Industrial acerola juice reversed diet-induced inhibition of hypothalamic complex I activity, whereas unripe and ripe preparations restored citrate synthase levels in a region-specific manner, with differential effects on mitochondrial function rather than a global increase in bioenergetic activity [76]. Not all respiratory alterations were normalized, reinforcing the notion of targeted mitochondrial adaptation. This selective modulation of the response suggests a partial restoration of cerebral energy homeostasis, potentially contributing to the maintenance of neuronal function under metabolic stress [76,78], which may be influenced by variation in phytochemical composition across juice preparations.

Fatigue during sustained physical effort provides a useful framework for examining bioenergetic adaptations. A weight-loaded swimming model was used to assess an arabinan-rich pectic polysaccharide isolated from acerola pulp after 28 days of oral supplementation (50–200 mg/kg), resulting in a prolonged time to exhaustion [75]. Treated groups showed higher post-exercise plasma glucose and triglyceride concentrations, together with increased lactate levels, indicating a shift in carbohydrate and lipid mobilization rather

 https://doi.org/10.3390/molecules31111792

than a reduction in metabolic demand [75]. At higher doses, mitochondrial respiratory capacity in permeabilized skeletal muscle fibers increased following adenosine diphosphate (ADP) stimulation, consistent with enhanced oxidative phosphorylation under high-energy requirements [75]. Glutathione levels were higher in the hippocampus, whereas lipid peroxidation and plasma markers of hepatic injury remained unchanged [75]. The dose-dependent nature of these responses, particularly the enhancement of mitochondrial respiratory capacity at higher doses, suggests that the metabolic effects of the arabinan-rich pectic polysaccharide are modulated by the level of supplementation, reflecting a threshold for bioenergetic adaptation under conditions of increased energy demand [75]. A coordinated pattern of peripheral and central adaptations emerges, with improved mitochondrial function in skeletal muscle and selective modulation of antioxidant status in the brain, supporting enhanced tolerance to metabolic stress during sustained exercise.

Acerola-derived matrices exert biological effects in experimental models beyond antioxidant and metabolic regulation, including gastrointestinal protection, modulation of genomic stability, neuroenergetic responses and improved fatigue-related performance [49,75,77,78]. Together, these observations indicate that the biological activity of acerola extends beyond a single physiological domain and involves multiple organ systems.

3.5. *Evidence from Human Studies*

Human studies evaluating the biological effects of acerola remain more limited than preclinical investigations and are mainly based on small-scale interventions. Available trials point to antioxidant-related effects, although results are not consistent between markers [24,25]. More recently, clinical investigations have explored additional physiological outcomes beyond global antioxidant capacity.

In a pilot trial involving 22 elite endurance athletes supplemented with 300 g/day of acerola pulp for three weeks, Vítek et al. [24] reported reductions in circulating immunoglobulins (including IgG subclasses and IgA), complement component C1q, and selected metabolic markers, whereas total antioxidant status and glutathione reductase activity remained unchanged. Phytochemical profiling of the pulp showed high vitamin C content and flavonoids such as quercitrin and aceronidin, which may be related to the biological effects reported [24]. The use of a highly trained population, together with the short intervention period and limited sample size, should be considered when interpreting these findings, as both baseline metabolic status and exercise-related adaptations may influence the outcomes.

The lack of variation in global redox indices may also be related to the limited specificity of broad antioxidant assays, particularly in individuals with already optimized redox homeostasis. The inclusion of more specific markers of oxidative damage, such as lipid peroxidation or protein oxidation products, together with quantitative cytokine profiling, could provide a clearer assessment of whether acerola directly influences redox balance or is more closely associated with inflammatory and metabolic adjustments. Larger placebo-controlled trials with longer follow-up periods are needed to strengthen causal interpretation and improve the clinical relevance of these observations.

Earlier pharmacokinetic evidence by Uchida et al. [25] demonstrated that ingestion of acerola juice containing 50 mg of ascorbic acid resulted in greater plasma exposure and lower urinary excretion compared with isolated ascorbic acid, supporting a matrix-dependent modulation of vitamin C bioavailability. However, the acute design and the absence of functional clinical endpoints limit inference regarding sustained physiological effects.

 https://doi.org/10.3390/molecules31111792

Most studies rely on fresh pulp or juice, whereas evidence on acerola by-products in human populations remains limited, despite their well-characterized phytochemical profile. This gap underscores the need for clinical trials using standardized by-product matrices and incorporating redox, inflammatory, and metabolic endpoints, allowing a clearer link between compositional data and clinically relevant outcomes.

A key limitation in current clinical research using acerola and its derived matrices as an intervention is the lack of standardized intervention protocols, particularly regarding dose, duration, and matrix composition. In addition, the absence of human studies specifically evaluating acerola by-products restricts translational interpretation, especially given their high phytochemical density and potential industrial application. Addressing these limitations will be essential to establishing the functional and clinical relevance of acerola-derived bioactives.

4. Critical Appraisal and Translational Considerations

Variability in the phytochemical composition of acerola remains a key challenge for interpreting its biological effects. Differences related to cultivars, fruit maturity, processing conditions, and the use of distinct matrices (e.g., fresh pulp, juice, by-products, extracts and isolated compounds) result in heterogeneous profiles of bioactive compounds, including ascorbic acid, polyphenols, and dietary fiber. This heterogeneity limits comparability across studies and complicates the interpretation of matrix-dependent effects. The lack of standardization in analytical methods used for compound identification and quantification further contributes to this limitation.

From a precision health perspective, foodomics and multi-omics approaches may provide an integrative framework for linking the characterization of acerola-derived matrices to their biological effects. In acerola and its by-products, high-throughput platforms such as metabolomics and proteomics may improve the profiling of bioactive compounds and their transformations during processing, thereby refining the characterization of phytochemical composition and functional potential, as discussed in Section 3.3. At the molecular and cellular levels, transcriptomics, proteomics, and metabolomics have supported the identification of molecular targets, signaling pathways, and gene–nutrient interactions involved in the response to these phytochemical matrices [29,82]. Integrative analyses may further clarify modulation of antioxidant and cytoprotective pathways, such as Nrf2/Keap1, as well as inflammatory pathways including NF-κB, although their application in acerola research remains limited [83].

These approaches may also help elucidate interactions between acerola bioactives and the gut microbiota, including shifts in microbial composition and metabolite production (e.g., short-chain fatty acids), which are recognized mediators of systemic effects [84–86]. Nevertheless, current evidence remains limited and largely indirect, and the specific contribution of acerola-derived compounds to microbiota-mediated responses has not yet been clearly established. The integration of omics data from both food matrices and biological responses may additionally support the identification of biomarkers related to efficacy, interindividual variability, and metabolic responsiveness; however, the validation of such markers in the context of acerola intake remains scarce [87].

In parallel, micro- and nanotechnology-based delivery systems are being explored to improve the stability and bioavailability of acerola-derived bioactives. Techniques such as nanoencapsulation and nanocarriers may protect labile compounds, including ascorbic acid and polyphenols, from degradation, enhance intestinal absorption, and allow controlled release and tissue-specific distribution [88,89]. Despite these advances, most evidence is derived from experimental systems, and their applicability in human

 https://doi.org/10.3390/molecules31111792

settings remains to be demonstrated. The combination of mechanistic insights from omics approaches with advanced delivery systems may represent a promising direction for future research, although its translational potential requires further validation in well-controlled studies [90,91].

Although the biological effects of acerola and its by-products are suggested by relatively consistent experimental evidence [30,44,52,54,71,72], important limitations remain. Uncertainties regarding bioavailability, incomplete characterization of bioactive fractions, and inconsistent dosage strategies limit the establishment of clear relationships between intake and systemic effects. A summary of the main biological effects reported for acerola pulp and its by-products in experimental and clinical settings is presented in Figure 3.

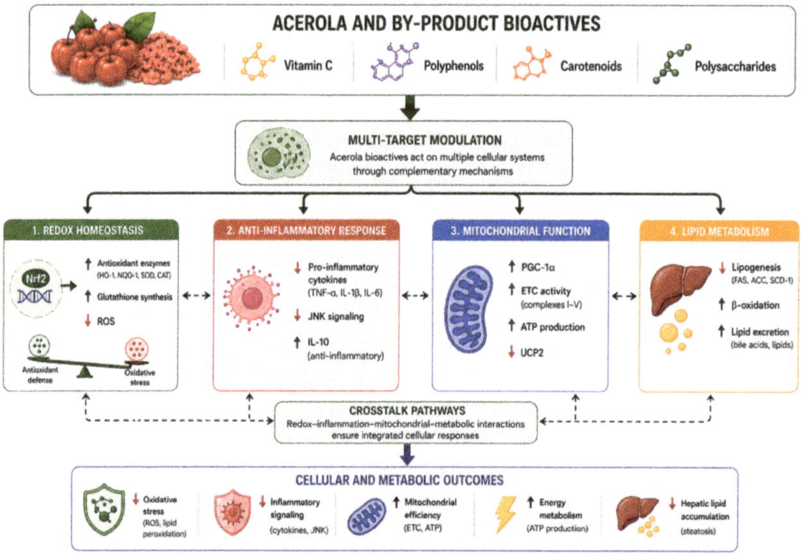

Figure 3. Molecular mechanisms underlying the biological effects of acerola and by-product bioactives, HO-1, Heme oxygenase-1; NQO-1, Quinone oxidoreductase-1; SOD, Superoxide dismutase; CAT, Catalase; ROS, Reactive oxygen species; PGC-1α, Peroxisome proliferator-activated receptor gamma coactivator-1α; ATP, Adenosine Triphosphate; UCP2, Uncoupling protein 2; TNF-α, Tumor necrosis factor alpha; IL-1β, Interleukin-1 beta; IL-6, Interleukin-6; JNK, Jun N-terminal kinases; IL-10, Interleukin-10, FAS, Fatty acid synthase; ACC, Acetyl-CoA carboxylase; SCD-1, Stearoyl-CoA desaturase-1; upward arrows (↑), indicates increase; downward arrows (↓), indicates decrease.

Most preclinical studies have been conducted in high-fat diet models of dyslipidemia and obesity [30,44,52,54,71,72]. Although these models are relevant for cardiometabolic research and frequently report antioxidant and metabolic effects, particularly along the enterohepatic axis, their predominance restricts broader pathophysiological interpretation. Evidence in other contexts, including cancer, microbiota-mediated responses, and neurodegenerative processes, remains relatively scarce, limiting the generalization of these findings across biological systems. Expanding experimental designs to include a wider range of biological conditions may help determine whether the observed effects are model-specific or reflect more general physiological responses.

Comparative studies between acerola pulp and its by-products remain scarce, particularly those evaluating whole matrices alongside isolated bioactive fractions. This distinction is relevant because differences in phytochemical density, fiber content, and matrix complex-

 https://doi.org/10.3390/molecules31111792

ity may influence bioavailability and biological responses. Future studies may benefit from head-to-head designs across systems, animal models, and human trials to clarify relative efficacy, dose equivalence, and underlying mechanisms.

Among tropical fruit residues, acerola by-products appear to be supported by a relatively more consistent body of preclinical evidence. By-products from fruits such as cajá (*Spondias mombin* L.), umbu (*Spondias tuberosa* Arruda), guava (*Psidium guajava*), and cashew apple (*Anacardium occidentale*) have shown antioxidant and metabolic potential [54,92–94]; however, most available studies remain restricted to compositional analyses or assays, with fewer investigations addressing systemic and tissue-level effects in vivo. This suggests that acerola has been more extensively explored in experimental settings, although clinical validation remains limited. While animal studies provide important mechanistic insights, their predominance, particularly in models of metabolic disorders, limits broader translational applicability. Greater standardization of study design and increased inclusion of human studies will be necessary to establish efficacy, safety, and optimal intake conditions.

The valorization of acerola by-products is increasingly aligned with sustainability and bioeconomy strategies. The recovery and use of pomace, peels, and seeds may help reduce agro-industrial waste while generating value-added ingredients for food, nutraceutical, and pharmaceutical applications. These practices are consistent with circular economy principles and may support the United Nations Sustainable Development Goals (SDGs), particularly those related to responsible production (SDG 12), innovation (SDG 9), and health (SDG 3). Given Brazil's leading role in acerola production, the use of these by-products may also contribute to regional development, especially in semi-arid areas, while encouraging the development of standardized bioactive ingredients. In this context, acerola by-products represent promising resources within more sustainable production systems rather than as organic waste.

This review has inherent limitations that should be acknowledged. Although based on a structured search, its narrative design does not allow for quantitative synthesis and may introduce selection bias. In addition, the heterogeneity of the available evidence and predominance of preclinical studies limit comparability and translational interpretation.

5. Conclusions and Future Perspectives

Acerola and its by-products constitute phytochemically complex matrices rich in ascorbic acid, hydroxycinnamic acids, flavonoids, carotenoids, and bioactive polysaccharides. Preclinical evidence suggests that these constituents may modulate redox homeostasis and inflammatory signaling pathways, potentially through coordinated antioxidant effects and the regulation of lipid and glucose metabolism. By integrating current evidence across compositional, experimental, and emerging translational approaches, this review summarizes the current understanding of acerola-derived matrices as sources of bioactive compounds. However, most of the available evidence derives from experimental models and is limited by variability in study design, lack of standardization, and the scarcity of direct comparative studies, which restricts translational interpretation.

Future studies should prioritize detailed phytochemical characterization, dose–response relationships, bioavailability assessment, and well-controlled clinical trials incorporating molecular and mechanistic biomarkers to better clarify efficacy and safety.

Current evidence suggests the promising but still preliminary biological relevance of acerola-derived matrices. Further advances in standardization and translational validation will be necessary to determine their applicability in technological and clinical contexts.

https://doi.org/10.3390/molecules31111792

Author Contributions: Conceptualization, J.d.S.A., J.M.D.d.A., L.C.S., J.C.C.S., K.S.B., L.R.R.A.C. and A.N.V.d.A.; methodology, J.d.S.A., J.M.D.d.A., L.C.S., J.C.C.S., K.S.B., L.R.R.A.C. and A.N.V.d.A.; validation, J.d.S.A., J.M.D.d.A. and A.N.V.d.A.; investigation, J.d.S.A., J.M.D.d.A., L.C.S., J.C.C.S., K.S.B., L.R.R.A.C. and A.N.V.d.A.; resources, J.d.S.A. and L.R.R.A.C.; data curation, J.d.S.A., J.M.D.d.A. and A.N.V.d.A.; writing—original draft preparation, J.d.S.A., J.M.D.d.A., L.C.S., J.C.C.S., K.S.B., L.R.R.A.C. and A.N.V.d.A.; writing—review and editing, J.d.S.A., J.M.D.d.A. and A.N.V.d.A.; visualization, J.d.S.A., J.M.D.d.A., L.C.S., J.C.C.S., K.S.B., L.R.R.A.C. and A.N.V.d.A.; supervision, J.d.S.A. and A.N.V.d.A.; project administration, J.d.S.A.; funding acquisition, J.d.S.A. and L.R.R.A.C. All authors have read and agreed to the published version of the manuscript.

Funding: This research was funded by the Coordenação de Aperfeiçoamento de Pessoal de Nível Superior (CAPES), Brazil, Finance Code 001 (doctoral scholarships to J.M.D.d.A., L.C.S. and J.C.C.S.), and by the Conselho Nacional de Desenvolvimento Científico e Tecnológico (CNPq), Grant No. 303939/2025-7 (Research Productivity Fellowship to J.d.S.A.).

Institutional Review Board Statement: Not applicable.

Informed Consent Statement: Not applicable.

Data Availability Statement: No new data were created or analyzed in this study. Data sharing is not applicable to this article.

Acknowledgments: The authors acknowledge Karolinska Institutet for supporting the Article Processing Charge (APC) through an institutional publishing agreement, which covered the open access publication fees.

Conflicts of Interest: The authors declare no conflicts of interest.

Abbreviations

ADP	Adenosine diphosphate
CAT	Catalase
DNA	Deoxyribonucleic acid
FAS	Fatty acid synthase
FRAP	Ferric reducing antioxidant power
GSH	Glutathione
GSSG	Oxidized glutathione
HFD	High-fat diet
HO-1	Heme oxygenase-1
IL-1β	Interleukin-1 beta
IL-10	Interleukin-10
JNK	Jun N-terminal kinases
MASLD	Metabolic dysfunction-associated steatotic liver disease
MDA	Malondialdehyde
ORAC	Oxygen radical absorbance capacity
Nrf2	Nuclear factor (erythroid-derived-2)-like 2
NQO-1	NAD(P)H:quinone oxidoreductase-1
PGC-1α	Peroxisome proliferator-activated receptor gamma coactivator-1α
PGE_2	Prostaglandin E2
SCD-1	Stearoyl-CoA desaturase-1
SOD	Superoxide dismutase
SREBP-1c	Sterol regulatory element-binding protein-1c
TAC	Total antioxidant capacity
TNF-α	Tumor necrosis factor alpha
UCP2	Uncoupling protein 2

 https://doi.org/10.3390/molecules31111792

References

1. Vega-Baudrit, J.R.; Camacho, M.; Batista-Menezes, D.; Corrales Ureña, Y.; Zúñiga, J.M.; Chacón, A.M.; Lecot, N.; Henríquez, L.C.; Lopretti, M. Acerola (*Malpighia* spp.) waste: A sustainable approach to nutraceutical, pharmaceutical, and energy applications. *Recycling* **2023**, *8*, 96. [CrossRef]
2. Silva, T.O.M.; Nogueira, J.P.; Rezende, Y.R.R.S.; Oliveira, C.S.D.; Narain, N. Bioactive compounds and antioxidants activities in the agroi-industrial residues of berries solvent and enzyme assisted extraction. *Food Sci. Technol.* **2022**, *42*, e61022. [CrossRef]
3. Teshome, E.; Teka, T.A.; Nandasiri, R.; Rout, J.R.; Harouna, D.V.; Astatkie, T.; Urugo, M.M. Fruit By-Products and Their Industrial Applications for Nutritional Benefits and Health Promotion: A comprehensivel Review. *Sustainability* **2023**, *15*, 7840. [CrossRef]
4. Miskinis, R.D.A.S.; Nascimento, L.Á.; Colussi, R. Bioactive compounds from acerola pomace: A review. *Food Chem.* **2023**, *404*, 134613. [CrossRef]
5. Silva, N.C.D.; Dourado, P.L.A.; Tosi, M.M. A review about acerola (*Malpighia emarginata* DC.) by-products as a promising raw material for the generation of green products. *Braz. J. Food Technol.* **2023**, *26*, e2023039. [CrossRef]
6. Rufino, M.D.S.M.; Pérez-Jiménez, J.; Tabernero, M.; Alves, R.E.; Brito, E.S.; Saura-Calixto, F. Acerola and cashew apple as sources of antioxidants and dietary fibre. *Int. J. Food Sci. Technol.* **2010**, *45*, 2227–2233. [CrossRef]
7. Martirosyan, D. Functional food science and bioactive compounds. *Bioact. Compd. Health Dis.* **2025**, *8*, 218–229. [CrossRef]
8. Laurindo, L.F.; Takeda, L.N.; Machado, N.M.; Otoboni, A.M.M.B.; Goulart, R.A.; Catharin, V.C.S.; Direito, R. Health benefits of acerola (*Malpighia* spp.) and its by-products: A comprehensive review of nutrient-rich composition, pharmacological potential, and industrial applications. *Food Biosci.* **2024**, *62*, 105422. [CrossRef]
9. Instituto Brasileiro de Geografia e Estatística (IBGE). Levantamento Sistemático da Produção Agrícola (LSPA)/Pesquisa Agrícola Municipal (PAM): Acerola: Area Harvested and Production (Brazil). Available online: https://www.ibge.gov.br/explica/producao-agropecuaria/acerola/br (accessed on 27 March 2026).
10. Mezadri, T.; Fernández-Pachón, M.S.; Villano, D.; García-Parrilla, M.C.; Troncoso, A.M. The acerola fruit: Composition, productive characteristics and economic importance. *Arch. Latinoam. Nutr.* **2006**, *56*, 101–109. [PubMed]
11. da Paz, M.E.A.; de Barcelos, S.C.; de Oliveira, V.C.; Pinto, M.B.; Sá, D.M.A.T. Impact of adding dehydrated acerola (*Malpighia emarginata* DC.) on the microbiological, colorimetric, and sensory characteristics of acerola ice cream. *Int. J. Gastron. Food Sci.* **2025**, *39*, 101101. [CrossRef]
12. Matuda, T.G.; Hoshino, L.M.; Ribeiro, E.P.; Tadini, C.C. The influence of polydextrose on freeze-dried unripe acerola (*Malpighia emarginata* DC.) by the concept of a state diagram. *J. Food Eng.* **2023**, *341*, 111349. [CrossRef]
13. Mezadri, T.; Villaño, D.; Fernández-Pachón, M.S.; García-Parrilla, M.C.; Troncoso, A.M. Antioxidant compounds and antioxidant activity in acerola (*Malpighia emarginata* DC.) fruits and derivatives. *J. Food Compos. Anal.* **2008**, *21*, 282–290. [CrossRef]
14. Belwal, T.; Devkota, H.P.; Hassan, H.A.; Ahluwalia, S.; Ramadan, M.F.; Mocan, A.; Atanasov, A.G. Phytopharmacology of acerola (*Malpighia* spp.) and its potential as functional food. *Trends Food Sci. Technol.* **2018**, *74*, 99–106. [CrossRef]
15. Freitas, E.E.S.; Dias, Ê.R.; Albuquerque, M.M.S.; Biondi, I.B.; Branco, C.R.C.; Cruz, R.S.; Camilloto, G.P. Antioxidant films based on poly(lactic acid) incorporated with crude extract from *Malpighia emarginata* DC. pomace for use in food packaging. *Packag. Technol. Sci.* **2024**, *37*, 39–50. [CrossRef]
16. Freitas, E.E.S.; Esteves, I.S.C.; Rodrigues, L.M.; Bomfim, A.S.; dos Santos, S.P.; Cruz, R.S.; Camilloto, G.P. Antioxidant films based on poly(vinyl alcohol) (PVOH) activated by a flavonoid-rich fraction from *Malpighia emarginata* DC. pomace in free and nanoencapsulated forms. *Food Packag. Shelf Life* **2025**, *52*, 101618. [CrossRef]
17. da Silva, N.C.; de Barros-Alexandrino, T.T.; Assis, O.B.G.; Martelli-Tosi, M. Extraction of phenolic compounds from acerola by-products using chitosan solution, encapsulation and application in extending the shelf-life of guava. *Food Chem.* **2021**, *354*, 129553. [CrossRef]
18. Gomes, A.C.A.; da Costa Lima, M.; de Oliveira, K.Á.R.; dos Santos Lima, M.; Magnani, M.; Câmara, M.P.S.; de Souza, E.L. Coatings with chitosan and phenolic-rich extract from acerola (*Malpighia emarginata* DC.) or jabuticaba (*Plinia jaboticaba* (Vell.) Berg) processing by-product to control rot caused by *Lasiodiplodia* spp. in papaya (*Carica papaya* L.) fruit. *Int. J. Food Microbiol.* **2020**, *331*, 108694. [CrossRef]
19. Dantas, S.C.D.M.; Pontes Júnior, S.M.D.; Medeiros, F.G.M.D.; Santos Junior, L.C.; Alsina, O.L.S.D.; Medeiros, M.D.F.D.D. Spouted-bed drying of acerola pulp (*Malpighia emarginata* DC.): Effects of adding milk and milk protein on process performance and characterization of dried fruit powders. *J. Food Process Eng.* **2019**, *42*, e13205. [CrossRef]
20. de Oliveira, S.D.; Araujo, C.M.; Borges, G.D.S.C.; dos Santos Lima, M.; Viera, V.B.; Garcia, E.F.; de Oliveira, M.E.G. Improvement in physicochemical characteristics, bioactive compounds and antioxidant activity of acerola (*Malpighia emarginata* DC.) and guava (*Psidium guajava* L.) fruit by-products fermented with potentially probiotic lactobacilli. *LWT* **2020**, *134*, 110200. [CrossRef]
21. Vieira, A.D.S.; Battistini, C.; Bedani, R.; Saad, S.M.I. Acerola by-product may improve the gastrointestinal resistance of probiotic strains in a plant-based fermented beverage. *LWT* **2021**, *141*, 110858. [CrossRef]

https://doi.org/10.3390/molecules31111792

22. Umezu, T.; Takanashi, M.; Murakami, Y.; Ohno, S.I.; Kanekura, K.; Sudo, K.; Kuroda, M. Acerola exosome-like nanovesicles to systemically deliver nucleic acid medicine via oral administration. *Mol. Ther. Methods Clin. Dev.* **2021**, *21*, 199–208. [CrossRef]
23. Olędzki, R.; Harasym, J. Acerola (*Malpighia emarginata*) anti-inflammatory activity—A review. *Int. J. Mol. Sci.* **2024**, *25*, 2089. [CrossRef] [PubMed]
24. Vitek, L.; Hajšlová, J.; Bátovský, M.; Woronyczova, J.; Posova, H.; Zelenkova, M.; Kalousová, M. The effect of acerola intake on metabolic and immunological parameters in elite athletes. *Plant Foods Hum. Nutr.* **2025**, *80*, 198. [CrossRef]
25. Uchida, E.; Kondo, Y.; Amano, A.; Aizawa, S.; Hanamura, T.; Aoki, H.; Nagamine, K.; Koizumi, T.; Maruyama, N.; Ishigami, A. Absorption and excretion of ascorbic acid alone and in acerola (*Malpighia emarginata*) juice: Comparison in healthy Japanese subjects. *Biol. Pharm. Bull.* **2011**, *34*, 1744–1747. [CrossRef]
26. Hu, Y.; Yin, F.; Liu, Z.; Xie, H.; Xu, Y.; Zhou, D.; Zhu, B. Acerola polysaccharides ameliorate high-fat diet-induced non-alcoholic fatty liver disease through reduction of lipogenesis and improvement of mitochondrial functions in mice. *Food Funct.* **2020**, *11*, 1037–1048. [CrossRef] [PubMed]
27. Dallazen, J.L.; Ciapparini, P.G.; Maria-Ferreira, D.; da Luz, B.B.; Klosterhoff, R.R.; Felipe, L.P.G.; Silva, B.J.G.; Cordeiro, L.M.C.; Paula Werner, M.F. Arabinan rich pectic polysaccharide fraction from *Malpighia emarginata* fruits alleviates inflammatory pain in mice. *Food Res. Int.* **2024**, *176*, 113743. [CrossRef]
28. Marques, T.R.; Caetano, A.A.; Cesar, P.H.S.; Braga, M.A.; Machado, G.H.A.; Sousa, R.V.; Corrêa, A.D. Antioxidant activity and hepatoprotective potential of lyophilized extract of acerola bagasse against CCl_4-induced hepatotoxicity in Wistar rats. *J. Food Biochem.* **2018**, *42*, e12670. [CrossRef]
29. Hasin, Y.; Seldin, M.; Lusis, A. Multi-omics approaches to disease. *Genome Biol.* **2017**, *18*, 83. [CrossRef]
30. Batista, K.S.; Soares, N.L.; Dorand, V.A.M.; Alves, A.F.; Lima, M.S.; Pereira, R.A.; Souza, E.L.; Magnani, M.; Persuhn, D.C.; Aquino, J.S. Acerola fruit by-product alleviates lipid, glucose, and inflammatory changes in the enterohepatic axis of rats fed a high-fat diet. *Food Chem.* **2023**, *403*, 134322. [CrossRef] [PubMed]
31. Vilvert, J.C.; Freitas, S.T.D.; Veloso, C.M.; Amaral, C.L.F. Genetic Diversity on Acerola Quality: A Systematic Review. *Braz. Arch. Biol. Technol.* **2024**, *67*, e24220490. [CrossRef]
32. Islas-Osuna, M.A.; Tiznado-Hernández, M.E. Acerola (*Malpighia emarginata* DC.). In *Postharvest Biology and Technology of Tropical and Subtropical Fruits: Fundamental Issues*; Yahia, E.M., Ed.; Woodhead Publishing: Cambridge, UK, 2011; Volume 1, pp. 315–361.
33. Whyte, T.A.; Goldson-Barnaby, A.; Bailey, D. Medicinal properties and food applications of *Malpighia emarginata* (acerola cherry). *Caribb. J. Sci. Technol.* **2023**, *11*, 15–20. [CrossRef]
34. Queiroga, V.P.; Gomes, J.P.; Mendes, N.V.B.; Queiroz, A.J.M.; Lima, D.C.; Figueiredo Neto, A.; Girão, E.G.; Albuquerque, E.M.B. Chapter 1: Sistema produtivo da acerola (*Malpighia emarginata* Sessé & Mociño ex DC). In *Acerola (Malpighia emarginata Sessé & Mociño ex DC.): Tecnologias de Plantio e Utilização*; Embrapa Mandioca e Fruticultura: Cruz das Almas, Brazil, 2023. Available online: https://ainfo.cnptia.embrapa.br/digital/bitstream/doc/1162484/1/Sistema-produtivio-acerola-2023.pdf (accessed on 13 February 2026).
35. EMPAER-PB; UFPB. *Paraíba se Torna Pioneira ao Lançar Acerola Mais Saborosa e Rentável*; EMPAER-PB: João Pessoa, Brazil, 2023. Available online: https://empaer.pb.gov.br/noticias/paraiba-se-torna-pioneira-ao-lancar-acerola-mais-saborosa-e-rentavel (accessed on 18 February 2026).
36. Carneiro Ferreira, I.; Pereira da Silva, V.; Vilvert, J.C.; de França Souza, F.; de Freitas, S.T.; dos Santos Lima, M. Brazilian varieties of acerola (*Malpighia emarginata* DC.) produced under tropical semi-arid conditions: Bioactive phenolic compounds, sugars, organic acids, and antioxidant capacity. *J. Food Biochem.* **2021**, *45*, e13829. [CrossRef]
37. Paiva, Y.F.; Figueirêdo, R.M.F.; Queiroz, A.J.M.; Amadeu, L.T.S.; Santos, F.S.; Reis, C.G.; Carvalho, A.J.A.; Lima, M.S.; Lima, A.G.B.; Gomes, J.P.; et al. Physicochemical aspects, bioactive compounds, phenolic profile and antioxidant activity of tropical red fruits and their blend. *Molecules* **2023**, *28*, 4866. [CrossRef] [PubMed]
38. de Oliveira Filho, J.G.; Estevam, E.B.B.; Lemes, R.S.; Prado, D.M.F.; Silva, V.P.; Silva, L.S.; Duarte, L.G.R.; Egea, M.B. Acerola (*Malpighia emarginata*) pulp: Characterization and stability of anthocyanins under different conditions. *Food Sci. Technol.* **2023**, *43*, e0000110. [CrossRef]
39. da Silva, L.M.R.; de Figueiredo, E.A.T.; Ricardo, N.M.P.S.; Vieira, I.G.P.; de Figueiredo, R.W.; Brasil, I.M.; Gomes, C.L. Quantification of bioactive compounds in pulps and by-products of tropical fruits from Brazil. *Food Chem.* **2014**, *143*, 398–404. [CrossRef] [PubMed]
40. Berni, P.; Campoli, S.S.; Negri, T.C.; Toledo, N.M.V.; Canniatti-Brazaca, S.G. Non-conventional tropical fruits: Characterization, antioxidant potential and carotenoid bioaccessibility. *Plant Foods Hum. Nutr.* **2019**, *74*, 141–148. [CrossRef]
41. Montero, I.F.; Saravia, S.A.M.; Santos, R.A.; dos Santos, R.C.; Marca, J.A.F.; da Costa, H.N.R. Nutrients in Amazonian fruit pulps with functional and pharmacological interest. *Afr. J. Pharm. Pharmacol.* **2020**, *14*, 118–127. [CrossRef]

42. Binh, H.Q.; Tram, P.N.; Diep, D.T.N. Bioactive compounds and strategy processing for acerola: A review. *CTU J. Innov. Sustain. Dev.* **2022**, *14*, 46–60. [CrossRef]
43. Silveira, A.G.; Ribeiro, L.B.; Sampaio, L.M.; Lima, A.R.; Alves Filho, E.G.; Ribeiro, P.R.; Oliveira, L.S. UPLC-QTOF/MS-MS based profiling coupled to antioxidant activity for evaluation of unripe acerola fruits from different clones. *J. Braz. Chem. Soc.* **2024**, *35*, e20230172. [CrossRef]
44. Rodríguez-Rodríguez, E.; Herrero-Lodares, C.; Sánchez-Prieto, M.; Olmedilla-Alonso, B.; Sánchez-Moreno, C.; De Ancos, B. Sustainable extraction methods of carotenoids from mango (*Mangifera indica* L. 'Kent') pulp: Ultrasound-assisted extraction and green solvents. *Food Chem.* **2024**, *450*, 139253. [CrossRef]
45. Lara-Abia, S.; Welti-Chanes, J.; Cano, M.P. Effect of ultrasound-assisted extraction of carotenoids from papaya (*Carica papaya* L. cv. Sweet Mary) using vegetable oils. *Molecules* **2022**, *27*, 638. [CrossRef]
46. Miranda, N.G.M.; Gomes, F.D.S.; Cabral, L.M.C.; Jesus, M.S.C.D. Thermal concentration effects of tomato pulp and carotenoids extraction from the residue. *Sci. Agric.* **2022**, *80*, e20220047. [CrossRef]
47. da Silva Pereira, A.C.; Wurlitzer, N.J.; Dionísio, A.P.; Soares, M.V.L.; Bastos, M.D.S.R.; Alves, R.E.; Brasil, I.M.M. Synergistic, additive and antagonistic effects of fruit mixtures on total antioxidant capacities and bioactive compounds in tropical fruit juices. *Arch. Latinoam. Nutr.* **2015**, *65*, 119–127.
48. Zheng, X.; Chen, Z.; Guo, Z.; Chen, M.; Xie, B.; Sun, Z.; Hu, K. Effect of novel processing techniques on the carotenoid release during the production of red guava juice. *Molecules* **2024**, *29*, 487. [CrossRef] [PubMed]
49. Delva, L.; Goodrich-Schneider, R. Antioxidant activity and antimicrobial properties of phenolic extracts from acerola (*Malpighia emarginata* DC.). *Int. J. Food Sci. Technol.* **2013**, *48*, 1048–1056. [CrossRef]
50. Cruz, R.G.; Beney, L.; Gervais, P.; Lira, S.P.; Vieira, T.M.F.S.; Dupont, S. Comparison of the antioxidant property of acerola extracts with synthetic antioxidants using an in vivo method with yeasts. *Food Chem.* **2019**, *277*, 698–705. [CrossRef]
51. Macedo Albuquerque Junior, N.; Silva, W.P.d.; Lima, D.A.; dos Santos Silva, R.; Moura, H.V.; Silva, A.P.d.F.; de Oliveira Carvalho, R.; Andrade, F.S.; Amadeu, L.T.S. Thermodynamic, bioactive, and antioxidant potential of green acerola residue as influenced by convective drying. *J. Therm. Anal. Calorim.* **2025**, *150*, 5213–5226. [CrossRef]
52. Batista, K.S.; Cavalcante, H.C.; Gomes, J.A.S.; Silva, L.A.; Cavalcanti, N.S.H.; Garcia, E.F.; Menezes, F.N.D.D.; Lima, T.A.S.; Souza, E.L.; Magnani, M.; et al. Effects of supplementation of tropical fruit processing by-products on lipid profile, retinol levels and intestinal function in Wistar rats. *An. Acad. Bras. Cienc.* **2023**, *95*, e20201684. [CrossRef] [PubMed]
53. Poletto, P.; Álvarez-Rivera, G.; López, G.D.; Borges, O.M.; Mendiola, J.A.; Ibáñez, E.; Cifuentes, A. Recovery of Ascorbic Acid, Phenolic Compounds and Carotenoids from Acerola By-Products: An Opportunity for Their Valorization. *LWT* **2021**, *146*, 111654. [CrossRef]
54. Batista, K.S.; Alves, A.F.; Lima, M.S.; Silva, L.A.; Lins, P.P.; Gomes, J.A.S.; Silva, A.S.; Toscano, L.T.; Meireles, B.R.L.A.; Cordeiro, A.M.T.M.; et al. Beneficial effects of consumption of acerola, cashew or guava processing by-products on intestinal health and lipid metabolism in dyslipidaemic female Wistar rats. *Br. J. Nutr.* **2018**, *119*, 30–41. [CrossRef]
55. Vargas-Ramella, M.; Ely, J.; Favaro-Trindade, C.S.; Dos Santos, B.A.; Campagnol, P.C.B.; Lorenzo, J.M. Analytical characterization and quantification of plant-derived antioxidants focusing on methods, limitations, and data comparability. *Food Anal. Methods* **2026**, *19*, 162. [CrossRef]
56. Usman, I.; Hussain, M.; Imran, A.; Afzaal, M.; Saeed, F.; Javed, M.; Saewan, S.A. Traditional and innovative approaches for the extraction of bioactive compounds. *Int. J. Food Prop.* **2022**, *25*, 1215–1233. [CrossRef]
57. Rezende, Y.R.R.S.; Nogueira, J.P.; Narain, N. Comparison and optimization of conventional and ultrasound-assisted extraction for bioactive compounds and antioxidant activity from agro-industrial acerola (*Malpighia emarginata* DC.) residue. *LWT* **2017**, *85*, 158–169. [CrossRef]
58. Ciccoritti, R.; Ciorba, R.; Ceccarelli, D.; Amoriello, M.; Amoriello, T. Phytochemical and Functional Properties of Fruit and Vegetable Processing By-Products. *Appl. Sci.* **2024**, *14*, 9172. [CrossRef]
59. Raczkowska, E.; Serek, P. Health-Promoting Properties and the Use of Fruit Pomace in the Food Industry—A Review. *Nutrients* **2024**, *16*, 2757. [CrossRef]
60. Hasan, M.M.; Islam, M.R.; Haque, A.R.; Kabir, M.R.; Khushe, K.J.; Hasan, S.M.K. Trends and challenges of fruit by-products utilization: Insights into safety, sensory, and benefits of the use for the development of innovative healthy food: A review. *Bioresour. Bioprocess.* **2024**, *11*, 10. [CrossRef]
61. da Silva Donadone, D.B.; de Castro França, I.A.; Silva, D.L.G.; Faria, M.G.I.; Ruiz, S.P.; Barros, B.C.B. Enhancement of Phenolic Recovery by Probe-Type Ultrasound-Assisted Extraction of Acerola By-Product and Evaluation of Antioxidant and Antibacterial Activities. *Appl. Sci.* **2025**, *15*, 9154. [CrossRef]

 https://doi.org/10.3390/molecules31111792

62. Xu, M.; Shen, C.; Zheng, H.; Xu, Y.; Xue, C.; Zhu, B.; Hu, J. Metabolomic analysis of acerola cherry (*Malpighia emarginata*) fruit during ripening development via UPLC-Q-TOF and contribution to the antioxidant activity. *Food Res. Int.* **2020**, *130*, 108915. [CrossRef]
63. Mesquita, P.C.; Rodrigues, L.G.G.; Mazzutti, S.; Ribeiro, P.R.V.; de Brito, E.S.; Lanza, M. Untargeted metabolomic profile of recovered bioactive compounds by subcritical water extraction of acerola (*Malpighia emarginata* DC.) pomace. *Food Chem.* **2022**, *397*, 133718. [CrossRef]
64. Badejo, A.A.; Fujikawa, Y.; Esaka, M. Gene expression of ascorbic acid biosynthesis related enzymes of the Smirnoff–Wheeler pathway in acerola (*Malpighia glabra*). *J. Plant Physiol.* **2009**, *166*, 652–660. [CrossRef]
65. dos Santos, C.P.; Batista, M.C.; da Cruz Saraiva, K.D.; Roque, A.L.M.; de Souza Miranda, R.; Alexandre e Silva, L.M.; Costa, J.H. Transcriptome analysis of acerola fruit ripening: Insights into ascorbate, ethylene, respiration, and softening metabolisms. *Plant Mol. Biol.* **2019**, *101*, 269–296. [CrossRef] [PubMed]
66. Li, M.; Li, D.; Feng, F.; Zhang, S.; Ma, F.; Cheng, L. Proteomic analysis reveals dynamic regulation of fruit development and sugar and acid accumulation in apple. *J. Exp. Bot.* **2016**, *67*, 5145–5157. [CrossRef]
67. Chan, Z. Proteomic responses of fruits to environmental stresses. *Front. Plant Sci.* **2013**, *3*, 311. [CrossRef]
68. Wu, X.; Gong, F.; Wang, W. Protein extraction from plant tissues for 2DE and its application in proteomic analysis. *Proteomics* **2014**, *14*, 645–658. [CrossRef]
69. Brajkovic, S.; Rugen, N.; Agius, C.; Berner, N.; Eckert, S.; Sakhteman, A.; Kuster, B. Getting ready for large-scale proteomics in crop plants. *Nutrients* **2023**, *15*, 783. [CrossRef] [PubMed]
70. Kilambi, H.V.; Manda, K.; Sanivarapu, H.; Maurya, V.K.; Sharma, R.; Sreelakshmi, Y. Shotgun proteomics of tomato fruits: Evaluation, optimization and validation of sample preparation methods and mass spectrometric parameters. *Front. Plant Sci.* **2016**, *7*, 969. [CrossRef] [PubMed]
71. Batista, K.S.; dos Santos Lima, M.; Alves, A.F.; Cavalcante, H.C.; de Souza, D.M.; de Oliveira, G.C.; Toscano, L.T.; Silva, A.S.; Rodrigues, J.F.; Meireles, B.R.L.A.; et al. Antioxidant potential of acerola by-product along the enterohepatic axis of rats fed a high-fat diet. *Food Res. Int.* **2023**, *173*, 113380. [CrossRef]
72. Batista, A.C.V.; Ribeiro, M.A.; Oliveira, K.A.; Freitas, P.A.; Santos, N.S.; Magalhães, L.A.; Magalhães, S.C.; Fonseca, S.G.C.; Aquino, J.S.; Souza, E.L.; et al. Effects of consumption of acerola, cashew and guava by-products on adiposity and redox homeostasis of adipose tissue in obese rats. *Clin. Nutr. ESPEN* **2021**, *43*, 283–289. [CrossRef]
73. Miller, G.D.; Ragalie-Carr, J.; Torres-Gonzalez, M. Perspective: Seeing the forest through the trees: The importance of food matrix in diet quality and human health. *Adv. Nutr.* **2023**, *14*, 363–365. [CrossRef]
74. Jin, M.Y.; Li, M.Y.; Huang, R.M.; Wu, X.Y.; Sun, Y.M.; Xu, Z.L. Structural features and anti-inflammatory properties of pectic polysaccharides: A review. *Trends Food Sci. Technol.* **2021**, *107*, 284–298. [CrossRef]
75. Klosterhoff, R.R.; Kanazawa, L.K.; Furlanetto, A.L.; Peixoto, J.V.; Corso, C.R.; Adami, E.R.; Iacomini, M.; Fogaça, R.T.H.; Acco, A.; Cadena, S.M.S.C.; et al. Anti-fatigue activity of an arabinan-rich pectin from acerola (*Malpighia emarginata*). *Int. J. Biol. Macromol.* **2018**, *109*, 1147–1153. [CrossRef]
76. Leffa, D.D.; Rezin, G.T.; Daumann, F.; Longaretti, L.M.; Dajori, A.L.F.; Gomes, L.M.; Silva, M.C.; Streck, E.L.; Andrade, V.M. Effects of acerola (*Malpighia emarginata* DC.) juice intake on brain energy metabolism of mice fed a cafeteria diet. *Mol. Neurobiol.* **2017**, *54*, 954–963. [CrossRef] [PubMed]
77. de Sousa Sabino, L.B.; Costa Gonzaga, M.L.; Siqueira Oliveira, L.; Duarte, A.S.G.; Silva, L.M.A.; Brito, E.S.; Figueiredo, R.W.; Silva, L.M.R.; Sousa, P.H.M. Polysaccharides from acerola, cashew apple, pineapple, mango and passion fruit co-products: Structure, cytotoxicity and gastroprotective effects. *Bioact. Carbohydr. Diet. Fibre* **2020**, *24*, 100228. [CrossRef]
78. Leffa, D.D.; Silva, J.; Daumann, F.; Dajori, A.L.F.; Longaretti, L.M.; Damiani, A.P.; Lira, F.S.; Campos, F.; Ferraz, A.B.F.; Corrêa, D.S.; et al. Corrective effects of acerola (*Malpighia emarginata* DC.) juice intake on biochemical and genotoxical parameters in mice fed on a high-fat diet. *Mutat. Res. Fundam. Mol. Mech. Mutagen.* **2014**, *770*, 144–152. [CrossRef]
79. Wang, X.Y.; Yin, J.Y.; Hu, J.L.; Nie, S.P.; Xie, M.Y. Gastroprotective polysaccharide from natural sources: Review on structure, mechanism, and structure–activity relationship. *Food Front.* **2022**, *3*, 560–591. [CrossRef]
80. Jomova, K.; Alomar, S.Y.; Nepovimova, E.; Kuca, K.; Valko, M. Heavy metals: Toxicity and human health effects. *Arch. Toxicol.* **2025**, *99*, 153–209. [CrossRef]
81. Jomova, K.; Alomar, S.Y.; Alwasel, S.H.; Nepovimova, E.; Kuca, K.; Valko, M. Several lines of antioxidant defense against oxidative stress: Antioxidant enzymes, nanomaterials with multiple enzyme-mimicking activities, and low-molecular-weight antioxidants. *Arch. Toxicol.* **2024**, *98*, 1323–1367. [CrossRef] [PubMed]
82. Rahimah, S.; Tallei, T.E.; Savitri, M.; Yamada, C.; Kim, H.J.; Choi, M.; Park, M.N.; Ophinni, Y.; Kim, B. Functional foods and dietary matrices for metabolic improvement: An integrative review of mechanisms, evidence, and future directions. *J. Funct. Foods* **2026**, *139*, 107240. [CrossRef]

 https://doi.org/10.3390/molecules31111792

83. Liu, T.; Zhang, L.; Joo, D.; Sun, S.-C. NF-κB signaling in inflammation. *Signal Transduct. Target. Ther.* **2017**, *2*, 17023. [CrossRef]
84. Wu, X.; Zha, J.; Koffas, M.A.G. Microbial production of bioactive chemicals for human health. *Curr. Opin. Food Sci.* **2020**, *32*, 9–16. [CrossRef]
85. Liu, S.H.; Yang, X.F.; Liang, L.; Song, B.B.; Song, X.M.; Yang, Y.J.; Alhoot, M.A. Regulatory mechanisms of the gut microbiota-short chain fatty acids signaling axis in slow transit constipation and progress in multi-target interventions. *Front. Microbiol.* **2025**, *16*, 1689597. [CrossRef]
86. Chen, N.; Chan, C.K.; Khan, F.U.; Makhijani, P.; Valentino, T.R.; Winer, S.; Luk, C.T.; Winer, D.A. Gut microbiota-derived metabolites as immune modulators in aging and age-related chronic inflammatory diseases. *Ageing Res. Rev.* **2026**, *116*, 103036. [CrossRef] [PubMed]
87. Ordovas, J.M.; Ferguson, L.R.; Tai, E.S.; Mathers, J.C. Personalised nutrition and health. *BMJ* **2018**, *361*, k2173. [CrossRef] [PubMed]
88. Martínez-Ballesta, M.; Gil-Izquierdo, Á.; García-Viguera, C.; Domínguez-Perles, R. Nanoparticles and controlled delivery for bioactive compounds: Outlining challenges for new "smart-foods" for health. *Foods* **2018**, *7*, 72. [CrossRef]
89. Pateiro, M.; Gómez, B.; Munekata, P.E.; Barba, F.J.; Putnik, P.; Kovačević, D.B.; Lorenzo, J.M. Nanoencapsulation of promising bioactive compounds to improve their absorption, stability, functionality and the appearance of the final food products. *Molecules* **2021**, *26*, 1547. [CrossRef]
90. Hao, M.; Tan, X.; Liu, K.; Xin, N. Nanoencapsulation of nutraceuticals: Enhancing stability and bioavailability in functional foods. *Front. Nutr.* **2025**, *12*, 1746176. [CrossRef]
91. de Góes Carneiro, A.P.; de Aguiar, A.L.L.; de Lima, A.C.S.; da Silva, L.M.R.; de Sousa, P.H.M.; de Figueiredo, R.W. Bioactive potential of nanoparticles of acerola byproduct (*Malpighia* sp. L): Bioaccessibility in nectar. *Res. Soc. Dev.* **2020**, *9*, e159996691. [CrossRef]
92. Lucena, T.L.C.; Batista, K.S.; Pinheiro, R.O.; Cavalcante, H.C.; Gomes, J.A.S.; Silva, L.A.; Lins, P.P.; Ferreira, F.S.; Lima, R.F.; Lima, M.S.; et al. Nutritional characterization, antioxidant, and lipid-Lowering effects of yellow mombin (*Spondias mombin*) supplemented to rats fed a high-fat diet. *Foods* **2022**, *11*, 3064. [CrossRef]
93. Gutierrez-Montiel, D.; Guerrero-Barrera, A.L.; Chávez-Vela, N.A.; Avelar-Gonzalez, F.J.; Ornelas-García, I.G. *Psidium guajava* L.: From byproduct and use in traditional Mexican medicine to antimicrobial agent. *Front. Nutr.* **2023**, *10*, 1108306. [CrossRef]
94. Zié, M.; Alabi, T.; Karamoko, G.; Blecker, C. Valorization of cashew apple bagasse in food application: Focus on the use and extraction of nutritional or bioactive compounds. *Food Humanit.* **2023**, *1*, 848–863. [CrossRef]

 https://doi.org/10.3390/molecules31111792

MDPI AG
Grosspeteranlage 5
4052 Basel
Switzerland
Tel.: +41 61 683 77 34

Molecules Editorial Office
E-mail: molecules@mdpi.com
www.mdpi.com/journal/molecules